# GNU Octave 4.0 Reference Manual 1/2

A catalogue record for this book is available from the Hong Kong Public Libraries.

Published in Hong Kong by Samurai Media Limited.

Email: info@samuraimedia.org

ISBN 978-988-8381-05-0

Copyright 1996, 1997, 1999, 2000, 2001, 2002, 2005, 2006, 2007, 2011, 2013, 2015 John W. Eaton.
This is the fourth edition of the Octave documentation, and is consistent with version 4.0.0 of Octave.
Permission is granted to make and distribute verbatim copies of this manual provided the copyright notice and this permission notice are preserved on all copies.
Permission is granted to copy and distribute modified versions of this manual under the conditions for verbatim copying, provided that the entire resulting derived work is distributed under the terms of a permission notice identical to this one.
Permission is granted to copy and distribute translations of this manual into another language, under the same conditions as for modified versions.
Portions of this document have been adapted from the gawk, readline, gcc, and C library manuals, published by the Free Software Foundation, Inc., 51 Franklin Street, Fifth Floor, Boston, MA 02110-13011307, USA.
Minor modifications for publication Copyright 2015 Samurai Media Limited.

Background Cover Image by https://www.flickr.com/people/webtreatsetc/

# Table of Contents

**Preface** .................................................................... 1
    Acknowledgements .................................................. 1
    Citing Octave in Publications ..................................... 5
    How You Can Contribute to Octave ................................. 5
    Distribution ....................................................... 6

**1 A Brief Introduction to Octave** ........................... 7
    1.1 Running Octave ................................................ 7
    1.2 Simple Examples ............................................... 7
        1.2.1 Elementary Calculations ................................. 7
        1.2.2 Creating a Matrix ....................................... 8
        1.2.3 Matrix Arithmetic ....................................... 8
        1.2.4 Solving Systems of Linear Equations ..................... 8
        1.2.5 Integrating Differential Equations ...................... 9
        1.2.6 Producing Graphical Output ............................. 10
        1.2.7 Editing What You Have Typed ........................... 10
        1.2.8 Help and Documentation ................................ 10
    1.3 Conventions .................................................. 11
        1.3.1 Fonts .................................................. 11
        1.3.2 Evaluation Notation .................................... 11
        1.3.3 Printing Notation ...................................... 12
        1.3.4 Error Messages ......................................... 12
        1.3.5 Format of Descriptions ................................. 12
            1.3.5.1 A Sample Function Description ................... 12
            1.3.5.2 A Sample Command Description .................... 13

**2 Getting Started** ............................................. 15
    2.1 Invoking Octave from the Command Line ....................... 15
        2.1.1 Command Line Options .................................. 15
        2.1.2 Startup Files .......................................... 19
    2.2 Quitting Octave .............................................. 19
    2.3 Commands for Getting Help ................................... 20
    2.4 Command Line Editing ........................................ 25
        2.4.1 Cursor Motion ......................................... 25
        2.4.2 Killing and Yanking .................................... 26
        2.4.3 Commands For Changing Text ............................ 27
        2.4.4 Letting Readline Type For You ......................... 27
        2.4.5 Commands For Manipulating The History ................. 28
        2.4.6 Customizing `readline` ................................. 32
        2.4.7 Customizing the Prompt ................................ 32
        2.4.8 Diary and Echo Commands ............................... 34
    2.5 How Octave Reports Errors .................................... 35

|  |  |
|---|---|
| 2.6 Executable Octave Programs | 36 |
| 2.7 Comments in Octave Programs | 37 |
|     2.7.1 Single Line Comments | 37 |
|     2.7.2 Block Comments | 37 |
|     2.7.3 Comments and the Help System | 38 |

## 3 Data Types .................................................. 39

|  |  |
|---|---|
| 3.1 Built-in Data Types | 39 |
|     3.1.1 Numeric Objects | 42 |
|     3.1.2 Missing Data | 43 |
|     3.1.3 String Objects | 43 |
|     3.1.4 Data Structure Objects | 43 |
|     3.1.5 Cell Array Objects | 44 |
| 3.2 User-defined Data Types | 44 |
| 3.3 Object Sizes | 44 |

## 4 Numeric Data Types ...................................... 47

|  |  |
|---|---|
| 4.1 Matrices | 48 |
|     4.1.1 Empty Matrices | 51 |
| 4.2 Ranges | 52 |
| 4.3 Single Precision Data Types | 53 |
| 4.4 Integer Data Types | 54 |
|     4.4.1 Integer Arithmetic | 56 |
| 4.5 Bit Manipulations | 57 |
| 4.6 Logical Values | 60 |
| 4.7 Promotion and Demotion of Data Types | 61 |
| 4.8 Predicates for Numeric Objects | 62 |

## 5 Strings ...................................................... 67

|  |  |
|---|---|
| 5.1 Escape Sequences in String Constants | 67 |
| 5.2 Character Arrays | 68 |
| 5.3 Creating Strings | 69 |
|     5.3.1 Concatenating Strings | 70 |
|     5.3.2 Converting Numerical Data to Strings | 73 |
| 5.4 Comparing Strings | 76 |
| 5.5 Manipulating Strings | 77 |
| 5.6 String Conversions | 91 |
| 5.7 Character Class Functions | 96 |

# 6 Data Containers .................................................... 99
## 6.1 Structures ...................................................... 99
### 6.1.1 Basic Usage and Examples ................................... 99
### 6.1.2 Structure Arrays ............................................ 103
### 6.1.3 Creating Structures ........................................ 104
### 6.1.4 Manipulating Structures .................................... 107
### 6.1.5 Processing Data in Structures .............................. 111
## 6.2 Cell Arrays .................................................... 112
### 6.2.1 Basic Usage of Cell Arrays ................................. 112
### 6.2.2 Creating Cell Arrays ....................................... 114
### 6.2.3 Indexing Cell Arrays ....................................... 116
### 6.2.4 Cell Arrays of Strings ..................................... 118
### 6.2.5 Processing Data in Cell Arrays ............................. 119
## 6.3 Comma Separated Lists .......................................... 120
### 6.3.1 Comma Separated Lists Generated from Cell Arrays ........... 121
### 6.3.2 Comma Separated Lists Generated from Structure Arrays ...... 122

# 7 Variables .......................................................... 123
## 7.1 Global Variables ................................................ 124
## 7.2 Persistent Variables ............................................ 126
## 7.3 Status of Variables ............................................. 127

# 8 Expressions ........................................................ 135
## 8.1 Index Expressions ............................................... 135
### 8.1.1 Advanced Indexing .......................................... 136
## 8.2 Calling Functions ............................................... 139
### 8.2.1 Call by Value .............................................. 140
### 8.2.2 Recursion .................................................. 141
## 8.3 Arithmetic Operators ............................................ 141
## 8.4 Comparison Operators ............................................ 145
## 8.5 Boolean Expressions ............................................. 146
### 8.5.1 Element-by-element Boolean Operators ....................... 146
### 8.5.2 Short-circuit Boolean Operators ............................ 148
## 8.6 Assignment Expressions .......................................... 149
## 8.7 Increment Operators ............................................. 152
## 8.8 Operator Precedence ............................................. 152

# 9 Evaluation ......................................................... 155
## 9.1 Calling a Function by its Name .................................. 155
## 9.2 Evaluation in a Different Context ............................... 157

## 10 Statements ... 159

- 10.1 The if Statement ... 159
- 10.2 The switch Statement ... 161
  - 10.2.1 Notes for the C Programmer ... 162
- 10.3 The while Statement ... 163
- 10.4 The do-until Statement ... 164
- 10.5 The for Statement ... 164
  - 10.5.1 Looping Over Structure Elements ... 165
- 10.6 The break Statement ... 166
- 10.7 The continue Statement ... 167
- 10.8 The unwind_protect Statement ... 168
- 10.9 The try Statement ... 168
- 10.10 Continuation Lines ... 169

## 11 Functions and Scripts ... 171

- 11.1 Introduction to Function and Script Files ... 171
- 11.2 Defining Functions ... 171
- 11.3 Multiple Return Values ... 174
- 11.4 Variable-length Argument Lists ... 182
- 11.5 Ignoring Arguments ... 184
- 11.6 Variable-length Return Lists ... 184
- 11.7 Returning from a Function ... 185
- 11.8 Default Arguments ... 186
- 11.9 Function Files ... 187
  - 11.9.1 Manipulating the Load Path ... 189
  - 11.9.2 Subfunctions ... 192
  - 11.9.3 Private Functions ... 192
  - 11.9.4 Nested Functions ... 192
  - 11.9.5 Overloading and Autoloading ... 195
  - 11.9.6 Function Locking ... 196
  - 11.9.7 Function Precedence ... 197
- 11.10 Script Files ... 198
- 11.11 Function Handles, Anonymous Functions, Inline Functions ... 199
  - 11.11.1 Function Handles ... 199
  - 11.11.2 Anonymous Functions ... 201
  - 11.11.3 Inline Functions ... 202
- 11.12 Commands ... 203
- 11.13 Organization of Functions Distributed with Octave ... 203

## 12 Errors and Warnings ... 205

- 12.1 Handling Errors ... 205
  - 12.1.1 Raising Errors ... 205
  - 12.1.2 Catching Errors ... 208
  - 12.1.3 Recovering From Errors ... 211
- 12.2 Handling Warnings ... 212
  - 12.2.1 Issuing Warnings ... 212
  - 12.2.2 Enabling and Disabling Warnings ... 218

## 13 Debugging ... 219
- 13.1 Entering Debug Mode ... 219
- 13.2 Leaving Debug Mode ... 220
- 13.3 Breakpoints ... 220
- 13.4 Debug Mode ... 223
- 13.5 Call Stack ... 224
- 13.6 Profiling ... 225
- 13.7 Profiler Example ... 227

## 14 Input and Output ... 231
- 14.1 Basic Input and Output ... 231
  - 14.1.1 Terminal Output ... 231
    - 14.1.1.1 Paging Screen Output ... 234
  - 14.1.2 Terminal Input ... 236
  - 14.1.3 Simple File I/O ... 238
    - 14.1.3.1 Saving Data on Unexpected Exits ... 247
- 14.2 C-Style I/O Functions ... 249
  - 14.2.1 Opening and Closing Files ... 249
  - 14.2.2 Simple Output ... 251
  - 14.2.3 Line-Oriented Input ... 252
  - 14.2.4 Formatted Output ... 253
  - 14.2.5 Output Conversion for Matrices ... 255
  - 14.2.6 Output Conversion Syntax ... 255
  - 14.2.7 Table of Output Conversions ... 256
  - 14.2.8 Integer Conversions ... 257
  - 14.2.9 Floating-Point Conversions ... 258
  - 14.2.10 Other Output Conversions ... 258
  - 14.2.11 Formatted Input ... 259
  - 14.2.12 Input Conversion Syntax ... 260
  - 14.2.13 Table of Input Conversions ... 261
  - 14.2.14 Numeric Input Conversions ... 262
  - 14.2.15 String Input Conversions ... 262
  - 14.2.16 Binary I/O ... 262
  - 14.2.17 Temporary Files ... 265
  - 14.2.18 End of File and Errors ... 266
  - 14.2.19 File Positioning ... 268

## 15 Plotting ... 271
- 15.1 Introduction to Plotting ... 271
- 15.2 High-Level Plotting ... 271
  - 15.2.1 Two-Dimensional Plots ... 271
    - 15.2.1.1 Axis Configuration ... 299
    - 15.2.1.2 Two-dimensional Function Plotting ... 301
    - 15.2.1.3 Two-dimensional Geometric Shapes ... 304
  - 15.2.2 Three-Dimensional Plots ... 305
    - 15.2.2.1 Aspect Ratio ... 321
    - 15.2.2.2 Three-dimensional Function Plotting ... 322

|          |           |                                                  |     |
|----------|-----------|--------------------------------------------------|-----|
|          | 15.2.2.3  | Three-dimensional Geometric Shapes               | 325 |
| 15.2.3   | Plot Annotations                                              | 327 |
| 15.2.4   | Multiple Plots on One Page                                    | 333 |
| 15.2.5   | Multiple Plot Windows                                         | 335 |
| 15.2.6   | Manipulation of Plot Objects                                  | 335 |
| 15.2.7   | Manipulation of Plot Windows                                  | 337 |
| 15.2.8   | Use of the `interpreter` Property                             | 341 |
| 15.2.9   | Printing and Saving Plots                                     | 344 |
| 15.2.10  | Interacting with Plots                                        | 350 |
| 15.2.11  | Test Plotting Functions                                       | 351 |
| 15.3 Graphics Data Structures                                              | 352 |
| 15.3.1   | Introduction to Graphics Structures                           | 352 |
| 15.3.2   | Graphics Objects                                              | 354 |
|          | 15.3.2.1  | Creating Graphics Objects                        | 354 |
|          | 15.3.2.2  | Handle Functions                                 | 357 |
| 15.3.3   | Graphics Object Properties                                    | 362 |
|          | 15.3.3.1  | Root Figure Properties                           | 362 |
|          | 15.3.3.2  | Figure Properties                                | 364 |
|          | 15.3.3.3  | Axes Properties                                  | 367 |
|          | 15.3.3.4  | Line Properties                                  | 372 |
|          | 15.3.3.5  | Text Properties                                  | 374 |
|          | 15.3.3.6  | Image Properties                                 | 376 |
|          | 15.3.3.7  | Patch Properties                                 | 377 |
|          | 15.3.3.8  | Surface Properties                               | 380 |
| 15.3.4   | Searching Properties                                          | 382 |
| 15.3.5   | Managing Default Properties                                   | 383 |
| 15.4 Advanced Plotting                                                     | 384 |
| 15.4.1   | Colors                                                        | 384 |
| 15.4.2   | Line Styles                                                   | 385 |
| 15.4.3   | Marker Styles                                                 | 385 |
| 15.4.4   | Callbacks                                                     | 385 |
| 15.4.5   | Application-defined Data                                      | 387 |
| 15.4.6   | Object Groups                                                 | 387 |
|          | 15.4.6.1  | Data Sources in Object Groups                    | 392 |
|          | 15.4.6.2  | Area Series                                      | 392 |
|          | 15.4.6.3  | Bar Series                                       | 393 |
|          | 15.4.6.4  | Contour Groups                                   | 394 |
|          | 15.4.6.5  | Error Bar Series                                 | 395 |
|          | 15.4.6.6  | Line Series                                      | 396 |
|          | 15.4.6.7  | Quiver Group                                     | 396 |
|          | 15.4.6.8  | Scatter Group                                    | 397 |
|          | 15.4.6.9  | Stair Group                                      | 398 |
|          | 15.4.6.10 | Stem Series                                      | 398 |
|          | 15.4.6.11 | Surface Group                                    | 399 |
| 15.4.7   | Graphics Toolkits                                             | 400 |
|          | 15.4.7.1  | Customizing Toolkit Behavior                     | 400 |

# 16 Matrix Manipulation ... 403
- 16.1 Finding Elements and Checking Conditions ... 403
- 16.2 Rearranging Matrices ... 407
- 16.3 Special Utility Matrices ... 416
- 16.4 Famous Matrices ... 424

# 17 Arithmetic ... 433
- 17.1 Exponents and Logarithms ... 433
- 17.2 Complex Arithmetic ... 435
- 17.3 Trigonometry ... 436
- 17.4 Sums and Products ... 440
- 17.5 Utility Functions ... 441
- 17.6 Special Functions ... 449
- 17.7 Rational Approximations ... 456
- 17.8 Coordinate Transformations ... 457
- 17.9 Mathematical Constants ... 458

# 18 Linear Algebra ... 463
- 18.1 Techniques Used for Linear Algebra ... 463
- 18.2 Basic Matrix Functions ... 463
- 18.3 Matrix Factorizations ... 470
- 18.4 Functions of a Matrix ... 481
- 18.5 Specialized Solvers ... 483

# 19 Vectorization and Faster Code Execution ... 487
- 19.1 Basic Vectorization ... 487
- 19.2 Broadcasting ... 489
  - 19.2.1 Broadcasting and Legacy Code ... 492
- 19.3 Function Application ... 492
- 19.4 Accumulation ... 497
- 19.5 JIT Compiler ... 499
- 19.6 Miscellaneous Techniques ... 500
- 19.7 Examples ... 502

# 20 Nonlinear Equations ... 503
- 20.1 Solvers ... 503
- 20.2 Minimizers ... 506

## 21 Diagonal and Permutation Matrices ... 509
- 21.1 Creating and Manipulating Diagonal/Permutation Matrices ... 509
  - 21.1.1 Creating Diagonal Matrices ... 510
  - 21.1.2 Creating Permutation Matrices ... 510
  - 21.1.3 Explicit and Implicit Conversions ... 511
- 21.2 Linear Algebra with Diagonal/Permutation Matrices ... 512
  - 21.2.1 Expressions Involving Diagonal Matrices ... 512
  - 21.2.2 Expressions Involving Permutation Matrices ... 513
- 21.3 Functions That Are Aware of These Matrices ... 514
  - 21.3.1 Diagonal Matrix Functions ... 514
  - 21.3.2 Permutation Matrix Functions ... 514
- 21.4 Examples of Usage ... 514
- 21.5 Differences in Treatment of Zero Elements ... 515

## 22 Sparse Matrices ... 517
- 22.1 Creation and Manipulation of Sparse Matrices ... 517
  - 22.1.1 Storage of Sparse Matrices ... 517
  - 22.1.2 Creating Sparse Matrices ... 518
  - 22.1.3 Finding Information about Sparse Matrices ... 524
  - 22.1.4 Basic Operators and Functions on Sparse Matrices ... 527
    - 22.1.4.1 Sparse Functions ... 527
    - 22.1.4.2 Return Types of Operators and Functions ... 528
    - 22.1.4.3 Mathematical Considerations ... 529
- 22.2 Linear Algebra on Sparse Matrices ... 538
- 22.3 Iterative Techniques Applied to Sparse Matrices ... 546
- 22.4 Real Life Example using Sparse Matrices ... 554

## 23 Numerical Integration ... 559
- 23.1 Functions of One Variable ... 559
- 23.2 Orthogonal Collocation ... 566
- 23.3 Functions of Multiple Variables ... 567

## 24 Differential Equations ... 571
- 24.1 Ordinary Differential Equations ... 571
- 24.2 Differential-Algebraic Equations ... 573

## 25 Optimization ... 583
- 25.1 Linear Programming ... 583
- 25.2 Quadratic Programming ... 589
- 25.3 Nonlinear Programming ... 591
- 25.4 Linear Least Squares ... 593

# 26 Statistics ... 597
- 26.1 Descriptive Statistics ... 597
- 26.2 Basic Statistical Functions ... 603
- 26.3 Statistical Plots ... 607
- 26.4 Correlation and Regression Analysis ... 608
- 26.5 Distributions ... 610
- 26.6 Tests ... 618
- 26.7 Random Number Generation ... 625

# 27 Sets ... 633
- 27.1 Set Operations ... 633

# 28 Polynomial Manipulations ... 637
- 28.1 Evaluating Polynomials ... 637
- 28.2 Finding Roots ... 638
- 28.3 Products of Polynomials ... 639
- 28.4 Derivatives / Integrals / Transforms ... 642
- 28.5 Polynomial Interpolation ... 643
- 28.6 Miscellaneous Functions ... 651

# 29 Interpolation ... 653
- 29.1 One-dimensional Interpolation ... 653
- 29.2 Multi-dimensional Interpolation ... 657

# 30 Geometry ... 661
- 30.1 Delaunay Triangulation ... 661
    - 30.1.1 Plotting the Triangulation ... 663
    - 30.1.2 Identifying Points in Triangulation ... 665
- 30.2 Voronoi Diagrams ... 668
- 30.3 Convex Hull ... 671
- 30.4 Interpolation on Scattered Data ... 673

# 31 Signal Processing ... 675

# 32 Image Processing ... 689
- 32.1 Loading and Saving Images ... 689
- 32.2 Displaying Images ... 695
- 32.3 Representing Images ... 697
- 32.4 Plotting on top of Images ... 705
- 32.5 Color Conversion ... 706

## 33 Audio Processing . . . . . . . . . . . . . . . . . . . . . . . . . . . . . . . . . . . . 709
### 33.1 Audio File Utilities . . . . . . . . . . . . . . . . . . . . . . . . . . . . . . . . . . . . . . . 709
### 33.2 Audio Device Information . . . . . . . . . . . . . . . . . . . . . . . . . . . . . . . . . 709
### 33.3 Audio Player . . . . . . . . . . . . . . . . . . . . . . . . . . . . . . . . . . . . . . . . . . . . . . 710
#### 33.3.1 Playback . . . . . . . . . . . . . . . . . . . . . . . . . . . . . . . . . . . . . . . . . . . 710
#### 33.3.2 Properties . . . . . . . . . . . . . . . . . . . . . . . . . . . . . . . . . . . . . . . . . . 711
### 33.4 Audio Recorder . . . . . . . . . . . . . . . . . . . . . . . . . . . . . . . . . . . . . . . . . . . 712
#### 33.4.1 Recording . . . . . . . . . . . . . . . . . . . . . . . . . . . . . . . . . . . . . . . . . 712
#### 33.4.2 Data Retrieval . . . . . . . . . . . . . . . . . . . . . . . . . . . . . . . . . . . . . 712
#### 33.4.3 Properties . . . . . . . . . . . . . . . . . . . . . . . . . . . . . . . . . . . . . . . . . . 713
### 33.5 Audio Data Processing . . . . . . . . . . . . . . . . . . . . . . . . . . . . . . . . . . . . 713

## 34 Object Oriented Programming . . . . . . . . . . . . . . . . . . . . . . . . 717
### 34.1 Creating a Class . . . . . . . . . . . . . . . . . . . . . . . . . . . . . . . . . . . . . . . . . . . 717
### 34.2 Manipulating Classes . . . . . . . . . . . . . . . . . . . . . . . . . . . . . . . . . . . . . . 719
### 34.3 Indexing Objects . . . . . . . . . . . . . . . . . . . . . . . . . . . . . . . . . . . . . . . . . . 722
#### 34.3.1 Defining Indexing And Indexed Assignment . . . . . . . . . . . 722
#### 34.3.2 Indexed Assignment Optimization . . . . . . . . . . . . . . . . . . . . 726
### 34.4 Overloading Objects . . . . . . . . . . . . . . . . . . . . . . . . . . . . . . . . . . . . . . . 727
#### 34.4.1 Function Overloading . . . . . . . . . . . . . . . . . . . . . . . . . . . . . . . 727
#### 34.4.2 Operator Overloading . . . . . . . . . . . . . . . . . . . . . . . . . . . . . . . 727
#### 34.4.3 Precedence of Objects . . . . . . . . . . . . . . . . . . . . . . . . . . . . . . 728
### 34.5 Inheritance and Aggregation . . . . . . . . . . . . . . . . . . . . . . . . . . . . . . . 730

## 35 GUI Development . . . . . . . . . . . . . . . . . . . . . . . . . . . . . . . . . . . . . 735
### 35.1 I/O Dialogs . . . . . . . . . . . . . . . . . . . . . . . . . . . . . . . . . . . . . . . . . . . . . . . 735
### 35.2 Progress Bar . . . . . . . . . . . . . . . . . . . . . . . . . . . . . . . . . . . . . . . . . . . . . . 736
### 35.3 UI Elements . . . . . . . . . . . . . . . . . . . . . . . . . . . . . . . . . . . . . . . . . . . . . . 737
### 35.4 GUI Utility Functions . . . . . . . . . . . . . . . . . . . . . . . . . . . . . . . . . . . . . 737
### 35.5 User-Defined Preferences . . . . . . . . . . . . . . . . . . . . . . . . . . . . . . . . . . 739

## 36 System Utilities . . . . . . . . . . . . . . . . . . . . . . . . . . . . . . . . . . . . . . . 743
### 36.1 Timing Utilities . . . . . . . . . . . . . . . . . . . . . . . . . . . . . . . . . . . . . . . . . . . 743
### 36.2 Filesystem Utilities . . . . . . . . . . . . . . . . . . . . . . . . . . . . . . . . . . . . . . . . 754
### 36.3 File Archiving Utilities . . . . . . . . . . . . . . . . . . . . . . . . . . . . . . . . . . . . 763
### 36.4 Networking Utilities . . . . . . . . . . . . . . . . . . . . . . . . . . . . . . . . . . . . . . 766
#### 36.4.1 FTP Objects . . . . . . . . . . . . . . . . . . . . . . . . . . . . . . . . . . . . . . . 766
#### 36.4.2 URL Manipulation . . . . . . . . . . . . . . . . . . . . . . . . . . . . . . . . . 768
#### 36.4.3 Base64 and Binary Data Transmission . . . . . . . . . . . . . . . 769
### 36.5 Controlling Subprocesses . . . . . . . . . . . . . . . . . . . . . . . . . . . . . . . . . . 769
### 36.6 Process, Group, and User IDs . . . . . . . . . . . . . . . . . . . . . . . . . . . . . 777
### 36.7 Environment Variables . . . . . . . . . . . . . . . . . . . . . . . . . . . . . . . . . . . . 778
### 36.8 Current Working Directory . . . . . . . . . . . . . . . . . . . . . . . . . . . . . . . . 778
### 36.9 Password Database Functions . . . . . . . . . . . . . . . . . . . . . . . . . . . . . 780
### 36.10 Group Database Functions . . . . . . . . . . . . . . . . . . . . . . . . . . . . . . . 781
### 36.11 System Information . . . . . . . . . . . . . . . . . . . . . . . . . . . . . . . . . . . . . . 782
### 36.12 Hashing Functions . . . . . . . . . . . . . . . . . . . . . . . . . . . . . . . . . . . . . . . 786

# 37 Java Interface ... 787
## 37.1 Java Interface Functions ... 787
## 37.2 Dialog Box Functions ... 793
## 37.3 FAQ - Frequently asked Questions ... 796
### 37.3.1 How to distinguish between Octave and Matlab? ... 796
### 37.3.2 How to make Java classes available to Octave? ... 796
### 37.3.3 How to create an instance of a Java class? ... 798
### 37.3.4 How can I handle memory limitations? ... 798
### 37.3.5 Which TeX symbols are implemented in dialog functions? ... 799

# 38 Packages ... 801
## 38.1 Installing and Removing Packages ... 801
## 38.2 Using Packages ... 805
## 38.3 Administrating Packages ... 805
## 38.4 Creating Packages ... 805
### 38.4.1 The DESCRIPTION File ... 807
### 38.4.2 The INDEX File ... 809
### 38.4.3 PKG_ADD and PKG_DEL Directives ... 810
### 38.4.4 Missing Components ... 810

# Appendix A  External Code Interface ... 813
## A.1 Oct-Files ... 814
### A.1.1 Getting Started with Oct-Files ... 814
### A.1.2 Matrices and Arrays in Oct-Files ... 817
### A.1.3 Character Strings in Oct-Files ... 820
### A.1.4 Cell Arrays in Oct-Files ... 822
### A.1.5 Structures in Oct-Files ... 822
### A.1.6 Sparse Matrices in Oct-Files ... 824
#### A.1.6.1 Array and Sparse Class Differences ... 824
#### A.1.6.2 Creating Sparse Matrices in Oct-Files ... 825
#### A.1.6.3 Using Sparse Matrices in Oct-Files ... 828
### A.1.7 Accessing Global Variables in Oct-Files ... 829
### A.1.8 Calling Octave Functions from Oct-Files ... 830
### A.1.9 Calling External Code from Oct-Files ... 831
### A.1.10 Allocating Local Memory in Oct-Files ... 834
### A.1.11 Input Parameter Checking in Oct-Files ... 834
### A.1.12 Exception and Error Handling in Oct-Files ... 835
### A.1.13 Documentation and Test of Oct-Files ... 837
## A.2 Mex-Files ... 838
### A.2.1 Getting Started with Mex-Files ... 838
### A.2.2 Working with Matrices and Arrays in Mex-Files ... 840
### A.2.3 Character Strings in Mex-Files ... 842
### A.2.4 Cell Arrays with Mex-Files ... 843
### A.2.5 Structures with Mex-Files ... 844
### A.2.6 Sparse Matrices with Mex-Files ... 846
### A.2.7 Calling Other Functions in Mex-Files ... 849
## A.3 Standalone Programs ... 850

## Appendix B  Test and Demo Functions . . . . . . . . . . . . . . . . . . 853
B.1  Test Functions . . . . . . . . . . . . . . . . . . . . . . . . . . . . . . . . . . . . . . . . . . . . . 853
B.2  Demonstration Functions . . . . . . . . . . . . . . . . . . . . . . . . . . . . . . . . . . . 860

## Appendix C  Tips and Standards . . . . . . . . . . . . . . . . . . . . . . . . 865
C.1  Writing Clean Octave Programs . . . . . . . . . . . . . . . . . . . . . . . . . . . . 865
C.2  Tips on Writing Comments . . . . . . . . . . . . . . . . . . . . . . . . . . . . . . . . 865
C.3  Conventional Headers for Octave Functions . . . . . . . . . . . . . . . . . 866
C.4  Tips for Documentation Strings . . . . . . . . . . . . . . . . . . . . . . . . . . . . 867

## Appendix D  Contributing Guidelines . . . . . . . . . . . . . . . . . . . 875
D.1  How to Contribute . . . . . . . . . . . . . . . . . . . . . . . . . . . . . . . . . . . . . . . . 875
D.2  Building the Development Sources . . . . . . . . . . . . . . . . . . . . . . . . . 875
D.3  Basics of Generating a Changeset . . . . . . . . . . . . . . . . . . . . . . . . . . 875
D.4  General Guidelines . . . . . . . . . . . . . . . . . . . . . . . . . . . . . . . . . . . . . . . 877
D.5  Octave Sources (m-files) . . . . . . . . . . . . . . . . . . . . . . . . . . . . . . . . . . 878
D.6  C++ Sources . . . . . . . . . . . . . . . . . . . . . . . . . . . . . . . . . . . . . . . . . . . . 879
D.7  Other Sources . . . . . . . . . . . . . . . . . . . . . . . . . . . . . . . . . . . . . . . . . . . 880

## Appendix E  Obsolete Functions . . . . . . . . . . . . . . . . . . . . . . . . 881

## Appendix F  Known Causes of Trouble . . . . . . . . . . . . . . . . . . 885
F.1  Actual Bugs We Haven't Fixed Yet . . . . . . . . . . . . . . . . . . . . . . . . . 885
F.2  Reporting Bugs . . . . . . . . . . . . . . . . . . . . . . . . . . . . . . . . . . . . . . . . . . 885
    F.2.1  Have You Found a Bug? . . . . . . . . . . . . . . . . . . . . . . . . . . . . 885
    F.2.2  Where to Report Bugs . . . . . . . . . . . . . . . . . . . . . . . . . . . . . 886
    F.2.3  How to Report Bugs . . . . . . . . . . . . . . . . . . . . . . . . . . . . . . . 886
    F.2.4  Sending Patches for Octave . . . . . . . . . . . . . . . . . . . . . . . . 887
F.3  How To Get Help with Octave . . . . . . . . . . . . . . . . . . . . . . . . . . . . . 888

## Appendix G  Installing Octave . . . . . . . . . . . . . . . . . . . . . . . . . . 889
G.1  Build Dependencies . . . . . . . . . . . . . . . . . . . . . . . . . . . . . . . . . . . . . . 889
    G.1.1  Obtaining the Dependencies Automatically . . . . . . . . . . . . . . . . . . 889
    G.1.2  Build Tools . . . . . . . . . . . . . . . . . . . . . . . . . . . . . . . . . . . . . . 889
    G.1.3  External Packages . . . . . . . . . . . . . . . . . . . . . . . . . . . . . . . . 890
G.2  Running Configure and Make . . . . . . . . . . . . . . . . . . . . . . . . . . . . . 892
G.3  Compiling Octave with 64-bit Indexing . . . . . . . . . . . . . . . . . . . . . 896
G.4  Installation Problems . . . . . . . . . . . . . . . . . . . . . . . . . . . . . . . . . . . . . 899

## Appendix H  Emacs Octave Support . . . . . . . . . . . . . . . . . . . . 903
H.1  Installing EOS . . . . . . . . . . . . . . . . . . . . . . . . . . . . . . . . . . . . . . . . . . . 903
H.2  Using Octave Mode . . . . . . . . . . . . . . . . . . . . . . . . . . . . . . . . . . . . . . 903
H.3  Running Octave from Within Emacs . . . . . . . . . . . . . . . . . . . . . . . 907
H.4  Using the Emacs Info Reader for Octave . . . . . . . . . . . . . . . . . . . 908

## Appendix I     Grammar and Parser .................... 911
### I.1   Keywords ........................................................... 911
### I.2   Parser .............................................................. 911

## Appendix J     GNU GENERAL PUBLIC LICENSE .... 913

## Concept Index ............................................... 925

## Function Index .............................................. 935

## Operator Index .............................................. 949

# Preface

Octave was originally intended to be companion software for an undergraduate-level textbook on chemical reactor design being written by James B. Rawlings of the University of Wisconsin-Madison and John G. Ekerdt of the University of Texas.

Clearly, Octave is now much more than just another 'courseware' package with limited utility beyond the classroom. Although our initial goals were somewhat vague, we knew that we wanted to create something that would enable students to solve realistic problems, and that they could use for many things other than chemical reactor design problems. We find that most students pick up the basics of Octave quickly, and are using it confidently in just a few hours.

Although it was originally intended to be used to teach reactor design, it has been used in several other undergraduate and graduate courses in the Chemical Engineering Department at the University of Texas, and the math department at the University of Texas has been using it for teaching differential equations and linear algebra as well. More recently, Octave has been used as the primary computational tool for teaching Stanford's online Machine Learning class (`ml-class.org`) taught by Andrew Ng. Tens of thousands of students participated in the course.

If you find Octave useful, please let us know. We are always interested to find out how Octave is being used.

Virtually everyone thinks that the name Octave has something to do with music, but it is actually the name of one of John W. Eaton's former professors who wrote a famous textbook on chemical reaction engineering, and who was also well known for his ability to do quick 'back of the envelope' calculations. We hope that this software will make it possible for many people to do more ambitious computations just as easily.

Everyone is encouraged to share this software with others under the terms of the GNU General Public License (see Appendix J [Copying], page 913). You are also encouraged to help make Octave more useful by writing and contributing additional functions for it, and by reporting any problems you may have.

## Acknowledgements

Many people have contributed to Octave's development. The following people have helped code parts of Octave or aided in various other ways (listed alphabetically).

| | | |
|---|---|---|
| Ben Abbott | Drew Abbot | Andy Adler |
| Adam H. Aitkenhead | Giles Anderson | Joel Andersson |
| Lachlan Andrew | Pedro Angelo | Muthiah Annamalai |
| Markus Appel | Branden Archer | Willem Atsma |
| Marco Atzeri | Shai Ayal | Roger Banks |
| Ben Barrowes | Alexander Barth | David Bateman |
| Heinz Bauschke | Julien Bect | Stefan Beller |
| Roman Belov | Markus Bergholz | Karl Berry |
| David Billinghurst | Don Bindner | Jakub Bogusz |
| Moritz Borgmann | Paul Boven | Richard Bovey |
| John Bradshaw | Marcus Brinkmann | Max Brister |
| Remy Bruno | Clemens Buchacher | Ansgar Burchard |

Marco Caliari              Daniel Calvelo              John C. Campbell
Juan Pablo Carbajal        Jean-Francois Cardoso       Joao Cardoso
Larrie Carr                David Castelow              Vincent Cautaerts
Clinton Chee               Albert Chin-A-Young         Carsten Clark
Catalin Codreanu           J. D. Cole                  Martin Costabel
Michael Creel              Richard Crozier             Jeff Cunningham
Martin Dalecki             Jacob Dawid                 Jorge Barros de Abreu
Carlo de Falco             Thomas D. Dean              Philippe Defert
Bill Denney                Fabian Deutsch              Christos Dimitrakakis
Pantxo Diribarne           Vivek Dogra                 John Donoghue
David M. Doolin            Carn Draug                  Pascal A. Dupuis
John W. Eaton              Dirk Eddelbuettel           Pieter Eendebak
Paul Eggert                Stephen Eglen               Peter Ekberg
Edmund Grimley Evans       Rolf Fabian                 Gunnar Farnebck
Massimiliano Fasi          Stephen Fegan               Ramon Garcia Fernandez
Torsten Finke              Colin Foster                Jose Daniel Munoz Frias
Brad Froehle               Castor Fu                   Eduardo Gallestey
Walter Gautschi            Klaus Gebhardt              Driss Ghaddab
Eugenio Gianniti           Nicolo Giorgetti            Arun Giridhar
Michael D. Godfrey         Michael Goffioul            Glenn Golden
Tomislav Goles             Keith Goodman               Brian Gough
Michael C. Grant           Steffen Groot               Etienne Grossmann
David Grundberg            Kyle Guinn                  Vaibhav Gupta
Peter Gustafson            Kai Habel                   Patrick Hcker
William P. Y. Hadisoeseno  Jaroslav Hajek              Benjamin Hall
Kim Hansen                 Sren Hauberg                Dave Hawthorne
Daniel Heiserer            Martin Helm                 Stefan Hepp
Martin Hepperle            Jordi Gutirrez Hermoso      Yozo Hida
Ryan Hinton                Roman Hodek                 A. Scottedward Hodel
Richard Allan Holcombe     Tom Holroyd                 David Hoover
Kurt Hornik                Christopher Hulbert         Cyril Humbert
John Hunt                  Teemu Ikonen                Alan W. Irwin
Allan Jacobs               Geoff Jacobsen              Vytautas Janauskas
Mats Jansson               Cai Jianming                Steven G. Johnson
Heikki Junes               Matthias Jschke             Atsushi Kajita
Jarkko Kaleva              Avinoam Kalma               Mohamed Kamoun
Lute Kamstra               Fotios Kasolis              Thomas Kasper
Joel Keay                  Mumit Khan                  Paul Kienzle
Aaron A. King              Erik Kjellson               Arno J. Klaassen
Alexander Klein            Geoffrey Knauth             Heine Kolltveit
Ken Kouno                  Kacper Kowalik              Daniel Kraft
Nir Krakauer               Aravindh Krishnamoorthy     Oyvind Kristiansen
Artem Krosheninnikov       Piotr Krzyzanowski          Volker Kuhlmann
Ilya Kurdyukov             Tetsuro Kurita              Philipp Kutin
Miroslaw Kwasniak          Rafael Laboissiere          Kai Labusch
Claude Lacoursiere         Walter Landry               Bill Lash
Dirk Laurie                Maurice LeBrun              Friedrich Leisch

# Preface

Johannes Leuschner
Timo Lindfors
David Livings
Massimo Lorenzin
Hoxide Ma
Jens-Uwe Mager
Alexander Mamonov
Axel Mathi
Christoph Mayer
Ronald van der Meer
Thorsten Meyer
Mike Miller
Antoine Moreau
Hannes Mller

Iain Murray
Philip Nienhuis
Rick Niles
Patrick Noffke
Michael O'Brien
Kai T. Ohlhus
Luis F. Ortiz
Scott Pakin
Gabriele Pannocchia
Per Persson
Danilo Piazzalunga
Robert Platt
Tom Poage
Ondrej Popp
Konstantinos Poulios
Pooja Rao
Balint Reczey
Michael Reifenberger
Jason Riedy
Petter Risholm
Peter Rosin
Mark van Rossum
Kevin Ruland
Olli Saarela
Radek Salac
Aleksej Saushev
Julian Schnidder
Sebastian Schubert
Thomas L. Scofield
Vanya Sergeev
Baylis Shanks

Thorsten Liebig
Benjamin Lindner
Sebastien Loisel
Emil Lucretiu
Colin Macdonald
Stefan Mahr
Ricardo Marranita
Makoto Matsumoto
Laurent Mazet
Jlio Hoffimann Mendes
Stefan Miereis
Serviscope Minor
Kai P. Mueller
Victor Munoz

Carmen Navarrete
Al Niessner
Takuji Nishimura
Eric Norum
Peter O'Gorman
Arno Onken
Carl Osterwisch
Jos Luis Garca Pallero
Sylvain Pelissier
Primozz Peterlin
Nicholas Piper
Hans Ekkehard Plesser
Nathan Podlich
Jef Poskanzer
Jarno Rajahalme
James B. Rawlings
Joshua Redstone
Jens Restemeier
E. Joshua Rigler
Matthew W. Roberts
Andrew Ross
Joe Rothweiler
Kristian Rumberg
Toni Saarela
Mike Sander
Alois Schlgl
Sebastian Schoeps
Lasse Schuirmann
Daniel J. Sebald
Marko Seric
Andriy Shinkarchuck

Jyh-miin Lin
Ross Lippert
Erik de Castro Lopo
Yi-Hong Lyu
James Macnicol
Rob Mahurin
Orestes Mas
Tatsuro Matsuoka
G. D. McBain
Ed Meyer
Petr Mikulik
Stefan Monnier
Armin Mller
PrasannaKumar Muralidharan

Todd Neal
Felipe G. Nievinski
Kai Noda
Krzesimir Nowak
Thorsten Ohl
Valentin Ortega-Clavero
Janne Olavi Paanajrvi
Jason Alan Palmer
Rolando Pereira
Jim Peterson
Elias Pipping
Sergey Plotnikov
Orion Poplawski
Francesco Potort
Eduardo Ramos
Eric S. Raymond
Lukas Reichlin
Anthony Richardson
Sander van Rijn
Dmitry Roshchin
Fabio Rossi
David Rrich
Ryan Rusaw
Juhani Saastamoinen
Ben Sapp
Michel D. Schmid
Nicol N. Schraudolph
Ludwig Schwardt
Dmitri A. Sergatskov
Ahsan Ali Shahid
Robert T. Short

| | | |
|---|---|---|
| Joseph P. Skudlarek | John Smith | Julius Smith |
| Shan G. Smith | Peter L. Sondergaard | Joerg Specht |
| Quentin H. Spencer | Christoph Spiel | David Spies |
| Richard Stallman | Russell Standish | Brett Stewart |
| Doug Stewart | Jonathan Stickel | Judd Storrs |
| Thomas Stuart | Ivan Sutoris | John Swensen |
| Daisuke Takago | Ariel Tankus | Falk Tannhuser |
| Duncan Temple Lang | Matthew Tenny | Kris Thielemans |
| Georg Thimm | Corey Thomasson | Olaf Till |
| Christophe Tournery | Thomas Treichl | Karsten Trulsen |
| David Turner | Frederick Umminger | Utkarsh Upadhyay |
| Stefan van der Walt | Peter Van Wieren | James R. Van Zandt |
| Risto Vanhanen | Gregory Vanuxem | Mihas Varantsou |
| Ivana Varekova | Sbastien Villemot | Marco Vitetta |
| Daniel Wagenaar | Thomas Walter | Andreas Weber |
| Olaf Weber | Thomas Weber | Rik Wehbring |
| Bob Weigel | Andreas Weingessel | Martin Weiser |
| Michael Weitzel | David Wells | Joachim Wiesemann |
| Fook Fah Yap | Sean Young | Johannes Zarl |
| Michael Zeising | Federico Zenith | Alex Zvoleff |

Special thanks to the following people and organizations for supporting the development of Octave:

- The United States Department of Energy, through grant number DE-FG02-04ER25635.
- Ashok Krishnamurthy, David Hudak, Juan Carlos Chaves, and Stanley C. Ahalt of the Ohio Supercomputer Center.
- The National Science Foundation, through grant numbers CTS-0105360, CTS-9708497, CTS-9311420, CTS-8957123, and CNS-0540147.
- The industrial members of the Texas-Wisconsin Modeling and Control Consortium (TWMCC).
- The Paul A. Elfers Endowed Chair in Chemical Engineering at the University of Wisconsin-Madison.
- Digital Equipment Corporation, for an equipment grant as part of their External Research Program.
- Sun Microsystems, Inc., for an Academic Equipment grant.
- International Business Machines, Inc., for providing equipment as part of a grant to the University of Texas College of Engineering.
- Texaco Chemical Company, for providing funding to continue the development of this software.
- The University of Texas College of Engineering, for providing a Challenge for Excellence Research Supplement, and for providing an Academic Development Funds grant.
- The State of Texas, for providing funding through the Texas Advanced Technology Program under Grant No. 003658-078.
- Noel Bell, Senior Engineer, Texaco Chemical Company, Austin Texas.

- John A. Turner, Group Leader, Continuum Dynamics (CCS-2), Los Alamos National Laboratory, for registering the `octave.org` domain name.
- James B. Rawlings, Professor, University of Wisconsin-Madison, Department of Chemical and Biological Engineering.
- Richard Stallman, for writing GNU.

This project would not have been possible without the GNU software used in and to produce Octave.

## Citing Octave in Publications

In view of the many contributions made by numerous developers over many years it is common courtesy to cite Octave in publications when it has been used during the course of research or the preparation of figures. The `citation` function can automatically generate a recommended citation text for Octave or any of its packages. See the help text below on how to use `citation`.

`citation` [Command]
`citation package` [Command]

    Display instructions for citing GNU Octave or its packages in publications.

    When called without an argument, display information on how to cite the core GNU Octave system.

    When given a package name *package*, display information on citing the specific named package. Note that some packages may not yet have instructions on how to cite them.

    The GNU Octave developers and its active community of package authors have invested a lot of time and effort in creating GNU Octave as it is today. Please give credit where credit is due and cite GNU Octave and its packages when you use them.

## How You Can Contribute to Octave

There are a number of ways that you can contribute to help make Octave a better system. Perhaps the most important way to contribute is to write high-quality code for solving new problems, and to make your code freely available for others to use. See Appendix D [Contributing Guidelines], page 875, for detailed information on contributing new code.

If you find Octave useful, consider providing additional funding to continue its development. Even a modest amount of additional funding could make a significant difference in the amount of time that is available for development and support.

Donations supporting Octave development may be made on the web at https://my.fsf.org/donate/working-together/octave. These donations also help to support the Free Software Foundation

If you'd prefer to pay by check or money order, you can do so by sending a check to the FSF at the following address:

    Free Software Foundation
    51 Franklin Street, Suite 500
    Boston, MA 02110-1335
    USA

If you pay by check, please be sure to write "GNU Octave" in the memo field of your check.

If you cannot provide funding or contribute code, you can still help make Octave better and more reliable by reporting any bugs you find and by offering suggestions for ways to improve Octave. See Appendix F [Trouble], page 885, for tips on how to write useful bug reports.

## Distribution

Octave is *free* software. This means that everyone is free to use it and free to redistribute it on certain conditions. Octave is not, however, in the public domain. It is copyrighted and there are restrictions on its distribution, but the restrictions are designed to ensure that others will have the same freedom to use and redistribute Octave that you have. The precise conditions can be found in the GNU General Public License that comes with Octave and that also appears in Appendix J [Copying], page 913.

To download a copy of Octave, please visit `http://www.octave.org/download.html`.

# 1 A Brief Introduction to Octave

GNU Octave is a high-level language, primarily intended for numerical computations. It is typically used for such problems as solving linear and nonlinear equations, numerical linear algebra, statistical analysis, and for performing other numerical experiments. It may also be used as a batch-oriented language for automated data processing.

Until recently GNU Octave provided a command-line interface only with graphical plots displayed in separate windows. However, by default the current version runs with a graphical user interface.

GNU Octave is freely redistributable software. You may redistribute it and/or modify it under the terms of the GNU General Public License as published by the Free Software Foundation. The GPL is included in this manual, see Appendix J [Copying], page 913.

This manual provides comprehensive documentation on how to install, run, use, and extend GNU Octave. Additional chapters describe how to report bugs and help contribute code.

This document corresponds to Octave version 4.0.0.

## 1.1 Running Octave

On most systems, Octave is started with the shell command 'octave'. This starts the graphical user interface (GUI). The central window in the GUI is the Octave command-line interface. In this window Octave displays an initial message and then a prompt indicating it is ready to accept input. If you have chosen the traditional command-line interface then only the command prompt appears in the same window that was running a shell. In any case, you can immediately begin typing Octave commands.

If you get into trouble, you can usually interrupt Octave by typing `Control-C` (written `C-c` for short). `C-c` gets its name from the fact that you type it by holding down CTRL and then pressing C. Doing this will normally return you to Octave's prompt.

To exit Octave, type `quit` or `exit` at the Octave prompt.

On systems that support job control, you can suspend Octave by sending it a `SIGTSTP` signal, usually by typing `C-z`.

## 1.2 Simple Examples

The following chapters describe all of Octave's features in detail, but before doing that, it might be helpful to give a sampling of some of its capabilities.

If you are new to Octave, we recommend that you try these examples to begin learning Octave by using it. Lines marked like so, 'octave:13>', are lines you type, ending each with a carriage return. Octave will respond with an answer, or by displaying a graph.

### 1.2.1 Elementary Calculations

Octave can easily be used for basic numerical calculations. Octave knows about arithmetic operations (+,-,*,/), exponentiation (^), natural logarithms/exponents (log, exp), and the trigonometric functions (sin, cos, ...). Moreover, Octave calculations work on real or imaginary numbers (i,j). In addition, some mathematical constants such as the base of the natural logarithm (e) and the ratio of a circle's circumference to its diameter (pi) are pre-defined.

For example, to verify Euler's Identity,

$$e^{i\pi} = -1$$

type the following which will evaluate to `-1` within the tolerance of the calculation.

```
octave:1> exp (i*pi)
```

### 1.2.2 Creating a Matrix

Vectors and matrices are the basic building blocks for numerical analysis. To create a new matrix and store it in a variable so that you can refer to it later, type the command

```
octave:1> A = [ 1, 1, 2; 3, 5, 8; 13, 21, 34 ]
```

Octave will respond by printing the matrix in neatly aligned columns. Octave uses a comma or space to separate entries in a row, and a semicolon or carriage return to separate one row from the next. Ending a command with a semicolon tells Octave not to print the result of the command. For example,

```
octave:2> B = rand (3, 2);
```

will create a 3 row, 2 column matrix with each element set to a random value between zero and one.

To display the value of a variable, simply type the name of the variable at the prompt. For example, to display the value stored in the matrix `B`, type the command

```
octave:3> B
```

### 1.2.3 Matrix Arithmetic

Octave has a convenient operator notation for performing matrix arithmetic. For example, to multiply the matrix `A` by a scalar value, type the command

```
octave:4> 2 * A
```

To multiply the two matrices `A` and `B`, type the command

```
octave:5> A * B
```

and to form the matrix product $A^T A$, type the command

```
octave:6> A' * A
```

### 1.2.4 Solving Systems of Linear Equations

Systems of linear equations are ubiquitous in numerical analysis. To solve the set of linear equations `Ax = b`, use the left division operator, '\':

```
x = A \ b
```

This is conceptually equivalent to $A^{-1}b$, but avoids computing the inverse of a matrix directly.

If the coefficient matrix is singular, Octave will print a warning message and compute a minimum norm solution.

A simple example comes from chemistry and the need to obtain balanced chemical equations. Consider the burning of hydrogen and oxygen to produce water.

$$H_2 + O_2 \rightarrow H_2O$$

The equation above is not accurate. The Law of Conservation of Mass requires that the number of molecules of each type balance on the left- and right-hand sides of the equation. Writing the variable overall reaction with individual equations for hydrogen and oxygen one finds:

$$x_1 H_2 + x_2 O_2 \rightarrow H_2 O$$

$$H: \quad 2x_1 + 0x_2 \rightarrow 2$$

$$O: \quad 0x_1 + 2x_2 \rightarrow 1$$

The solution in Octave is found in just three steps.

```
octave:1> A = [ 2, 0; 0, 2 ];
octave:2> b = [ 2; 1 ];
octave:3> x = A \ b
```

## 1.2.5 Integrating Differential Equations

Octave has built-in functions for solving nonlinear differential equations of the form

$$\frac{dx}{dt} = f(x, t), \qquad x(t = t_0) = x_0$$

For Octave to integrate equations of this form, you must first provide a definition of the function $f(x, t)$. This is straightforward, and may be accomplished by entering the function body directly on the command line. For example, the following commands define the right-hand side function for an interesting pair of nonlinear differential equations. Note that while you are entering a function, Octave responds with a different prompt, to indicate that it is waiting for you to complete your input.

```
octave:1> function xdot = f (x, t)
>
>   r = 0.25;
>   k = 1.4;
>   a = 1.5;
>   b = 0.16;
>   c = 0.9;
>   d = 0.8;
>
>   xdot(1) = r*x(1)*(1 - x(1)/k) - a*x(1)*x(2)/(1 + b*x(1));
>   xdot(2) = c*a*x(1)*x(2)/(1 + b*x(1)) - d*x(2);
>
> endfunction
```

Given the initial condition

```
octave:2> x0 = [1; 2];
```

and the set of output times as a column vector (note that the first output time corresponds to the initial condition given above)

```
octave:3> t = linspace (0, 50, 200)';
```
it is easy to integrate the set of differential equations:
```
octave:4> x = lsode ("f", x0, t);
```
The function `lsode` uses the Livermore Solver for Ordinary Differential Equations, described in A. C. Hindmarsh, *ODEPACK, a Systematized Collection of ODE Solvers*, in: Scientific Computing, R. S. Stepleman et al. (Eds.), North-Holland, Amsterdam, 1983, pages 55–64.

### 1.2.6 Producing Graphical Output

To display the solution of the previous example graphically, use the command
```
octave:1> plot (t, x)
```
If you are using a graphical user interface, Octave will automatically create a separate window to display the plot.

To save a plot once it has been displayed on the screen, use the print command. For example,
```
print -dpdf foo.pdf
```
will create a file called 'foo.pdf' that contains a rendering of the current plot in Portable Document Format. The command
```
help print
```
explains more options for the `print` command and provides a list of additional output file formats.

### 1.2.7 Editing What You Have Typed

At the Octave prompt, you can recall, edit, and reissue previous commands using Emacs- or vi-style editing commands. The default keybindings use Emacs-style commands. For example, to recall the previous command, press `Control-p` (written `C-p` for short). Doing this will normally bring back the previous line of input. `C-n` will bring up the next line of input, `C-b` will move the cursor backward on the line, `C-f` will move the cursor forward on the line, etc.

A complete description of the command line editing capability is given in this manual, see Section 2.4 [Command Line Editing], page 25.

### 1.2.8 Help and Documentation

Octave has an extensive help facility. The same documentation that is available in printed form is also available from the Octave prompt, because both forms of the documentation are created from the same input file.

In order to get good help you first need to know the name of the command that you want to use. The name of this function may not always be obvious, but a good place to start is to type `help --list`. This will show you all the operators, keywords, built-in functions, and loadable functions available in the current session of Octave. An alternative is to search the documentation using the `lookfor` function (described in Section 2.3 [Getting Help], page 20).

Once you know the name of the function you wish to use, you can get more help on the function by simply including the name as an argument to help. For example,

```
help plot
```
will display the help text for the `plot` function.

Octave sends output that is too long to fit on one screen through a pager like `less` or `more`. Type a RET to advance one line, a SPC to advance one page, and Q to quit the pager.

The part of Octave's help facility that allows you to read the complete text of the printed manual from within Octave normally uses a separate program called Info. When you invoke Info you will be put into a menu driven program that contains the entire Octave manual. Help for using Info is provided in this manual, see Section 2.3 [Getting Help], page 20.

## 1.3 Conventions

This section explains the notational conventions that are used in this manual. You may want to skip this section and refer back to it later.

### 1.3.1 Fonts

Examples of Octave code appear in this font or form: `svd (a)`. Names that represent variables or function arguments appear in this font or form: *first-number*. Commands that you type at the shell prompt appear in this font or form: 'octave --no-init-file'. Commands that you type at the Octave prompt sometimes appear in this font or form: `foo --bar --baz`. Specific keys on your keyboard appear in this font or form: RET.

### 1.3.2 Evaluation Notation

In the examples in this manual, results from expressions that you evaluate are indicated with '$\Rightarrow$'. For example:

```
sqrt (2)
     ⇒ 1.4142
```

You can read this as "`sqrt (2)` evaluates to 1.4142".

In some cases, matrix values that are returned by expressions are displayed like this

```
[1, 2; 3, 4] == [1, 3; 2, 4]
     ⇒ [ 1, 0; 0, 1 ]
```

and in other cases, they are displayed like this

```
eye (3)
     ⇒   1  0  0
         0  1  0
         0  0  1
```

in order to clearly show the structure of the result.

Sometimes to help describe one expression, another expression is shown that produces identical results. The exact equivalence of expressions is indicated with '$\equiv$'. For example:

```
rot90 ([1, 2; 3, 4], -1)
≡
rot90 ([1, 2; 3, 4], 3)
≡
rot90 ([1, 2; 3, 4], 7)
```

### 1.3.3 Printing Notation

Many of the examples in this manual print text when they are evaluated. In this manual the printed text resulting from an example is indicated by '⊣'. The value that is returned by evaluating the expression is displayed with '⇒' (`1` in the next example) and follows on a separate line.

```
printf ("foo %s\n", "bar")
    ⊣ foo bar
    ⇒ 1
```

### 1.3.4 Error Messages

Some examples signal errors. This normally displays an error message on your terminal. Error messages are shown on a line beginning with `error:`.

```
fieldnames ([1, 2; 3, 4])
error: fieldnames: Invalid input argument
```

### 1.3.5 Format of Descriptions

Functions and commands are described in this manual in a uniform format. The first line of a description contains the name of the item followed by its arguments, if any. The category—function, command, or whatever—is printed next to the right margin. If there are multiple ways to invoke the function then each allowable form is listed.

The description follows on succeeding lines, sometimes with examples.

#### 1.3.5.1 A Sample Function Description

In a function description, the name of the function being described appears first. It is followed on the same line by a list of parameters. The names used for the parameters are also used in the body of the description.

After all of the calling forms have been enumerated, the next line is a concise one-sentence summary of the function.

After the summary there may be documentation on the inputs and outputs, examples of function usage, notes about the algorithm used, and references to related functions.

Here is a description of an imaginary function `foo`:

**foo** (*x*) [Function File]
**foo** (*x*, *y*) [Function File]
**foo** (*x*, *y*, ...) [Function File]
> The function `foo` subtracts *x* from *y*, then adds the remaining arguments to the result.
>
> If *y* is not supplied, then the number 19 is used by default.
>
> Example:
>
>     foo (1, [3, 5], 3, 9)
>          ⇒ [ 14, 16 ]
>     foo (5)
>          ⇒ 14
>
> More generally,
>
>     foo (w, x, y, ...)
>     ≡
>     x - w + y + ...
>
> **See also:** bar

Any parameter whose name contains the name of a type (e.g., *integer* or *matrix*) is expected to be of that type. Parameters named *object* may be of any type. Parameters with other sorts of names (e.g., *new_file*) are discussed specifically in the description of the function. In some sections, features common to parameters of several functions are described at the beginning.

Functions in Octave may be defined in several different ways. The category name for functions may include a tag that indicates the way that the function is defined. These additional tags include

Function File
: The function described is defined using Octave commands stored in a text file. See Section 11.9 [Function Files], page 187.

Built-in Function
: The function described is written in a language like C++, C, or Fortran, and is part of the compiled Octave binary.

Loadable Function
: The function described is written in a language like C++, C, or Fortran. On systems that support dynamic linking of user-supplied functions, it may be automatically linked while Octave is running, but only if it is needed. See Appendix A [External Code Interface], page 813.

Mapping Function
: The function described works element-by-element for matrix and vector arguments.

### 1.3.5.2 A Sample Command Description

Command descriptions have a format similar to function descriptions, except that the word 'Function' is replaced by 'Command'. Commands are functions that may be called without surrounding their arguments in parentheses. For example, here is the description for Octave's `diary` command:

`diary`                                                                [Command]
`diary` *on*                                                           [Command]
`diary` *off*                                                          [Command]
`diary` *filename*                                                     [Command]

Record a list of all commands *and* the output they produce, mixed together just as they appear on the terminal.

Valid options are:

on      Start recording a session in a file called '`diary`' in the current working directory.

off     Stop recording the session in the diary file.

*filename*   Record the session in the file named *filename*.

With no arguments, `diary` toggles the current diary state.

**See also:** history.

# 2 Getting Started

This chapter explains some of Octave's basic features, including how to start an Octave session, get help at the command prompt, edit the command line, and write Octave programs that can be executed as commands from your shell.

## 2.1 Invoking Octave from the Command Line

Normally, Octave is used interactively by running the program 'octave' without any arguments. Once started, Octave reads commands from the terminal until you tell it to exit.

You can also specify the name of a file on the command line, and Octave will read and execute the commands from the named file and then exit when it is finished.

You can further control how Octave starts by using the command-line options described in the next section, and Octave itself can remind you of the options available. Type 'octave --help' to display all available options and briefly describe their use ('octave -h' is a shorter equivalent).

### 2.1.1 Command Line Options

Here is a complete list of the command line options that Octave accepts.

`--built-in-docstrings-file` *filename*
: Specify the name of the file containing documentation strings for the built-in functions of Octave. This value is normally correct and should only need to specified in extraordinary situations.

`--debug`
`-d`
: Enter parser debugging mode. Using this option will cause Octave's parser to print a lot of information about the commands it reads, and is probably only useful if you are actually trying to debug the parser.

`--debug-jit`
: Enable JIT compiler debugging and tracing.

`--doc-cache-file` *filename*
: Specify the name of the doc cache file to use. The value of *filename* specified on the command line will override any value of `OCTAVE_DOC_CACHE_FILE` found in the environment, but not any commands in the system or user startup files that use the `doc_cache_file` function.

`--echo-commands`
`-x`
: Echo commands as they are executed.

`--eval` *code*
: Evaluate *code* and exit when finished unless '`--persist`' is also specified.

`--exec-path` *path*
: Specify the path to search for programs to run. The value of *path* specified on the command line will override any value of `OCTAVE_EXEC_PATH` found in the environment, but not any commands in the system or user startup files that set the built-in variable `EXEC_PATH`.

`--force-gui`
: Force the graphical user interface (GUI) to start.

`--help`
`-h`
`-?`
: Print short help message and exit.

`--image-path` *path*
: Add path to the head of the search path for images. The value of *path* specified on the command line will override any value of `OCTAVE_IMAGE_PATH` found in the environment, but not any commands in the system or user startup files that set the built-in variable `IMAGE_PATH`.

`--info-file` *filename*
: Specify the name of the info file to use. The value of *filename* specified on the command line will override any value of `OCTAVE_INFO_FILE` found in the environment, but not any commands in the system or user startup files that use the `info_file` function.

`--info-program` *program*
: Specify the name of the info program to use. The value of *program* specified on the command line will override any value of `OCTAVE_INFO_PROGRAM` found in the environment, but not any commands in the system or user startup files that use the `info_program` function.

`--interactive`
`-i`
: Force interactive behavior. This can be useful for running Octave via a remote shell command or inside an Emacs shell buffer. For another way to run Octave within Emacs, see Appendix H [Emacs Octave Support], page 903.

`--jit-compiler`
: Enable the JIT compiler used for accelerating loops.

`--line-editing`
: Force readline use for command-line editing.

`--no-gui`
: Disable the graphical user interface (GUI) and use the command line interface (CLI) instead.

`--no-history`
`-H`
: Disable recording of command-line history.

`--no-init-file`
: Don't read the initialization files '`~/.octaverc`' and '`.octaverc`'.

`--no-init-path`
: Don't initialize the search path for function files to include default locations.

`--no-line-editing`
: Disable command-line editing.

`--no-site-file`
: Don't read the site-wide '`octaverc`' initialization files.

`--no-window-system`
`-W`    Disable use of a windowing system including graphics. This forces a strictly terminal-only environment.

`--norc`
`-f`    Don't read any of the system or user initialization files at startup. This is equivalent to using both of the options '`--no-init-file`' and '`--no-site-file`'.

`--path` *path*
`-p` *path*    Add path to the head of the search path for function files. The value of *path* specified on the command line will override any value of `OCTAVE_PATH` found in the environment, but not any commands in the system or user startup files that set the internal load path through one of the path functions.

`--persist`
         Go to interactive mode after '`--eval`' or reading from a file named on the command line.

`--silent`
`--quiet`
`-q`    Don't print the usual greeting and version message at startup.

`--texi-macros-file` *filename*
         Specify the name of the file containing Texinfo macros for use by makeinfo.

`--traditional`
`--braindead`
         For compatibility with MATLAB, set initial values for user preferences to the following values

```
                PS1                             = ">> "
                PS2                             = ""
                beep_on_error                   = true
                confirm_recursive_rmdir         = false
                crash_dumps_octave_core         = false
                disable_diagonal_matrix         = true
                disable_permutation_matrix      = true
                disable_range                   = true
                fixed_point_format              = true
                history_timestamp_format_string = "%%-- %D %I:%M %p --%%"
                page_screen_output              = false
                print_empty_dimensions          = false
                save_default_options            = "-mat-binary"
                struct_levels_to_print          = 0
```

and disable the following warnings

```
                Octave:abbreviated-property-match
                Octave:fopen-file-in-path
                Octave:function-name-clash
                Octave:load-file-in-path
                Octave:possible-matlab-short-circuit-operator
```

Note that this does not enable the `Octave:language-extension` warning, which you might want if you want to be told about writing code that works in Octave but not MATLAB (see [warning], page 212, [warning_ids], page 214).

`--verbose`
`-V`      Turn on verbose output.

`--version`
`-v`      Print the program version number and exit.

`file`    Execute commands from *file*. Exit when done unless '`--persist`' is also specified.

Octave also includes several functions which return information about the command line, including the number of arguments and all of the options.

`argv ()`                                                                  [Built-in Function]

Return the command line arguments passed to Octave.

For example, if you invoked Octave using the command

```
octave --no-line-editing --silent
```

`argv` would return a cell array of strings with the elements '`--no-line-editing`' and '`--silent`'.

If you write an executable Octave script, `argv` will return the list of arguments passed to the script. See Section 2.6 [Executable Octave Programs], page 36, for an example of how to create an executable Octave script.

`program_name ()`                                                          [Built-in Function]

Return the last component of the value returned by `program_invocation_name`.

**See also:** [program_invocation_name], page 18.

`program_invocation_name ()`                                               [Built-in Function]

Return the name that was typed at the shell prompt to run Octave.

If executing a script from the command line (e.g., `octave foo.m`) or using an executable Octave script, the program name is set to the name of the script. See Section 2.6 [Executable Octave Programs], page 36, for an example of how to create an executable Octave script.

**See also:** [program_name], page 18.

Here is an example of using these functions to reproduce the command line which invoked Octave.

```
printf ("%s", program_name ());
arg_list = argv ();
for i = 1:nargin
  printf (" %s", arg_list{i});
endfor
printf ("\n");
```

See Section 6.2.3 [Indexing Cell Arrays], page 116, for an explanation of how to retrieve objects from cell arrays, and Section 11.2 [Defining Functions], page 171, for information about the variable `nargin`.

## 2.1.2 Startup Files

When Octave starts, it looks for commands to execute from the files in the following list. These files may contain any valid Octave commands, including function definitions.

*octave-home*/share/octave/site/m/startup/octaverc

> where *octave-home* is the directory in which Octave is installed (the default is '/usr/local'). This file is provided so that changes to the default Octave environment can be made globally for all users at your site for all versions of Octave you have installed. Care should be taken when making changes to this file since all users of Octave at your site will be affected. The default file may be overridden by the environment variable OCTAVE_SITE_INITFILE.

*octave-home*/share/octave/*version*/m/startup/octaverc

> where *octave-home* is the directory in which Octave is installed (the default is '/usr/local'), and *version* is the version number of Octave. This file is provided so that changes to the default Octave environment can be made globally for all users of a particular version of Octave. Care should be taken when making changes to this file since all users of Octave at your site will be affected. The default file may be overridden by the environment variable OCTAVE_VERSION_INITFILE.

~/.octaverc

> This file is used to make personal changes to the default Octave environment.

.octaverc

> This file can be used to make changes to the default Octave environment for a particular project. Octave searches for this file in the current directory after it reads '~/.octaverc'. Any use of the cd command in the '~/.octaverc' file will affect the directory where Octave searches for '.octaverc'.
>
> If you start Octave in your home directory, commands from the file '~/.octaverc' will only be executed once.

A message will be displayed as each of the startup files is read if you invoke Octave with the '--verbose' option but without the '--silent' option.

## 2.2 Quitting Octave

Shutdown is initiated with the exit or quit commands (they are equivalent). Similar to startup, Octave has a shutdown process that can be customized by user script files. During shutdown Octave will search for the script file 'finish.m' in the function load path. Commands to save all workspace variables or cleanup temporary files may be placed there. Additional functions to execute on shutdown may be registered with atexit.

| | |
|---|---|
| exit | [Built-in Function] |
| exit (*status*) | [Built-in Function] |
| quit | [Built-in Function] |
| quit (*status*) | [Built-in Function] |

> Exit the current Octave session.
>
> If the optional integer value *status* is supplied, pass that value to the operating system as Octave's exit status. The default value is zero.

When exiting, Octave will attempt to run the m-file 'finish.m' if it exists. User commands to save the workspace or clean up temporary files may be placed in that file. Alternatively, another m-file may be scheduled to run using `atexit`.

**See also:** [atexit], page 20.

`atexit (fcn)` [Built-in Function]
`atexit (fcn, flag)` [Built-in Function]

Register a function to be called when Octave exits.

For example,

```
function last_words ()
  disp ("Bye bye");
endfunction
atexit ("last_words");
```

will print the message "Bye bye" when Octave exits.

The additional argument *flag* will register or unregister *fcn* from the list of functions to be called when Octave exits. If *flag* is true, the function is registered, and if *flag* is false, it is unregistered. For example, after registering the function `last_words` above,

```
atexit ("last_words", false);
```

will remove the function from the list and Octave will not call `last_words` when it exits.

Note that `atexit` only removes the first occurrence of a function from the list, so if a function was placed in the list multiple times with `atexit`, it must also be removed from the list multiple times.

**See also:** [quit], page 19.

## 2.3 Commands for Getting Help

The entire text of this manual is available from the Octave prompt via the command *doc*. In addition, the documentation for individual user-written functions and variables is also available via the `help` command. This section describes the commands used for reading the manual and the documentation strings for user-supplied functions and variables. See Section 11.9 [Function Files], page 187, for more information about how to document the functions you write.

`help name` [Command]
`help --list` [Command]
`help .` [Command]
`help` [Command]

Display the help text for *name*.

For example, the command `help help` prints a short message describing the `help` command.

Given the single argument `--list`, list all operators, keywords, built-in functions, and loadable functions available in the current session of Octave.

Given the single argument ., list all operators available in the current session of Octave.

If invoked without any arguments, `help` display instructions on how to access help from the command line.

The help command can provide information about most operators, for example `help +`, but not the comma and semicolon characters which are used by the Octave interpreter as command separators. For help on either of these type `help comma` or `help semicolon`.

**See also:** [doc], page 21, [lookfor], page 21, [which], page 132, [info], page 22.

doc *function_name* [Command]
doc [Command]

Display documentation for the function *function_name* directly from an online version of the printed manual, using the GNU Info browser.

If invoked without an argument, the manual is shown from the beginning.

For example, the command `doc rand` starts the GNU Info browser at the `rand` node in the online version of the manual.

Once the GNU Info browser is running, help for using it is available using the command `C-h`.

**See also:** [help], page 20.

lookfor *str* [Command]
lookfor *-all str* [Command]
[*fcn, help1str*] = lookfor (*str*) [Function File]
[*fcn, help1str*] = lookfor ("*-all*", *str*) [Function File]

Search for the string *str* in the documentation of all functions in the current function search path.

By default, `lookfor` looks for *str* in just the first sentence of the help string for each function found. The entire help text of each function can be searched by using the "`-all`" argument. All searches are case insensitive.

When called with no output arguments, `lookfor` prints the list of matching functions to the terminal. Otherwise, the output argument *fcns* contains the function names and *help1str* contains the first sentence from the help string of each function.

Programming Note: The ability of `lookfor` to correctly identify the first sentence of the help text is dependent on the format of the function's help. All Octave core functions are correctly formatted, but the same can not be guaranteed for external packages and user-supplied functions. Therefore, the use of the "`-all`" argument may be necessary to find related functions that are not a part of Octave.

The speed of lookup is greatly enhanced by having a cached documentation file. See `doc_cache_create` for more information.

**See also:** [help], page 20, [doc], page 21, [which], page 132, [path], page 190, [doc_cache_create], page 24.

To see what is new in the current release of Octave, use the `news` function.

`news`                                                                          [Command]
`news` *package*                                                                [Command]
> Display the current NEWS file for Octave or an installed package.
>
> When called without an argument, display the NEWS file for Octave.
>
> When given a package name *package*, display the current NEWS file for that package.
>
> **See also:** [ver], page 784, [pkg], page 801.

`info ()`                                                                       [Function File]
> Display contact information for the GNU Octave community.

`warranty ()`                                                                   [Built-in Function]
> Describe the conditions for copying and distributing Octave.

The following functions can be used to change which programs are used for displaying the documentation, and where the documentation can be found.

`val = info_file ()`                                                            [Built-in Function]
`old_val = info_file (`*new_val*`)`                                             [Built-in Function]
`info_file (`*new_val*`, "`*local*`")`                                          [Built-in Function]
> Query or set the internal variable that specifies the name of the Octave info file.
>
> The default value is '*octave-home*/info/octave.info', in which *octave-home* is the root directory of the Octave installation. The default value may be overridden by the environment variable `OCTAVE_INFO_FILE`, or the command line argument '`--info-file FNAME`'.
>
> When called from inside a function with the `"local"` option, the variable is changed locally for the function and any subroutines it calls. The original variable value is restored when exiting the function.
>
> **See also:** [info_program], page 22, [doc], page 21, [help], page 20, [makeinfo_program], page 23.

`val = info_program ()`                                                         [Built-in Function]
`old_val = info_program (`*new_val*`)`                                          [Built-in Function]
`info_program (`*new_val*`, "`*local*`")`                                       [Built-in Function]
> Query or set the internal variable that specifies the name of the info program to run.
>
> The default value is '*octave-home*/libexec/octave/*version*/exec/*arch*/info' in which *octave-home* is the root directory of the Octave installation, *version* is the Octave version number, and *arch* is the system type (for example, `i686-pc-linux-gnu`). The default value may be overridden by the environment variable `OCTAVE_INFO_PROGRAM`, or the command line argument '`--info-program NAME`'.
>
> When called from inside a function with the `"local"` option, the variable is changed locally for the function and any subroutines it calls. The original variable value is restored when exiting the function.
>
> **See also:** [info_file], page 22, [doc], page 21, [help], page 20, [makeinfo_program], page 23.

Chapter 2: Getting Started    23

`val = makeinfo_program ()`    [Built-in Function]
`old_val = makeinfo_program (new_val)`    [Built-in Function]
`makeinfo_program (new_val, "local")`    [Built-in Function]

> Query or set the internal variable that specifies the name of the program that Octave runs to format help text containing Texinfo markup commands.
>
> The default value is `makeinfo`.
>
> When called from inside a function with the `"local"` option, the variable is changed locally for the function and any subroutines it calls. The original variable value is restored when exiting the function.
>
> **See also:** [texi_macros_file], page 23, [info_file], page 22, [info_program], page 22, [doc], page 21, [help], page 20.

`val = texi_macros_file ()`    [Built-in Function]
`old_val = texi_macros_file (new_val)`    [Built-in Function]
`texi_macros_file (new_val, "local")`    [Built-in Function]

> Query or set the internal variable that specifies the name of the file containing Texinfo macros that are prepended to documentation strings before they are passed to makeinfo.
>
> The default value is '*octave-home*/share/octave/*version*/etc/macros.texi', in which *octave-home* is the root directory of the Octave installation, and *version* is the Octave version number. The default value may be overridden by the environment variable `OCTAVE_TEXI_MACROS_FILE`, or the command line argument '--texi-macros-file FNAME'.
>
> When called from inside a function with the `"local"` option, the variable is changed locally for the function and any subroutines it calls. The original variable value is restored when exiting the function.
>
> **See also:** [makeinfo_program], page 23.

`val = doc_cache_file ()`    [Built-in Function]
`old_val = doc_cache_file (new_val)`    [Built-in Function]
`doc_cache_file (new_val, "local")`    [Built-in Function]

> Query or set the internal variable that specifies the name of the Octave documentation cache file.
>
> A cache file significantly improves the performance of the `lookfor` command. The default value is '*octave-home*/share/octave/*version*/etc/doc-cache', in which *octave-home* is the root directory of the Octave installation, and *version* is the Octave version number. The default value may be overridden by the environment variable `OCTAVE_DOC_CACHE_FILE`, or the command line argument '--doc-cache-file FNAME'.
>
> When called from inside a function with the `"local"` option, the variable is changed locally for the function and any subroutines it calls. The original variable value is restored when exiting the function.
>
> **See also:** [doc_cache_create], page 24, [lookfor], page 21, [info_program], page 22, [doc], page 21, [help], page 20, [makeinfo_program], page 23.
>
> **See also:** [lookfor], page 21.

`val = built_in_docstrings_file ()` [Built-in Function]
`old_val = built_in_docstrings_file (`*`new_val`*`)` [Built-in Function]
`built_in_docstrings_file (`*`new_val`*`, "`*`local`*`")` [Built-in Function]

Query or set the internal variable that specifies the name of the file containing docstrings for built-in Octave functions.

The default value is '*octave-home*/share/octave/*version*/etc/built-in-docstrings', in which *octave-home* is the root directory of the Octave installation, and *version* is the Octave version number. The default value may be overridden by the environment variable `OCTAVE_BUILT_IN_DOCSTRINGS_FILE`, or the command line argument '`--built-in-docstrings-file FNAME`'.

Note: This variable is only used when Octave is initializing itself. Modifying it during a running session of Octave will have no effect.

`val = suppress_verbose_help_message ()` [Built-in Function]
`old_val = suppress_verbose_help_message (`*`new_val`*`)` [Built-in Function]
`suppress_verbose_help_message (`*`new_val`*`, "`*`local`*`")` [Built-in Function]

Query or set the internal variable that controls whether Octave will add additional help information to the end of the output from the `help` command and usage messages for built-in commands.

When called from inside a function with the `"local"` option, the variable is changed locally for the function and any subroutines it calls. The original variable value is restored when exiting the function.

The following functions are principally used internally by Octave for generating the documentation. They are documented here for completeness and because they may occasionally be useful for users.

`doc_cache_create (`*`out_file`*`, `*`directory`*`)` [Function File]
`doc_cache_create (`*`out_file`*`)` [Function File]
`doc_cache_create ()` [Function File]

Generate documentation cache for all functions in *directory*.

A documentation cache is generated for all functions in *directory* which may be a single string or a cell array of strings. The cache is used to speed up the function `lookfor`.

The cache is saved in the file *out_file* which defaults to the value '`doc-cache`' if not given.

If no directory is given (or it is the empty matrix), a cache for built-in operators, etc. is generated.

**See also:** [doc_cache_file], page 23, [lookfor], page 21, [path], page 190.

`[`*`text, format`*`] = get_help_text (`*`name`*`)` [Built-in Function]

Return the raw help text of function *name*.

The raw help text is returned in *text* and the format in *format* The format is a string which is one of `"texinfo"`, `"html"`, or `"plain text"`.

**See also:** [get_help_text_from_file], page 25.

# Chapter 2: Getting Started

[*text*, *format*] = get_help_text_from_file (*fname*)      [Built-in Function]

> Return the raw help text from the file *fname*.
>
> The raw help text is returned in *text* and the format in *format* The format is a string which is one of "texinfo", "html", or "plain text".
>
> **See also:** [get_help_text], page 24.

*text* = get_first_help_sentence (*name*)      [Function File]
*text* = get_first_help_sentence (*name*, *max_len*)      [Function File]
[*text*, *status*] = get_first_help_sentence (...)      [Function File]

> Return the first sentence of a function's help text.
>
> The first sentence is defined as the text after the function declaration until either the first period (".") or the first appearance of two consecutive newlines ("\n\n"). The text is truncated to a maximum length of *max_len*, which defaults to 80.
>
> The optional output argument *status* returns the status reported by `makeinfo`. If only one output argument is requested, and *status* is nonzero, a warning is displayed.
>
> As an example, the first sentence of this help text is
>
> ```
> get_first_help_sentence ("get_first_help_sentence")
>    ⊣ ans = Return the first sentence of a function's help text.
> ```

## 2.4 Command Line Editing

Octave uses the GNU Readline library to provide an extensive set of command-line editing and history features. Only the most common features are described in this manual. In addition, all of the editing functions can be bound to different key strokes at the user's discretion. This manual assumes no changes from the default Emacs bindings. See the GNU Readline Library manual for more information on customizing Readline and for a complete feature list.

To insert printing characters (letters, digits, symbols, etc.), simply type the character. Octave will insert the character at the cursor and advance the cursor forward.

Many of the command-line editing functions operate using control characters. For example, the character `Control-a` moves the cursor to the beginning of the line. To type `C-a`, hold down CTRL and then press A. In the following sections, control characters such as `Control-a` are written as `C-a`.

Another set of command-line editing functions use Meta characters. To type `M-u`, hold down the META key and press U. Depending on the keyboard, the META key may be labeled ALT or even WINDOWS. If your terminal does not have a META key, you can still type Meta characters using two-character sequences starting with `ESC`. Thus, to enter `M-u`, you would type ESC U. The `ESC` character sequences are also allowed on terminals with real Meta keys. In the following sections, Meta characters such as `Meta-u` are written as `M-u`.

### 2.4.1 Cursor Motion

The following commands allow you to position the cursor.

`C-b`      Move back one character.

`C-f`      Move forward one character.

BACKSPACE
: Delete the character to the left of the cursor.

`DEL`     Delete the character underneath the cursor.

`C-d`     Delete the character underneath the cursor.

`M-f`     Move forward a word.

`M-b`     Move backward a word.

`C-a`     Move to the start of the line.

`C-e`     Move to the end of the line.

`C-l`     Clear the screen, reprinting the current line at the top.

`C-_`
`C-/`     Undo the last action. You can undo all the way back to an empty line.

`M-r`     Undo all changes made to this line. This is like typing the 'undo' command enough times to get back to the beginning.

The above table describes the most basic possible keystrokes that you need in order to do editing of the input line. On most terminals, you can also use the left and right arrow keys in place of `C-f` and `C-b` to move forward and backward.

Notice how `C-f` moves forward a character, while `M-f` moves forward a word. It is a loose convention that control keystrokes operate on characters while meta keystrokes operate on words.

The function `clc` will allow you to clear the screen from within Octave programs.

`clc ()` [Built-in Function]
`home ()` [Built-in Function]
: Clear the terminal screen and move the cursor to the upper left corner.

### 2.4.2 Killing and Yanking

*Killing* text means to delete the text from the line, but to save it away for later use, usually by *yanking* it back into the line. If the description for a command says that it 'kills' text, then you can be sure that you can get the text back in a different (or the same) place later.

Here is the list of commands for killing text.

`C-k`     Kill the text from the current cursor position to the end of the line.

`M-d`     Kill from the cursor to the end of the current word, or if between words, to the end of the next word.

`M-DEL`   Kill from the cursor to the start of the previous word, or if between words, to the start of the previous word.

`C-w`     Kill from the cursor to the previous whitespace. This is different than `M-DEL` because the word boundaries differ.

And, here is how to *yank* the text back into the line. Yanking means to copy the most-recently-killed text from the kill buffer.

# Chapter 2: Getting Started

C-y          Yank the most recently killed text back into the buffer at the cursor.

M-y          Rotate the kill-ring, and yank the new top. You can only do this if the prior command is C-y or M-y.

When you use a kill command, the text is saved in a *kill-ring*. Any number of consecutive kills save all of the killed text together, so that when you yank it back, you get it in one clean sweep. The kill ring is not line specific; the text that you killed on a previously typed line is available to be yanked back later, when you are typing another line.

## 2.4.3 Commands For Changing Text

The following commands can be used for entering characters that would otherwise have a special meaning (e.g., TAB, C-q, etc.), or for quickly correcting typing mistakes.

C-q  
C-v          Add the next character that you type to the line verbatim. This is how to insert things like C-q for example.

M-TAB      Insert a tab character.

C-t          Drag the character before the cursor forward over the character at the cursor, also moving the cursor forward. If the cursor is at the end of the line, then transpose the two characters before it.

M-t          Drag the word behind the cursor past the word in front of the cursor moving the cursor over that word as well.

M-u          Uppercase the characters following the cursor to the end of the current (or following) word, moving the cursor to the end of the word.

M-l          Lowercase the characters following the cursor to the end of the current (or following) word, moving the cursor to the end of the word.

M-c          Uppercase the character following the cursor (or the beginning of the next word if the cursor is between words), moving the cursor to the end of the word.

## 2.4.4 Letting Readline Type For You

The following commands allow Octave to complete command and variable names for you.

TAB          Attempt to do completion on the text before the cursor. Octave can complete the names of commands and variables.

M-?          List the possible completions of the text before the cursor.

*val* = completion_append_char ()          [Built-in Function]  
*old_val* = completion_append_char (*new_val*)          [Built-in Function]  
completion_append_char (*new_val*, "*local*")          [Built-in Function]

     Query or set the internal character variable that is appended to successful command-line completion attempts.

     The default value is " " (a single space).

     When called from inside a function with the "`local`" option, the variable is changed locally for the function and any subroutines it calls. The original variable value is restored when exiting the function.

`completion_matches (`*hint*`)` [Built-in Function]

    Generate possible completions given *hint*.

    This function is provided for the benefit of programs like Emacs which might be controlling Octave and handling user input. The current command number is not incremented when this function is called. This is a feature, not a bug.

### 2.4.5 Commands For Manipulating The History

Octave normally keeps track of the commands you type so that you can recall previous commands to edit or execute them again. When you exit Octave, the most recent commands you have typed, up to the number specified by the variable `history_size`, are saved in a file. When Octave starts, it loads an initial list of commands from the file named by the variable `history_file`.

Here are the commands for simple browsing and searching the history list.

LFD  
RET     Accept the current line regardless of where the cursor is. If the line is non-empty, add it to the history list. If the line was a history line, then restore the history line to its original state.

C-p     Move 'up' through the history list.

C-n     Move 'down' through the history list.

M-<     Move to the first line in the history.

M->     Move to the end of the input history, i.e., the line you are entering!

C-r     Search backward starting at the current line and moving 'up' through the history as necessary. This is an incremental search.

C-s     Search forward starting at the current line and moving 'down' through the history as necessary.

On most terminals, you can also use the up and down arrow keys in place of C-p and C-n to move through the history list.

In addition to the keyboard commands for moving through the history list, Octave provides three functions for viewing, editing, and re-running chunks of commands from the history list.

`history` [Command]  
`history` *opt1* ... [Command]  
`h = history ()` [Built-in Function]  
`h = history (`*opt1*`, ...)` [Built-in Function]

    If invoked with no arguments, `history` displays a list of commands that you have executed.

    Valid options are:

    *n*  
    -*n*     Display only the most recent *n* lines of history.

    -c     Clear the history list.

Chapter 2: Getting Started 29

> `-q`  Don't number the displayed lines of history. This is useful for cutting and pasting commands using the X Window System.
>
> `-r file`  Read the file *file*, appending its contents to the current history list. If the name is omitted, use the default history file (normally '`~/.octave_hist`').
>
> `-w file`  Write the current history to the file *file*. If the name is omitted, use the default history file (normally '`~/.octave_hist`').

For example, to display the five most recent commands that you have typed without displaying line numbers, use the command *history -q 5*.

If invoked with a single output argument, the history will be saved to that argument as a cell string and will not be output to screen.

**See also:** [edit_history], page 29, [run_history], page 29.

`edit_history`  [Command]
`edit_history cmd_number`  [Command]
`edit_history first last`  [Command]

Edit the history list using the editor named by the variable `EDITOR`.

The commands to be edited are first copied to a temporary file. When you exit the editor, Octave executes the commands that remain in the file. It is often more convenient to use `edit_history` to define functions rather than attempting to enter them directly on the command line. The block of commands is executed as soon as you exit the editor. To avoid executing any commands, simply delete all the lines from the buffer before leaving the editor.

When invoked with no arguments, edit the previously executed command; With one argument, edit the specified command *cmd_number*; With two arguments, edit the list of commands between *first* and *last*. Command number specifiers may also be negative where -1 refers to the most recently executed command. The following are equivalent and edit the most recently executed command.

```
edit_history
edit_history -1
```

When using ranges, specifying a larger number for the first command than the last command reverses the list of commands before they are placed in the buffer to be edited.

**See also:** [run_history], page 29, [history], page 28.

`run_history`  [Command]
`run_history cmd_number`  [Command]
`run_history first last`  [Command]

Run commands from the history list.

When invoked with no arguments, run the previously executed command;

With one argument, run the specified command *cmd_number*;

With two arguments, run the list of commands between *first* and *last*. Command number specifiers may also be negative where -1 refers to the most recently executed command. For example, the command

```
run_history
    OR
run_history -1
```
executes the most recent command again. The command
```
run_history 13 169
```
executes commands 13 through 169.

Specifying a larger number for the first command than the last command reverses the list of commands before executing them. For example:
```
disp (1)
disp (2)
run_history -1 -2
⇒
  2
  1
```

See also: [edit_history], page 29, [history], page 28.

Octave also allows you customize the details of when, where, and how history is saved.

`val = history_save ()`         [Built-in Function]
`old_val = history_save (new_val)`         [Built-in Function]
`history_save (new_val, "local")`         [Built-in Function]

    Query or set the internal variable that controls whether commands entered on the command line are saved in the history file.

    When called from inside a function with the `"local"` option, the variable is changed locally for the function and any subroutines it calls. The original variable value is restored when exiting the function.

    See also: [history_control], page 30, [history_file], page 31, [history_size], page 31, [history_timestamp_format_string], page 31.

`val = history_control ()`         [Built-in Function]
`old_val = history_control (new_val)`         [Built-in Function]

    Query or set the internal variable that specifies how commands are saved to the history list.

    The default value is an empty character string, but may be overridden by the environment variable `OCTAVE_HISTCONTROL`.

    The value of `history_control` is a colon-separated list of values controlling how commands are saved on the history list. If the list of values includes `ignorespace`, lines which begin with a space character are not saved in the history list. A value of `ignoredups` causes lines matching the previous history entry to not be saved. A value of `ignoreboth` is shorthand for `ignorespace` and `ignoredups`. A value of `erasedups` causes all previous lines matching the current line to be removed from the history list before that line is saved. Any value not in the above list is ignored. If `history_control` is the empty string, all commands are saved on the history list, subject to the value of `history_save`.

    See also: [history_file], page 31, [history_size], page 31, [history_timestamp_format_string], page 31, [history_save], page 30.

Chapter 2: Getting Started    31

*val* = history_file ()    [Built-in Function]
*old_val* = history_file (*new_val*)    [Built-in Function]
: Query or set the internal variable that specifies the name of the file used to store command history.

  The default value is '~/.octave_hist', but may be overridden by the environment variable OCTAVE_HISTFILE.

  **See also:** [history_size], page 31, [history_save], page 30, [history_timestamp_format_string], page 31.

*val* = history_size ()    [Built-in Function]
*old_val* = history_size (*new_val*)    [Built-in Function]
: Query or set the internal variable that specifies how many entries to store in the history file.

  The default value is 1000, but may be overridden by the environment variable OCTAVE_HISTSIZE.

  **See also:** [history_file], page 31, [history_timestamp_format_string], page 31, [history_save], page 30.

*val* = history_timestamp_format_string ()    [Built-in Function]
*old_val* = history_timestamp_format_string (*new_val*)    [Built-in Function]
history_timestamp_format_string (*new_val*, "*local*")    [Built-in Function]
: Query or set the internal variable that specifies the format string for the comment line that is written to the history file when Octave exits.

  The format string is passed to **strftime**. The default value is

  ```
  "# Octave VERSION, %a %b %d %H:%M:%S %Y %Z <USER@HOST>"
  ```

  When called from inside a function with the "local" option, the variable is changed locally for the function and any subroutines it calls. The original variable value is restored when exiting the function.

  **See also:** [strftime], page 745, [history_file], page 31, [history_size], page 31, [history_save], page 30.

*val* = EDITOR ()    [Built-in Function]
*old_val* = EDITOR (*new_val*)    [Built-in Function]
EDITOR (*new_val*, "*local*")    [Built-in Function]
: Query or set the internal variable that specifies the default text editor.

  The default value is taken from the environment variable EDITOR when Octave starts. If the environment variable is not initialized, EDITOR will be set to "emacs".

  When called from inside a function with the "local" option, the variable is changed locally for the function and any subroutines it calls. The original variable value is restored when exiting the function.

  **See also:** [edit], page 187, [edit_history], page 29.

## 2.4.6 Customizing `readline`

Octave uses the GNU Readline library for command-line editing and history features. Readline is very flexible and can be modified through a configuration file of commands (See the GNU Readline library for the exact command syntax). The default configuration file is normally '`~/.inputrc`'.

Octave provides two commands for initializing Readline and thereby changing the command line behavior.

`readline_read_init_file (file)`     [Built-in Function]
    Read the readline library initialization file *file*.

    If *file* is omitted, read the default initialization file (normally '`~/.inputrc`').

    See Section "Readline Init File" in *GNU Readline Library*, for details.

    **See also:** [readline_re_read_init_file], page 32.

`readline_re_read_init_file ()`     [Built-in Function]
    Re-read the last readline library initialization file that was read.

    See Section "Readline Init File" in *GNU Readline Library*, for details.

    **See also:** [readline_read_init_file], page 32.

## 2.4.7 Customizing the Prompt

The following variables are available for customizing the appearance of the command-line prompts. Octave allows the prompt to be customized by inserting a number of backslash-escaped special characters that are decoded as follows:

'`\t`'     The time.

'`\d`'     The date.

'`\n`'     Begins a new line by printing the equivalent of a carriage return followed by a line feed.

'`\s`'     The name of the program (usually just '`octave`').

'`\w`'     The current working directory.

'`\W`'     The basename of the current working directory.

'`\u`'     The username of the current user.

'`\h`'     The hostname, up to the first '.'.

'`\H`'     The hostname.

'`\#`'     The command number of this command, counting from when Octave starts.

'`\!`'     The history number of this command. This differs from '`\#`' by the number of commands in the history list when Octave starts.

'`\$`'     If the effective UID is 0, a '`#`', otherwise a '`$`'.

'`\nnn`'     The character whose character code in octal is *nnn*.

'`\\`'     A backslash.

Chapter 2: Getting Started                                                                 33

*val* = PS1 ()                                                          [Built-in Function]
*old_val* = PS1 (*new_val*)                                             [Built-in Function]
PS1 (*new_val*, "*local*")                                              [Built-in Function]

    Query or set the primary prompt string.

    When executing interactively, Octave displays the primary prompt when it is ready to read a command.

    The default value of the primary prompt string is `"octave:\#> "`. To change it, use a command like

        PS1 ("\\u@\\H> ")

which will result in the prompt 'boris@kremvax> ' for the user 'boris' logged in on the host 'kremvax.kgb.su'. Note that two backslashes are required to enter a backslash into a double-quoted character string. See Chapter 5 [Strings], page 67.

    You can also use ANSI escape sequences if your terminal supports them. This can be useful for coloring the prompt. For example,

        PS1 ("\\[\\033[01;31m\\]\\s:\\#> \\[\\033[0m\\]")

will give the default Octave prompt a red coloring.

    When called from inside a function with the `"local"` option, the variable is changed locally for the function and any subroutines it calls. The original variable value is restored when exiting the function.

    **See also:** [PS2], page 33, [PS4], page 33.

*val* = PS2 ()                                                          [Built-in Function]
*old_val* = PS2 (*new_val*)                                             [Built-in Function]
PS2 (*new_val*, "*local*")                                              [Built-in Function]

    Query or set the secondary prompt string.

    The secondary prompt is printed when Octave is expecting additional input to complete a command. For example, if you are typing a `for` loop that spans several lines, Octave will print the secondary prompt at the beginning of each line after the first. The default value of the secondary prompt string is `"> "`.

    When called from inside a function with the `"local"` option, the variable is changed locally for the function and any subroutines it calls. The original variable value is restored when exiting the function.

    **See also:** [PS1], page 32, [PS4], page 33.

*val* = PS4 ()                                                          [Built-in Function]
*old_val* = PS4 (*new_val*)                                             [Built-in Function]
PS4 (*new_val*, "*local*")                                              [Built-in Function]

    Query or set the character string used to prefix output produced when echoing commands is enabled.

    The default value is `"+ "`. See Section 2.4.8 [Diary and Echo Commands], page 34, for a description of echoing commands.

    When called from inside a function with the `"local"` option, the variable is changed locally for the function and any subroutines it calls. The original variable value is restored when exiting the function.

**See also:** [echo], page 34, [echo_executing_commands], page 34, [PS1], page 32, [PS2], page 33.

## 2.4.8 Diary and Echo Commands

Octave's diary feature allows you to keep a log of all or part of an interactive session by recording the input you type and the output that Octave produces in a separate file.

`diary`                           [Command]
`diary` *on*                      [Command]
`diary` *off*                     [Command]
`diary` *filename*                [Command]

> Record a list of all commands *and* the output they produce, mixed together just as they appear on the terminal.
>
> Valid options are:
>
> on       Start recording a session in a file called '`diary`' in the current working directory.
>
> off      Stop recording the session in the diary file.
>
> *filename*      Record the session in the file named *filename*.
>
> With no arguments, `diary` toggles the current diary state.
>
> **See also:** [history], page 28.

Sometimes it is useful to see the commands in a function or script as they are being evaluated. This can be especially helpful for debugging some kinds of problems.

`echo`                            [Command]
`echo` *on*                       [Command]
`echo` *off*                      [Command]
`echo` *on all*                   [Command]
`echo` *off all*                  [Command]

> Control whether commands are displayed as they are executed.
>
> Valid options are:
>
> on      Enable echoing of commands as they are executed in script files.
>
> off      Disable echoing of commands as they are executed in script files.
>
> on all      Enable echoing of commands as they are executed in script files and functions.
>
> off all      Disable echoing of commands as they are executed in script files and functions.
>
> With no arguments, `echo` toggles the current echo state.

`val = echo_executing_commands ()`                         [Built-in Function]
`old_val = echo_executing_commands (`*new_val*`)`          [Built-in Function]
`echo_executing_commands (`*new_val*`, "`*local*`")`       [Built-in Function]

> Query or set the internal variable that controls the echo state.
>
> It may be the sum of the following values:

Chapter 2: Getting Started

| | |
|---|---|
| 1 | Echo commands read from script files. |
| 2 | Echo commands from functions. |
| 4 | Echo commands read from command line. |

More than one state can be active at once. For example, a value of 3 is equivalent to the command *echo on all*.

The value of `echo_executing_commands` may be set by the *echo* command or the command line option '--echo-commands'.

When called from inside a function with the "local" option, the variable is changed locally for the function and any subroutines it calls. The original variable value is restored when exiting the function.

## 2.5 How Octave Reports Errors

Octave reports two kinds of errors for invalid programs.

A *parse error* occurs if Octave cannot understand something you have typed. For example, if you misspell a keyword,

```
octave:13> function y = f (x) y = x***2; endfunction
```

Octave will respond immediately with a message like this:

```
parse error:

  syntax error

>>> function y = f (x) y = x***2; endfunction
                               ^
```

For most parse errors, Octave uses a caret ('^') to mark the point on the line where it was unable to make sense of your input. In this case, Octave generated an error message because the keyword for exponentiation (**) was misspelled. It marked the error at the third '*' because the code leading up to this was correct but the final '*' was not understood.

Another class of error message occurs at evaluation time. These errors are called *run-time errors*, or sometimes *evaluation errors*, because they occur when your program is being *run*, or *evaluated*. For example, if after correcting the mistake in the previous function definition, you type

```
octave:13> f ()
```

Octave will respond with

```
error: 'x' undefined near line 1 column 24
error: called from:
error:    f at line 1, column 22
```

This error message has several parts, and gives quite a bit of information to help you locate the source of the error. The messages are generated from the point of the innermost error, and provide a traceback of enclosing expressions and function calls.

In the example above, the first line indicates that a variable named 'x' was found to be undefined near line 1 and column 24 of some function or expression. For errors occurring within functions, lines are counted from the beginning of the file containing the function definition. For errors occurring outside of an enclosing function, the line number indicates the input line number, which is usually displayed in the primary prompt string.

The second and third lines of the error message indicate that the error occurred within the function f. If the function f had been called from within another function, for example, g, the list of errors would have ended with one more line:

```
error:    g at line 1, column 17
```

These lists of function calls make it fairly easy to trace the path your program took before the error occurred, and to correct the error before trying again.

## 2.6 Executable Octave Programs

Once you have learned Octave, you may want to write self-contained Octave scripts, using the '#!' script mechanism. You can do this on GNU systems and on many Unix systems[1].

Self-contained Octave scripts are useful when you want to write a program which users can invoke without knowing that the program is written in the Octave language. Octave scripts are also used for batch processing of data files. Once an algorithm has been developed and tested in the interactive portion of Octave, it can be committed to an executable script and used again and again on new data files.

As a trivial example of an executable Octave script, you might create a text file named 'hello', containing the following lines:

```
#! octave-interpreter-name -qf
# a sample Octave program
printf ("Hello, world!\n");
```

(where *octave-interpreter-name* should be replaced with the full path and name of your Octave binary). Note that this will only work if '#!' appears at the very beginning of the file. After making the file executable (with the chmod command on Unix systems), you can simply type:

```
hello
```

at the shell, and the system will arrange to run Octave as if you had typed:

```
octave hello
```

The line beginning with '#!' lists the full path and filename of an interpreter to be run, and an optional initial command line argument to pass to that interpreter. The operating system then runs the interpreter with the given argument and the full argument list of the executed program. The first argument in the list is the full file name of the Octave executable. The rest of the argument list will either be options to Octave, or data files, or both. The '-qf' options are usually specified in stand-alone Octave programs to prevent them from printing the normal startup message, and to keep them from behaving differently depending on the contents of a particular user's '~/.octaverc' file. See Section 2.1 [Invoking Octave from the Command Line], page 15.

Note that some operating systems may place a limit on the number of characters that are recognized after '#!'. Also, the arguments appearing in a '#!' line are parsed differently by various shells/systems. The majority of them group all the arguments together in one string and pass it to the interpreter as a single argument. In this case, the following script:

---

[1] The '#!' mechanism works on Unix systems derived from Berkeley Unix, System V Release 4, and some System V Release 3 systems.

```
#! octave-interpreter-name -q -f # comment
```

is equivalent to typing at the command line:

```
octave "-q -f # comment"
```

which will produce an error message. Unfortunately, it is not possible for Octave to determine whether it has been called from the command line or from a '#!' script, so some care is needed when using the '#!' mechanism.

Note that when Octave is started from an executable script, the built-in function `argv` returns a cell array containing the command line arguments passed to the executable Octave script, not the arguments passed to the Octave interpreter on the '#!' line of the script. For example, the following program will reproduce the command line that was used to execute the script, not '-qf'.

```
#! /bin/octave -qf
printf ("%s", program_name ());
arg_list = argv ();
for i = 1:nargin
  printf (" %s", arg_list{i});
endfor
printf ("\n");
```

## 2.7 Comments in Octave Programs

A *comment* is some text that is included in a program for the sake of human readers, and which is NOT an executable part of the program. Comments can explain what the program does, and how it works. Nearly all programming languages have provisions for comments, because programs are typically hard to understand without them.

### 2.7.1 Single Line Comments

In the Octave language, a comment starts with either the sharp sign character, '#', or the percent symbol '%' and continues to the end of the line. Any text following the sharp sign or percent symbol is ignored by the Octave interpreter and not executed. The following example shows whole line and partial line comments.

```
function countdown
  # Count down for main rocket engines
  disp (3);
  disp (2);
  disp (1);
  disp ("Blast Off!");   # Rocket leaves pad
endfunction
```

### 2.7.2 Block Comments

Entire blocks of code can be commented by enclosing the code between matching '#{' and '#}' or '%{' and '%}' markers. For example,

```
function quick_countdown
  # Count down for main rocket engines
  disp (3);
 #{
  disp (2);
  disp (1);
 #}
  disp ("Blast Off!");   # Rocket leaves pad
endfunction
```

will produce a very quick countdown from '3' to "Blast Off" as the lines "disp (2);" and "disp (1);" won't be executed.

The block comment markers must appear alone as the only characters on a line (excepting whitespace) in order to be parsed correctly.

### 2.7.3 Comments and the Help System

The `help` command (see Section 2.3 [Getting Help], page 20) is able to find the first block of comments in a function and return those as a documentation string. This means that the same commands used to get help on built-in functions are available for properly formatted user-defined functions. For example, after defining the function `f` below,

```
function xdot = f (x, t)

# usage: f (x, t)
#
# This function defines the right-hand
# side functions for a set of nonlinear
# differential equations.

  r = 0.25;
  ...
endfunction
```

the command *help f* produces the output

```
usage: f (x, t)

This function defines the right-hand
side functions for a set of nonlinear
differential equations.
```

Although it is possible to put comment lines into keyboard-composed, throw-away Octave programs, it usually isn't very useful because the purpose of a comment is to help you or another person understand the program at a later time.

The `help` parser currently only recognizes single line comments (see Section 2.7.1 [Single Line Comments], page 37) and not block comments for the initial help text.

# 3 Data Types

All versions of Octave include a number of built-in data types, including real and complex scalars and matrices, character strings, a data structure type, and an array that can contain all data types.

It is also possible to define new specialized data types by writing a small amount of C++ code. On some systems, new data types can be loaded dynamically while Octave is running, so it is not necessary to recompile all of Octave just to add a new type. See Appendix A [External Code Interface], page 813, for more information about Octave's dynamic linking capabilities. Section 3.2 [User-defined Data Types], page 44 describes what you must do to define a new data type for Octave.

`typeinfo ()` [Built-in Function]
`typeinfo (expr)` [Built-in Function]

> Return the type of the expression *expr*, as a string.
>
> If *expr* is omitted, return a cell array of strings containing all the currently installed data types.
>
> **See also:** [class], page 39, [isa], page 39.

## 3.1 Built-in Data Types

The standard built-in data types are real and complex scalars and matrices, ranges, character strings, a data structure type, and cell arrays. Additional built-in data types may be added in future versions. If you need a specialized data type that is not currently provided as a built-in type, you are encouraged to write your own user-defined data type and contribute it for distribution in a future release of Octave.

The data type of a variable can be determined and changed through the use of the following functions.

`classname = class (obj)` [Function File]
`class (s, id)` [Function File]
`class (s, id, p, ...)` [Function File]

> Return the class of the object *obj*, or create a class with fields from structure *s* and name (string) *id*.
>
> Additional arguments name a list of parent classes from which the new class is derived.
>
> **See also:** [typeinfo], page 39, [isa], page 39.

`isa (obj, classname)` [Function File]

> Return true if *obj* is an object from the class *classname*.
>
> *classname* may also be one of the following class categories:
>
> `"float"`    Floating point value comprising classes `"double"` and `"single"`.
>
> `"integer"`
> > Integer value comprising classes (u)int8, (u)int16, (u)int32, (u)int64.
>
> `"numeric"`
> > Numeric value comprising either a floating point or integer value.

If *classname* is a cell array of string, a logical array of the same size is returned, containing true for each class to which *obj* belongs to.

**See also:** [class], page 39, [typeinfo], page 39.

cast (*val*, "*type*")                                                                                  [Function File]
Convert *val* to data type *type*.

*val* must be one of the numeric classes:

```
"double"
"single"
"logical"
"char"
"int8"
"int16"
"int32"
"int64"
"uint8"
"uint16"
"uint32"
"uint64"
```

The value *val* may be modified to fit within the range of the new type.

Examples:

```
cast (-5, "uint8")
    ⇒ 0
cast (300, "int8")
    ⇒ 127
```

**See also:** [typecast], page 40, [int8], page 54, [uint8], page 55, [int16], page 55, [uint16], page 55, [int32], page 55, [uint32], page 55, [int64], page 55, [uint64], page 55, [double], page 47, [single], page 53, [logical], page 60, [char], page 71, [class], page 39, [typeinfo], page 39.

*y* = typecast (*x*, "*class*")                                                           [Built-in Function]
Return a new array *y* resulting from interpreting the data of *x* in memory as data of the numeric class *class*.

Both the class of *x* and *class* must be one of the built-in numeric classes:

Chapter 3: Data Types

```
"logical"
"char"
"int8"
"int16"
"int32"
"int64"
"uint8"
"uint16"
"uint32"
"uint64"
"double"
"single"
"double complex"
"single complex"
```

the last two are only used with *class*; they indicate that a complex-valued result is requested. Complex arrays are stored in memory as consecutive pairs of real numbers. The sizes of integer types are given by their bit counts. Both logical and char are typically one byte wide; however, this is not guaranteed by C++. If your system is IEEE conformant, single and double will be 4 bytes and 8 bytes wide, respectively. `"logical"` is not allowed for *class*.

If the input is a row vector, the return value is a row vector, otherwise it is a column vector.

If the bit length of *x* is not divisible by that of *class*, an error occurs.

An example of the use of typecast on a little-endian machine is

```
x = uint16 ([1, 65535]);
typecast (x, "uint8")
⇒ [   1,   0, 255, 255]
```

**See also:** [cast], page 40, [bitpack], page 41, [bitunpack], page 42, [swapbytes], page 41.

swapbytes (*x*) [Function File]

Swap the byte order on values, converting from little endian to big endian and vice versa.

For example:

```
swapbytes (uint16 (1:4))
⇒ [   256   512   768   1024]
```

**See also:** [typecast], page 40, [cast], page 40.

*y* = bitpack (*x*, *class*) [Built-in Function]

Return a new array *y* resulting from interpreting the logical array *x* as raw bit patterns for data of the numeric class *class*.

*class* must be one of the built-in numeric classes:

```
"double"
"single"
"double complex"
"single complex"
"char"
"int8"
"int16"
"int32"
"int64"
"uint8"
"uint16"
"uint32"
"uint64"
```

The number of elements of *x* should be divisible by the bit length of *class*. If it is not, excess bits are discarded. Bits come in increasing order of significance, i.e., `x(1)` is bit 0, `x(2)` is bit 1, etc.

The result is a row vector if *x* is a row vector, otherwise it is a column vector.

**See also:** [bitunpack], page 42, [typecast], page 40.

*y* = bitunpack (*x*)                                                       [Built-in Function]

Return a logical array *y* corresponding to the raw bit patterns of *x*.

*x* must belong to one of the built-in numeric classes:

```
"double"
"single"
"char"
"int8"
"int16"
"int32"
"int64"
"uint8"
"uint16"
"uint32"
"uint64"
```

The result is a row vector if *x* is a row vector; otherwise, it is a column vector.

**See also:** [bitpack], page 41, [typecast], page 40.

### 3.1.1 Numeric Objects

Octave's built-in numeric objects include real, complex, and integer scalars and matrices. All built-in floating point numeric data is currently stored as double precision numbers. On systems that use the IEEE floating point format, values in the range of approximately $2.2251 \times 10^{-308}$ to $1.7977 \times 10^{308}$ can be stored, and the relative precision is approximately $2.2204 \times 10^{-16}$. The exact values are given by the variables `realmin`, `realmax`, and `eps`, respectively.

Matrix objects can be of any size, and can be dynamically reshaped and resized. It is easy to extract individual rows, columns, or submatrices using a variety of powerful indexing features. See Section 8.1 [Index Expressions], page 135.

Chapter 3: Data Types                                                              43

See Chapter 4 [Numeric Data Types], page 47, for more information.

## 3.1.2 Missing Data

It is possible to represent missing data explicitly in Octave using `NA` (short for "Not Available"). Missing data can only be represented when data is represented as floating point numbers. In this case missing data is represented as a special case of the representation of NaN.

`NA`                                                                  [Built-in Function]
`NA (n)`                                                              [Built-in Function]
`NA (n, m)`                                                           [Built-in Function]
`NA (n, m, k, ...)`                                                   [Built-in Function]
`NA (..., class)`                                                     [Built-in Function]

   Return a scalar, matrix, or N-dimensional array whose elements are all equal to the special constant used to designate missing values.

   Note that NA always compares not equal to NA (NA != NA). To find NA values, use the `isna` function.

   When called with no arguments, return a scalar with the value 'NA'.

   When called with a single argument, return a square matrix with the dimension specified.

   When called with more than one scalar argument the first two arguments are taken as the number of rows and columns and any further arguments specify additional matrix dimensions.

   The optional argument *class* specifies the return type and may be either `"double"` or `"single"`.

   **See also:** [isna], page 43.

`isna (x)`                                                             [Mapping Function]
   Return a logical array which is true where the elements of *x* are NA (missing) values and false where they are not.

   For example:
```
        isna ([13, Inf, NA, NaN])
             ⇒ [ 0, 0, 1, 0 ]
```
   **See also:** [isnan], page 404, [isinf], page 404, [isfinite], page 405.

## 3.1.3 String Objects

A character string in Octave consists of a sequence of characters enclosed in either double-quote or single-quote marks. Internally, Octave currently stores strings as matrices of characters. All the indexing operations that work for matrix objects also work for strings.

See Chapter 5 [Strings], page 67, for more information.

## 3.1.4 Data Structure Objects

Octave's data structure type can help you to organize related objects of different types. The current implementation uses an associative array with indices limited to strings, but the syntax is more like C-style structures.

See Section 6.1 [Structures], page 99, for more information.

### 3.1.5 Cell Array Objects

A Cell Array in Octave is general array that can hold any number of different data types.

See Section 6.2 [Cell Arrays], page 112, for more information.

## 3.2 User-defined Data Types

Someday I hope to expand this to include a complete description of Octave's mechanism for managing user-defined data types. Until this feature is documented here, you will have to make do by reading the code in the 'ov.h', 'ops.h', and related files from Octave's 'src' directory.

## 3.3 Object Sizes

The following functions allow you to determine the size of a variable or expression. These functions are defined for all objects. They return −1 when the operation doesn't make sense. For example, Octave's data structure type doesn't have rows or columns, so the `rows` and `columns` functions return −1 for structure arguments.

`ndims (a)` [Built-in Function]

> Return the number of dimensions of a.
>
> For any array, the result will always be greater than or equal to 2. Trailing singleton dimensions are not counted.
>
>     ndims (ones (4, 1, 2, 1))
>        ⇒ 3
>
> See also: [size], page 45.

`columns (a)` [Built-in Function]

> Return the number of columns of a.
>
> See also: [rows], page 44, [size], page 45, [length], page 45, [numel], page 44, [isscalar], page 63, [isvector], page 63, [ismatrix], page 63.

`rows (a)` [Built-in Function]

> Return the number of rows of a.
>
> See also: [columns], page 44, [size], page 45, [length], page 45, [numel], page 44, [isscalar], page 63, [isvector], page 63, [ismatrix], page 63.

`numel (a)` [Built-in Function]
`numel (a, idx1, idx2, ...)` [Built-in Function]

> Return the number of elements in the object a.
>
> Optionally, if indices idx1, idx2, ... are supplied, return the number of elements that would result from the indexing
>
>     a(idx1, idx2, ...)
>
> Note that the indices do not have to be numerical. For example,
>
>     a = 1;
>     b = ones (2, 3);
>     numel (a, b)

will return 6, as this is the number of ways to index with *b*.

This method is also called when an object appears as lvalue with cs-list indexing, i.e., `object{...}` or `object(...).field`.

**See also:** [size], page 45.

`length (a)`       [Built-in Function]
: Return the length of the object *a*.

    The length is 0 for empty objects, 1 for scalars, and the number of elements for vectors. For matrix objects, the length is the number of rows or columns, whichever is greater (this odd definition is used for compatibility with MATLAB).

    **See also:** [numel], page 44, [size], page 45.

`size (a)`       [Built-in Function]
`size (a, dim)`       [Built-in Function]
: Return the number of rows and columns of *a*.

    With one input argument and one output argument, the result is returned in a row vector. If there are multiple output arguments, the number of rows is assigned to the first, and the number of columns to the second, etc. For example:

    ```
    size ([1, 2; 3, 4; 5, 6])
        ⇒ [ 3, 2 ]

    [nr, nc] = size ([1, 2; 3, 4; 5, 6])
        ⇒ nr = 3
        ⇒ nc = 2
    ```

    If given a second argument, `size` will return the size of the corresponding dimension. For example,

    ```
    size ([1, 2; 3, 4; 5, 6], 2)
        ⇒ 2
    ```

    returns the number of columns in the given matrix.

    **See also:** [numel], page 44, [ndims], page 44, [length], page 45, [rows], page 44, [columns], page 44.

`isempty (a)`       [Built-in Function]
: Return true if *a* is an empty matrix (any one of its dimensions is zero).

    **See also:** [isnull], page 45, [isa], page 39.

`isnull (x)`       [Built-in Function]
: Return true if *x* is a special null matrix, string, or single quoted string.

    Indexed assignment with such a value on the right-hand side should delete array elements. This function should be used when overloading indexed assignment for user-defined classes instead of `isempty`, to distinguish the cases:

    `A(I) = []`     This should delete elements if I is nonempty.

    `X = []; A(I) = X`
    : This should give an error if I is nonempty.

    **See also:** [isempty], page 45, [isindex], page 138.

`sizeof (val)` [Built-in Function]
: Return the size of *val* in bytes.

    **See also:** [whos], page 128.

`size_equal (a, b, ...)` [Built-in Function]
: Return true if the dimensions of all arguments agree.

    Trailing singleton dimensions are ignored. When called with a single or no argument `size_equal` returns true.

    **See also:** [size], page 45, [numel], page 44, [ndims], page 44.

`squeeze (x)` [Built-in Function]
: Remove singleton dimensions from *x* and return the result.

    Note that for compatibility with MATLAB, all objects have a minimum of two dimensions and row vectors are left unchanged.

    **See also:** [reshape], page 410.

# 4 Numeric Data Types

A *numeric constant* may be a scalar, a vector, or a matrix, and it may contain complex values.

The simplest form of a numeric constant, a scalar, is a single number that can be an integer, a decimal fraction, a number in scientific (exponential) notation, or a complex number. Note that by default numeric constants are represented within Octave in double-precision floating point format (complex constants are stored as pairs of double-precision floating point values). It is, however, possible to represent real integers as described in Section 4.4 [Integer Data Types], page 54. Here are some examples of real-valued numeric constants, which all have the same value:

```
105
1.05e+2
1050e-1
```

To specify complex constants, you can write an expression of the form

```
3 + 4i
3.0 + 4.0i
0.3e1 + 40e-1i
```

all of which are equivalent. The letter 'i' in the previous example stands for the pure imaginary constant, defined as $\sqrt{-1}$.

For Octave to recognize a value as the imaginary part of a complex constant, a space must not appear between the number and the 'i'. If it does, Octave will print an error message, like this:

```
octave:13> 3 + 4 i

parse error:

  syntax error

>>> 3 + 4 i
```

You may also use 'j', 'I', or 'J' in place of the 'i' above. All four forms are equivalent.

**double** (*x*)     [Built-in Function]
    Convert *x* to double precision type.

    **See also:** [single], page 53.

**complex** (*x*)     [Built-in Function]
**complex** (*re*, *im*)     [Built-in Function]
    Return a complex value from real arguments.

    With 1 real argument *x*, return the complex result `x + 0i`.

    With 2 real arguments, return the complex result `re + im`. `complex` can often be more convenient than expressions such as `a + i*b`. For example:

```
complex ([1, 2], [3, 4])
    ⇒ [ 1 + 3i   2 + 4i ]
```

**See also:** [real], page 436, [imag], page 435, [iscomplex], page 63, [abs], page 435, [arg], page 435.

## 4.1 Matrices

It is easy to define a matrix of values in Octave. The size of the matrix is determined automatically, so it is not necessary to explicitly state the dimensions. The expression

```
a = [1, 2; 3, 4]
```

results in the matrix

$$a = \begin{bmatrix} 1 & 2 \\ 3 & 4 \end{bmatrix}$$

Elements of a matrix may be arbitrary expressions, provided that the dimensions all make sense when combining the various pieces. For example, given the above matrix, the expression

```
[ a, a ]
```

produces the matrix

```
ans =

  1  2  1  2
  3  4  3  4
```

but the expression

```
[ a, 1 ]
```

produces the error

```
error: number of rows must match (1 != 2) near line 13, column 6
```

(assuming that this expression was entered as the first thing on line 13, of course).

Inside the square brackets that delimit a matrix expression, Octave looks at the surrounding context to determine whether spaces and newline characters should be converted into element and row separators, or simply ignored, so an expression like

```
a = [ 1 2
      3 4 ]
```

will work. However, some possible sources of confusion remain. For example, in the expression

```
[ 1 - 1 ]
```

the '-' is treated as a binary operator and the result is the scalar 0, but in the expression

```
[ 1 -1 ]
```

the '-' is treated as a unary operator and the result is the vector [ 1, -1 ]. Similarly, the expression

```
[ sin (pi) ]
```

will be parsed as

# Chapter 4: Numeric Data Types

```
[ sin, (pi) ]
```

and will result in an error since the `sin` function will be called with no arguments. To get around this, you must omit the space between `sin` and the opening parenthesis, or enclose the expression in a set of parentheses:

```
[ (sin (pi)) ]
```

Whitespace surrounding the single quote character ('', used as a transpose operator and for delimiting character strings) can also cause confusion. Given `a = 1`, the expression

```
[ 1 a' ]
```

results in the single quote character being treated as a transpose operator and the result is the vector `[ 1, 1 ]`, but the expression

```
[ 1 a ' ]
```

produces the error message

```
parse error:

  syntax error

>>> [ 1 a ' ]
```

because not doing so would cause trouble when parsing the valid expression

```
[ a 'foo' ]
```

For clarity, it is probably best to always use commas and semicolons to separate matrix elements and rows.

The maximum number of elements in a matrix is fixed when Octave is compiled. The allowable number can be queried with the function `sizemax`. Note that other factors, such as the amount of memory available on your machine, may limit the maximum size of matrices to something smaller.

`sizemax ()` [Built-in Function]

Return the largest value allowed for the size of an array.

If Octave is compiled with 64-bit indexing, the result is of class int64, otherwise it is of class int32. The maximum array size is slightly smaller than the maximum value allowable for the relevant class as reported by `intmax`.

**See also:** [intmax], page 55.

When you type a matrix or the name of a variable whose value is a matrix, Octave responds by printing the matrix in with neatly aligned rows and columns. If the rows of the matrix are too large to fit on the screen, Octave splits the matrix and displays a header before each section to indicate which columns are being displayed. You can use the following variables to control the format of the output.

`val = output_max_field_width ()` [Built-in Function]
`old_val = output_max_field_width (new_val)` [Built-in Function]
`output_max_field_width (new_val, "local")` [Built-in Function]

Query or set the internal variable that specifies the maximum width of a numeric output field.

When called from inside a function with the "local" option, the variable is changed locally for the function and any subroutines it calls. The original variable value is restored when exiting the function.

**See also:** [format], page 232, [fixed_point_format], page 51, [output_precision], page 50.

`val = output_precision ()`  [Built-in Function]
`old_val = output_precision (new_val)`  [Built-in Function]
`output_precision (new_val, "local")`  [Built-in Function]

Query or set the internal variable that specifies the minimum number of significant figures to display for numeric output.

When called from inside a function with the "local" option, the variable is changed locally for the function and any subroutines it calls. The original variable value is restored when exiting the function.

**See also:** [format], page 232, [fixed_point_format], page 51, [output_max_field_width], page 49.

It is possible to achieve a wide range of output styles by using different values of `output_precision` and `output_max_field_width`. Reasonable combinations can be set using the `format` function. See Section 14.1 [Basic Input and Output], page 231.

`val = split_long_rows ()`  [Built-in Function]
`old_val = split_long_rows (new_val)`  [Built-in Function]
`split_long_rows (new_val, "local")`  [Built-in Function]

Query or set the internal variable that controls whether rows of a matrix may be split when displayed to a terminal window.

If the rows are split, Octave will display the matrix in a series of smaller pieces, each of which can fit within the limits of your terminal width and each set of rows is labeled so that you can easily see which columns are currently being displayed. For example:

```
octave:13> rand (2,10)
ans =

Columns 1 through 6:

  0.75883  0.93290  0.40064  0.43818  0.94958  0.16467
  0.75697  0.51942  0.40031  0.61784  0.92309  0.40201

Columns 7 through 10:

  0.90174  0.11854  0.72313  0.73326
  0.44672  0.94303  0.56564  0.82150
```

When called from inside a function with the "local" option, the variable is changed locally for the function and any subroutines it calls. The original variable value is restored when exiting the function.

**See also:** [format], page 232.

# Chapter 4: Numeric Data Types

Octave automatically switches to scientific notation when values become very large or very small. This guarantees that you will see several significant figures for every value in a matrix. If you would prefer to see all values in a matrix printed in a fixed point format, you can set the built-in variable `fixed_point_format` to a nonzero value. But doing so is not recommended, because it can produce output that can easily be misinterpreted.

*val* = fixed_point_format ()  [Built-in Function]
*old_val* = fixed_point_format (*new_val*)  [Built-in Function]
fixed_point_format (*new_val*, "*local*")  [Built-in Function]

> Query or set the internal variable that controls whether Octave will use a scaled format to print matrix values.
>
> The scaled format prints a scaling factor on the first line of output chosen such that the largest matrix element can be written with a single leading digit. For example:
>
> ```
> logspace (1, 7, 5)'
> ans =
> 
>    1.0e+07  *
> 
>    0.00000
>    0.00003
>    0.00100
>    0.03162
>    1.00000
> ```
>
> Notice that the first value appears to be 0 when it is actually 1. Because of the possibility for confusion you should be careful about enabling `fixed_point_format`.
>
> When called from inside a function with the `"local"` option, the variable is changed locally for the function and any subroutines it calls. The original variable value is restored when exiting the function.
>
> **See also:** [format], page 232, [output_max_field_width], page 49, [output_precision], page 50.

## 4.1.1 Empty Matrices

A matrix may have one or both dimensions zero, and operations on empty matrices are handled as described by Carl de Boor in *An Empty Exercise*, SIGNUM, Volume 25, pages 2-6, 1990 and C. N. Nett and W. M. Haddad, in *A System-Theoretic Appropriate Realization of the Empty Matrix Concept*, IEEE Transactions on Automatic Control, Volume 38, Number 5, May 1993. Briefly, given a scalar $s$, an $m \times n$ matrix $M_{m \times n}$, and an $m \times n$ empty matrix $[]_{m \times n}$ (with either one or both dimensions equal to zero), the following are true:

$$s \cdot []_{m \times n} = []_{m \times n} \cdot s = []_{m \times n}$$
$$[]_{m \times n} + []_{m \times n} = []_{m \times n}$$
$$[]_{0 \times m} \cdot M_{m \times n} = []_{0 \times n}$$
$$M_{m \times n} \cdot []_{n \times 0} = []_{m \times 0}$$
$$[]_{m \times 0} \cdot []_{0 \times n} = 0_{m \times n}$$

By default, dimensions of the empty matrix are printed along with the empty matrix symbol, '[]'. The built-in variable `print_empty_dimensions` controls this behavior.

`val = print_empty_dimensions ()` [Built-in Function]
`old_val = print_empty_dimensions (`*`new_val`*`)` [Built-in Function]
`print_empty_dimensions (`*`new_val`*`, "`*`local`*`")` [Built-in Function]

> Query or set the internal variable that controls whether the dimensions of empty matrices are printed along with the empty matrix symbol, '`[]`'.
>
> For example, the expression
>
>     zeros (3, 0)
>
> will print
>
>     ans = [](3x0)
>
> When called from inside a function with the `"local"` option, the variable is changed locally for the function and any subroutines it calls. The original variable value is restored when exiting the function.
>
> **See also:** [format], page 232.

Empty matrices may also be used in assignment statements as a convenient way to delete rows or columns of matrices. See Section 8.6 [Assignment Expressions], page 149.

When Octave parses a matrix expression, it examines the elements of the list to determine whether they are all constants. If they are, it replaces the list with a single matrix constant.

## 4.2 Ranges

A *range* is a convenient way to write a row vector with evenly spaced elements. A range expression is defined by the value of the first element in the range, an optional value for the increment between elements, and a maximum value which the elements of the range will not exceed. The base, increment, and limit are separated by colons (the ':' character) and may contain any arithmetic expressions and function calls. If the increment is omitted, it is assumed to be 1. For example, the range

    1 : 5

defines the set of values '`[ 1, 2, 3, 4, 5 ]`', and the range

    1 : 3 : 5

defines the set of values '`[ 1, 4 ]`'.

Although a range constant specifies a row vector, Octave does *not* normally convert range constants to vectors unless it is necessary to do so. This allows you to write a constant like '1 : 10000' without using 80,000 bytes of storage on a typical 32-bit workstation.

A common example of when it does become necessary to convert ranges into vectors occurs when they appear within a vector (i.e., inside square brackets). For instance, whereas

    x = 0 : 0.1 : 1;

defines x to be a variable of type **range** and occupies 24 bytes of memory, the expression

    y = [ 0 : 0.1 : 1];

defines y to be of type **matrix** and occupies 88 bytes of memory.

This space saving optimization may be disabled using the function *disable_range*.

*val* = disable_range ()                                               [Built-in Function]
*old_val* = disable_range (*new_val*)                                  [Built-in Function]
disable_range (*new_val*, "*local*")                                   [Built-in Function]

> Query or set the internal variable that controls whether ranges are stored in a special space-efficient format.
>
> The default value is true. If this option is disabled Octave will store ranges as full matrices.
>
> When called from inside a function with the "local" option, the variable is changed locally for the function and any subroutines it calls. The original variable value is restored when exiting the function.
>
> **See also:** [disable_diagonal_matrix], page 509, [disable_permutation_matrix], page 509.

Note that the upper (or lower, if the increment is negative) bound on the range is not always included in the set of values, and that ranges defined by floating point values can produce surprising results because Octave uses floating point arithmetic to compute the values in the range. If it is important to include the endpoints of a range and the number of elements is known, you should use the linspace function instead (see Section 16.3 [Special Utility Matrices], page 416).

When adding a scalar to a range, subtracting a scalar from it (or subtracting a range from a scalar) and multiplying by scalar, Octave will attempt to avoid unpacking the range and keep the result as a range, too, if it can determine that it is safe to do so. For instance, doing

        a = 2*(1:1e7) - 1;

will produce the same result as '1:2:2e7-1', but without ever forming a vector with ten million elements.

Using zero as an increment in the colon notation, as '1:0:1' is not allowed, because a division by zero would occur in determining the number of range elements. However, ranges with zero increment (i.e., all elements equal) are useful, especially in indexing, and Octave allows them to be constructed using the built-in function *ones*. Note that because a range must be a row vector, 'ones (1, 10)' produces a range, while 'ones (10, 1)' does not.

When Octave parses a range expression, it examines the elements of the expression to determine whether they are all constants. If they are, it replaces the range expression with a single range constant.

## 4.3 Single Precision Data Types

Octave includes support for single precision data types, and most of the functions in Octave accept single precision values and return single precision answers. A single precision variable is created with the single function.

single (*x*)                                                           [Built-in Function]

> Convert *x* to single precision type.
>
> **See also:** [double], page 47.

for example:
```
sngl = single (rand (2, 2))
    ⇒ sngl =
        0.37569   0.92982
        0.11962   0.50876
class (sngl)
    ⇒ single
```
Many functions can also return single precision values directly. For example
```
ones (2, 2, "single")
zeros (2, 2, "single")
eye (2, 2,  "single")
rand (2, 2, "single")
NaN (2, 2, "single")
NA (2, 2, "single")
Inf (2, 2, "single")
```
will all return single precision matrices.

## 4.4 Integer Data Types

Octave supports integer matrices as an alternative to using double precision. It is possible to use both signed and unsigned integers represented by 8, 16, 32, or 64 bits. It should be noted that most computations require floating point data, meaning that integers will often change type when involved in numeric computations. For this reason integers are most often used to store data, and not for calculations.

In general most integer matrices are created by casting existing matrices to integers. The following example shows how to cast a matrix into 32 bit integers.
```
float = rand (2, 2)
    ⇒ float = 0.37569   0.92982
              0.11962   0.50876
integer = int32 (float)
    ⇒ integer = 0   1
                0   1
```
As can be seen, floating point values are rounded to the nearest integer when converted.

**isinteger (x)** [Built-in Function]
Return true if x is an integer object (int8, uint8, int16, etc.).

Note that `isinteger (14)` is false because numeric constants in Octave are double precision floating point values.

See also: [isfloat], page 62, [ischar], page 68, [islogical], page 62, [isnumeric], page 62, [isa], page 39.

**int8 (x)** [Built-in Function]
Convert x to 8-bit integer type.

See also: [uint8], page 55, [int16], page 55, [uint16], page 55, [int32], page 55, [uint32], page 55, [int64], page 55, [uint64], page 55.

## Chapter 4: Numeric Data Types

**uint8 (*x*)**     [Built-in Function]

    Convert *x* to unsigned 8-bit integer type.

    **See also:** [int8], page 54, [int16], page 55, [uint16], page 55, [int32], page 55, [uint32], page 55, [int64], page 55, [uint64], page 55.

**int16 (*x*)**     [Built-in Function]

    Convert *x* to 16-bit integer type.

    **See also:** [int8], page 54, [uint8], page 55, [uint16], page 55, [int32], page 55, [uint32], page 55, [int64], page 55, [uint64], page 55.

**uint16 (*x*)**     [Built-in Function]

    Convert *x* to unsigned 16-bit integer type.

    **See also:** [int8], page 54, [uint8], page 55, [int16], page 55, [int32], page 55, [uint32], page 55, [int64], page 55, [uint64], page 55.

**int32 (*x*)**     [Built-in Function]

    Convert *x* to 32-bit integer type.

    **See also:** [int8], page 54, [uint8], page 55, [int16], page 55, [uint16], page 55, [uint32], page 55, [int64], page 55, [uint64], page 55.

**uint32 (*x*)**     [Built-in Function]

    Convert *x* to unsigned 32-bit integer type.

    **See also:** [int8], page 54, [uint8], page 55, [int16], page 55, [uint16], page 55, [int32], page 55, [int64], page 55, [uint64], page 55.

**int64 (*x*)**     [Built-in Function]

    Convert *x* to 64-bit integer type.

    **See also:** [int8], page 54, [uint8], page 55, [int16], page 55, [uint16], page 55, [int32], page 55, [uint32], page 55, [uint64], page 55.

**uint64 (*x*)**     [Built-in Function]

    Convert *x* to unsigned 64-bit integer type.

    **See also:** [int8], page 54, [uint8], page 55, [int16], page 55, [uint16], page 55, [int32], page 55, [uint32], page 55, [int64], page 55.

**intmax (*type*)**     [Built-in Function]

    Return the largest integer that can be represented in an integer type.

    The variable *type* can be

        `int8`      signed 8-bit integer.

        `int16`      signed 16-bit integer.

        `int32`      signed 32-bit integer.

        `int64`      signed 64-bit integer.

        `uint8`      unsigned 8-bit integer.

        `uint16`      unsigned 16-bit integer.

`uint32`    unsigned 32-bit integer.

`uint64`    unsigned 64-bit integer.

The default for *type* is `int32`.

**See also:** [intmin], page 56, [flintmax], page 56, [bitmax], page 58.

`intmin (type)` [Built-in Function]

Return the smallest integer that can be represented in an integer type.

The variable *type* can be

`int8`      signed 8-bit integer.

`int16`     signed 16-bit integer.

`int32`     signed 32-bit integer.

`int64`     signed 64-bit integer.

`uint8`     unsigned 8-bit integer.

`uint16`    unsigned 16-bit integer.

`uint32`    unsigned 32-bit integer.

`uint64`    unsigned 64-bit integer.

The default for *type* is `int32`.

**See also:** [intmax], page 55, [flintmax], page 56, [bitmax], page 58.

`flintmax ()` [Built-in Function]
`flintmax ("double")` [Built-in Function]
`flintmax ("single")` [Built-in Function]

Return the largest integer that can be represented consecutively in a floating point value.

The default class is `"double"`, but `"single"` is a valid option. On IEEE-754 compatible systems, `flintmax` is $2^53$ for `"double"` and $2^24$ for `"single"`.

**See also:** [bitmax], page 58, [intmax], page 55, [realmax], page 461, [realmin], page 461.

### 4.4.1 Integer Arithmetic

While many numerical computations can't be carried out in integers, Octave does support basic operations like addition and multiplication on integers. The operators `+`, `-`, `.*`, and `./` work on integers of the same type. So, it is possible to add two 32 bit integers, but not to add a 32 bit integer and a 16 bit integer.

When doing integer arithmetic one should consider the possibility of underflow and overflow. This happens when the result of the computation can't be represented using the chosen integer type. As an example it is not possible to represent the result of $10 - 20$ when using unsigned integers. Octave makes sure that the result of integer computations is the integer that is closest to the true result. So, the result of $10 - 20$ when using unsigned integers is zero.

When doing integer division Octave will round the result to the nearest integer. This is different from most programming languages, where the result is often floored to the nearest integer. So, the result of `int32 (5) ./ int32 (8)` is 1.

Chapter 4: Numeric Data Types                                                 57

`idivide (x, y, op)`                                                 [Function File]
> Integer division with different rounding rules.
>
> The standard behavior of integer division such as `a ./ b` is to round the result to the nearest integer. This is not always the desired behavior and `idivide` permits integer element-by-element division to be performed with different treatment for the fractional part of the division as determined by the *op* flag. *op* is a string with one of the values:
>
> `"fix"`   Calculate `a ./ b` with the fractional part rounded towards zero.
>
> `"round"` Calculate `a ./ b` with the fractional part rounded towards the nearest integer.
>
> `"floor"` Calculate `a ./ b` with the fractional part rounded towards negative infinity.
>
> `"ceil"`  Calculate `a ./ b` with the fractional part rounded towards positive infinity.
>
> If *op* is not given it defaults to `"fix"`. An example demonstrating these rounding rules is
>
> ```
> idivide (int8 ([-3, 3]), int8 (4), "fix")
>   ⇒ int8 ([0, 0])
> idivide (int8 ([-3, 3]), int8 (4), "round")
>   ⇒ int8 ([-1, 1])
> idivide (int8 ([-3, 3]), int8 (4), "floor")
>   ⇒ int8 ([-1, 0])
> idivide (int8 ([-3, 3]), int8 (4), "ceil")
>   ⇒ int8 ([0, 1])
> ```
>
> **See also:** [ldivide], page 143, [rdivide], page 144.

## 4.5 Bit Manipulations

Octave provides a number of functions for the manipulation of numeric values on a bit by bit basis. The basic functions to set and obtain the values of individual bits are `bitset` and `bitget`.

`C = bitset (A, n)`                                                  [Function File]
`C = bitset (A, n, val)`                                             [Function File]
> Set or reset bit(s) *n* of the unsigned integers in *A*.
>
> *val* = 0 resets and *val* = 1 sets the bits. The least significant bit is $n = 1$. All variables must be the same size or scalars.
>
> ```
> dec2bin (bitset (10, 1))
>   ⇒ 1011
> ```
>
> **See also:** [bitand], page 58, [bitor], page 58, [bitxor], page 59, [bitget], page 57, [bitcmp], page 59, [bitshift], page 59, [bitmax], page 58.

`c = bitget (A, n)`                                                  [Function File]
> Return the status of bit(s) *n* of the unsigned integers in *A*.
>
> The least significant bit is $n = 1$.

```
bitget (100, 8:-1:1)
⇒  0  1  1  0  0  1  0  0
```

**See also:** [bitand], page 58, [bitor], page 58, [bitxor], page 59, [bitset], page 57, [bitcmp], page 59, [bitshift], page 59, [bitmax], page 58.

The arguments to all of Octave's bitwise operations can be scalar or arrays, except for `bitcmp`, whose *k* argument must a scalar. In the case where more than one argument is an array, then all arguments must have the same shape, and the bitwise operator is applied to each of the elements of the argument individually. If at least one argument is a scalar and one an array, then the scalar argument is duplicated. Therefore

```
bitget (100, 8:-1:1)
```

is the same as

```
bitget (100 * ones (1, 8), 8:-1:1)
```

It should be noted that all values passed to the bit manipulation functions of Octave are treated as integers. Therefore, even though the example for `bitset` above passes the floating point value 10, it is treated as the bits [1, 0, 1, 0] rather than the bits of the native floating point format representation of 10.

As the maximum value that can be represented by a number is important for bit manipulation, particularly when forming masks, Octave supplies the function `bitmax`.

**bitmax ()**                                                                                         [Built-in Function]
**bitmax ("*double*")**                                                  [Built-in Function]
**bitmax ("*single*")**                                                 [Built-in Function]

Return the largest integer that can be represented within a floating point value.

The default class is `"double"`, but `"single"` is a valid option. On IEEE-754 compatible systems, `bitmax` is $2^{53} - 1$ for `"double"` and $2^{24} - 1$ for `"single"`.

**See also:** [flintmax], page 56, [intmax], page 55, [realmax], page 461, [realmin], page 461.

This is the double precision version of the function `intmax`, previously discussed.

Octave also includes the basic bitwise 'and', 'or', and 'exclusive or' operators.

**bitand (*x*, *y*)**                                                          [Built-in Function]

Return the bitwise AND of non-negative integers.

*x*, *y* must be in the range [0,bitmax]

**See also:** [bitor], page 58, [bitxor], page 59, [bitset], page 57, [bitget], page 57, [bitcmp], page 59, [bitshift], page 59, [bitmax], page 58.

**bitor (*x*, *y*)**                                                             [Built-in Function]

Return the bitwise OR of non-negative integers.

*x*, *y* must be in the range [0,bitmax]

**See also:** [bitor], page 58, [bitxor], page 59, [bitset], page 57, [bitget], page 57, [bitcmp], page 59, [bitshift], page 59, [bitmax], page 58.

Chapter 4: Numeric Data Types 59

**bitxor (x, y)** [Built-in Function]
> Return the bitwise XOR of non-negative integers.
>
> x, y must be in the range [0,bitmax]
>
> **See also:** [bitand], page 58, [bitor], page 58, [bitset], page 57, [bitget], page 57, [bitcmp], page 59, [bitshift], page 59, [bitmax], page 58.

The bitwise 'not' operator is a unary operator that performs a logical negation of each of the bits of the value. For this to make sense, the mask against which the value is negated must be defined. Octave's bitwise 'not' operator is `bitcmp`.

**bitcmp (A, k)** [Function File]
> Return the $k$-bit complement of integers in $A$.
>
> If k is omitted k = log2 (bitmax) + 1 is assumed.
>
> ```
> bitcmp (7,4)
>    ⇒ 8
> dec2bin (11)
>    ⇒ 1011
> dec2bin (bitcmp (11, 6))
>    ⇒ 110100
> ```
>
> **See also:** [bitand], page 58, [bitor], page 58, [bitxor], page 59, [bitset], page 57, [bitget], page 57, [bitcmp], page 59, [bitshift], page 59, [bitmax], page 58.

Octave also includes the ability to left-shift and right-shift values bitwise.

**bitshift (a, k)** [Built-in Function]
**bitshift (a, k, n)** [Built-in Function]
> Return a $k$ bit shift of $n$-digit unsigned integers in $a$.
>
> A positive $k$ leads to a left shift; A negative value to a right shift.
>
> If n is omitted it defaults to log2(bitmax)+1. n must be in the range [1,log2(bitmax)+1] usually [1,33].
>
> ```
> bitshift (eye (3), 1)
> ⇒
> 2 0 0
> 0 2 0
> 0 0 2
>
> bitshift (10, [-2, -1, 0, 1, 2])
>    ⇒ 2   5   10   20   40
> ```
>
> **See also:** [bitand], page 58, [bitor], page 58, [bitxor], page 59, [bitset], page 57, [bitget], page 57, [bitcmp], page 59, [bitmax], page 58.

Bits that are shifted out of either end of the value are lost. Octave also uses arithmetic shifts, where the sign bit of the value is kept during a right shift. For example:

```
bitshift (-10, -1)
   ⇒ -5
bitshift (int8 (-1), -1)
   ⇒ -1
```

Note that `bitshift (int8 (-1), -1)` is -1 since the bit representation of -1 in the `int8` data type is [1, 1, 1, 1, 1, 1, 1, 1].

## 4.6 Logical Values

Octave has built-in support for logical values, i.e., variables that are either `true` or `false`. When comparing two variables, the result will be a logical value whose value depends on whether or not the comparison is true.

The basic logical operations are `&`, `|`, and `!`, which correspond to "Logical And", "Logical Or", and "Logical Negation". These operations all follow the usual rules of logic.

It is also possible to use logical values as part of standard numerical calculations. In this case `true` is converted to 1, and `false` to 0, both represented using double precision floating point numbers. So, the result of `true*22 - false/6` is 22.

Logical values can also be used to index matrices and cell arrays. When indexing with a logical array the result will be a vector containing the values corresponding to `true` parts of the logical array. The following example illustrates this.

```
data = [ 1, 2; 3, 4 ];
idx = (data <= 2);
data(idx)
     ⇒ ans = [ 1; 2 ]
```

Instead of creating the `idx` array it is possible to replace `data(idx)` with `data( data <= 2 )` in the above code.

Logical values can also be constructed by casting numeric objects to logical values, or by using the `true` or `false` functions.

**logical** (*x*)                                                              [Built-in Function]

Convert the numeric object *x* to logical type.

Any nonzero values will be converted to true (1) while zero values will be converted to false (0). The non-numeric value NaN cannot be converted and will produce an error.

Compatibility Note: Octave accepts complex values as input, whereas MATLAB issues an error.

**See also:** [double], page 47, [single], page 53, [char], page 71.

**true** (*x*)                                                                 [Built-in Function]
**true** (*n, m*)                                                              [Built-in Function]
**true** (*n, m, k, ...*)                                                      [Built-in Function]

Return a matrix or N-dimensional array whose elements are all logical 1.

If invoked with a single scalar integer argument, return a square matrix of the specified size.

If invoked with two or more scalar integer arguments, or a vector of integer values, return an array with given dimensions.

**See also:** [false], page 61.

`false (x)`                                              [Built-in Function]
`false (n, m)`                                           [Built-in Function]
`false (n, m, k, ...)`                                   [Built-in Function]

> Return a matrix or N-dimensional array whose elements are all logical 0.
>
> If invoked with a single scalar integer argument, return a square matrix of the specified size.
>
> If invoked with two or more scalar integer arguments, or a vector of integer values, return an array with given dimensions.
>
> **See also:** [true], page 60.

## 4.7 Promotion and Demotion of Data Types

Many operators and functions can work with mixed data types. For example,

```
uint8 (1) + 1
    ⇒ 2
```

where the above operator works with an 8-bit integer and a double precision value and returns an 8-bit integer value. Note that the type is demoted to an 8-bit integer, rather than promoted to a double precision value as might be expected. The reason is that if Octave promoted values in expressions like the above with all numerical constants would need to be explicitly cast to the appropriate data type like

```
uint8 (1) + uint8 (1)
    ⇒ 2
```

which becomes difficult for the user to apply uniformly and might allow hard to find bugs to be introduced. The same applies to single precision values where a mixed operation such as

```
single (1) + 1
    ⇒ 2
```

returns a single precision value. The mixed operations that are valid and their returned data types are

| Mixed Operation     | Result  |
|---------------------|---------|
| double OP single    | single  |
| double OP integer   | integer |
| double OP char      | double  |
| double OP logical   | double  |
| single OP integer   | integer |
| single OP char      | single  |
| single OP logical   | single  |

The same logic applies to functions with mixed arguments such as

```
min (single (1), 0)
    ⇒ 0
```

where the returned value is single precision.

In the case of mixed type indexed assignments, the type is not changed. For example,

```
x = ones (2, 2);
x(1, 1) = single (2)
   ⇒ x = 2   1
         1   1
```

where x remains of the double precision type.

## 4.8 Predicates for Numeric Objects

Since the type of a variable may change during the execution of a program, it can be necessary to do type checking at run-time. Doing this also allows you to change the behavior of a function depending on the type of the input. As an example, this naive implementation of abs returns the absolute value of the input if it is a real number, and the length of the input if it is a complex number.

```
function a = abs (x)
  if (isreal (x))
    a = sign (x) .* x;
  elseif (iscomplex (x))
    a = sqrt (real(x).^2 + imag(x).^2);
  endif
endfunction
```

The following functions are available for determining the type of a variable.

**isnumeric (*x*)** [Built-in Function]

Return true if *x* is a numeric object, i.e., an integer, real, or complex array.

Logical and character arrays are not considered to be numeric.

**See also:** [isinteger], page 54, [isfloat], page 62, [isreal], page 62, [iscomplex], page 63, [islogical], page 62, [ischar], page 68, [iscell], page 113, [isstruct], page 107, [isa], page 39.

**islogical (*x*)** [Built-in Function]
**isbool (*x*)** [Built-in Function]

Return true if *x* is a logical object.

**See also:** [isfloat], page 62, [isinteger], page 54, [ischar], page 68, [isnumeric], page 62, [isa], page 39.

**isfloat (*x*)** [Built-in Function]

Return true if *x* is a floating-point numeric object.

Objects of class double or single are floating-point objects.

**See also:** [isinteger], page 54, [ischar], page 68, [islogical], page 62, [isnumeric], page 62, [isa], page 39.

**isreal (*x*)** [Built-in Function]

Return true if *x* is a non-complex matrix or scalar.

For compatibility with MATLAB, this includes logical and character matrices.

**See also:** [iscomplex], page 63, [isnumeric], page 62, [isa], page 39.

Chapter 4: Numeric Data Types 63

**iscomplex (x)** [Built-in Function]

Return true if *x* is a complex-valued numeric object.

**See also:** [isreal], page 62, [isnumeric], page 62, [islogical], page 62, [ischar], page 68, [isfloat], page 62, [isa], page 39.

**ismatrix (a)** [Built-in Function]

Return true if *a* is a 2-D array.

**See also:** [isscalar], page 63, [isvector], page 63, [iscell], page 113, [isstruct], page 107, [issparse], page 524, [isa], page 39.

**isvector (x)** [Function File]

Return true if *x* is a vector.

A vector is a 2-D array where one of the dimensions is equal to 1. As a consequence a 1x1 array, or scalar, is also a vector.

**See also:** [isscalar], page 63, [ismatrix], page 63, [size], page 45, [rows], page 44, [columns], page 44, [length], page 45.

**isrow (x)** [Function File]

Return true if *x* is a row vector 1xN with non-negative N.

**See also:** [iscolumn], page 63, [isscalar], page 63, [isvector], page 63, [ismatrix], page 63.

**iscolumn (x)** [Function File]

Return true if *x* is a column vector Nx1 with non-negative N.

**See also:** [isrow], page 63, [isscalar], page 63, [isvector], page 63, [ismatrix], page 63.

**isscalar (x)** [Built-in Function]

Return true if *x* is a scalar.

**See also:** [isvector], page 63, [ismatrix], page 63.

**issquare (x)** [Function File]

Return true if *x* is a square matrix.

**See also:** [isscalar], page 63, [isvector], page 63, [ismatrix], page 63, [size], page 45.

**issymmetric (A)** [Function File]
**issymmetric (A, *tol*)** [Function File]

Return true if *A* is a symmetric matrix within the tolerance specified by *tol*.

The default tolerance is zero (uses faster code).

Matrix *A* is considered symmetric if `norm (A - A.', Inf) / norm (A, Inf) < tol`.

**See also:** [ishermitian], page 63, [isdefinite], page 64.

**ishermitian (A)** [Function File]
**ishermitian (A, *tol*)** [Function File]

Return true if *A* is Hermitian within the tolerance specified by *tol*.

The default tolerance is zero (uses faster code).

Matrix *A* is considered symmetric if `norm (A - A', Inf) / norm (A, Inf) < tol`.

**See also:** [issymmetric], page 63, [isdefinite], page 64.

**isdefinite** (*A*) [Function File]
**isdefinite** (*A*, *tol*) [Function File]
    Return 1 if *A* is symmetric positive definite within the tolerance specified by *tol* or 0 if *A* is symmetric positive semidefinite. Otherwise, return -1.

    If *tol* is omitted, use a tolerance of `100 * eps * norm (A, "fro")`

    **See also:** [issymmetric], page 63, [ishermitian], page 63.

**isbanded** (*A*, *lower*, *upper*) [Function File]
    Return true if *A* is a matrix with entries confined between *lower* diagonals below the main diagonal and *upper* diagonals above the main diagonal.

    *lower* and *upper* must be non-negative integers.

    **See also:** [isdiag], page 64, [istril], page 64, [istriu], page 64, [bandwidth], page 464.

**isdiag** (*A*) [Function File]
    Return true if *A* is a diagonal matrix.

    **See also:** [isbanded], page 64, [istril], page 64, [istriu], page 64, [diag], page 415, [bandwidth], page 464.

**istril** (*A*) [Function File]
    Return true if *A* is a lower triangular matrix.

    A lower triangular matrix has nonzero entries only on the main diagonal and below.

    **See also:** [istriu], page 64, [isbanded], page 64, [isdiag], page 64, [tril], page 414, [bandwidth], page 464.

**istriu** (*A*) [Function File]
    Return true if *A* is an upper triangular matrix.

    An upper triangular matrix has nonzero entries only on the main diagonal and above.

    **See also:** [isdiag], page 64, [isbanded], page 64, [istril], page 64, [triu], page 414, [bandwidth], page 464.

**isprime** (*x*) [Function File]
    Return a logical array which is true where the elements of *x* are prime numbers and false where they are not.

    A prime number is conventionally defined as a positive integer greater than 1 (e.g., 2, 3, ...) which is divisible only by itself and 1. Octave extends this definition to include both negative integers and complex values. A negative integer is prime if its positive counterpart is prime. This is equivalent to `isprime (abs (x))`.

    If `class (x)` is complex, then primality is tested in the domain of Gaussian integers (http://en.wikipedia.org/wiki/Gaussian_integer). Some non-complex integers are prime in the ordinary sense, but not in the domain of Gaussian integers. For example, $5 = (1 + 2i) * (1 - 2i)$ shows that 5 is not prime because it has a factor other than itself and 1. Exercise caution when testing complex and real values together in the same matrix.

    Examples:

# Chapter 4: Numeric Data Types

```
isprime (1:6)
    ⇒ [0, 1, 1, 0, 1, 0]
isprime ([i, 2, 3, 5])
    ⇒ [0, 0, 1, 0]
```

Programming Note: `isprime` is appropriate if the maximum value in x is not too large (< 1e15). For larger values special purpose factorization code should be used.

Compatibility Note: *matlab* does not extend the definition of prime numbers and will produce an error if given negative or complex inputs.

**See also:** [primes], page 448, [factor], page 447, [gcd], page 447, [lcm], page 447.

If instead of knowing properties of variables, you wish to know which variables are defined and to gather other information about the workspace itself, see Section 7.3 [Status of Variables], page 127.

# 5 Strings

A *string constant* consists of a sequence of characters enclosed in either double-quote or single-quote marks. For example, both of the following expressions

    "parrot"
    'parrot'

represent the string whose contents are 'parrot'. Strings in Octave can be of any length.

Since the single-quote mark is also used for the transpose operator (see Section 8.3 [Arithmetic Ops], page 141) but double-quote marks have no other purpose in Octave, it is best to use double-quote marks to denote strings.

Strings can be concatenated using the notation for defining matrices. For example, the expression

    [ "foo" , "bar" , "baz" ]

produces the string whose contents are 'foobarbaz'. See Chapter 4 [Numeric Data Types], page 47, for more information about creating matrices.

## 5.1 Escape Sequences in String Constants

In double-quoted strings, the backslash character is used to introduce *escape sequences* that represent other characters. For example, '\n' embeds a newline character in a double-quoted string and '\"' embeds a double quote character. In single-quoted strings, backslash is not a special character. Here is an example showing the difference:

    toascii ("\n")
        ⇒ 10
    toascii ('\n')
        ⇒ [ 92 110 ]

Here is a table of all the escape sequences used in Octave (within double quoted strings). They are the same as those used in the C programming language.

| | |
|---|---|
| \\ | Represents a literal backslash, '\'. |
| \" | Represents a literal double-quote character, '"'. |
| \' | Represents a literal single-quote character, '''. |
| \0 | Represents the null character, control-@, ASCII code 0. |
| \a | Represents the "alert" character, control-g, ASCII code 7. |
| \b | Represents a backspace, control-h, ASCII code 8. |
| \f | Represents a formfeed, control-l, ASCII code 12. |
| \n | Represents a newline, control-j, ASCII code 10. |
| \r | Represents a carriage return, control-m, ASCII code 13. |
| \t | Represents a horizontal tab, control-i, ASCII code 9. |
| \v | Represents a vertical tab, control-k, ASCII code 11. |
| \nnn | Represents the octal value *nnn*, where *nnn* are one to three digits between 0 and 7. For example, the code for the ASCII ESC (escape) character is '\033'. |

\xhh…     Represents the hexadecimal value *hh*, where *hh* are hexadecimal digits ('0' through '9' and either 'A' through 'F' or 'a' through 'f'). Like the same construct in ANSI C, the escape sequence continues until the first non-hexadecimal digit is seen. However, using more than two hexadecimal digits produces undefined results.

In a single-quoted string there is only one escape sequence: you may insert a single quote character using two single quote characters in succession. For example,

```
'I can''t escape'
    ⇒ I can't escape
```

In scripts the two different string types can be distinguished if necessary by using `is_dq_string` and `is_sq_string`.

**is_dq_string** (*x*)                                                     [Built-in Function]
    Return true if *x* is a double-quoted character string.

    **See also:** [is_sq_string], page 68, [ischar], page 68.

**is_sq_string** (*x*)                                                     [Built-in Function]
    Return true if *x* is a single-quoted character string.

    **See also:** [is_dq_string], page 68, [ischar], page 68.

## 5.2 Character Arrays

The string representation used by Octave is an array of characters, so internally the string `"dddddddddd"` is actually a row vector of length 10 containing the value 100 in all places (100 is the ASCII code of `"d"`). This lends itself to the obvious generalization to character matrices. Using a matrix of characters, it is possible to represent a collection of same-length strings in one variable. The convention used in Octave is that each row in a character matrix is a separate string, but letting each column represent a string is equally possible.

The easiest way to create a character matrix is to put several strings together into a matrix.

```
collection = [ "String #1"; "String #2" ];
```

This creates a 2-by-9 character matrix.

The function `ischar` can be used to test if an object is a character matrix.

**ischar** (*x*)                                                           [Built-in Function]
    Return true if *x* is a character array.

    **See also:** [isfloat], page 62, [isinteger], page 54, [islogical], page 62, [isnumeric], page 62, [iscellstr], page 119, [isa], page 39.

To test if an object is a string (i.e., a character vector and not a character matrix) you can use the `ischar` function in combination with the `isvector` function as in the following example:

```
ischar (collection)
    ⇒ 1

ischar (collection) && isvector (collection)
    ⇒ 0

ischar ("my string") && isvector ("my string")
    ⇒ 1
```

One relevant question is, what happens when a character matrix is created from strings of different length. The answer is that Octave puts blank characters at the end of strings shorter than the longest string. It is possible to use a different character than the blank character using the `string_fill_char` function.

*val* = `string_fill_char` ()                               [Built-in Function]
*old_val* = `string_fill_char` (*new_val*)                  [Built-in Function]
`string_fill_char` (*new_val*, "*local*")                   [Built-in Function]

> Query or set the internal variable used to pad all rows of a character matrix to the same length.
>
> The value must be a single character and the default is " " (a single space). For example:
>
> ```
> string_fill_char ("X");
> [ "these"; "are"; "strings" ]
>     ⇒  "theseXX"
>        "areXXXX"
>        "strings"
> ```
>
> When called from inside a function with the `"local"` option, the variable is changed locally for the function and any subroutines it calls. The original variable value is restored when exiting the function.

This shows a problem with character matrices. It simply isn't possible to represent strings of different lengths. The solution is to use a cell array of strings, which is described in Section 6.2.4 [Cell Arrays of Strings], page 118.

## 5.3 Creating Strings

The easiest way to create a string is, as illustrated in the introduction, to enclose a text in double-quotes or single-quotes. It is however possible to create a string without actually writing a text. The function `blanks` creates a string of a given length consisting only of blank characters (ASCII code 32).

`blanks` (*n*)                                              [Function File]

> Return a string of *n* blanks.
>
> For example:

```
blanks (10);
whos ans
    ⇒
    Attr Name      Size                    Bytes  Class
    ==== ====      ====                    =====  =====
         ans       1x10                       10  char
```

See also: [repmat], page 417.

## 5.3.1 Concatenating Strings

Strings can be concatenated using matrix notation (see Chapter 5 [Strings], page 67, Section 5.2 [Character Arrays], page 68) which is often the most natural method. For example:

```
fullname = [fname ".txt"];
email = ["<" user "@" domain ">"];
```

In each case it is easy to see what the final string will look like. This method is also the most efficient. When using matrix concatenation the parser immediately begins joining the strings without having to process the overhead of a function call and the input validation of the associated function.

Nevertheless, there are several other functions for concatenating string objects which can be useful in specific circumstances: **char**, **strvcat**, **strcat**, and **cstrcat**. Finally, the general purpose concatenation functions can be used: see [cat], page 408, [horzcat], page 409, and [vertcat], page 409.

- All string concatenation functions except **cstrcat** convert numerical input into character data by taking the corresponding ASCII character for each element, as in the following example:

    ```
    char ([98, 97, 110, 97, 110, 97])
        ⇒ banana
    ```

- **char** and **strvcat** concatenate vertically, while **strcat** and **cstrcat** concatenate horizontally. For example:

    ```
    char ("an apple", "two pears")
        ⇒ an apple
          two pears

    strcat ("oc", "tave", " is", " good", " for you")
        ⇒ octave is good for you
    ```

- **char** generates an empty row in the output for each empty string in the input. **strvcat**, on the other hand, eliminates empty strings.

    ```
    char ("orange", "green", "", "red")
        ⇒ orange
          green

          red
    ```

# Chapter 5: Strings

```
strvcat ("orange", "green", "", "red")
    ⇒ orange
      green
      red
```

- All string concatenation functions except `cstrcat` also accept cell array data (see Section 6.2 [Cell Arrays], page 112). `char` and `strvcat` convert cell arrays into character arrays, while `strcat` concatenates within the cells of the cell arrays:

```
char ({"red", "green", "", "blue"})
    ⇒ red
      green

      blue

strcat ({"abc"; "ghi"}, {"def"; "jkl"})
    ⇒
      {
        [1,1] = abcdef
        [2,1] = ghijkl
      }
```

- `strcat` removes trailing white space in the arguments (except within cell arrays), while `cstrcat` leaves white space untouched. Both kinds of behavior can be useful as can be seen in the examples:

```
strcat (["dir1";"directory2"], ["/";"/"], ["file1";"file2"])
    ⇒ dir1/file1
      directory2/file2

cstrcat (["thirteen apples"; "a banana"], [" 5$";" 1$"])
    ⇒ thirteen apples 5$
      a banana        1$
```

Note that in the above example for `cstrcat`, the white space originates from the internal representation of the strings in a string array (see Section 5.2 [Character Arrays], page 68).

**char** (*x*)  [Built-in Function]
**char** (*x*, ...)  [Built-in Function]
**char** (*s1*, *s2*, ...)  [Built-in Function]
**char** (*cell_array*)  [Built-in Function]

Create a string array from one or more numeric matrices, character matrices, or cell arrays.

Arguments are concatenated vertically. The returned values are padded with blanks as needed to make each row of the string array have the same length. Empty input strings are significant and will concatenated in the output.

For numerical input, each element is converted to the corresponding ASCII character. A range error results if an input is outside the ASCII range (0-255).

For cell arrays, each element is concatenated separately. Cell arrays converted through `char` can mostly be converted back with `cellstr`. For example:

```
char ([97, 98, 99], "", {"98", "99", 100}, "str1", ["ha", "lf"])
    ⇒ ["abc  "
       "     "
       "98   "
       "99   "
       "d    "
       "str1 "
       "half "]
```

**See also:** [strvcat], page 72, [cellstr], page 119.

**strvcat** (*x*)                                                                                               [Built-in Function]
**strvcat** (*x, ...*)                                                                      [Built-in Function]
**strvcat** (*s1, s2, ...*)                                                   [Built-in Function]
**strvcat** (*cell_array*)                                             [Built-in Function]

Create a character array from one or more numeric matrices, character matrices, or cell arrays.

Arguments are concatenated vertically. The returned values are padded with blanks as needed to make each row of the string array have the same length. Unlike `char`, empty strings are removed and will not appear in the output.

For numerical input, each element is converted to the corresponding ASCII character. A range error results if an input is outside the ASCII range (0-255).

For cell arrays, each element is concatenated separately. Cell arrays converted through `strvcat` can mostly be converted back with `cellstr`. For example:

```
strvcat ([97, 98, 99], "", {"98", "99", 100}, "str1", ["ha", "lf"])
    ⇒ ["abc  "
       "98   "
       "99   "
       "d    "
       "str1 "
       "half "]
```

**See also:** [char], page 71, [strcat], page 72, [cstrcat], page 73.

**strcat** (*s1, s2, ...*)                                                           [Function File]

Return a string containing all the arguments concatenated horizontally.

If the arguments are cell strings, `strcat` returns a cell string with the individual cells concatenated. For numerical input, each element is converted to the corresponding ASCII character. Trailing white space for any character string input is eliminated before the strings are concatenated. Note that cell string values do **not** have whitespace trimmed.

For example:

```
strcat ("|", " leading space is preserved", "|")
    ⇒ | leading space is preserved|
strcat ("|", "trailing space is eliminated ", "|")
    ⇒ |trailing space is eliminated|
```

Chapter 5: Strings                                                              73

```
strcat ("homogeneous space |", "   ", "| is also eliminated")
     ⇒ homogeneous space || is also eliminated
s = [ "ab"; "cde" ];
strcat (s, s, s)
    ⇒
        "ababab    "
        "cdecdecde"
s = { "ab"; "cd " };
strcat (s, s, s)
    ⇒
        {
          [1,1] = ababab
          [2,1] = cd cd cd
        }
```

**See also:** [cstrcat], page 73, [char], page 71, [strvcat], page 72.

cstrcat (*s1*, *s2*, ...)                                      [Function File]
Return a string containing all the arguments concatenated horizontally with trailing white space preserved.

For example:
```
cstrcat ("ab   ", "cd")
     ⇒ "ab   cd"
s = [ "ab"; "cde" ];
cstrcat (s, s, s)
     ⇒ "ab ab ab "
        "cdecdecde"
```

**See also:** [strcat], page 72, [char], page 71, [strvcat], page 72.

## 5.3.2 Converting Numerical Data to Strings

Apart from the string concatenation functions (see Section 5.3.1 [Concatenating Strings], page 70) which cast numerical data to the corresponding ASCII characters, there are several functions that format numerical data as strings. `mat2str` and `num2str` convert real or complex matrices, while `int2str` converts integer matrices. `int2str` takes the real part of complex values and round fractional values to integer. A more flexible way to format numerical data as strings is the `sprintf` function (see Section 14.2.4 [Formatted Output], page 253, [sprintf], page 254).

s = mat2str (*x*, *n*)                                         [Function File]
s = mat2str (*x*, *n*, "*class*")                              [Function File]
Format real, complex, and logical matrices as strings.

The returned string may be used to reconstruct the original matrix by using the `eval` function.

The precision of the values is given by *n*. If *n* is a scalar then both real and imaginary parts of the matrix are printed to the same precision. Otherwise `n(1)` defines the precision of the real part and n(2) defines the precision of the imaginary part. The default for $n$ is 15.

If the argument "class" is given then the class of x is included in the string in such a way that eval will result in the construction of a matrix of the same class.

```
mat2str ([ -1/3 + i/7; 1/3 - i/7 ], [4 2])
    ⇒ "[-0.3333+0.14i;0.3333-0.14i]"

mat2str ([ -1/3 +i/7; 1/3 -i/7 ], [4 2])
    ⇒ "[-0.3333+0i 0+0.14i;0.3333+0i -0-0.14i]"

mat2str (int16 ([1 -1]), "class")
    ⇒ "int16([1 -1])"

mat2str (logical (eye (2)))
    ⇒ "[true false;false true]"

isequal (x, eval (mat2str (x)))
    ⇒ 1
```

**See also:** [sprintf], page 254, [num2str], page 74, [int2str], page 75.

num2str (*x*)        [Function File]
num2str (*x, precision*)        [Function File]
num2str (*x, format*)        [Function File]

Convert a number (or array) to a string (or a character array).

The optional second argument may either give the number of significant digits (*precision*) to be used in the output or a format template string (*format*) as in `sprintf` (see Section 14.2.4 [Formatted Output], page 253). `num2str` can also process complex numbers.

Examples:

Chapter 5: Strings

```
num2str (123.456)
    ⇒ "123.46"

num2str (123.456, 4)
    ⇒ "123.5"

s = num2str ([1, 1.34; 3, 3.56], "%5.1f")
    ⇒ s =
      1.0  1.3
      3.0  3.6
whos s
    ⇒
      Attr Name        Size                     Bytes  Class
      ==== ====        ====                     =====  =====
           s           2x8                         16  char

num2str (1.234 + 27.3i)
    ⇒ "1.234+27.3i"
```

Notes:

For MATLAB compatibility, leading spaces are stripped before returning the string.

The **num2str** function is not very flexible. For better control over the results, use **sprintf** (see Section 14.2.4 [Formatted Output], page 253).

For complex x, the format string may only contain one output conversion specification and nothing else. Otherwise, results will be unpredictable.

**See also:** [sprintf], page 254, [int2str], page 75, [mat2str], page 73.

**int2str (n)** [Function File]

Convert an integer (or array of integers) to a string (or a character array).

```
int2str (123)
    ⇒ "123"

s = int2str ([1, 2, 3; 4, 5, 6])
    ⇒ s =
      1  2  3
      4  5  6

whos s
    ⇒
      Attr Name        Size                     Bytes  Class
      ==== ====        ====                     =====  =====
           s           2x7                         14  char
```

This function is not very flexible. For better control over the results, use **sprintf** (see Section 14.2.4 [Formatted Output], page 253).

**See also:** [sprintf], page 254, [num2str], page 74, [mat2str], page 73.

## 5.4 Comparing Strings

Since a string is a character array, comparisons between strings work element by element as the following example shows:

```
GNU = "GNU's Not UNIX";
spaces = (GNU == " ")
    ⇒ spaces =
        0 0 0 0 0 1 0 0 0 1 0 0 0 0
```

To determine if two strings are identical it is necessary to use the `strcmp` function. It compares complete strings and is case sensitive. `strncmp` compares only the first N characters (with N given as a parameter). `strcmpi` and `strncmpi` are the corresponding functions for case-insensitive comparison.

**strcmp** (*s1*, *s2*)                                                                                     [Built-in Function]

Return 1 if the character strings *s1* and *s2* are the same, and 0 otherwise.

If either *s1* or *s2* is a cell array of strings, then an array of the same size is returned, containing the values described above for every member of the cell array. The other argument may also be a cell array of strings (of the same size or with only one element), char matrix or character string.

**Caution:** For compatibility with MATLAB, Octave's strcmp function returns 1 if the character strings are equal, and 0 otherwise. This is just the opposite of the corresponding C library function.

**See also:** [strcmpi], page 76, [strncmp], page 76, [strncmpi], page 77.

**strncmp** (*s1*, *s2*, *n*)                                                                         [Built-in Function]

Return 1 if the first *n* characters of strings *s1* and *s2* are the same, and 0 otherwise.

```
strncmp ("abce", "abcd", 3)
    ⇒ 1
```

If either *s1* or *s2* is a cell array of strings, then an array of the same size is returned, containing the values described above for every member of the cell array. The other argument may also be a cell array of strings (of the same size or with only one element), char matrix or character string.

```
strncmp ("abce", {"abcd", "bca", "abc"}, 3)
    ⇒ [1, 0, 1]
```

**Caution:** For compatibility with MATLAB, Octave's strncmp function returns 1 if the character strings are equal, and 0 otherwise. This is just the opposite of the corresponding C library function.

**See also:** [strncmpi], page 77, [strcmp], page 76, [strcmpi], page 76.

**strcmpi** (*s1*, *s2*)                                                                          [Built-in Function]

Return 1 if the character strings *s1* and *s2* are the same, disregarding case of alphabetic characters, and 0 otherwise.

If either *s1* or *s2* is a cell array of strings, then an array of the same size is returned, containing the values described above for every member of the cell array. The other argument may also be a cell array of strings (of the same size or with only one element), char matrix or character string.

Chapter 5: Strings 77

**Caution:** For compatibility with MATLAB, Octave's strcmp function returns 1 if the character strings are equal, and 0 otherwise. This is just the opposite of the corresponding C library function.

**Caution:** National alphabets are not supported.

**See also:** [strcmp], page 76, [strncmp], page 76, [strncmpi], page 77.

strncmpi (*s1*, *s2*, *n*)                               [Built-in Function]
Return 1 if the first *n* character of *s1* and *s2* are the same, disregarding case of alphabetic characters, and 0 otherwise.

If either *s1* or *s2* is a cell array of strings, then an array of the same size is returned, containing the values described above for every member of the cell array. The other argument may also be a cell array of strings (of the same size or with only one element), char matrix or character string.

**Caution:** For compatibility with MATLAB, Octave's strncmpi function returns 1 if the character strings are equal, and 0 otherwise. This is just the opposite of the corresponding C library function.

**Caution:** National alphabets are not supported.

**See also:** [strncmp], page 76, [strcmp], page 76, [strcmpi], page 76.

## 5.5 Manipulating Strings

Octave supports a wide range of functions for manipulating strings. Since a string is just a matrix, simple manipulations can be accomplished using standard operators. The following example shows how to replace all blank characters with underscores.

```
quote = ...
  "First things first, but not necessarily in that order";
quote( quote == " " ) = "_"
⇒ quote =
    First_things_first,_but_not_necessarily_in_that_order
```

For more complex manipulations, such as searching, replacing, and general regular expressions, the following functions come with Octave.

deblank (*s*)                                              [Function File]
Remove trailing whitespace and nulls from *s*.

If *s* is a matrix, *deblank* trims each row to the length of longest string. If *s* is a cell array of strings, operate recursively on each string element.

Examples:

```
deblank ("    abc   ")
    ⇒ "    abc"

deblank ([" abc   "; "  def   "])
    ⇒ [" abc  " ; "  def"]
```

**See also:** [strtrim], page 78.

`strtrim (s)` [Function File]

Remove leading and trailing whitespace from s.

If s is a matrix, *strtrim* trims each row to the length of longest string. If s is a cell array of strings, operate recursively on each string element.

For example:

```
strtrim ("    abc  ")
    ⇒ "abc"

strtrim ([" abc   "; "  def   "])
    ⇒ ["abc  " ; "  def"]
```

See also: [deblank], page 77.

`strtrunc (s, n)` [Function File]

Truncate the character string s to length n.

If s is a character matrix, then the number of columns is adjusted.

If s is a cell array of strings, then the operation is performed on each cell element and the new cell array is returned.

`findstr (s, t)` [Function File]
`findstr (s, t, overlap)` [Function File]

Return the vector of all positions in the longer of the two strings s and t where an occurrence of the shorter of the two starts.

If the optional argument *overlap* is true (default), the returned vector can include overlapping positions. For example:

```
findstr ("ababab", "a")
    ⇒ [1, 3, 5];
findstr ("abababa", "aba", 0)
    ⇒ [1, 5]
```

**Caution:** `findstr` is scheduled for deprecation. Use `strfind` in all new code.

See also: [strfind], page 79, [strmatch], page 80, [strcmp], page 76, [strncmp], page 76, [strcmpi], page 76, [strncmpi], page 77, [find], page 405.

`idx = strchr (str, chars)` [Function File]
`idx = strchr (str, chars, n)` [Function File]
`idx = strchr (str, chars, n, direction)` [Function File]
`[i, j] = strchr (...)` [Function File]

Search for the string *str* for occurrences of characters from the set *chars*.

The return value(s), as well as the n and *direction* arguments behave identically as in `find`.

This will be faster than using regexp in most cases.

See also: [find], page 405.

`index (s, t)` [Function File]
`index (s, t, direction)` [Function File]

Return the position of the first occurrence of the string t in the string s, or 0 if no occurrence is found.

Chapter 5: Strings 79

*s* may also be a string array or cell array of strings.

For example:

    index ("Teststring", "t")
       ⇒ 4

If *direction* is `"first"`, return the first element found. If *direction* is `"last"`, return the last element found.

**See also:** [find], page 405, [rindex], page 79.

**rindex (*s*, *t*)**                                                        [Function File]
Return the position of the last occurrence of the character string *t* in the character string *s*, or 0 if no occurrence is found.

*s* may also be a string array or cell array of strings.

For example:

    rindex ("Teststring", "t")
       ⇒ 6

The `rindex` function is equivalent to `index` with *direction* set to `"last"`.

**See also:** [find], page 405, [index], page 78.

*idx* = **strfind** (*str*, *pattern*)                                         [Built-in Function]
*idx* = **strfind** (*cellstr*, *pattern*)                                     [Built-in Function]
*idx* = **strfind** (..., "*overlaps*", *val*)                            [Built-in Function]
Search for *pattern* in the string *str* and return the starting index of every such occurrence in the vector *idx*.

If there is no such occurrence, or if *pattern* is longer than *str*, or if *pattern* itself is empty, then *idx* is the empty array [].

The optional argument `"overlaps"` determines whether the pattern can match at every position in *str* (true), or only for unique occurrences of the complete pattern (false). The default is true.

If a cell array of strings *cellstr* is specified then *idx* is a cell array of vectors, as specified above.

Examples:

```
strfind ("abababa", "aba")
    ⇒ [1, 3, 5]

strfind ("abababa", "aba", "overlaps", false)
    ⇒ [1, 5]

strfind ({"abababa", "bebebe", "ab"}, "aba")
    ⇒
      {
        [1,1] =

           1   3   5

        [1,2] = [](1x0)
        [1,3] = [](1x0)
      }
```

**See also:** [findstr], page 78, [strmatch], page 80, [regexp], page 87, [regexpi], page 89, [find], page 405.

*str* = **strjoin** (*cstr*)     [Function File]
*str* = **strjoin** (*cstr*, *delimiter*)     [Function File]

Join the elements of the cell string array, *cstr*, into a single string.

If no *delimiter* is specified, the elements of *cstr* are separated by a space.

If *delimiter* is specified as a string, the cell string array is joined using the string. Escape sequences are supported.

If *delimiter* is a cell string array whose length is one less than *cstr*, then the elements of *cstr* are joined by interleaving the cell string elements of *delimiter*. Escape sequences are not supported.

```
strjoin ({'Octave','Scilab','Lush','Yorick'}, '*')
    ⇒ 'Octave*Scilab*Lush*Yorick'
```

**See also:** [strsplit], page 81.

**strmatch** (*s*, *A*)     [Function File]
**strmatch** (*s*, *A*, "*exact*")     [Function File]

Return indices of entries of *A* which begin with the string *s*.

The second argument *A* must be a string, character matrix, or a cell array of strings.

If the third argument "**exact**" is not given, then *s* only needs to match *A* up to the length of *s*. Trailing spaces and nulls in *s* and *A* are ignored when matching.

For example:

Chapter 5: Strings 81

>       strmatch ("apple", "apple juice")
>            ⇒ 1
>
>       strmatch ("apple", ["apple     "; "apple juice"; "an apple"])
>            ⇒ [1; 2]
>
>       strmatch ("apple", ["apple     "; "apple juice"; "an apple"], "exact")
>            ⇒ [1]

**Caution:** `strmatch` is scheduled for deprecation. Use `strncmp` (normal case), or `strcmp` ("exact" case), or `regexp` in all new code.

**See also:** [strfind], page 79, [findstr], page 78, [strcmp], page 76, [strncmp], page 76, [strcmpi], page 76, [strncmpi], page 77, [find], page 405.

[tok, rem] = strtok (str)                                                               [Function File]
[tok, rem] = strtok (str, delim)                                            [Function File]

Find all characters in the string *str* up to, but not including, the first character which is in the string *delim*.

*str* may also be a cell array of strings in which case the function executes on every individual string and returns a cell array of tokens and remainders.

Leading delimiters are ignored. If *delim* is not specified, whitespace is assumed.

If *rem* is requested, it contains the remainder of the string, starting at the first delimiter.

Examples:

>       strtok ("this is the life")
>            ⇒ "this"
>
>       [tok, rem] = strtok ("14*27+31", "+-*/")
>            ⇒
>              tok = 14
>              rem = *27+31

**See also:** [index], page 78, [strsplit], page 81, [strchr], page 78, [isspace], page 97.

[cstr] = strsplit (str)                                                                        [Function File]
[cstr] = strsplit (str, del)                                                           [Function File]
[cstr] = strsplit (..., name, value)                                   [Function File]
[cstr, matches] = strsplit (...)                                           [Function File]

Split the string *str* using the delimiters specified by *del* and return a cell string array of substrings.

If a delimiter is not specified the string is split at whitespace {" ", "\f", "\n", "\r", "\t", "\v"}. Otherwise, the delimiter, *del* must be a string or cell array of strings. By default, consecutive delimiters in the input string *s* are collapsed into one resulting in a single split.

Supported *name*/*value* pair arguments are:

- *collapsedelimiters* which may take the value of `true` (default) or `false`.

- *delimitertype* which may take the value of `"simple"` (default) or `"regularexpression"`. A simple delimiter matches the text exactly as written. Otherwise, the syntax for regular expressions outlined in `regexp` is used.

The optional second output, *matches*, returns the delimiters which were matched in the original string.

Examples with simple delimiters:

```
strsplit ("a b c")
    ⇒
        {
          [1,1] = a
          [1,2] = b
          [1,3] = c
        }

strsplit ("a,b,c", ",")
    ⇒
        {
          [1,1] = a
          [1,2] = b
          [1,3] = c
        }

strsplit ("a foo b,bar c", {" ", ",", "foo", "bar"})
    ⇒
        {
          [1,1] = a
          [1,2] = b
          [1,3] = c
        }

strsplit ("a,,b, c", {",", " "}, "collapsedelimiters", false)
    ⇒
        {
          [1,1] = a
          [1,2] =
          [1,3] = b
          [1,4] =
          [1,5] = c
        }
```

Examples with regularexpression delimiters:

```
strsplit ("a foo b,bar c", ',|\s|foo|bar', "delimitertype", "regularexpression")
    ⇒
    {
          [1,1] = a
          [1,2] = b
          [1,3] = c
```

# Chapter 5: Strings

```
        }
        strsplit ("a,,b, c", '[, ]', "collapsedelimiters", false, "delimitertype", "regularexpression")
        ⇒
        {
                [1,1] = a
                [1,2] =
                [1,3] = b
                [1,4] =
                [1,5] = c
        }
        strsplit ("a,\t,b, c", {',', '\s'}, "delimitertype", "regularexpression")
        ⇒
        {
                [1,1] = a
                [1,2] = b
                [1,3] = c
        }
        strsplit ("a,\t,b, c", {',', ' ', '\t'}, "collapsedelimiters", false)
        ⇒
        {
                [1,1] = a
                [1,2] =
                [1,3] =
                [1,4] = b
                [1,5] =
                [1,6] = c
        }
```

**See also:** [ostrsplit], page 83, [strjoin], page 80, [strtok], page 81, [regexp], page 87.

[*cstr*] = ostrsplit (*s*, *sep*)      [Function File]
[*cstr*] = ostrsplit (*s*, *sep*, *strip_empty*)      [Function File]

Split the string *s* using one or more separators *sep* and return a cell array of strings.

Consecutive separators and separators at boundaries result in empty strings, unless *strip_empty* is true. The default value of *strip_empty* is false.

2-D character arrays are split at separators and at the original column boundaries.

Example:

```
ostrsplit ("a,b,c", ",")
    ⇒
        {
          [1,1] = a
          [1,2] = b
          [1,3] = c
        }

ostrsplit (["a,b" ; "cde"], ",")
    ⇒
        {
          [1,1] = a
          [1,2] = b
          [1,3] = cde
        }
```

**See also:** [strsplit], page 81, [strtok], page 81.

[a, ...] = strread (*str*)     [Function File]
[a, ...] = strread (*str*, *format*)     [Function File]
[a, ...] = strread (*str*, *format*, *format_repeat*)     [Function File]
[a, ...] = strread (*str*, *format*, *prop1*, *value1*, ...)     [Function File]
[a, ...] = strread (*str*, *format*, *format_repeat*, *prop1*,     [Function File]
    *value1*, ...)

Read data from a string.

The string *str* is split into words that are repeatedly matched to the specifiers in *format*. The first word is matched to the first specifier, the second to the second specifier and so forth. If there are more words than specifiers, the process is repeated until all words have been processed.

The string *format* describes how the words in *str* should be parsed. It may contain any combination of the following specifiers:

%s        The word is parsed as a string.

%f
%n        The word is parsed as a number and converted to double.

%d
%u        The word is parsed as a number and converted to int32.

'%*', '%*f', '%*s'

       The word is skipped.

       For %s and %d, %f, %n, %u and the associated %*s ... specifiers an optional width can be specified as %Ns, etc. where N is an integer > 1. For %f, format specifiers like %N.Mf are allowed.

literals    In addition the format may contain literal character strings; these will be skipped during reading.

Parsed word corresponding to the first specifier are returned in the first output argument and likewise for the rest of the specifiers.

By default, *format* is "%f", meaning that numbers are read from *str*. This will do if *str* contains only numeric fields.

For example, the string

```
str = "\
Bunny Bugs    5.5\n\
Duck Daffy    -7.5e-5\n\
Penguin Tux   6"
```

can be read using

```
[a, b, c] = strread (str, "%s %s %f");
```

Optional numeric argument *format_repeat* can be used for limiting the number of items read:

-1          (default) read all of the string until the end.

N           Read N times *nargout* items. 0 (zero) is an acceptable value for *format_repeat*.

The behavior of `strread` can be changed via property-value pairs. The following properties are recognized:

"commentstyle"
: Parts of *str* are considered comments and will be skipped. *value* is the comment style and can be any of the following.

    - "shell" Everything from # characters to the nearest end-of-line is skipped.
    - "c" Everything between /* and */ is skipped.
    - "c++" Everything from // characters to the nearest end-of-line is skipped.
    - "matlab" Everything from % characters to the nearest end-of-line is skipped.
    - user-supplied. Two options: (1) One string, or 1x1 cell string: Skip everything to the right of it; (2) 2x1 cell string array: Everything between the left and right strings is skipped.

"delimiter"
: Any character in *value* will be used to split *str* into words (default value = any whitespace).

"emptyvalue":
: Value to return for empty numeric values in non-whitespace delimited data. The default is NaN. When the data type does not support NaN (int32 for example), then default is zero.

"multipledelimsasone"
: Treat a series of consecutive delimiters, without whitespace in between, as a single delimiter. Consecutive delimiter series need not be vertically "aligned".

"treatasempty"
: Treat single occurrences (surrounded by delimiters or whitespace) of the string(s) in *value* as missing values.

"returnonerror"
: If *value* true (1, default), ignore read errors and return normally. If false (0), return an error.

"whitespace"
: Any character in *value* will be interpreted as whitespace and trimmed; the string defining whitespace must be enclosed in double quotes for proper processing of special characters like "\t". The default value for whitespace is " \b\r\n\t" (note the space). Unless whitespace is set to "" (empty) AND at least one "%s" format conversion specifier is supplied, a space is always part of whitespace.

When the number of words in *str* doesn't match an exact multiple of the number of format conversion specifiers, strread's behavior depends on the last character of *str*:

last character = "\n"
: Data columns are padded with empty fields or Nan so that all columns have equal length

last character is not "\n"
: Data columns are not padded; strread returns columns of unequal length

**See also:** [textscan], page 246, [textread], page 244, [load], page 241, [dlmread], page 244, [fscanf], page 259.

*newstr* = **strrep** (*str*, *ptn*, *rep*)      [Built-in Function]
*newstr* = **strrep** (*cellstr*, *ptn*, *rep*)      [Built-in Function]
*newstr* = **strrep** (..., "*overlaps*", *val*)      [Built-in Function]
: Replace all occurrences of the pattern *ptn* in the string *str* with the string *rep* and return the result.

The optional argument "overlaps" determines whether the pattern can match at every position in *str* (true), or only for unique occurrences of the complete pattern (false). The default is true.

*s* may also be a cell array of strings, in which case the replacement is done for each element and a cell array is returned.

Example:

```
strrep ("This is a test string", "is", "&%$")
  ⇒  "Th&%$ &%$ a test string"
```

**See also:** [regexprep], page 90, [strfind], page 79, [findstr], page 78.

**substr** (*s*, *offset*)      [Function File]
**substr** (*s*, *offset*, *len*)      [Function File]
: Return the substring of *s* which starts at character number *offset* and is *len* characters long.

Position numbering for offsets begins with 1. If *offset* is negative, extraction starts that far from the end of the string.

Chapter 5: Strings

If *len* is omitted, the substring extends to the end of *s*. A negative value for *len* extracts to within *len* characters of the end of the string

Examples:

```
substr ("This is a test string", 6, 9)
    ⇒ "is a test"
substr ("This is a test string", -11)
    ⇒ "test string"
substr ("This is a test string", -11, -7)
    ⇒ "test"
```

This function is patterned after the equivalent function in Perl.

[*s*, *e*, *te*, *m*, *t*, *nm*, *sp*] = regexp (*str*, *pat*)           [Built-in Function]
[...] = regexp (*str*, *pat*, "*opt1*", ...)           [Built-in Function]

Regular expression string matching.

Search for *pat* in *str* and return the positions and substrings of any matches, or empty values if there are none.

The matched pattern *pat* can include any of the standard regex operators, including:

.           Match any character

\* + ? {}     Repetition operators, representing

      \*           Match zero or more times

      +           Match one or more times

      ?           Match zero or one times

      {*n*}       Match exactly *n* times

      {*n*,}      Match *n* or more times

      {*m*,*n*}    Match between *m* and *n* times

[...] [^...]

          List operators. The pattern will match any character listed between "[" and "]". If the first character is "^" then the pattern is inverted and any character except those listed between brackets will match.

          Escape sequences defined below can also be used inside list operators. For example, a template for a floating point number might be [-+.\d]+.

() (?:)      Grouping operator. The first form, parentheses only, also creates a token.

|           Alternation operator. Match one of a choice of regular expressions. The alternatives must be delimited by the grouping operator () above.

^ $        Anchoring operators. Requires pattern to occur at the start (^) or end ($) of the string.

In addition, the following escaped characters have special meaning.

\d         Match any digit

\D         Match any non-digit

| | |
|---|---|
| `\s` | Match any whitespace character |
| `\S` | Match any non-whitespace character |
| `\w` | Match any word character |
| `\W` | Match any non-word character |
| `\<` | Match the beginning of a word |
| `\>` | Match the end of a word |
| `\B` | Match within a word |

Implementation Note: For compatibility with MATLAB, escape sequences in *pat* (e.g., `"\n"` => newline) are expanded even when *pat* has been defined with single quotes. To disable expansion use a second backslash before the escape sequence (e.g., `"\\n"`) or use the `regexptranslate` function.

The outputs of `regexp` default to the order given below

| | |
|---|---|
| *s* | The start indices of each matching substring |
| *e* | The end indices of each matching substring |
| *te* | The extents of each matched token surrounded by (...) in *pat* |
| *m* | A cell array of the text of each match |
| *t* | A cell array of the text of each token matched |
| *nm* | A structure containing the text of each matched named token, with the name being used as the fieldname. A named token is denoted by (?<name>...). |
| *sp* | A cell array of the text not returned by match, i.e., what remains if you split the string based on *pat*. |

Particular output arguments, or the order of the output arguments, can be selected by additional *opt* arguments. These are strings and the correspondence between the output arguments and the optional argument are

| | |
|---|---|
| `'start'` | *s* |
| `'end'` | *e* |
| `'tokenExtents'` | *te* |
| `'match'` | *m* |
| `'tokens'` | *t* |
| `'names'` | *nm* |
| `'split'` | *sp* |

Additional arguments are summarized below.

'once'   Return only the first occurrence of the pattern.

'matchcase'
    Make the matching case sensitive. (default)

    Alternatively, use (?-i) in the pattern.

Chapter 5: Strings

'ignorecase'
: Ignore case when matching the pattern to the string.

  Alternatively, use (?i) in the pattern.

'stringanchors'
: Match the anchor characters at the beginning and end of the string. (default)

  Alternatively, use (?-m) in the pattern.

'lineanchors'
: Match the anchor characters at the beginning and end of the line.

  Alternatively, use (?m) in the pattern.

'dotall'
: The pattern . matches all characters including the newline character. (default)

  Alternatively, use (?s) in the pattern.

'dotexceptnewline'
: The pattern . matches all characters except the newline character.

  Alternatively, use (?-s) in the pattern.

'literalspacing'
: All characters in the pattern, including whitespace, are significant and are used in pattern matching. (default)

  Alternatively, use (?-x) in the pattern.

'freespacing'
: The pattern may include arbitrary whitespace and also comments beginning with the character '#'.

  Alternatively, use (?x) in the pattern.

'noemptymatch'
: Zero-length matches are not returned. (default)

'emptymatch'
: Return zero-length matches.

  regexp ('a', 'b*', 'emptymatch') returns [1 2] because there are zero or more 'b' characters at positions 1 and end-of-string.

**See also:** [regexpi], page 89, [strfind], page 79, [regexprep], page 90.

[s, e, te, m, t, nm, sp] = regexpi (str, pat)  [Built-in Function]
[...] = regexpi (str, pat, "opt1", ...)  [Built-in Function]
: Case insensitive regular expression string matching.

Search for *pat* in *str* and return the positions and substrings of any matches, or empty values if there are none. See [regexp], page 87, for details on the syntax of the search pattern.

**See also:** [regexp], page 87.

`outstr = regexprep (string, pat, repstr)`     [Built-in Function]
`outstr = regexprep (string, pat, repstr, "opt1", ...)`     [Built-in Function]

> Replace occurrences of pattern *pat* in *string* with *repstr*.
>
> The pattern is a regular expression as documented for `regexp`. See [regexp], page 87.
>
> The replacement string may contain $i, which substitutes for the ith set of parentheses in the match string. For example,
>
> > `regexprep ("Bill Dunn", '(\w+) (\w+)', '$2, $1')`
>
> returns "Dunn, Bill"
>
> Options in addition to those of `regexp` are
>
> 'once'     Replace only the first occurrence of *pat* in the result.
>
> 'warnings'
> > This option is present for compatibility but is ignored.
>
> Implementation Note: For compatibility with MATLAB, escape sequences in *pat* (e.g., "\n" => newline) are expanded even when *pat* has been defined with single quotes. To disable expansion use a second backslash before the escape sequence (e.g., "\\n") or use the `regexptranslate` function.
>
> **See also:** [regexp], page 87, [regexpi], page 89, [strrep], page 86.

`regexptranslate (op, s)`     [Function File]

> Translate a string for use in a regular expression.
>
> This may include either wildcard replacement or special character escaping.
>
> The behavior is controlled by *op* which can take the following values
>
> "wildcard"
> > The wildcard characters ., *, and ? are replaced with wildcards that are appropriate for a regular expression. For example:
> >
> > > `regexptranslate ("wildcard", "*.m")`
> > > ⇒ ".*\.m"
>
> "escape"     The characters $.?[], that have special meaning for regular expressions are escaped so that they are treated literally. For example:
>
> > `regexptranslate ("escape", "12.5")`
> > ⇒ "12\.5"
>
> **See also:** [regexp], page 87, [regexpi], page 89, [regexprep], page 90.

`untabify (t)`     [Function File]
`untabify (t, tw)`     [Function File]
`untabify (t, tw, deblank)`     [Function File]

> Replace TAB characters in *t* with spaces.
>
> The input, *t*, may be either a 2-D character array, or a cell array of character strings. The output is the same class as the input.
>
> The tab width is specified by *tw*, and defaults to eight.
>
> If the optional argument *deblank* is true, then the spaces will be removed from the end of the character data.

The following example reads a file and writes an untabified version of the same file with trailing spaces stripped.

```
fid = fopen ("tabbed_script.m");
text = char (fread (fid, "uchar")');
fclose (fid);
fid = fopen ("untabified_script.m", "w");
text = untabify (strsplit (text, "\n"), 8, true);
fprintf (fid, "%s\n", text{:});
fclose (fid);
```

**See also:** [strjust], page 94, [strsplit], page 81, [deblank], page 77.

## 5.6 String Conversions

Octave supports various kinds of conversions between strings and numbers. As an example, it is possible to convert a string containing a hexadecimal number to a floating point number.

```
hex2dec ("FF")
    ⇒ 255
```

**bin2dec (s)** [Function File]

Return the decimal number corresponding to the binary number represented by the string s.

For example:

```
bin2dec ("1110")
    ⇒ 14
```

Spaces are ignored during conversion and may be used to make the binary number more readable.

```
bin2dec ("1000 0001")
    ⇒ 129
```

If s is a string matrix, return a column vector with one converted number per row of s; Invalid rows evaluate to NaN.

If s is a cell array of strings, return a column vector with one converted number per cell element in s.

**See also:** [dec2bin], page 91, [base2dec], page 92, [hex2dec], page 92.

**dec2bin (d, len)** [Function File]

Return a binary number corresponding to the non-negative integer d, as a string of ones and zeros.

For example:

```
dec2bin (14)
    ⇒ "1110"
```

If d is a matrix or cell array, return a string matrix with one row per element in d, padded with leading zeros to the width of the largest value.

The optional second argument, len, specifies the minimum number of digits in the result.

**See also:** [bin2dec], page 91, [dec2base], page 92, [dec2hex], page 92.

`dec2hex (`*`d, len`*`)` [Function File]
> Return the hexadecimal string corresponding to the non-negative integer *d*.
>
> For example:
>
>     dec2hex (2748)
>        ⇒ "ABC"
>
> If *d* is a matrix or cell array, return a string matrix with one row per element in *d*, padded with leading zeros to the width of the largest value.
>
> The optional second argument, *len*, specifies the minimum number of digits in the result.
>
> **See also:** [hex2dec], page 92, [dec2base], page 92, [dec2bin], page 91.

`hex2dec (`*`s`*`)` [Function File]
> Return the integer corresponding to the hexadecimal number represented by the string *s*.
>
> For example:
>
>     hex2dec ("12B")
>        ⇒ 299
>     hex2dec ("12b")
>        ⇒ 299
>
> If *s* is a string matrix, return a column vector with one converted number per row of *s*; Invalid rows evaluate to NaN.
>
> If *s* is a cell array of strings, return a column vector with one converted number per cell element in *s*.
>
> **See also:** [dec2hex], page 92, [base2dec], page 92, [bin2dec], page 91.

`dec2base (`*`d, base`*`)` [Function File]
`dec2base (`*`d, base, len`*`)` [Function File]
> Return a string of symbols in base *base* corresponding to the non-negative integer *d*.
>
>     dec2base (123, 3)
>        ⇒ "11120"
>
> If *d* is a matrix or cell array, return a string matrix with one row per element in *d*, padded with leading zeros to the width of the largest value.
>
> If *base* is a string then the characters of *base* are used as the symbols for the digits of *d*. Space (' ') may not be used as a symbol.
>
>     dec2base (123, "aei")
>        ⇒ "eeeia"
>
> The optional third argument, *len*, specifies the minimum number of digits in the result.
>
> **See also:** [base2dec], page 92, [dec2bin], page 91, [dec2hex], page 92.

`base2dec (`*`s, base`*`)` [Function File]
> Convert *s* from a string of digits in base *base* to a decimal integer (base 10).

Chapter 5: Strings                                                                                   93

>     base2dec ("11120", 3)
>         ⇒ 123

If s is a string matrix, return a column vector with one value per row of s. If a row contains invalid symbols then the corresponding value will be NaN.

If s is a cell array of strings, return a column vector with one value per cell element in s.

If base is a string, the characters of base are used as the symbols for the digits of s. Space (' ') may not be used as a symbol.

>     base2dec ("yyyzx", "xyz")
>         ⇒ 123

**See also:** [dec2base], page 92, [bin2dec], page 91, [hex2dec], page 92.

**s = num2hex (n)**                                                              [Built-in Function]

Typecast a double or single precision number or vector to a 8 or 16 character hexadecimal string of the IEEE 754 representation of the number.

For example:

>     num2hex ([-1, 1, e, Inf])
>         ⇒ "bff0000000000000
>             3ff0000000000000
>             4005bf0a8b145769
>             7ff0000000000000"

If the argument n is a single precision number or vector, the returned string has a length of 8. For example:

>     num2hex (single ([-1, 1, e, Inf]))
>         ⇒ "bf800000
>             3f800000
>             402df854
>             7f800000"

**See also:** [hex2num], page 93, [hex2dec], page 92, [dec2hex], page 92.

**n = hex2num (s)**                                                              [Built-in Function]
**n = hex2num (s, class)**                                                       [Built-in Function]

Typecast the 16 character hexadecimal character string to an IEEE 754 double precision number.

If fewer than 16 characters are given the strings are right padded with '0' characters.

Given a string matrix, hex2num treats each row as a separate number.

>     hex2num (["4005bf0a8b145769"; "4024000000000000"])
>         ⇒ [2.7183; 10.000]

The optional argument class can be passed as the string "single" to specify that the given string should be interpreted as a single precision number. In this case, s should be an 8 character hexadecimal string. For example:

>     hex2num (["402df854"; "41200000"], "single")
>         ⇒ [2.7183; 10.000]

**See also:** [num2hex], page 93, [hex2dec], page 92, [dec2hex], page 92.

**str2double** (*s*)                                                    [Built-in Function]

Convert a string to a real or complex number.

The string must be in one of the following formats where a and b are real numbers and the complex unit is 'i' or 'j':

- a + bi
- a + b*i
- a + i*b
- bi + a
- b*i + a
- i*b + a

If present, a and/or b are of the form [+-]d[,.]d[[eE][+-]d] where the brackets indicate optional arguments and 'd' indicates zero or more digits. The special input values Inf, NaN, and NA are also accepted.

*s* may be a character string, character matrix, or cell array. For character arrays the conversion is repeated for every row, and a double or complex array is returned. Empty rows in *s* are deleted and not returned in the numeric array. For cell arrays each character string element is processed and a double or complex array of the same dimensions as *s* is returned.

For unconvertible scalar or character string input **str2double** returns a NaN. Similarly, for character array input **str2double** returns a NaN for any row of *s* that could not be converted. For a cell array, **str2double** returns a NaN for any element of *s* for which conversion fails. Note that numeric elements in a mixed string/numeric cell array are not strings and the conversion will fail for these elements and return NaN.

**str2double** can replace **str2num**, and it avoids the security risk of using **eval** on unknown data.

**See also:** [str2num], page 95.

**strjust** (*s*)                                                        [Function File]
**strjust** (*s, pos*)                                                   [Function File]

Return the text, *s*, justified according to *pos*, which may be "left", "center", or "right".

If *pos* is omitted it defaults to "right".

Null characters are replaced by spaces. All other character data are treated as non-white space.

Example:

```
strjust (["a"; "ab"; "abc"; "abcd"])
     ⇒
        "   a"
        "  ab"
        " abc"
        "abcd"
```

**See also:** [deblank], page 77, [strrep], page 86, [strtrim], page 78, [untabify], page 90.

# Chapter 5: Strings

`x = str2num (s)`                                                                                       [Function File]
`[x, state] = str2num (s)`                                                     [Function File]

    Convert the string (or character array) $s$ to a number (or an array).

    Examples:

```
str2num ("3.141596")
     ⇒ 3.141596

str2num (["1, 2, 3"; "4, 5, 6"])
     ⇒ 1 2 3
       4 5 6
```

    The optional second output, *state*, is logically true when the conversion is successful. If the conversion fails the numeric output, $x$, is empty and *state* is false.

    **Caution:** As `str2num` uses the `eval` function to do the conversion, `str2num` will execute any code contained in the string $s$. Use `str2double` for a safer and faster conversion.

    For cell array of strings use `str2double`.

    **See also:** [str2double], page 94, [eval], page 155.

`toascii (s)`                                                         [Mapping Function]

    Return ASCII representation of $s$ in a matrix.

    For example:

```
toascii ("ASCII")
     ⇒ [ 65, 83, 67, 73, 73 ]
```

    **See also:** [char], page 71.

`tolower (s)`                                                   [Mapping Function]
`lower (s)`                                                    [Mapping Function]

    Return a copy of the string or cell string $s$, with each uppercase character replaced by the corresponding lowercase one; non-alphabetic characters are left unchanged.

    For example:

```
tolower ("MiXeD cAsE 123")
     ⇒ "mixed case 123"
```

    **See also:** [toupper], page 95.

`toupper (s)`                                                   [Mapping Function]
`upper (s)`                                                   [Mapping Function]

    Return a copy of the string or cell string $s$, with each lowercase character replaced by the corresponding uppercase one; non-alphabetic characters are left unchanged.

    For example:

```
toupper ("MiXeD cAsE 123")
     ⇒ "MIXED CASE 123"
```

    **See also:** [tolower], page 95.

`do_string_escapes (string)` [Built-in Function]
> Convert escape sequences in *string* to the characters they represent.
>
> Escape sequences begin with a leading backslash ('\') followed by 1–3 characters (.e.g., `"\n"` => newline).
>
> **See also:** [undo_string_escapes], page 96.

`undo_string_escapes (s)` [Built-in Function]
> Convert special characters in strings back to their escaped forms.
>
> For example, the expression
>
>     bell = "\a";
>
> assigns the value of the alert character (control-g, ASCII code 7) to the string variable `bell`. If this string is printed, the system will ring the terminal bell (if it is possible). This is normally the desired outcome. However, sometimes it is useful to be able to print the original representation of the string, with the special characters replaced by their escape sequences. For example,
>
>     octave:13> undo_string_escapes (bell)
>     ans = \a
>
> replaces the unprintable alert character with its printable representation.
>
> **See also:** [do_string_escapes], page 96.

## 5.7 Character Class Functions

Octave also provides the following character class test functions patterned after the functions in the standard C library. They all operate on string arrays and return matrices of zeros and ones. Elements that are nonzero indicate that the condition was true for the corresponding character in the string array. For example:

    isalpha ("!Q@WERT^Y&")
        ⇒ [ 0, 1, 0, 1, 1, 1, 1, 0, 1, 0 ]

`isalnum (s)` [Mapping Function]
> Return a logical array which is true where the elements of *s* are letters or digits and false where they are not.
>
> This is equivalent to (`isalpha (s) | isdigit (s)`).
>
> **See also:** [isalpha], page 96, [isdigit], page 97, [ispunct], page 97, [isspace], page 97, [iscntrl], page 97.

`isalpha (s)` [Mapping Function]
> Return a logical array which is true where the elements of *s* are letters and false where they are not.
>
> This is equivalent to (`islower (s) | isupper (s)`).
>
> **See also:** [isdigit], page 97, [ispunct], page 97, [isspace], page 97, [iscntrl], page 97, [isalnum], page 96, [islower], page 97, [isupper], page 97.

Chapter 5: Strings

**isletter** (*s*)                                                                      [Function File]

Return a logical array which is true where the elements of *s* are letters and false where they are not.

This is an alias for the `isalpha` function.

**See also:** [isalpha], page 96, [isdigit], page 97, [ispunct], page 97, [isspace], page 97, [iscntrl], page 97, [isalnum], page 96.

**islower** (*s*)                                                                     [Mapping Function]

Return a logical array which is true where the elements of *s* are lowercase letters and false where they are not.

**See also:** [isupper], page 97, [isalpha], page 96, [isletter], page 97, [isalnum], page 96.

**isupper** (*s*)                                                                    [Mapping Function]

Return a logical array which is true where the elements of *s* are uppercase letters and false where they are not.

**See also:** [islower], page 97, [isalpha], page 96, [isletter], page 97, [isalnum], page 96.

**isdigit** (*s*)                                                                     [Mapping Function]

Return a logical array which is true where the elements of *s* are decimal digits (0-9) and false where they are not.

**See also:** [isxdigit], page 97, [isalpha], page 96, [isletter], page 97, [ispunct], page 97, [isspace], page 97, [iscntrl], page 97.

**isxdigit** (*s*)                                                                  [Mapping Function]

Return a logical array which is true where the elements of *s* are hexadecimal digits (0-9 and a-fA-F).

**See also:** [isdigit], page 97.

**ispunct** (*s*)                                                                   [Mapping Function]

Return a logical array which is true where the elements of *s* are punctuation characters and false where they are not.

**See also:** [isalpha], page 96, [isdigit], page 97, [isspace], page 97, [iscntrl], page 97.

**isspace** (*s*)                                                                   [Mapping Function]

Return a logical array which is true where the elements of *s* are whitespace characters (space, formfeed, newline, carriage return, tab, and vertical tab) and false where they are not.

**See also:** [iscntrl], page 97, [ispunct], page 97, [isalpha], page 96, [isdigit], page 97.

**iscntrl** (*s*)                                                                   [Mapping Function]

Return a logical array which is true where the elements of *s* are control characters and false where they are not.

**See also:** [ispunct], page 97, [isspace], page 97, [isalpha], page 96, [isdigit], page 97.

**isgraph** (*s*)                                                                   [Mapping Function]

Return a logical array which is true where the elements of *s* are printable characters (but not the space character) and false where they are not.

**See also:** [isprint], page 98.

`isprint (s)` [Mapping Function]

> Return a logical array which is true where the elements of s are printable characters (including the space character) and false where they are not.
>
> **See also:** [isgraph], page 97.

`isascii (s)` [Mapping Function]

> Return a logical array which is true where the elements of s are ASCII characters (in the range 0 to 127 decimal) and false where they are not.

`isstrprop (str, prop)` [Function File]

> Test character string properties.
>
> For example:
>
>     isstrprop ("abc123", "alpha")
>       ⇒ [1, 1, 1, 0, 0, 0]
>
> If str is a cell array, `isstrpop` is applied recursively to each element of the cell array.
>
> Numeric arrays are converted to character strings.
>
> The second argument prop must be one of
>
> `"alpha"`  True for characters that are alphabetic (letters).
>
> `"alnum"`
> `"alphanum"`
> True for characters that are alphabetic or digits.
>
> `"lower"`  True for lowercase letters.
>
> `"upper"`  True for uppercase letters.
>
> `"digit"`  True for decimal digits (0-9).
>
> `"xdigit"` True for hexadecimal digits (a-fA-F0-9).
>
> `"space"`
> `"wspace"` True for whitespace characters (space, formfeed, newline, carriage return, tab, vertical tab).
>
> `"punct"`  True for punctuation characters (printing characters except space or letter or digit).
>
> `"cntrl"`  True for control characters.
>
> `"graph"`
> `"graphic"`
> True for printing characters except space.
>
> `"print"`  True for printing characters including space.
>
> `"ascii"`  True for characters that are in the range of ASCII encoding.
>
> **See also:** [isalpha], page 96, [isalnum], page 96, [islower], page 97, [isupper], page 97, [isdigit], page 97, [isxdigit], page 97, [isspace], page 97, [ispunct], page 97, [iscntrl], page 97, [isgraph], page 97, [isprint], page 98, [isascii], page 98.

# 6 Data Containers

Octave includes support for two different mechanisms to contain arbitrary data types in the same variable. Structures, which are C-like, and are indexed with named fields, and cell arrays, where each element of the array can have a different data type and or shape. Multiple input arguments and return values of functions are organized as another data container, the comma separated list.

## 6.1 Structures

Octave includes support for organizing data in structures. The current implementation uses an associative array with indices limited to strings, but the syntax is more like C-style structures.

### 6.1.1 Basic Usage and Examples

Here are some examples of using data structures in Octave.

Elements of structures can be of any value type. For example, the three expressions

```
x.a = 1;
x.b = [1, 2; 3, 4];
x.c = "string";
```

create a structure with three elements. The '.' character separates the structure name from the field name and indicates to Octave that this variable is a structure. To print the value of the structure you can type its name, just as for any other variable:

```
x
    ⇒ x =
    {
      a = 1
      b =

        1  2
        3  4

      c = string
    }
```

Note that Octave may print the elements in any order.

Structures may be copied just like any other variable:

```
y = x
    ⇒ y =
    {
      a = 1
      b =

        1 2
        3 4

      c = string
    }
```

Since structures are themselves values, structure elements may reference other structures. The following statements change the value of the element b of the structure x to be a data structure containing the single element d, which has a value of 3.

```
x.b.d = 3;
x.b
    ⇒ ans =
    {
      d = 3
    }

x
    ⇒ x =
    {
      a = 1
      b =
      {
        d = 3
      }

      c = string
    }
```

Note that when Octave prints the value of a structure that contains other structures, only a few levels are displayed. For example:

```
a.b.c.d.e = 1;
a
    ⇒ a =
        {
          b =
          {
            c =
            {
              1x1 struct array containing the fields:

              d: 1x1 struct
            }
          }
        }
```

This prevents long and confusing output from large deeply nested structures. The number of levels to print for nested structures may be set with the function `struct_levels_to_print`, and the function `print_struct_array_contents` may be used to enable printing of the contents of structure arrays.

`val = struct_levels_to_print ()` [Built-in Function]
`old_val = struct_levels_to_print (new_val)` [Built-in Function]
`struct_levels_to_print (new_val, "local")` [Built-in Function]
> Query or set the internal variable that specifies the number of structure levels to display.
>
> When called from inside a function with the `"local"` option, the variable is changed locally for the function and any subroutines it calls. The original variable value is restored when exiting the function.
>
> **See also:** [print_struct_array_contents], page 101.

`val = print_struct_array_contents ()` [Built-in Function]
`old_val = print_struct_array_contents (new_val)` [Built-in Function]
`print_struct_array_contents (new_val, "local")` [Built-in Function]
> Query or set the internal variable that specifies whether to print struct array contents.
>
> If true, values of struct array elements are printed. This variable does not affect scalar structures whose elements are always printed. In both cases, however, printing will be limited to the number of levels specified by *struct_levels_to_print*.
>
> When called from inside a function with the `"local"` option, the variable is changed locally for the function and any subroutines it calls. The original variable value is restored when exiting the function.
>
> **See also:** [struct_levels_to_print], page 101.

Functions can return structures. For example, the following function separates the real and complex parts of a matrix and stores them in two elements of the same structure variable.

```
function y = f (x)
  y.re = real (x);
  y.im = imag (x);
endfunction
```

When called with a complex-valued argument, `f` returns the data structure containing the real and imaginary parts of the original function argument.

```
f (rand (2) + rand (2) * I)
    ⇒ ans =
      {
        im =

          0.26475  0.14828
          0.18436  0.83669

        re =

          0.040239  0.242160
          0.238081  0.402523

      }
```

Function return lists can include structure elements, and they may be indexed like any other variable. For example:

```
[ x.u, x.s(2:3,2:3), x.v ] = svd ([1, 2; 3, 4]);
x
    ⇒ x =
      {
        u =

          -0.40455  -0.91451
          -0.91451   0.40455

        s =

          0.00000  0.00000  0.00000
          0.00000  5.46499  0.00000
          0.00000  0.00000  0.36597

        v =

          -0.57605   0.81742
          -0.81742  -0.57605

      }
```

It is also possible to cycle through all the elements of a structure in a loop, using a special form of the `for` statement (see Section 10.5.1 [Looping Over Structure Elements], page 165).

## 6.1.2 Structure Arrays

A structure array is a particular instance of a structure, where each of the fields of the structure is represented by a cell array. Each of these cell arrays has the same dimensions. Conceptually, a structure array can also be seen as an array of structures with identical fields. An example of the creation of a structure array is

```
x(1).a = "string1";
x(2).a = "string2";
x(1).b = 1;
x(2).b = 2;
```

which creates a 2-by-1 structure array with two fields. Another way to create a structure array is with the **struct** function (see Section 6.1.3 [Creating Structures], page 104). As previously, to print the value of the structure array, you can type its name:

```
x
    ⇒ x =
    {
      1x2 struct array containing the fields:

        a
        b
    }
```

Individual elements of the structure array can be returned by indexing the variable like `x(1)`, which returns a structure with two fields:

```
x(1)
    ⇒ ans =
    {
      a = string1
      b = 1
    }
```

Furthermore, the structure array can return a comma separated list of field values (see Section 6.3 [Comma Separated Lists], page 120), if indexed by one of its own field names. For example:

```
x.a
    ⇒
      ans = string1
      ans = string2
```

Here is another example, using this comma separated list on the left-hand side of an assignment:

```
[x.a] = deal ("new string1", "new string2");
x(1).a
    ⇒ ans = new string1
x(2).a
    ⇒ ans = new string2
```

Just as for numerical arrays, it is possible to use vectors as indices (see Section 8.1 [Index Expressions], page 135):

```
x(3:4) = x(1:2);
[x([1,3]).a] = deal ("other string1", "other string2");
x.a
```
⇒
```
    ans = other string1
    ans = new string2
    ans = other string2
    ans = new string2
```
The function `size` will return the size of the structure. For the example above
```
size (x)
```
⇒ ans =

   1   4

Elements can be deleted from a structure array in a similar manner to a numerical array, by assigning the elements to an empty matrix. For example
```
in = struct ("call1", {x, Inf, "last"},
             "call2", {x, Inf, "first"})
```
⇒ in =
```
    {
        1x3 struct array containing the fields:

            call1
            call2
    }
```
```
in(1) = [];
in.call1
```
⇒
```
    ans = Inf
    ans = last
```

### 6.1.3 Creating Structures

Besides the index operator ".", Octave can use dynamic naming "(var)" or the `struct` function to create structures. Dynamic naming uses the string value of a variable as the field name. For example:
```
a = "field2";
x.a = 1;
x.(a) = 2;
x
```
⇒ x =
```
    {
        a =  1
        field2 =  2
    }
```
Dynamic indexing also allows you to use arbitrary strings, not merely valid Octave identifiers (note that this does not work on MATLAB):

Chapter 6: Data Containers

```
a = "long field with spaces (and funny char$)";
x.a = 1;
x.(a) = 2;
x
    ⇒ x =
      {
        a =  1
        long field with spaces (and funny char$) =  2
      }
```

The warning id `Octave:language-extension` can be enabled to warn about this usage. See [warning_ids], page 214.

More realistically, all of the functions that operate on strings can be used to build the correct field name before it is entered into the data structure.

```
names = ["Bill"; "Mary"; "John"];
ages  = [37; 26; 31];
for i = 1:rows (names)
  database.(names(i,:)) = ages(i);
endfor
database
    ⇒ database =
       {
         Bill =  37
         Mary =  26
         John =  31
       }
```

The third way to create structures is the **struct** command. **struct** takes pairs of arguments, where the first argument in the pair is the fieldname to include in the structure and the second is a scalar or cell array, representing the values to include in the structure or structure array. For example:

```
struct ("field1", 1, "field2", 2)
⇒ ans =
     {
       field1 =  1
       field2 =  2
     }
```

If the values passed to **struct** are a mix of scalar and cell arrays, then the scalar arguments are expanded to create a structure array with a consistent dimension. For example:

```
s = struct ("field1", {1, "one"}, "field2", {2, "two"},
        "field3", 3);
s.field1
    ⇒
        ans =  1
        ans = one

s.field2
    ⇒
        ans =  2
        ans = two

s.field3
    ⇒
        ans =  3
        ans =  3
```

If you want to create a struct which contains a cell array as an individual field, you must wrap it in another cell array as shown in the following example:

```
struct ("field1", {{1, "one"}}, "field2", 2)
    ⇒ ans =
      {
        field1 =

        {
          [1,1] =  1
          [1,2] = one
        }

        field2 =  2
      }
```

**s = struct ()**           [Built-in Function]
**s = struct (*field1*, *value1*, *field2*, *value2*, ...)**           [Built-in Function]
**s = struct (*obj*)**           [Built-in Function]

Create a scalar or array structure and initialize its values.

The *field1*, *field2*, ... variables are strings specifying the names of the fields and the *value1*, *value2*, ... variables can be of any type.

If the values are cell arrays, create a structure array and initialize its values. The dimensions of each cell array of values must match. Singleton cells and non-cell values are repeated so that they fill the entire array. If the cells are empty, create an empty structure array with the specified field names.

If the argument is an object, return the underlying struct.

Observe that the syntax is optimized for struct **arrays**. Consider the following examples:

Chapter 6: Data Containers 107

```
struct ("foo", 1)
   ⇒ scalar structure containing the fields:
     foo =  1

struct ("foo", {})
   ⇒ 0x0 struct array containing the fields:
     foo

struct ("foo", { {} })
   ⇒ scalar structure containing the fields:
     foo = {}(0x0)

struct ("foo", {1, 2, 3})
   ⇒ 1x3 struct array containing the fields:
     foo
```

The first case is an ordinary scalar struct—one field, one value. The second produces an empty struct array with one field and no values, since being passed an empty cell array of struct array values. When the value is a cell array containing a single entry, this becomes a scalar struct with that single entry as the value of the field. That single entry happens to be an empty cell array.

Finally, if the value is a non-scalar cell array, then **struct** produces a struct **array**.

**See also:** [cell2struct], page 120, [fieldnames], page 107, [getfield], page 109, [setfield], page 108, [rmfield], page 109, [isfield], page 108, [orderfields], page 109, [isstruct], page 107, [structfun], page 496.

The function `isstruct` can be used to test if an object is a structure or a structure array.

`isstruct (x)` [Built-in Function]
: Return true if x is a structure or a structure array.

**See also:** [ismatrix], page 63, [iscell], page 113, [isa], page 39.

## 6.1.4 Manipulating Structures

Other functions that can manipulate the fields of a structure are given below.

`numfields (s)` [Built-in Function]
: Return the number of fields of the structure s.

**See also:** [fieldnames], page 107.

`names = fieldnames (struct)` [Function File]
`names = fieldnames (obj)` [Function File]
`names = fieldnames (javaobj)` [Function File]
`names = fieldnames ("jclassname")` [Function File]
: Return a cell array of strings with the names of the fields in the specified input.

When the input is a structure *struct*, the names are the elements of the structure.

When the input is an Octave object *obj*, the names are the public properties of the object.

When the input is a Java object *javaobj* or Java classname *jclassname* the name are the public data elements of the object or class.

**See also:** [numfields], page 107, [isfield], page 108, [orderfields], page 109, [struct], page 106, [methods], page 718.

isfield (*x*, "*name*")                                                               [Built-in Function]
isfield (*x*, *name*)                                                                  [Built-in Function]

Return true if the *x* is a structure and it includes an element named *name*.

If *name* is a cell array of strings then a logical array of equal dimension is returned.

**See also:** [fieldnames], page 107.

*sout* = setfield (*s*, *field*, *val*)                                             [Function File]
*sout* = setfield (*s*, *sidx1*, *field1*, *fidx1*, *sidx2*, *field2*,       [Function File]
    *fidx2*, ..., *val*)

Return a *copy* of the structure *s* with the field member *field* set to the value *val*.

For example:

```
s = struct ();
s = setfield (s, "foo bar", 42);
```

This is equivalent to

```
s.("foo bar") = 42;
```

Note that ordinary structure syntax s.foo bar = 42 cannot be used here, as the field name is not a valid Octave identifier because of the space character. Using arbitrary strings for field names is incompatible with MATLAB, and this usage will emit a warning if the warning ID Octave:language-extension is enabled. See [XREFwarning_ids], page 214.

With the second calling form, set a field of a structure array. The input *sidx* selects an element of the structure array, *field* specifies the field name of the selected element, and *fidx* selects which element of the field (in the case of an array or cell array). The *sidx*, *field*, and *fidx* inputs can be repeated to address nested structure array elements. The structure array index and field element index must be cell arrays while the field name must be a string.

For example:

```
s = struct ("baz", 42);
setfield (s, {1}, "foo", {1}, "bar", 54)
⇒
  ans =
    scalar structure containing the fields:
      baz = 42
      foo =
        scalar structure containing the fields:
          bar = 54
```

The example begins with an ordinary scalar structure to which a nested scalar structure is added. In all cases, if the structure index *sidx* is not specified it defaults to

Chapter 6: Data Containers 109

1 (scalar structure). Thus, the example above could be written more concisely as
`setfield (s, "foo", "bar", 54)`

Finally, an example with nested structure arrays:

```
sa.foo = 1;
sa = setfield (sa, {2}, "bar", {3}, "baz", {1, 4}, 5);
sa(2).bar(3)
⇒
  ans =
    scalar structure containing the fields:
      baz =  0   0   0   5
```

Here *sa* is a structure array whose field at elements 1 and 2 is in turn another structure array whose third element is a simple scalar structure. The terminal scalar structure has a field which contains a matrix value.

Note that the same result as in the above example could be achieved by:

```
sa.foo = 1;
sa(2).bar(3).baz(1,4) = 5
```

**See also:** [getfield], page 109, [rmfield], page 109, [orderfields], page 109, [isfield], page 108, [fieldnames], page 107, [isstruct], page 107, [struct], page 106.

*val* = **getfield** (*s*, *field*)  [Function File]
*val* = **getfield** (*s*, *sidx1*, *field1*, *fidx1*, ...)  [Function File]

Get the value of the field named *field* from a structure or nested structure *s*.

If *s* is a structure array then *sidx* selects an element of the structure array, *field* specifies the field name of the selected element, and *fidx* selects which element of the field (in the case of an array or cell array). See `setfield` for a more complete description of the syntax.

**See also:** [setfield], page 108, [rmfield], page 109, [orderfields], page 109, [isfield], page 108, [fieldnames], page 107, [isstruct], page 107, [struct], page 106.

*sout* = **rmfield** (*s*, "*f*")  [Built-in Function]
*sout* = **rmfield** (*s*, *f*)  [Built-in Function]

Return a *copy* of the structure (array) *s* with the field *f* removed.

If *f* is a cell array of strings or a character array, remove each of the named fields.

**See also:** [orderfields], page 109, [fieldnames], page 107, [isfield], page 108.

*sout*] = **orderfields** (*s1*)  [Function File]
*sout*] = **orderfields** (*s1*, *s2*)  [Function File]
*sout*] = **orderfields** (*s1*, {*cellstr*})  [Function File]
*sout*] = **orderfields** (*s1*, *p*)  [Function File]
[*sout*, *p*] = **orderfields** (...)  [Function File]

Return a *copy* of *s1* with fields arranged alphabetically, or as specified by the second input.

Given one input struct *s1*, arrange field names alphabetically.

If a second struct argument is given, arrange field names in *s1* as they appear in *s2*. The second argument may also specify the order in a cell array of strings *cellstr*. The second argument may also be a permutation vector.

The optional second output argument *p* is the permutation vector which converts the original name order to the new name order.

Examples:

```
s = struct ("d", 4, "b", 2, "a", 1, "c", 3);
t1 = orderfields (s)
     ⇒ t1 =
       {
         a = 1
         b = 2
         c = 3
         d = 4
       }
t = struct ("d", {}, "c", {}, "b", {}, "a", {});
t2 = orderfields (s, t)
     ⇒ t2 =
       {
         d = 4
         c = 3
         b = 2
         a = 1
       }
t3 = orderfields (s, [3, 2, 4, 1])
     ⇒ t3 =
       {
         a = 1
         b = 2
         c = 3
         d = 4
       }
[t4, p] = orderfields (s, {"d", "c", "b", "a"})
     ⇒ t4 =
       {
         d = 4
         c = 3
         b = 2
         a = 1
       }
       p =
          1
          4
          2
          3
```

**See also:** [fieldnames], page 107, [getfield], page 109, [setfield], page 108, [rmfield], page 109, [isfield], page 108, [isstruct], page 107, [struct], page 106.

Chapter 6: Data Containers 111

**substruct** (*type*, *subs*, ...)  [Function File]
  Create a subscript structure for use with **subsref** or **subsasgn**.

  For example:

```
    idx = substruct ("()", {3, ":"})
       ⇒
         idx =
         {
           type = ()
           subs =
           {
             [1,1] = 3
             [1,2] = :
           }
         }
    x = [1, 2, 3;
         4, 5, 6;
         7, 8, 9];
    subsref (x, idx)
       ⇒ 7  8  9
```

  **See also:** [subsref], page 722, [subsasgn], page 724.

## 6.1.5 Processing Data in Structures

The simplest way to process data in a structure is within a `for` loop (see Section 10.5.1 [Looping Over Structure Elements], page 165). A similar effect can be achieved with the **structfun** function, where a user defined function is applied to each field of the structure. See [structfun], page 496.

Alternatively, to process the data in a structure, the structure might be converted to another type of container before being treated.

*c* = **struct2cell** (*s*)  [Built-in Function]
  Create a new cell array from the objects stored in the struct object.

  If $f$ is the number of fields in the structure, the resulting cell array will have a dimension vector corresponding to [f size(s)]. For example:

```
s = struct ("name", {"Peter", "Hannah", "Robert"},
            "age", {23, 16, 3});
c = struct2cell (s)
    ⇒ c = {2x1x3 Cell Array}
c(1,1,:)(:)
    ⇒
       {
         [1,1] = Peter
         [2,1] = Hannah
         [3,1] = Robert
       }
c(2,1,:)(:)
    ⇒
       {
         [1,1] = 23
         [2,1] = 16
         [3,1] = 3
       }
```

See also: [cell2struct], page 120, [fieldnames], page 107.

## 6.2 Cell Arrays

It can be both necessary and convenient to store several variables of different size or type in one variable. A cell array is a container class able to do just that. In general cell arrays work just like $N$-dimensional arrays with the exception of the use of '{' and '}' as allocation and indexing operators.

### 6.2.1 Basic Usage of Cell Arrays

As an example, the following code creates a cell array containing a string and a 2-by-2 random matrix

```
c = {"a string", rand(2, 2)};
```

To access the elements of a cell array, it can be indexed with the { and } operators. Thus, the variable created in the previous example can be indexed like this:

```
c{1}
    ⇒ ans = a string
```

As with numerical arrays several elements of a cell array can be extracted by indexing with a vector of indexes

```
c{1:2}
    ⇒ ans = a string
    ⇒ ans =

          0.593993   0.627732
          0.377037   0.033643
```

The indexing operators can also be used to insert or overwrite elements of a cell array. The following code inserts the scalar 3 on the third place of the previously created cell array

Chapter 6: Data Containers                                                                    113

```
    c{3} = 3
      ⇒ c =
        {
          [1,1] = a string
          [1,2] =

             0.593993   0.627732
             0.377037   0.033643

          [1,3] =   3
        }
```

Details on indexing cell arrays are explained in Section 6.2.3 [Indexing Cell Arrays], page 116.

In general nested cell arrays are displayed hierarchically as in the previous example. In some circumstances it makes sense to reference them by their index, and this can be performed by the `celldisp` function.

**celldisp** (*c*)                                                                    [Function File]
**celldisp** (*c*, *name*)                                                            [Function File]

Recursively display the contents of a cell array.

By default the values are displayed with the name of the variable c. However, this name can be replaced with the variable *name*. For example:

```
    c = {1, 2, {31, 32}};
    celldisp (c, "b")
      ⇒
        b{1} =
        1
        b{2} =
        2
        b{3}{1} =
        31
        b{3}{2} =
        32
```

**See also:** [disp], page 231.

To test if an object is a cell array, use the `iscell` function. For example:

```
    iscell (c)
      ⇒ ans = 1

    iscell (3)
      ⇒ ans = 0
```

**iscell** (*x*)                                                                    [Built-in Function]

Return true if *x* is a cell array object.

**See also:** [ismatrix], page 63, [isstruct], page 107, [iscellstr], page 119, [isa], page 39.

### 6.2.2 Creating Cell Arrays

The introductory example (see Section 6.2.1 [Basic Usage of Cell Arrays], page 112) showed how to create a cell array containing currently available variables. In many situations, however, it is useful to create a cell array and then fill it with data.

The `cell` function returns a cell array of a given size, containing empty matrices. This function is similar to the `zeros` function for creating new numerical arrays. The following example creates a 2-by-2 cell array containing empty matrices

```
c = cell (2,2)
     ⇒ c =
     {
       [1,1] = [](0x0)
       [2,1] = [](0x0)
       [1,2] = [](0x0)
       [2,2] = [](0x0)
     }
```

Just like numerical arrays, cell arrays can be multi-dimensional. The `cell` function accepts any number of positive integers to describe the size of the returned cell array. It is also possible to set the size of the cell array through a vector of positive integers. In the following example two cell arrays of equal size are created, and the size of the first one is displayed

```
c1 = cell (3, 4, 5);
c2 = cell ( [3, 4, 5] );
size (c1)
     ⇒ ans =
         3   4   5
```

As can be seen, the [size], page 45 function also works for cell arrays. As do other functions describing the size of an object, such as [length], page 45, [numel], page 44, [rows], page 44, and [columns], page 44.

| | |
|---|---|
| `cell (n)` | [Built-in Function] |
| `cell (m, n)` | [Built-in Function] |
| `cell (m, n, k, ...)` | [Built-in Function] |
| `cell ([m n ...])` | [Built-in Function] |

    Create a new cell array object.

    If invoked with a single scalar integer argument, return a square NxN cell array. If invoked with two or more scalar integer arguments, or a vector of integer values, return an array with the given dimensions.

    **See also:** [cellstr], page 119, [mat2cell], page 115, [num2cell], page 114, [struct2cell], page 111.

As an alternative to creating empty cell arrays, and then filling them, it is possible to convert numerical arrays into cell arrays using the `num2cell`, `mat2cell` and `cellslices` functions.

Chapter 6: Data Containers    115

$C$ = num2cell ($A$)  [Built-in Function]
$C$ = num2cell ($A$, *dim*)  [Built-in Function]

Convert the numeric matrix $A$ to a cell array.

If *dim* is defined, the value $C$ is of dimension 1 in this dimension and the elements of $A$ are placed into $C$ in slices. For example:

```
num2cell ([1,2;3,4])
    ⇒
        {
          [1,1] =  1
          [2,1] =  3
          [1,2] =  2
          [2,2] =  4
        }
num2cell ([1,2;3,4],1)
    ⇒
        {
          [1,1] =
             1
             3
          [1,2] =
             2
             4
        }
```

**See also:** [mat2cell], page 115.

$C$ = mat2cell ($A$, $m$, $n$)  [Built-in Function]
$C$ = mat2cell ($A$, $d1$, $d2$, ...)  [Built-in Function]
$C$ = mat2cell ($A$, $r$)  [Built-in Function]

Convert the matrix $A$ to a cell array.

If $A$ is 2-D, then it is required that sum ($m$) == size ($A$, 1) and sum ($n$) == size ($A$, 2). Similarly, if $A$ is multi-dimensional and the number of dimensional arguments is equal to the dimensions of $A$, then it is required that sum ($di$) == size ($A$, $i$).

Given a single dimensional argument $r$, the other dimensional arguments are assumed to equal size ($A$,$i$).

An example of the use of mat2cell is

```
mat2cell (reshape (1:16,4,4), [3,1], [3,1])
    ⇒
    {
       [1,1] =

          1   5   9
          2   6  10
          3   7  11

       [2,1] =
```

```
        4    8   12

   [1,2] =

      13
      14
      15

   [2,2] = 16
 }
```

**See also:** [num2cell], page 114, [cell2mat], page 119.

`sl = cellslices (x, lb, ub, dim)`                              [Built-in Function]

Given an array *x*, this function produces a cell array of slices from the array determined by the index vectors *lb*, *ub*, for lower and upper bounds, respectively.

In other words, it is equivalent to the following code:

```
n = length (lb);
sl = cell (1, n);
for i = 1:length (lb)
   sl{i} = x(:,...,lb(i):ub(i),...,:);
endfor
```

The position of the index is determined by *dim*. If not specified, slicing is done along the first non-singleton dimension.

**See also:** [cell2mat], page 119, [cellindexmat], page 118, [cellfun], page 494.

### 6.2.3 Indexing Cell Arrays

As shown in see Section 6.2.1 [Basic Usage of Cell Arrays], page 112 elements can be extracted from cell arrays using the '{' and '}' operators. If you want to extract or access subarrays which are still cell arrays, you need to use the '(' and ')' operators. The following example illustrates the difference:

```
c = {"1", "2", "3"; "x", "y", "z"; "4", "5", "6"};
c{2,3}
    ⇒ ans = z

c(2,3)
    ⇒ ans =
       {
          [1,1] = z
       }
```

So with '{}' you access elements of a cell array, while with '()' you access a sub array of a cell array.

Using the '(' and ')' operators, indexing works for cell arrays like for multi-dimensional arrays. As an example, all the rows of the first and third column of a cell array can be set to 0 with the following command:

```
c(:, [1, 3]) = {0}
   ⇒ =
     {
       [1,1] = 0
       [2,1] = 0
       [3,1] = 0
       [1,2] = 2
       [2,2] =  10
       [3,2] =  20
       [1,3] = 0
       [2,3] = 0
       [3,3] = 0
     }
```

Note, that the above can also be achieved like this:

```
c(:, [1, 3]) = 0;
```

Here, the scalar '0' is automatically promoted to cell array '{0}' and then assigned to the subarray of c.

To give another example for indexing cell arrays with '()', you can exchange the first and the second row of a cell array as in the following command:

```
c = {1, 2, 3; 4, 5, 6};
c([1, 2], :) = c([2, 1], :)
   ⇒ =
     {
       [1,1] =  4
       [2,1] =  1
       [1,2] =  5
       [2,2] =  2
       [1,3] =  6
       [2,3] =  3
     }
```

Accessing multiple elements of a cell array with the '{' and '}' operators will result in a comma-separated list of all the requested elements (see Section 6.3 [Comma Separated Lists], page 120). Using the '{' and '}' operators the first two rows in the above example can be swapped back like this:

```
[c{[1,2], :}] = deal (c{[2, 1], :})
   ⇒ =
     {
       [1,1] =  1
       [2,1] =  4
       [1,2] =  2
       [2,2] =  5
       [1,3] =  3
       [2,3] =  6
     }
```

As for struct arrays and numerical arrays, the empty matrix '[]' can be used to delete elements from a cell array:

```
x = {"1", "2"; "3", "4"};
x(1, :) = []
    ⇒ x =
        {
          [1,1] = 3
          [1,2] = 4
        }
```

The following example shows how to just remove the contents of cell array elements but not delete the space for them:

```
x = {"1", "2"; "3", "4"};
x{1, :} = []
    ⇒ x =
        {
          [1,1] = [](0x0)
          [2,1] = 3
          [1,2] = [](0x0)
          [2,2] = 4
        }
```

The indexing operations operate on the cell array and not on the objects within the cell array. By contrast, `cellindexmat` applies matrix indexing to the objects within each cell array entry and returns the requested values.

$y$ = `cellindexmat` ($x$, *varargin*)                                                         [Built-in Function]
    Perform indexing of matrices in a cell array.

    Given a cell array of matrices x, this function computes

```
Y = cell (size (X));
for i = 1:numel (X)
  Y{i} = X{i}(varargin{:});
endfor
```

**See also:** [cellslices], page 116, [cellfun], page 494.

### 6.2.4 Cell Arrays of Strings

One common use of cell arrays is to store multiple strings in the same variable. It is also possible to store multiple strings in a character matrix by letting each row be a string. This, however, introduces the problem that all strings must be of equal length. Therefore, it is recommended to use cell arrays to store multiple strings. For cases, where the character matrix representation is required for an operation, there are several functions that convert a cell array of strings to a character array and back. `char` and `strvcat` convert cell arrays to a character array (see Section 5.3.1 [Concatenating Strings], page 70), while the function `cellstr` converts a character array to a cell array of strings:

Chapter 6: Data Containers                                                                                              119

```
a = ["hello"; "world"];
c = cellstr (a)
    ⇒ c =
      {
        [1,1] = hello
        [2,1] = world
      }
```

*cstr* = cellstr (*strmat*)                                                                    [Built-in Function]

    Create a new cell array object from the elements of the string array *strmat*.

    Each row of *strmat* becomes an element of *cstr*. Any trailing spaces in a row are deleted before conversion.

    To convert back from a cellstr to a character array use `char`.

    **See also:** [cell], page 114, [char], page 71.

One further advantage of using cell arrays to store multiple strings is that most functions for string manipulations included with Octave support this representation. As an example, it is possible to compare one string with many others using the `strcmp` function. If one of the arguments to this function is a string and the other is a cell array of strings, each element of the cell array will be compared to the string argument:

```
c = {"hello", "world"};
strcmp ("hello", c)
    ⇒ ans =
         1    0
```

The following string functions support cell arrays of strings: `char`, `strvcat`, `strcat` (see Section 5.3.1 [Concatenating Strings], page 70), `strcmp`, `strncmp`, `strcmpi`, `strncmpi` (see Section 5.4 [Comparing Strings], page 76), `str2double`, `deblank`, `strtrim`, `strtrunc`, `strfind`, `strmatch`, , `regexp`, `regexpi` (see Section 5.5 [Manipulating Strings], page 77) and `str2double` (see Section 5.6 [String Conversions], page 91).

    The function `iscellstr` can be used to test if an object is a cell array of strings.

iscellstr (*cell*)                                                                              [Built-in Function]

    Return true if every element of the cell array *cell* is a character string.

    **See also:** [ischar], page 68.

## 6.2.5 Processing Data in Cell Arrays

Data that is stored in a cell array can be processed in several ways depending on the actual data. The simplest way to process that data is to iterate through it using one or more `for` loops. The same idea can be implemented more easily through the use of the `cellfun` function that calls a user-specified function on all elements of a cell array. See [cellfun], page 494.

An alternative is to convert the data to a different container, such as a matrix or a data structure. Depending on the data this is possible using the `cell2mat` and `cell2struct` functions.

`m = cell2mat (c)` [Function File]
    Convert the cell array *c* into a matrix by concatenating all elements of *c* into a hyperrectangle.

    Elements of *c* must be numeric, logical, or char matrices; or cell arrays; or structs; and `cat` must be able to concatenate them together.

    **See also:** [mat2cell], page 115, [num2cell], page 114.

`cell2struct (cell, fields)` [Built-in Function]
`cell2struct (cell, fields, dim)` [Built-in Function]
    Convert *cell* to a structure.

    The number of fields in *fields* must match the number of elements in *cell* along dimension *dim*, that is `numel (fields) == size (cell, dim)`. If *dim* is omitted, a value of 1 is assumed.

```
A = cell2struct ({"Peter", "Hannah", "Robert";
                  185, 170, 168},
                 {"Name","Height"}, 1);
A(1)
   ⇒
     {
       Name   = Peter
       Height = 185
     }
```

    **See also:** [struct2cell], page 111, [cell2mat], page 119, [struct], page 106.

## 6.3 Comma Separated Lists

Comma separated lists[1] are the basic argument type to all Octave functions - both for input and return arguments. In the example

    `max (a, b)`

'a, b' is a comma separated list. Comma separated lists can appear on both the right and left hand side of an assignment. For example

```
x = [1 0 1 0 0 1 1; 0 0 0 0 0 0 7];
[i, j] = find (x, 2, "last");
```

Here, 'x, 2, "last"' is a comma separated list constituting the input arguments of `find`. `find` returns a comma separated list of output arguments which is assigned element by element to the comma separated list '*i, j*'.

    Another example of where comma separated lists are used is in the creation of a new array with `[]` (see Section 4.1 [Matrices], page 48) or the creation of a cell array with `{}` (see Section 6.2.1 [Basic Usage of Cell Arrays], page 112). In the expressions

```
a = [1, 2, 3, 4];
c = {4, 5, 6, 7};
```

both '1, 2, 3, 4' and '4, 5, 6, 7' are comma separated lists.

---

[1] Comma-separated lists are also sometimes informally referred to as *cs-lists*.

# Chapter 6: Data Containers

Comma separated lists cannot be directly manipulated by the user. However, both structure arrays and cell arrays can be converted into comma separated lists, and thus used in place of explicitly written comma separated lists. This feature is useful in many ways, as will be shown in the following subsections.

## 6.3.1 Comma Separated Lists Generated from Cell Arrays

As has been mentioned above (see Section 6.2.3 [Indexing Cell Arrays], page 116), elements of a cell array can be extracted into a comma separated list with the { and } operators. By surrounding this list with [ and ], it can be concatenated into an array. For example:

```
a = {1, [2, 3], 4, 5, 6};
b = [a{1:4}]
    ⇒ b =
       1   2   3   4   5
```

Similarly, it is possible to create a new cell array containing cell elements selected with {}. By surrounding the list with '{' and '}' a new cell array will be created, as the following example illustrates:

```
a = {1, rand(2, 2), "three"};
b = { a{ [1, 3] } }
    ⇒ b =
       {
         [1,1] =  1
         [1,2] = three
       }
```

Furthermore, cell elements (accessed by {}) can be passed directly to a function. The list of elements from the cell array will be passed as an argument list to a given function as if it is called with the elements as individual arguments. The two calls to `printf` in the following example are identical but the latter is simpler and can handle cell arrays of an arbitrary size:

```
c = {"GNU", "Octave", "is", "Free", "Software"};
printf ("%s ", c{1}, c{2}, c{3}, c{4}, c{5});
     ⊣ GNU Octave is Free Software
printf ("%s ", c{:});
     ⊣ GNU Octave is Free Software
```

If used on the left-hand side of an assignment, a comma separated list generated with {} can be assigned to. An example is

```
in{1} = [10, 20, 30, 40, 50, 60, 70, 80, 90];
in{2} = inf;
in{3} = "last";
in{4} = "first";
out = cell (4, 1);
[out{1:3}] = find (in{1 : 3});
[out{4:6}] = find (in{[1, 2, 4]})
    ⇒ out =
      {
        [1,1] = 1
        [2,1] = 9
        [3,1] = 90
        [4,1] = 1
        [3,1] = 1
        [4,1] = 10
      }
```

## 6.3.2 Comma Separated Lists Generated from Structure Arrays

Structure arrays can equally be used to create comma separated lists. This is done by addressing one of the fields of a structure array. For example:

```
x = ceil (randn (10, 1));
in = struct ("call1", {x, 3, "last"},
             "call2", {x, inf, "first"});
out = struct ("call1", cell (2, 1), "call2", cell (2, 1));
[out.call1] = find (in.call1);
[out.call2] = find (in.call2);
```

# 7 Variables

Variables let you give names to values and refer to them later. You have already seen variables in many of the examples. The name of a variable must be a sequence of letters, digits and underscores, but it may not begin with a digit. Octave does not enforce a limit on the length of variable names, but it is seldom useful to have variables with names longer than about 30 characters. The following are all valid variable names

```
x
x15
__foo_bar_baz__
fucnrdthsucngtagdjb
```

However, names like `__foo_bar_baz__` that begin and end with two underscores are understood to be reserved for internal use by Octave. You should not use them in code you write, except to access Octave's documented internal variables and built-in symbolic constants.

Case is significant in variable names. The symbols `a` and `A` are distinct variables.

A variable name is a valid expression by itself. It represents the variable's current value. Variables are given new values with *assignment operators* and *increment operators*. See Section 8.6 [Assignment Expressions], page 149.

There is one built-in variable with a special meaning. The `ans` variable always contains the result of the last computation, where the output wasn't assigned to any variable. The code `a = cos (pi)` will assign the value -1 to the variable `a`, but will not change the value of `ans`. However, the code `cos (pi)` will set the value of `ans` to -1.

Variables in Octave do not have fixed types, so it is possible to first store a numeric value in a variable and then to later use the same name to hold a string value in the same program. Variables may not be used before they have been given a value. Doing so results in an error.

**ans**  [Automatic Variable]

The most recently computed result that was not explicitly assigned to a variable.

For example, after the expression

```
3^2 + 4^2
```

is evaluated, the value returned by `ans` is 25.

**isvarname (*name*)**  [Built-in Function]

Return true if *name* is a valid variable name.

**See also:** [iskeyword], page 911, [exist], page 130, [who], page 128.

**varname = genvarname (*str*)**  [Function File]
**varname = genvarname (*str*, *exclusions*)**  [Function File]

Create valid unique variable name(s) from *str*.

If *str* is a cellstr, then a unique variable is created for each cell in *str*.

```
genvarname ({"foo", "foo"})
   ⇒
     {
       [1,1] = foo
       [1,2] = foo1
     }
```

If *exclusions* is given, then the variable(s) will be unique to each other and to *exclusions* (*exclusions* may be either a string or a cellstr).

```
x = 3.141;
genvarname ("x", who ())
    ⇒ x1
```

Note that the result is a char array or cell array of strings, not the variables themselves. To define a variable, `eval()` can be used. The following trivial example sets x to 42.

```
name = genvarname ("x");
eval ([name " = 42"]);
    ⇒ x =  42
```

This can be useful for creating unique struct field names.

```
x = struct ();
for i = 1:3
  x.(genvarname ("a", fieldnames (x))) = i;
endfor
    ⇒ x =
       {
         a  =  1
         a1 =  2
         a2 =  3
       }
```

Since variable names may only contain letters, digits, and underscores, `genvarname` will replace any sequence of disallowed characters with an underscore. Also, variables may not begin with a digit; in this case an 'x' is added before the variable name.

Variable names beginning and ending with two underscores "__" are valid, but they are used internally by Octave and should generally be avoided; therefore, `genvarname` will not generate such names.

`genvarname` will also ensure that returned names do not clash with keywords such as `"for"` and `"if"`. A number will be appended if necessary. Note, however, that this does **not** include function names such as `"sin"`. Such names should be included in *exclusions* if necessary.

**See also:** [isvarname], page 123, [iskeyword], page 911, [exist], page 130, [who], page 128, [tempname], page 266, [eval], page 155.

`namelengthmax ()` [Function File]

Return the MATLAB compatible maximum variable name length.

Octave is capable of storing strings up to $2^{31} - 1$ in length. However for MATLAB compatibility all variable, function, and structure field names should be shorter than the length returned by `namelengthmax`. In particular, variables stored to a MATLAB file format ('*.mat') will have their names truncated to this length.

## 7.1 Global Variables

A variable that has been declared *global* may be accessed from within a function body without having to pass it as a formal parameter.

A variable may be declared global using a `global` declaration statement. The following statements are all global declarations.

```
global a
global a b
global c = 2
global d = 3 e f = 5
```

A global variable may only be initialized once in a `global` statement. For example, after executing the following code

```
global gvar = 1
global gvar = 2
```

the value of the global variable `gvar` is 1, not 2. Issuing a 'clear gvar' command does not change the above behavior, but 'clear all' does.

It is necessary declare a variable as global within a function body in order to access it. For example,

```
global x
function f ()
  x = 1;
endfunction
f ()
```

does *not* set the value of the global variable `x` to 1. In order to change the value of the global variable `x`, you must also declare it to be global within the function body, like this

```
function f ()
  global x;
  x = 1;
endfunction
```

Passing a global variable in a function parameter list will make a local copy and not modify the global value. For example, given the function

```
function f (x)
  x = 0
endfunction
```

and the definition of `x` as a global variable at the top level,

```
global x = 13
```

the expression

```
f (x)
```

will display the value of `x` from inside the function as 0, but the value of `x` at the top level remains unchanged, because the function works with a *copy* of its argument.

**isglobal (*name*)**                                            [Built-in Function]

Return true if *name* is a globally visible variable.

For example:

```
global x
isglobal ("x")
  ⇒ 1
```

**See also:** [isvarname], page 123, [exist], page 130.

## 7.2 Persistent Variables

A variable that has been declared *persistent* within a function will retain its contents in memory between subsequent calls to the same function. The difference between persistent variables and global variables is that persistent variables are local in scope to a particular function and are not visible elsewhere.

The following example uses a persistent variable to create a function that prints the number of times it has been called.

```
function count_calls ()
  persistent calls = 0;
  printf ("'count_calls' has been called %d times\n",
          ++calls);
endfunction

for i = 1:3
  count_calls ();
endfor

⊣ 'count_calls' has been called 1 times
⊣ 'count_calls' has been called 2 times
⊣ 'count_calls' has been called 3 times
```

As the example shows, a variable may be declared persistent using a `persistent` declaration statement. The following statements are all persistent declarations.

```
persistent a
persistent a b
persistent c = 2
persistent d = 3 e f = 5
```

The behavior of persistent variables is equivalent to the behavior of static variables in C.

Like global variables, a persistent variable may only be initialized once. For example, after executing the following code

```
persistent pvar = 1
persistent pvar = 2
```

the value of the persistent variable `pvar` is 1, not 2.

If a persistent variable is declared but not initialized to a specific value, it will contain an empty matrix. So, it is also possible to initialize a persistent variable by checking whether it is empty, as the following example illustrates.

```
function count_calls ()
  persistent calls;
  if (isempty (calls))
    calls = 0;
  endif
  printf ("'count_calls' has been called %d times\n",
          ++calls);
endfunction
```

Chapter 7: Variables                                                          127

This implementation behaves in exactly the same way as the previous implementation of `count_calls`.

The value of a persistent variable is kept in memory until it is explicitly cleared. Assuming that the implementation of `count_calls` is saved on disk, we get the following behavior.

```
for i = 1:2
  count_calls ();
endfor
⊣ 'count_calls' has been called 1 times
⊣ 'count_calls' has been called 2 times

clear
for i = 1:2
  count_calls ();
endfor
⊣ 'count_calls' has been called 3 times
⊣ 'count_calls' has been called 4 times

clear all
for i = 1:2
  count_calls ();
endfor
⊣ 'count_calls' has been called 1 times
⊣ 'count_calls' has been called 2 times

clear count_calls
for i = 1:2
  count_calls ();
endfor
⊣ 'count_calls' has been called 1 times
⊣ 'count_calls' has been called 2 times
```

That is, the persistent variable is only removed from memory when the function containing the variable is removed. Note that if the function definition is typed directly into the Octave prompt, the persistent variable will be cleared by a simple `clear` command as the entire function definition will be removed from memory. If you do not want a persistent variable to be removed from memory even if the function is cleared, you should use the `mlock` function (see Section 11.9.6 [Function Locking], page 196).

## 7.3 Status of Variables

When creating simple one-shot programs it can be very convenient to see which variables are available at the prompt. The function `who` and its siblings `whos` and `whos_line_format` will show different information about what is in memory, as the following shows.

```
str = "A random string";
who -variables
     ⊣ *** local user variables:
     ⊣
     ⊣ __nargin__   str
```

**who**                                                                         [Command]
**who** *pattern* ...                                                           [Command]
**who** *option pattern* ...                                                    [Command]
**C = who** ("*pattern*", ...)                                                  [Command]

    List currently defined variables matching the given patterns.

    Valid pattern syntax is the same as described for the `clear` command. If no patterns are supplied, all variables are listed.

    By default, only variables visible in the local scope are displayed.

    The following are valid options, but may not be combined.

    `global`    List variables in the global scope rather than the current scope.

    `-regexp`    The patterns are considered to be regular expressions when matching the variables to display. The same pattern syntax accepted by the `regexp` function is used.

    `-file`    The next argument is treated as a filename. All variables found within the specified file are listed. No patterns are accepted when reading variables from a file.

    If called as a function, return a cell array of defined variable names matching the given patterns.

    **See also:** [whos], page 128, [isglobal], page 125, [isvarname], page 123, [exist], page 130, [regexp], page 87.

**whos**                                                                        [Command]
**whos** *pattern* ...                                                          [Command]
**whos** *option pattern* ...                                                   [Command]
**S = whos** ("*pattern*", ...)                                         [Built-in Function]

    Provide detailed information on currently defined variables matching the given patterns.

    Options and pattern syntax are the same as for the `who` command.

    Extended information about each variable is summarized in a table with the following default entries.

    Attr    Attributes of the listed variable. Possible attributes are:

        blank    Variable in local scope

        a    Automatic variable. An automatic variable is one created by the interpreter, for example **argn**.

        c    Variable of complex type.

        f    Formal parameter (function argument).

## Chapter 7: Variables

g        Variable with global scope.

p        Persistent variable.

Name        The name of the variable.

Size        The logical size of the variable. A scalar is 1x1, a vector is 1xN or Nx1, a 2-D matrix is MxN.

Bytes        The amount of memory currently used to store the variable.

Class        The class of the variable. Examples include double, single, char, uint16, cell, and struct.

The table can be customized to display more or less information through the function `whos_line_format`.

If `whos` is called as a function, return a struct array of defined variable names matching the given patterns. Fields in the structure describing each variable are: name, size, bytes, class, global, sparse, complex, nesting, persistent.

**See also:** [who], page 128, [whos_line_format], page 129.

---

`val = whos_line_format ()`        [Built-in Function]
`old_val = whos_line_format (new_val)`        [Built-in Function]
`whos_line_format (new_val, "local")`        [Built-in Function]

Query or set the format string used by the command `whos`.

A full format string is:

    `%[modifier]<command>[:width[:left-min[:balance]]];`

The following command sequences are available:

%a        Prints attributes of variables (g=global, p=persistent, f=formal parameter, a=automatic variable).

%b        Prints number of bytes occupied by variables.

%c        Prints class names of variables.

%e        Prints elements held by variables.

%n        Prints variable names.

%s        Prints dimensions of variables.

%t        Prints type names of variables.

Every command may also have an alignment modifier:

l        Left alignment.

r        Right alignment (default).

c        Column-aligned (only applicable to command %s).

The `width` parameter is a positive integer specifying the minimum number of columns used for printing. No maximum is needed as the field will auto-expand as required.

The parameters `left-min` and `balance` are only available when the column-aligned modifier is used with the command '%s'. `balance` specifies the column number within the field width which will be aligned between entries. Numbering starts from 0 which indicates the leftmost column. `left-min` specifies the minimum field width to the left of the specified balance column.

The default format is:

`" %a:4; %ln:6; %cs:16:6:1; %rb:12; %lc:-1;\n"`

When called from inside a function with the `"local"` option, the variable is changed locally for the function and any subroutines it calls. The original variable value is restored when exiting the function.

**See also:** [whos], page 128.

Instead of displaying which variables are in memory, it is possible to determine if a given variable is available. That way it is possible to alter the behavior of a program depending on the existence of a variable. The following example illustrates this.

```
if (! exist ("meaning", "var"))
  disp ("The program has no 'meaning'");
endif
```

*c* = exist (*name*) [Built-in Function]
*c* = exist (*name*, *type*) [Built-in Function]

Check for the existence of *name* as a variable, function, file, directory, or class.

The return code *c* is one of

| | |
|---|---|
| 1 | *name* is a variable. |
| 2 | *name* is an absolute file name, an ordinary file in Octave's `path`, or (after appending '.m') a function file in Octave's `path`. |
| 3 | *name* is a '.oct' or '.mex' file in Octave's `path`. |
| 5 | *name* is a built-in function. |
| 7 | *name* is a directory. |
| 103 | *name* is a function not associated with a file (entered on the command line). |
| 0 | *name* does not exist. |

If the optional argument *type* is supplied, check only for symbols of the specified type. Valid types are

| | |
|---|---|
| `"var"` | Check only for variables. |
| `"builtin"` | Check only for built-in functions. |
| `"dir"` | Check only for directories. |
| `"file"` | Check only for files and directories. |
| `"class"` | Check only for classes. (Note: This option is accepted, but not currently implemented) |

Chapter 7: Variables

If no type is given, and there are multiple possible matches for name, `exist` will return a code according to the following priority list: variable, built-in function, oct-file, directory, file, class.

`exist` returns 2 if a regular file called *name* is present in Octave's search path. If you want information about other types of files not on the search path you should use some combination of the functions `file_in_path` and `stat` instead.

**See also:** [file_in_loadpath], page 191, [file_in_path], page 760, [dir_in_loadpath], page 191, [stat], page 757.

Usually Octave will manage the memory, but sometimes it can be practical to remove variables from memory manually. This is usually needed when working with large variables that fill a substantial part of the memory. On a computer that uses the IEEE floating point format, the following program allocates a matrix that requires around 128 MB memory.

```
large_matrix = zeros (4000, 4000);
```

Since having this variable in memory might slow down other computations, it can be necessary to remove it manually from memory. The `clear` function allows this.

`clear` [*options*] *pattern* ...                                                           [Command]

Delete the names matching the given patterns from the symbol table.

The pattern may contain the following special characters:

?          Match any single character.

*          Match zero or more characters.

[ *list* ]   Match the list of characters specified by *list*. If the first character is ! or ^, match all characters except those specified by *list*. For example, the pattern '[a-zA-Z]' will match all lowercase and uppercase alphabetic characters.

For example, the command

```
clear foo b*r
```

clears the name `foo` and all names that begin with the letter `b` and end with the letter `r`.

If `clear` is called without any arguments, all user-defined variables (local and global) are cleared from the symbol table.

If `clear` is called with at least one argument, only the visible names matching the arguments are cleared. For example, suppose you have defined a function `foo`, and then hidden it by performing the assignment `foo = 2`. Executing the command *clear foo* once will clear the variable definition and restore the definition of `foo` as a function. Executing *clear foo* a second time will clear the function definition.

The following options are available in both long and short form

-all, -a    Clear all local and global user-defined variables and all functions from the symbol table.

-exclusive, -x

Clear the variables that don't match the following pattern.

-functions, -f
: Clear the function names and the built-in symbols names.

-global, -g
: Clear global symbol names.

-variables, -v
: Clear local variable names.

-classes, -c
: Clears the class structure table and clears all objects.

-regexp, -r
: The arguments are treated as regular expressions as any variables that match will be cleared.

With the exception of `exclusive`, all long options can be used without the dash as well.

**See also:** [who], page 128, [whos], page 128, [exist], page 130.

pack ()                                                                                                  [Function File]
: Consolidate workspace memory in MATLAB.

    This function is provided for compatibility, but does nothing in Octave.

    **See also:** [clear], page 131.

Information about a function or variable such as its location in the file system can also be acquired from within Octave. This is usually only useful during development of programs, and not within a program.

type name ...                                                                                            [Command]
type -q name ...                                                                                         [Command]
text = type ("name", ...)                                                                                [Function File]
: Display the contents of name which may be a file, function (m-file), variable, operator, or keyword.

    type normally prepends a header line describing the category of name such as function or variable; The '-q' option suppresses this behavior.

    If no output variable is used the contents are displayed on screen. Otherwise, a cell array of strings is returned, where each element corresponds to the contents of each requested function.

which name ...                                                                                           [Command]
: Display the type of each name.

    If name is defined from a function file, the full name of the file is also displayed.

    **See also:** [help], page 20, [lookfor], page 21.

what                                                                                                     [Command]
what dir                                                                                                 [Command]
w = what (dir)                                                                                           [Function File]
: List the Octave specific files in directory dir.

If *dir* is not specified then the current directory is used.

If a return argument is requested, the files found are returned in the structure *w*. The structure contains the following fields:

| | |
|---|---|
| path | Full path to directory *dir* |
| m | Cell array of m-files |
| mat | Cell array of mat files |
| mex | Cell array of mex files |
| oct | Cell array of oct files |
| mdl | Cell array of mdl files |
| slx | Cell array of slx files |
| p | Cell array of p-files |
| classes | Cell array of class directories ('`@classname/`') |
| packages | Cell array of package directories ('`+pkgname/`') |

Compatibility Note: Octave does not support mdl, slx, and p files; nor does it support package directories. `what` will always return an empty list for these categories.

**See also:** [which], page 132, [ls], page 779, [exist], page 130.

# 8 Expressions

Expressions are the basic building block of statements in Octave. An expression evaluates to a value, which you can print, test, store in a variable, pass to a function, or assign a new value to a variable with an assignment operator.

An expression can serve as a statement on its own. Most other kinds of statements contain one or more expressions which specify data to be operated on. As in other languages, expressions in Octave include variables, array references, constants, and function calls, as well as combinations of these with various operators.

## 8.1 Index Expressions

An *index expression* allows you to reference or extract selected elements of a matrix or vector.

Indices may be scalars, vectors, ranges, or the special operator ':', which may be used to select entire rows or columns.

Vectors are indexed using a single index expression. Matrices (2-D) and higher multi-dimensional arrays are indexed using either one index or $N$ indices where $N$ is the dimension of the array. When using a single index expression to index 2-D or higher data the elements of the array are taken in column-first order (like Fortran).

The output from indexing assumes the dimensions of the index expression. For example:

```
a(2)         # result is a scalar
a(1:2)       # result is a row vector
a([1; 2])    # result is a column vector
```

As a special case, when a colon is used as a single index, the output is a column vector containing all the elements of the vector or matrix. For example:

```
a(:)         # result is a column vector
a(:)'        # result is a row vector
```

The above two code idioms are often used in place of **reshape** when a simple vector, rather than an arbitrarily sized array, is needed.

Given the matrix

```
a = [1, 2; 3, 4]
```

all of the following expressions are equivalent and select the first row of the matrix.

```
a(1, [1, 2])   # row 1, columns 1 and 2
a(1, 1:2)      # row 1, columns in range 1-2
a(1, :)        # row 1, all columns
```

In index expressions the keyword **end** automatically refers to the last entry for a particular dimension. This magic index can also be used in ranges and typically eliminates the needs to call **size** or **length** to gather array bounds before indexing. For example:

```
a = [1, 2, 3, 4];

a(1:end/2)        # first half of a => [1, 2]
a(end + 1) = 5;   # append element
a(end) = [];      # delete element
a(1:2:end)        # odd elements of a => [1, 3]
a(2:2:end)        # even elements of a => [2, 4]
a(end:-1:1)       # reversal of a => [4, 3, 2 , 1]
```

### 8.1.1 Advanced Indexing

An array with 'n' dimensions can be indexed using 'm' indices. More generally, the set of index tuples determining the result is formed by the Cartesian product of the index vectors (or ranges or scalars).

For the ordinary and most common case, m == n, and each index corresponds to its respective dimension. If m < n and every index is less than the size of the array in the $i^{th}$ dimension, m(i) < n(i), then the index expression is padded with trailing singleton dimensions ([ones (m-n, 1)]). If m < n but one of the indices m(i) is outside the size of the current array, then the last n-m+1 dimensions are folded into a single dimension with an extent equal to the product of extents of the original dimensions. This is easiest to understand with an example.

```
a = reshape (1:8, 2, 2, 2)   # Create 3-D array
a =

ans(:,:,1) =

   1   3
   2   4

ans(:,:,2) =

   5   7
   6   8

a(2,1,2);    # Case (m == n): ans = 6
a(2,1);      # Case (m < n), idx within array:
             # equivalent to a(2,1,1), ans = 2
a(2,4);      # Case (m < n), idx outside array:
             # Dimension 2 & 3 folded into new dimension of size 2x2 = 4
             # Select 2nd row, 4th element of [2, 4, 6, 8], ans = 8
```

One advanced use of indexing is to create arrays filled with a single value. This can be done by using an index of ones on a scalar value. The result is an object with the dimensions of the index expression and every element equal to the original scalar. For example, the following statements

```
a = 13;
a(ones (1, 4))
```

produce a vector whose four elements are all equal to 13.

Chapter 8: Expressions 137

Similarly, by indexing a scalar with two vectors of ones it is possible to create a matrix. The following statements

```
a = 13;
a(ones (1, 2), ones (1, 3))
```

create a 2x3 matrix with all elements equal to 13.

The last example could also be written as

```
13(ones (2, 3))
```

It is more efficient to use indexing rather than the code construction `scalar * ones (N, M, ...)` because it avoids the unnecessary multiplication operation. Moreover, multiplication may not be defined for the object to be replicated whereas indexing an array is always defined. The following code shows how to create a 2x3 cell array from a base unit which is not itself a scalar.

```
{"Hello"}(ones (2, 3))
```

It should be, noted that `ones (1, n)` (a row vector of ones) results in a range (with zero increment). A range is stored internally as a starting value, increment, end value, and total number of values; hence, it is more efficient for storage than a vector or matrix of ones whenever the number of elements is greater than 4. In particular, when 'r' is a row vector, the expressions

```
r(ones (1, n), :)
r(ones (n, 1), :)
```

will produce identical results, but the first one will be significantly faster, at least for 'r' and 'n' large enough. In the first case the index is held in compressed form as a range which allows Octave to choose a more efficient algorithm to handle the expression.

A general recommendation, for a user unaware of these subtleties, is to use the function **repmat** for replicating smaller arrays into bigger ones.

A second use of indexing is to speed up code. Indexing is a fast operation and judicious use of it can reduce the requirement for looping over individual array elements which is a slow operation.

Consider the following example which creates a 10-element row vector $a$ containing the values $a_i = \sqrt{i}$.

```
for i = 1:10
  a(i) = sqrt (i);
endfor
```

It is quite inefficient to create a vector using a loop like this. In this case, it would have been much more efficient to use the expression

```
a = sqrt (1:10);
```

which avoids the loop entirely.

In cases where a loop cannot be avoided, or a number of values must be combined to form a larger matrix, it is generally faster to set the size of the matrix first (pre-allocate storage), and then insert elements using indexing commands. For example, given a matrix a,

```
[nr, nc] = size (a);
x = zeros (nr, n * nc);
for i = 1:n
  x(:,(i-1)*nc+1:i*nc) = a;
endfor
```

is considerably faster than

```
x = a;
for i = 1:n-1
  x = [x, a];
endfor
```

because Octave does not have to repeatedly resize the intermediate result.

*ind* = sub2ind (*dims*, *i*, *j*)                                               [Function File]
*ind* = sub2ind (*dims*, *s1*, *s2*, ..., *sN*)                               [Function File]

    Convert subscripts to a linear index.

    The following example shows how to convert the two-dimensional index (2,3) of a 3-by-3 matrix to a linear index. The matrix is linearly indexed moving from one column to next, filling up all rows in each column.

```
linear_index = sub2ind ([3, 3], 2, 3)
    ⇒ 8
```

    **See also:** [ind2sub], page 138.

[*s1*, *s2*, ..., *sN*] = ind2sub (*dims*, *ind*)                                 [Function File]

    Convert a linear index to subscripts.

    The following example shows how to convert the linear index 8 in a 3-by-3 matrix into a subscript. The matrix is linearly indexed moving from one column to next, filling up all rows in each column.

```
[r, c] = ind2sub ([3, 3], 8)
    ⇒ r = 2
    ⇒ c = 3
```

    **See also:** [sub2ind], page 138.

isindex (*ind*)                                                                                    [Built-in Function]
isindex (*ind*, *n*)                                                                     [Built-in Function]

    Return true if *ind* is a valid index.

    Valid indices are either positive integers (although possibly of real data type), or logical arrays.

    If present, *n* specifies the maximum extent of the dimension to be indexed. When possible the internal result is cached so that subsequent indexing using *ind* will not perform the check again.

    Implementation Note: Strings are first converted to double values before the checks for valid indices are made. Unless a string contains the NULL character "\0", it will always be a valid index.

Chapter 8: Expressions                                                                 139

*val* = allow_noninteger_range_as_index ()                          [Built-in Function]
*old_val* = allow_noninteger_range_as_index (*new_val*)             [Built-in Function]
allow_noninteger_range_as_index (*new_val*, "*local*")              [Built-in Function]

> Query or set the internal variable that controls whether non-integer ranges are allowed as indices.
>
> This might be useful for MATLAB compatibility; however, it is still not entirely compatible because MATLAB treats the range expression differently in different contexts.
>
> When called from inside a function with the "local" option, the variable is changed locally for the function and any subroutines it calls. The original variable value is restored when exiting the function.

## 8.2 Calling Functions

A *function* is a name for a particular calculation. Because it has a name, you can ask for it by name at any point in the program. For example, the function `sqrt` computes the square root of a number.

A fixed set of functions are *built-in*, which means they are available in every Octave program. The `sqrt` function is one of these. In addition, you can define your own functions. See Chapter 11 [Functions and Scripts], page 171, for information about how to do this.

The way to use a function is with a *function call* expression, which consists of the function name followed by a list of *arguments* in parentheses. The arguments are expressions which give the raw materials for the calculation that the function will do. When there is more than one argument, they are separated by commas. If there are no arguments, you can omit the parentheses, but it is a good idea to include them anyway, to clearly indicate that a function call was intended. Here are some examples:

```
sqrt (x^2 + y^2)      # One argument
ones (n, m)           # Two arguments
rand ()               # No arguments
```

Each function expects a particular number of arguments. For example, the `sqrt` function must be called with a single argument, the number to take the square root of:

```
sqrt (argument)
```

Some of the built-in functions take a variable number of arguments, depending on the particular usage, and their behavior is different depending on the number of arguments supplied.

Like every other expression, the function call has a value, which is computed by the function based on the arguments you give it. In this example, the value of `sqrt (argument)` is the square root of the argument. A function can also have side effects, such as assigning the values of certain variables or doing input or output operations.

Unlike most languages, functions in Octave may return multiple values. For example, the following statement

```
[u, s, v] = svd (a)
```

computes the singular value decomposition of the matrix `a` and assigns the three result matrices to `u`, `s`, and `v`.

The left side of a multiple assignment expression is itself a list of expressions, and is allowed to be a list of variable names or index expressions. See also Section 8.1 [Index Expressions], page 135, and Section 8.6 [Assignment Ops], page 149.

## 8.2.1 Call by Value

In Octave, unlike Fortran, function arguments are passed by value, which means that each argument in a function call is evaluated and assigned to a temporary location in memory before being passed to the function. There is currently no way to specify that a function parameter should be passed by reference instead of by value. This means that it is impossible to directly alter the value of a function parameter in the calling function. It can only change the local copy within the function body. For example, the function

```
function f (x, n)
  while (n-- > 0)
    disp (x);
  endwhile
endfunction
```

displays the value of the first argument $n$ times. In this function, the variable $n$ is used as a temporary variable without having to worry that its value might also change in the calling function. Call by value is also useful because it is always possible to pass constants for any function parameter without first having to determine that the function will not attempt to modify the parameter.

The caller may use a variable as the expression for the argument, but the called function does not know this: it only knows what value the argument had. For example, given a function called as

```
foo = "bar";
fcn (foo)
```

you should not think of the argument as being "the variable `foo`." Instead, think of the argument as the string value, `"bar"`.

Even though Octave uses pass-by-value semantics for function arguments, values are not copied unnecessarily. For example,

```
x = rand (1000);
f (x);
```

does not actually force two 1000 by 1000 element matrices to exist *unless* the function `f` modifies the value of its argument. Then Octave must create a copy to avoid changing the value outside the scope of the function `f`, or attempting (and probably failing!) to modify the value of a constant or the value of a temporary result.

# Chapter 8: Expressions

## 8.2.2 Recursion

With some restrictions[1], recursive function calls are allowed. A *recursive function* is one which calls itself, either directly or indirectly. For example, here is an inefficient[2] way to compute the factorial of a given integer:

```
function retval = fact (n)
  if (n > 0)
    retval = n * fact (n-1);
  else
    retval = 1;
  endif
endfunction
```

This function is recursive because it calls itself directly. It eventually terminates because each time it calls itself, it uses an argument that is one less than was used for the previous call. Once the argument is no longer greater than zero, it does not call itself, and the recursion ends.

The built-in variable `max_recursion_depth` specifies a limit to the recursion depth and prevents Octave from recursing infinitely.

*val* = **max_recursion_depth** ()  [Built-in Function]
*old_val* = **max_recursion_depth** (*new_val*)  [Built-in Function]
**max_recursion_depth** (*new_val*, "*local*")  [Built-in Function]

> Query or set the internal limit on the number of times a function may be called recursively.
>
> If the limit is exceeded, an error message is printed and control returns to the top level.
>
> When called from inside a function with the `"local"` option, the variable is changed locally for the function and any subroutines it calls. The original variable value is restored when exiting the function.

## 8.3 Arithmetic Operators

The following arithmetic operators are available, and work on scalars and matrices. The element-by-element operators and functions broadcast (see Section 19.2 [Broadcasting], page 489).

*x* + *y*    Addition. If both operands are matrices, the number of rows and columns must both agree, or they must be broadcastable to the same shape.

*x* .+ *y*   Element-by-element addition. This operator is equivalent to +.

*x* - *y*    Subtraction. If both operands are matrices, the number of rows and columns of both must agree, or they must be broadcastable to the same shape.

---

[1] Some of Octave's functions are implemented in terms of functions that cannot be called recursively. For example, the ODE solver `lsode` is ultimately implemented in a Fortran subroutine that cannot be called recursively, so `lsode` should not be called either directly or indirectly from within the user-supplied function that `lsode` requires. Doing so will result in an error.

[2] It would be much better to use `prod (1:n)`, or `gamma (n+1)` instead, after first checking to ensure that the value `n` is actually a positive integer.

| | |
|---|---|
| x .- y | Element-by-element subtraction. This operator is equivalent to -. |
| x * y | Matrix multiplication. The number of columns of x must agree with the number of rows of y, or they must be broadcastable to the same shape. |
| x .* y | Element-by-element multiplication. If both operands are matrices, the number of rows and columns must both agree, or they must be broadcastable to the same shape. |
| x / y | Right division. This is conceptually equivalent to the expression<br>    `(inverse (y') * x')'`<br>but it is computed without forming the inverse of y'.<br>If the system is not square, or if the coefficient matrix is singular, a minimum norm solution is computed. |
| x ./ y | Element-by-element right division. |
| x \ y | Left division. This is conceptually equivalent to the expression<br>    `inverse (x) * y`<br>but it is computed without forming the inverse of x.<br>If the system is not square, or if the coefficient matrix is singular, a minimum norm solution is computed. |
| x .\ y | Element-by-element left division. Each element of y is divided by each corresponding element of x. |
| x ^ y<br>x ** y | Power operator. If x and y are both scalars, this operator returns x raised to the power y. If x is a scalar and y is a square matrix, the result is computed using an eigenvalue expansion. If x is a square matrix, the result is computed by repeated multiplication if y is an integer, and by an eigenvalue expansion if y is not an integer. An error results if both x and y are matrices.<br>The implementation of this operator needs to be improved. |
| x .^ y<br>x .** y | Element-by-element power operator. If both operands are matrices, the number of rows and columns must both agree, or they must be broadcastable to the same shape. If several complex results are possible, the one with smallest non-negative argument (angle) is taken. This rule may return a complex root even when a real root is also possible. Use `realpow`, `realsqrt`, `cbrt`, or `nthroot` if a real result is preferred. |
| -x | Negation. |
| +x | Unary plus. This operator has no effect on the operand. |
| x' | Complex conjugate transpose. For real arguments, this operator is the same as the transpose operator. For complex arguments, this operator is equivalent to the expression<br>    `conj (x.')` |
| x.' | Transpose. |

Chapter 8: Expressions

Note that because Octave's element-by-element operators begin with a '.', there is a possible ambiguity for statements like

    1./m

because the period could be interpreted either as part of the constant or as part of the operator. To resolve this conflict, Octave treats the expression as if you had typed

    (1) ./ m

and not

    (1.) / m

Although this is inconsistent with the normal behavior of Octave's lexer, which usually prefers to break the input into tokens by preferring the longest possible match at any given point, it is more useful in this case.

**ctranspose** (x)                                              [Built-in Function]

    Return the complex conjugate transpose of x.

    This function and x' are equivalent.

    **See also:** [transpose], page 144.

**ldivide** (x, y)                                              [Built-in Function]

    Return the element-by-element left division of x and y.

    This function and x .\ y are equivalent.

    **See also:** [rdivide], page 144, [mldivide], page 143, [times], page 144, [plus], page 144.

**minus** (x, y)                                                [Built-in Function]

    This function and x - y are equivalent.

    **See also:** [plus], page 144, [uminus], page 144.

**mldivide** (x, y)                                             [Built-in Function]

    Return the matrix left division of x and y.

    This function and x \ y are equivalent.

    **See also:** [mrdivide], page 143, [ldivide], page 143, [rdivide], page 144.

**mpower** (x, y)                                               [Built-in Function]

    Return the matrix power operation of x raised to the y power.

    This function and x ^ y are equivalent.

    **See also:** [power], page 144, [mtimes], page 143, [plus], page 144, [minus], page 143.

**mrdivide** (x, y)                                             [Built-in Function]

    Return the matrix right division of x and y.

    This function and x / y are equivalent.

    **See also:** [mldivide], page 143, [rdivide], page 144, [plus], page 144, [minus], page 143.

**mtimes** (x, y)                                               [Built-in Function]
**mtimes** (x1, x2, ...)                                        [Built-in Function]

    Return the matrix multiplication product of inputs.

    This function and x * y are equivalent. If more arguments are given, the multiplication is applied cumulatively from left to right:

$$(\ldots((\texttt{x1} * \texttt{x2}) * \texttt{x3}) * \ldots)$$

At least one argument is required.

**See also:** [times], page 144, [plus], page 144, [minus], page 143, [rdivide], page 144, [mrdivide], page 143, [mldivide], page 143, [mpower], page 143.

`plus (x, y)` [Built-in Function]
`plus (x1, x2, ...)` [Built-in Function]

This function and `x + y` are equivalent.

If more arguments are given, the summation is applied cumulatively from left to right:

$$(\ldots((\texttt{x1} + \texttt{x2}) + \texttt{x3}) + \ldots)$$

At least one argument is required.

**See also:** [minus], page 143, [uplus], page 145.

`power (x, y)` [Built-in Function]

Return the element-by-element operation of x raised to the y power.

This function and `x .^ y` are equivalent.

If several complex results are possible, returns the one with smallest non-negative argument (angle). Use `realpow`, `realsqrt`, `cbrt`, or `nthroot` if a real result is preferred.

**See also:** [mpower], page 143, [realpow], page 434, [realsqrt], page 434, [cbrt], page 434, [nthroot], page 434.

`rdivide (x, y)` [Built-in Function]

Return the element-by-element right division of x and y.

This function and `x ./ y` are equivalent.

**See also:** [ldivide], page 143, [mrdivide], page 143, [times], page 144, [plus], page 144.

`times (x, y)` [Built-in Function]
`times (x1, x2, ...)` [Built-in Function]

Return the element-by-element multiplication product of inputs.

This function and `x .* y` are equivalent. If more arguments are given, the multiplication is applied cumulatively from left to right:

$$(\ldots((\texttt{x1} .* \texttt{x2}) .* \texttt{x3}) .* \ldots)$$

At least one argument is required.

**See also:** [mtimes], page 143, [rdivide], page 144.

`transpose (x)` [Built-in Function]

Return the transpose of x.

This function and `x.'` are equivalent.

**See also:** [ctranspose], page 143.

`uminus (x)` [Built-in Function]

This function and `- x` are equivalent.

**See also:** [uplus], page 145, [minus], page 143.

Chapter 8: Expressions                                                    145

**uplus** (*x*)                                              [Built-in Function]
> This function and + x are equivalent.
>
> **See also:** [uminus], page 144, [plus], page 144, [minus], page 143.

## 8.4 Comparison Operators

*Comparison operators* compare numeric values for relationships such as equality. They are written using *relational operators*.

All of Octave's comparison operators return a value of 1 if the comparison is true, or 0 if it is false. For matrix values, they all work on an element-by-element basis. Broadcasting rules apply. See Section 19.2 [Broadcasting], page 489. For example:

```
[1, 2; 3, 4] == [1, 3; 2, 4]
    ⇒   1  0
        0  1
```

According to broadcasting rules, if one operand is a scalar and the other is a matrix, the scalar is compared to each element of the matrix in turn, and the result is the same size as the matrix.

*x* < *y*        True if *x* is less than *y*.

*x* <= *y*       True if *x* is less than or equal to *y*.

*x* == *y*       True if *x* is equal to *y*.

*x* >= *y*       True if *x* is greater than or equal to *y*.

*x* > *y*        True if *x* is greater than *y*.

*x* != *y*
*x* ~= *y*       True if *x* is not equal to *y*.

For complex numbers, the following ordering is defined: *z1* < *z2* if and only if

```
abs (z1) < abs (z2)
|| (abs (z1) == abs (z2) && arg (z1) < arg (z2))
```

This is consistent with the ordering used by *max*, *min* and *sort*, but is not consistent with MATLAB, which only compares the real parts.

String comparisons may also be performed with the **strcmp** function, not with the comparison operators listed above. See Chapter 5 [Strings], page 67.

**eq** (*x*, *y*)                                            [Built-in Function]
> Return true if the two inputs are equal.
>
> This function is equivalent to x == y.
>
> **See also:** [ne], page 146, [isequal], page 146, [le], page 146, [ge], page 145, [gt], page 146, [ne], page 146, [lt], page 146.

**ge** (*x*, *y*)                                            [Built-in Function]
> This function is equivalent to x >= y.
>
> **See also:** [le], page 146, [eq], page 145, [gt], page 146, [ne], page 146, [lt], page 146.

`gt (x, y)`                                                            [Built-in Function]

    This function is equivalent to `x > y`.

    **See also:** [le], page 146, [eq], page 145, [ge], page 145, [ne], page 146, [lt], page 146.

`isequal (x1, x2, ...)`                                                [Function File]

    Return true if all of *x1*, *x2*, ... are equal.

    **See also:** [isequaln], page 146.

`isequaln (x1, x2, ...)`                                               [Function File]

    Return true if all of *x1*, *x2*, ... are equal under the additional assumption that NaN == NaN (no comparison of NaN placeholders in dataset).

    **See also:** [isequal], page 146.

`le (x, y)`                                                            [Built-in Function]

    This function is equivalent to `x <= y`.

    **See also:** [eq], page 145, [ge], page 145, [gt], page 146, [ne], page 146, [lt], page 146.

`lt (x, y)`                                                            [Built-in Function]

    This function is equivalent to `x < y`.

    **See also:** [le], page 146, [eq], page 145, [ge], page 145, [gt], page 146, [ne], page 146.

`ne (x, y)`                                                            [Built-in Function]

    Return true if the two inputs are not equal.

    This function is equivalent to `x != y`.

    **See also:** [eq], page 145, [isequal], page 146, [le], page 146, [ge], page 145, [lt], page 146.

## 8.5 Boolean Expressions

### 8.5.1 Element-by-element Boolean Operators

An *element-by-element boolean expression* is a combination of comparison expressions using the boolean operators "or" ('|'), "and" ('&'), and "not" ('!'), along with parentheses to control nesting. The truth of the boolean expression is computed by combining the truth values of the corresponding elements of the component expressions. A value is considered to be false if it is zero, and true otherwise.

Element-by-element boolean expressions can be used wherever comparison expressions can be used. They can be used in `if` and `while` statements. However, a matrix value used as the condition in an `if` or `while` statement is only true if *all* of its elements are nonzero.

Like comparison operations, each element of an element-by-element boolean expression also has a numeric value (1 if true, 0 if false) that comes into play if the result of the boolean expression is stored in a variable, or used in arithmetic.

Here are descriptions of the three element-by-element boolean operators.

`boolean1 & boolean2`

    Elements of the result are true if both corresponding elements of *boolean1* and *boolean2* are true.

Chapter 8: Expressions                                                                147

*boolean1* | *boolean2*
> Elements of the result are true if either of the corresponding elements of *boolean1* or *boolean2* is true.

! *boolean*
~ *boolean*
> Each element of the result is true if the corresponding element of *boolean* is false.

These operators work on an element-by-element basis. For example, the expression

    [1, 0; 0, 1] & [1, 0; 2, 3]

returns a two by two identity matrix.

For the binary operators, broadcasting rules apply. See Section 19.2 [Broadcasting], page 489. In particular, if one of the operands is a scalar and the other a matrix, the operator is applied to the scalar and each element of the matrix.

For the binary element-by-element boolean operators, both subexpressions *boolean1* and *boolean2* are evaluated before computing the result. This can make a difference when the expressions have side effects. For example, in the expression

    a & b++

the value of the variable *b* is incremented even if the variable *a* is zero.

This behavior is necessary for the boolean operators to work as described for matrix-valued operands.

z = and (*x*, *y*)                                                       [Built-in Function]
z = and (*x1*, *x2*, ...)                                                [Built-in Function]
> Return the logical AND of *x* and *y*.
>
> This function is equivalent to the operator syntax x & y. If more than two arguments are given, the logical AND is applied cumulatively from left to right:
>
>     (...((x1 & x2) & x3) & ...)
>
> At least one argument is required.
>
> **See also:** [or], page 147, [not], page 147, [xor], page 403.

z = not (*x*)                                                            [Built-in Function]
> Return the logical NOT of *x*.
>
> This function is equivalent to the operator syntax ! x.
>
> **See also:** [and], page 147, [or], page 147, [xor], page 403.

z = or (*x*, *y*)                                                        [Built-in Function]
z = or (*x1*, *x2*, ...)                                                 [Built-in Function]
> Return the logical OR of *x* and *y*.
>
> This function is equivalent to the operator syntax x | y. If more than two arguments are given, the logical OR is applied cumulatively from left to right:
>
>     (...((x1 | x2) | x3) | ...)
>
> At least one argument is required.
>
> **See also:** [and], page 147, [not], page 147, [xor], page 403.

## 8.5.2 Short-circuit Boolean Operators

Combined with the implicit conversion to scalar values in `if` and `while` conditions, Octave's element-by-element boolean operators are often sufficient for performing most logical operations. However, it is sometimes desirable to stop evaluating a boolean expression as soon as the overall truth value can be determined. Octave's *short-circuit* boolean operators work this way.

*boolean1* `&&` *boolean2*

> The expression *boolean1* is evaluated and converted to a scalar using the equivalent of the operation `all (`*boolean1*`(:))`. If it is false, the result of the overall expression is 0. If it is true, the expression *boolean2* is evaluated and converted to a scalar using the equivalent of the operation `all (`*boolean1*`(:))`. If it is true, the result of the overall expression is 1. Otherwise, the result of the overall expression is 0.
>
> **Warning:** there is one exception to the rule of evaluating `all (`*boolean1*`(:))`, which is when `boolean1` is the empty matrix. The truth value of an empty matrix is always `false` so `[] && true` evaluates to `false` even though `all ([])` is `true`.

*boolean1* `||` *boolean2*

> The expression *boolean1* is evaluated and converted to a scalar using the equivalent of the operation `all (`*boolean1*`(:))`. If it is true, the result of the overall expression is 1. If it is false, the expression *boolean2* is evaluated and converted to a scalar using the equivalent of the operation `all (`*boolean1*`(:))`. If it is true, the result of the overall expression is 1. Otherwise, the result of the overall expression is 0.
>
> **Warning:** the truth value of an empty matrix is always `false`, see the previous list item for details.

The fact that both operands may not be evaluated before determining the overall truth value of the expression can be important. For example, in the expression

```
a && b++
```

the value of the variable *b* is only incremented if the variable *a* is nonzero.

This can be used to write somewhat more concise code. For example, it is possible write

```
function f (a, b, c)
  if (nargin > 2 && ischar (c))
    ...
```

instead of having to use two `if` statements to avoid attempting to evaluate an argument that doesn't exist. For example, without the short-circuit feature, it would be necessary to write

```
function f (a, b, c)
  if (nargin > 2)
    if (ischar (c))
      ...
```

Writing

Chapter 8: Expressions 149

```
function f (a, b, c)
  if (nargin > 2 & ischar (c))
    ...
```

would result in an error if `f` were called with one or two arguments because Octave would be forced to try to evaluate both of the operands for the operator '&'.

MATLAB has special behavior that allows the operators '&' and '|' to short-circuit when used in the truth expression for `if` and `while` statements. Octave also behaves the same way by default, though the use of the '&' and '|' operators in this way is strongly discouraged. Instead, you should use the '&&' and '||' operators that always have short-circuit behavior.

Finally, the ternary operator (?:) is not supported in Octave. If short-circuiting is not important, it can be replaced by the `ifelse` function.

**merge** (*mask*, *tval*, *fval*)  [Built-in Function]
**ifelse** (*mask*, *tval*, *fval*)  [Built-in Function]

Merge elements of *true_val* and *false_val*, depending on the value of *mask*.

If *mask* is a logical scalar, the other two arguments can be arbitrary values. Otherwise, *mask* must be a logical array, and *tval*, *fval* should be arrays of matching class, or cell arrays. In the scalar mask case, *tval* is returned if *mask* is true, otherwise *fval* is returned.

In the array mask case, both *tval* and *fval* must be either scalars or arrays with dimensions equal to *mask*. The result is constructed as follows:

```
result(mask) = tval(mask);
result(! mask) = fval(! mask);
```

*mask* can also be arbitrary numeric type, in which case it is first converted to logical.

**See also:** [logical], page 60, [diff], page 404.

## 8.6 Assignment Expressions

An *assignment* is an expression that stores a new value into a variable. For example, the following expression assigns the value 1 to the variable `z`:

```
z = 1
```

After this expression is executed, the variable `z` has the value 1. Whatever old value `z` had before the assignment is forgotten. The '=' sign is called an *assignment operator*.

Assignments can store string values also. For example, the following expression would store the value "this food is good" in the variable `message`:

```
thing = "food"
predicate = "good"
message = [ "this " , thing , " is " , predicate ]
```

(This also illustrates concatenation of strings.)

Most operators (addition, concatenation, and so on) have no effect except to compute a value. If you ignore the value, you might as well not use the operator. An assignment operator is different. It does produce a value, but even if you ignore the value, the assignment still makes itself felt through the alteration of the variable. We call this a *side effect*.

The left-hand operand of an assignment need not be a variable (see Chapter 7 [Variables], page 123). It can also be an element of a matrix (see Section 8.1 [Index Expressions], page 135) or a list of return values (see Section 8.2 [Calling Functions], page 139). These are all called *lvalues*, which means they can appear on the left-hand side of an assignment operator. The right-hand operand may be any expression. It produces the new value which the assignment stores in the specified variable, matrix element, or list of return values.

It is important to note that variables do *not* have permanent types. The type of a variable is simply the type of whatever value it happens to hold at the moment. In the following program fragment, the variable `foo` has a numeric value at first, and a string value later on:

```
octave:13> foo = 1
foo = 1
octave:13> foo = "bar"
foo = bar
```

When the second assignment gives `foo` a string value, the fact that it previously had a numeric value is forgotten.

Assignment of a scalar to an indexed matrix sets all of the elements that are referenced by the indices to the scalar value. For example, if `a` is a matrix with at least two columns,

```
a(:, 2) = 5
```

sets all the elements in the second column of `a` to 5.

Assigning an empty matrix '`[]`' works in most cases to allow you to delete rows or columns of matrices and vectors. See Section 4.1.1 [Empty Matrices], page 51. For example, given a 4 by 5 matrix A, the assignment

```
A (3, :) = []
```

deletes the third row of A, and the assignment

```
A (:, 1:2:5) = []
```

deletes the first, third, and fifth columns.

An assignment is an expression, so it has a value. Thus, `z = 1` as an expression has the value 1. One consequence of this is that you can write multiple assignments together:

```
x = y = z = 0
```

stores the value 0 in all three variables. It does this because the value of `z = 0`, which is 0, is stored into `y`, and then the value of `y = z = 0`, which is 0, is stored into `x`.

This is also true of assignments to lists of values, so the following is a valid expression

```
[a, b, c] = [u, s, v] = svd (a)
```

that is exactly equivalent to

```
[u, s, v] = svd (a)
a = u
b = s
c = v
```

In expressions like this, the number of values in each part of the expression need not match. For example, the expression

```
[a, b] = [u, s, v] = svd (a)
```

is equivalent to

# Chapter 8: Expressions

```
[u, s, v] = svd (a)
a = u
b = s
```

The number of values on the left side of the expression can, however, not exceed the number of values on the right side. For example, the following will produce an error.

```
[a, b, c, d] = [u, s, v] = svd (a);
⊣ error: element number 4 undefined in return list
```

The symbol ~ may be used as a placeholder in the list of lvalues, indicating that the corresponding return value should be ignored and not stored anywhere:

```
[~, s, v] = svd (a);
```

This is cleaner and more memory efficient than using a dummy variable. The `nargout` value for the right-hand side expression is not affected. If the assignment is used as an expression, the return value is a comma-separated list with the ignored values dropped.

A very common programming pattern is to increment an existing variable with a given value, like this

```
a = a + 2;
```

This can be written in a clearer and more condensed form using the += operator

```
a += 2;
```

Similar operators also exist for subtraction (-=), multiplication (*=), and division (/=). An expression of the form

*expr1 op= expr2*

is evaluated as

*expr1 = (expr1) op (expr2)*

where *op* can be either +, -, *, or /, as long as *expr2* is a simple expression with no side effects. If *expr2* also contains an assignment operator, then this expression is evaluated as

*temp = expr2*
*expr1 = (expr1) op temp*

where *temp* is a placeholder temporary value storing the computed result of evaluating *expr2*. So, the expression

```
a *= b+1
```

is evaluated as

```
a = a * (b+1)
```

and *not*

```
a = a * b + 1
```

You can use an assignment anywhere an expression is called for. For example, it is valid to write x != (y = 1) to set y to 1 and then test whether x equals 1. But this style tends to make programs hard to read. Except in a one-shot program, you should rewrite it to get rid of such nesting of assignments. This is never very hard.

## 8.7 Increment Operators

*Increment operators* increase or decrease the value of a variable by 1. The operator to increment a variable is written as '++'. It may be used to increment a variable either before or after taking its value.

For example, to pre-increment the variable x, you would write ++x. This would add one to x and then return the new value of x as the result of the expression. It is exactly the same as the expression x = x + 1.

To post-increment a variable x, you would write x++. This adds one to the variable x, but returns the value that x had prior to incrementing it. For example, if x is equal to 2, the result of the expression x++ is 2, and the new value of x is 3.

For matrix and vector arguments, the increment and decrement operators work on each element of the operand.

Here is a list of all the increment and decrement expressions.

++x    This expression increments the variable x. The value of the expression is the *new* value of x. It is equivalent to the expression x = x + 1.

--x    This expression decrements the variable x. The value of the expression is the *new* value of x. It is equivalent to the expression x = x - 1.

x++    This expression causes the variable x to be incremented. The value of the expression is the *old* value of x.

x--    This expression causes the variable x to be decremented. The value of the expression is the *old* value of x.

## 8.8 Operator Precedence

*Operator precedence* determines how operators are grouped, when different operators appear close by in one expression. For example, '*' has higher precedence than '+'. Thus, the expression a + b * c means to multiply b and c, and then add a to the product (i.e., a + (b * c)).

You can overrule the precedence of the operators by using parentheses. You can think of the precedence rules as saying where the parentheses are assumed if you do not write parentheses yourself. In fact, it is wise to use parentheses whenever you have an unusual combination of operators, because other people who read the program may not remember what the precedence is in this case. You might forget as well, and then you too could make a mistake. Explicit parentheses will help prevent any such mistake.

When operators of equal precedence are used together, the leftmost operator groups first, except for the assignment operators, which group in the opposite order. Thus, the expression a - b + c groups as (a - b) + c, but the expression a = b = c groups as a = (b = c).

The precedence of prefix unary operators is important when another operator follows the operand. For example, -x^2 means -(x^2), because '-' has lower precedence than '^'.

Here is a table of the operators in Octave, in order of decreasing precedence. Unless noted, all operators group left to right.

# Chapter 8: Expressions

function call and array indexing, cell array indexing, and structure element indexing
> '()' '{}' '.'

postfix increment, and postfix decrement
> '++' '--'
>> These operators group right to left.

transpose and exponentiation
> ''' '.'' '^' '**' '.^' '.**'

unary plus, unary minus, prefix increment, prefix decrement, and logical "not"
> '+' '-' '++' '--' '~' '!'

multiply and divide
> '*' '/' '\' '.\' '.*' './'

add, subtract
> '+' '-'

colon    ':'

relational
> '<' '<=' '==' '>=' '>' '!=' '~='

element-wise "and"
> '&'

element-wise "or"
> '|'

logical "and"
> '&&'

logical "or"
> '||'

assignment
> '=' '+=' '-=' '*=' '/=' '\=' '^=' '.*=' './=' '.\=' '.^=' '|=' '&='
>> These operators group right to left.

# 9 Evaluation

Normally, you evaluate expressions simply by typing them at the Octave prompt, or by asking Octave to interpret commands that you have saved in a file.

Sometimes, you may find it necessary to evaluate an expression that has been computed and stored in a string, which is exactly what the **eval** function lets you do.

**eval** (*try*) [Built-in Function]
**eval** (*try, catch*) [Built-in Function]

Parse the string *try* and evaluate it as if it were an Octave program.

If execution fails, evaluate the optional string *catch*.

The string *try* is evaluated in the current context, so any results remain available after **eval** returns.

The following example creates the variable $A$ with the approximate value of 3.1416 in the current workspace.

```
eval ("A = acos(-1);");
```

If an error occurs during the evaluation of *try* then the *catch* string is evaluated, as the following example shows:

```
eval ('error ("This is a bad example");',
      'printf ("This error occurred:\n%s\n", lasterr ());');
     ⊣ This error occurred:
        This is a bad example
```

Programming Note: if you are only using **eval** as an error-capturing mechanism, rather than for the execution of arbitrary code strings, Consider using try/catch blocks or unwind_protect/unwind_protect_cleanup blocks instead. These techniques have higher performance and don't introduce the security considerations that the evaluation of arbitrary code does.

**See also:** [evalin], page 158.

## 9.1 Calling a Function by its Name

The **feval** function allows you to call a function from a string containing its name. This is useful when writing a function that needs to call user-supplied functions. The **feval** function takes the name of the function to call as its first argument, and the remaining arguments are given to the function.

The following example is a simple-minded function using **feval** that finds the root of a user-supplied function of one variable using Newton's method.

```
      function result = newtroot (fname, x)

      # usage: newtroot (fname, x)
      #
      #   fname : a string naming a function f(x).
      #   x     : initial guess

        delta = tol = sqrt (eps);
```

```
  maxit = 200;
  fx = feval (fname, x);
  for i = 1:maxit
    if (abs (fx) < tol)
      result = x;
      return;
    else
      fx_new = feval (fname, x + delta);
      deriv = (fx_new - fx) / delta;
      x = x - fx / deriv;
      fx = fx_new;
    endif
  endfor

  result = x;

endfunction
```

Note that this is only meant to be an example of calling user-supplied functions and should not be taken too seriously. In addition to using a more robust algorithm, any serious code would check the number and type of all the arguments, ensure that the supplied function really was a function, etc. See Section 4.8 [Predicates for Numeric Objects], page 62, for a list of predicates for numeric objects, and see Section 7.3 [Status of Variables], page 127, for a description of the `exist` function.

`feval (name, ...)` [Built-in Function]

Evaluate the function named *name*.

Any arguments after the first are passed as inputs to the named function. For example,

```
feval ("acos", -1)
     ⇒ 3.1416
```

calls the function `acos` with the argument '-1'.

The function `feval` can also be used with function handles of any sort (see Section 11.11.1 [Function Handles], page 199). Historically, `feval` was the only way to call user-supplied functions in strings, but function handles are now preferred due to the cleaner syntax they offer. For example,

```
f = @exp;
feval (f, 1)
     ⇒ 2.7183
f (1)
     ⇒ 2.7183
```

are equivalent ways to call the function referred to by *f*. If it cannot be predicted beforehand whether *f* is a function handle, function name in a string, or inline function then `feval` can be used instead.

A similar function `run` exists for calling user script files, that are not necessarily on the user path

# Chapter 9: Evaluation

**run** *script*     [Command]
**run** ("*script*")     [Function File]

    Run *script* in the current workspace.

    Scripts which reside in directories specified in Octave's load path, and which end with the extension '".m"', can be run simply by typing their name. For scripts not located on the load path, use `run`.

    The file name *script* can be a bare, fully qualified, or relative filename and with or without a file extension. If no extension is specified, Octave will first search for a script with the '".m"' extension before falling back to the script name without an extension.

    Implementation Note: If *script* includes a path component, then `run` first changes the working directory to the directory where *script* is found. Next, the script is executed. Finally, `run` returns to the original working directory unless *script* has specifically changed directories.

    **See also:** [path], page 190, [addpath], page 189, [source], page 199.

## 9.2 Evaluation in a Different Context

Before you evaluate an expression you need to substitute the values of the variables used in the expression. These are stored in the symbol table. Whenever the interpreter starts a new function it saves the current symbol table and creates a new one, initializing it with the list of function parameters and a couple of predefined variables such as `nargin`. Expressions inside the function use the new symbol table.

Sometimes you want to write a function so that when you call it, it modifies variables in your own context. This allows you to use a pass-by-name style of function, which is similar to using a pointer in programming languages such as C.

Consider how you might write `save` and `load` as m-files. For example:

```
function create_data
  x = linspace (0, 10, 10);
  y = sin (x);
  save mydata x y
endfunction
```

With `evalin`, you could write `save` as follows:

```
function save (file, name1, name2)
  f = open_save_file (file);
  save_var (f, name1, evalin ("caller", name1));
  save_var (f, name2, evalin ("caller", name2));
endfunction
```

Here, 'caller' is the `create_data` function and `name1` is the string `"x"`, which evaluates simply as the value of x.

You later want to load the values back from `mydata` in a different context:

```
function process_data
  load mydata
  ... do work ...
endfunction
```

With `assignin`, you could write `load` as follows:
```
function load (file)
  f = open_load_file (file);
  [name, val] = load_var (f);
  assignin ("caller", name, val);
  [name, val] = load_var (f);
  assignin ("caller", name, val);
endfunction
```
Here, 'caller' is the `process_data` function.

You can set and use variables at the command prompt using the context 'base' rather than 'caller'.

These functions are rarely used in practice. One example is the `fail` ('code', 'pattern') function which evaluates 'code' in the caller's context and checks that the error message it produces matches the given pattern. Other examples such as `save` and `load` are written in C++ where all Octave variables are in the 'caller' context and `evalin` is not needed.

**evalin** (*context*, *try*)  [Built-in Function]
**evalin** (*context*, *try*, *catch*)  [Built-in Function]

    Like `eval`, except that the expressions are evaluated in the context *context*, which may be either "caller" or "base".

    **See also:** [eval], page 155, [assignin], page 158.

**assignin** (*context*, *varname*, *value*)  [Built-in Function]

    Assign *value* to *varname* in context *context*, which may be either "base" or "caller".

    **See also:** [evalin], page 158.

# 10 Statements

Statements may be a simple constant expression or a complicated list of nested loops and conditional statements.

*Control statements* such as if, while, and so on control the flow of execution in Octave programs. All the control statements start with special keywords such as if and while, to distinguish them from simple expressions. Many control statements contain other statements; for example, the if statement contains another statement which may or may not be executed.

Each control statement has a corresponding *end* statement that marks the end of the control statement. For example, the keyword endif marks the end of an if statement, and endwhile marks the end of a while statement. You can use the keyword end anywhere a more specific end keyword is expected, but using the more specific keywords is preferred because if you use them, Octave is able to provide better diagnostics for mismatched or missing end tokens.

The list of statements contained between keywords like if or while and the corresponding end statement is called the *body* of a control statement.

## 10.1 The if Statement

The if statement is Octave's decision-making statement. There are three basic forms of an if statement. In its simplest form, it looks like this:

```
if (condition)
  then-body
endif
```

*condition* is an expression that controls what the rest of the statement will do. The *then-body* is executed only if *condition* is true.

The condition in an if statement is considered true if its value is nonzero, and false if its value is zero. If the value of the conditional expression in an if statement is a vector or a matrix, it is considered true only if it is non-empty and *all* of the elements are nonzero.

The second form of an if statement looks like this:

```
if (condition)
  then-body
else
  else-body
endif
```

If *condition* is true, *then-body* is executed; otherwise, *else-body* is executed.

Here is an example:

```
if (rem (x, 2) == 0)
  printf ("x is even\n");
else
  printf ("x is odd\n");
endif
```

In this example, if the expression `rem (x, 2) == 0` is true (that is, the value of `x` is divisible by 2), then the first `printf` statement is evaluated, otherwise the second `printf` statement is evaluated.

The third and most general form of the `if` statement allows multiple decisions to be combined in a single statement. It looks like this:

```
if (condition)
  then-body
elseif (condition)
  elseif-body
else
  else-body
endif
```

Any number of `elseif` clauses may appear. Each condition is tested in turn, and if one is found to be true, its corresponding *body* is executed. If none of the conditions are true and the `else` clause is present, its body is executed. Only one `else` clause may appear, and it must be the last part of the statement.

In the following example, if the first condition is true (that is, the value of `x` is divisible by 2), then the first `printf` statement is executed. If it is false, then the second condition is tested, and if it is true (that is, the value of `x` is divisible by 3), then the second `printf` statement is executed. Otherwise, the third `printf` statement is performed.

```
if (rem (x, 2) == 0)
  printf ("x is even\n");
elseif (rem (x, 3) == 0)
  printf ("x is odd and divisible by 3\n");
else
  printf ("x is odd\n");
endif
```

Note that the `elseif` keyword must not be spelled `else if`, as is allowed in Fortran. If it is, the space between the `else` and `if` will tell Octave to treat this as a new `if` statement within another `if` statement's `else` clause. For example, if you write

```
if (c1)
  body-1
else if (c2)
  body-2
endif
```

Octave will expect additional input to complete the first `if` statement. If you are using Octave interactively, it will continue to prompt you for additional input. If Octave is reading this input from a file, it may complain about missing or mismatched `end` statements, or, if you have not used the more specific `end` statements (`endif`, `endfor`, etc.), it may simply produce incorrect results, without producing any warning messages.

It is much easier to see the error if we rewrite the statements above like this,

```
if (c1)
  body-1
else
  if (c2)
    body-2
  endif
```

using the indentation to show how Octave groups the statements. See Chapter 11 [Functions and Scripts], page 171.

## 10.2 The switch Statement

It is very common to take different actions depending on the value of one variable. This is possible using the `if` statement in the following way

```
if (X == 1)
  do_something ();
elseif (X == 2)
  do_something_else ();
else
  do_something_completely_different ();
endif
```

This kind of code can however be very cumbersome to both write and maintain. To overcome this problem Octave supports the `switch` statement. Using this statement, the above example becomes

```
switch (X)
  case 1
    do_something ();
  case 2
    do_something_else ();
  otherwise
    do_something_completely_different ();
endswitch
```

This code makes the repetitive structure of the problem more explicit, making the code easier to read, and hence maintain. Also, if the variable X should change its name, only one line would need changing compared to one line per case when `if` statements are used.

The general form of the `switch` statement is

```
switch (expression)
  case label
    command_list
  case label
    command_list
  ...

  otherwise
    command_list
endswitch
```

where *label* can be any expression. However, duplicate *label* values are not detected, and only the *command_list* corresponding to the first match will be executed. For the `switch` statement to be meaningful at least one `case label command_list` clause must be present, while the `otherwise command_list` clause is optional.

If *label* is a cell array the corresponding *command_list* is executed if *any* of the elements of the cell array match *expression*. As an example, the following program will print 'Variable is either 6 or 7'.

```
A = 7;
switch (A)
  case { 6, 7 }
    printf ("variable is either 6 or 7\n");
  otherwise
    printf ("variable is neither 6 nor 7\n");
endswitch
```

As with all other specific `end` keywords, `endswitch` may be replaced by `end`, but you can get better diagnostics if you use the specific forms.

One advantage of using the `switch` statement compared to using `if` statements is that the *label*s can be strings. If an `if` statement is used it is *not* possible to write

```
if (X == "a string") # This is NOT valid
```

since a character-to-character comparison between X and the string will be made instead of evaluating if the strings are equal. This special-case is handled by the `switch` statement, and it is possible to write programs that look like this

```
switch (X)
  case "a string"
    do_something
    ...
endswitch
```

### 10.2.1 Notes for the C Programmer

The `switch` statement is also available in the widely used C programming language. There are, however, some differences between the statement in Octave and C

- Cases are exclusive, so they don't 'fall through' as do the cases in the `switch` statement of the C language.
- The *command_list* elements are not optional. Making the list optional would have meant requiring a separator between the label and the command list. Otherwise, things like

```
switch (foo)
  case (1) -2
    ...
```

would produce surprising results, as would

```
switch (foo)
  case (1)
  case (2)
    doit ();
    ...
```

particularly for C programmers. If `doit()` should be executed if *foo* is either 1 or 2, the above code should be written with a cell array like this

```
switch (foo)
  case { 1, 2 }
    doit ();
  ...
```

## 10.3 The while Statement

In programming, a *loop* means a part of a program that is (or at least can be) executed two or more times in succession.

The `while` statement is the simplest looping statement in Octave. It repeatedly executes a statement as long as a condition is true. As with the condition in an `if` statement, the condition in a `while` statement is considered true if its value is nonzero, and false if its value is zero. If the value of the conditional expression in a `while` statement is a vector or a matrix, it is considered true only if it is non-empty and *all* of the elements are nonzero.

Octave's `while` statement looks like this:

```
while (condition)
  body
endwhile
```

Here *body* is a statement or list of statements that we call the *body* of the loop, and *condition* is an expression that controls how long the loop keeps running.

The first thing the `while` statement does is test *condition*. If *condition* is true, it executes the statement *body*. After *body* has been executed, *condition* is tested again, and if it is still true, *body* is executed again. This process repeats until *condition* is no longer true. If *condition* is initially false, the body of the loop is never executed.

This example creates a variable `fib` that contains the first ten elements of the Fibonacci sequence.

```
fib = ones (1, 10);
i = 3;
while (i <= 10)
  fib (i) = fib (i-1) + fib (i-2);
  i++;
endwhile
```

Here the body of the loop contains two statements.

The loop works like this: first, the value of `i` is set to 3. Then, the `while` tests whether `i` is less than or equal to 10. This is the case when `i` equals 3, so the value of the `i`-th element of `fib` is set to the sum of the previous two values in the sequence. Then the `i++` increments the value of `i` and the loop repeats. The loop terminates when `i` reaches 11.

A newline is not required between the condition and the body; but using one makes the program clearer unless the body is very simple.

## 10.4 The do-until Statement

The `do-until` statement is similar to the `while` statement, except that it repeatedly executes a statement until a condition becomes true, and the test of the condition is at the end of the loop, so the body of the loop is always executed at least once. As with the condition in an `if` statement, the condition in a `do-until` statement is considered true if its value is nonzero, and false if its value is zero. If the value of the conditional expression in a `do-until` statement is a vector or a matrix, it is considered true only if it is non-empty and *all* of the elements are nonzero.

Octave's `do-until` statement looks like this:

```
do
  body
until (condition)
```

Here *body* is a statement or list of statements that we call the *body* of the loop, and *condition* is an expression that controls how long the loop keeps running.

This example creates a variable `fib` that contains the first ten elements of the Fibonacci sequence.

```
fib = ones (1, 10);
i = 2;
do
  i++;
  fib (i) = fib (i-1) + fib (i-2);
until (i == 10)
```

A newline is not required between the `do` keyword and the body; but using one makes the program clearer unless the body is very simple.

## 10.5 The for Statement

The `for` statement makes it more convenient to count iterations of a loop. The general form of the `for` statement looks like this:

```
for var = expression
  body
endfor
```

where *body* stands for any statement or list of statements, *expression* is any valid expression, and *var* may take several forms. Usually it is a simple variable name or an indexed variable. If the value of *expression* is a structure, *var* may also be a vector with two elements. See Section 10.5.1 [Looping Over Structure Elements], page 165, below.

The assignment expression in the `for` statement works a bit differently than Octave's normal assignment statement. Instead of assigning the complete result of the expression, it assigns each column of the expression to *var* in turn. If *expression* is a range, a row vector, or a scalar, the value of *var* will be a scalar each time the loop body is executed. If *var* is a column vector or a matrix, *var* will be a column vector each time the loop body is executed.

The following example shows another way to create a vector containing the first ten elements of the Fibonacci sequence, this time using the `for` statement:

# Chapter 10: Statements

```
fib = ones (1, 10);
for i = 3:10
  fib (i) = fib (i-1) + fib (i-2);
endfor
```

This code works by first evaluating the expression 3:10, to produce a range of values from 3 to 10 inclusive. Then the variable i is assigned the first element of the range and the body of the loop is executed once. When the end of the loop body is reached, the next value in the range is assigned to the variable i, and the loop body is executed again. This process continues until there are no more elements to assign.

Within Octave is it also possible to iterate over matrices or cell arrays using the `for` statement. For example consider

```
disp ("Loop over a matrix")
for i = [1,3;2,4]
  i
endfor
disp ("Loop over a cell array")
for i = {1,"two";"three",4}
  i
endfor
```

In this case the variable i takes on the value of the columns of the matrix or cell matrix. So the first loop iterates twice, producing two column vectors [1;2], followed by [3;4], and likewise for the loop over the cell array. This can be extended to loops over multi-dimensional arrays. For example:

```
a = [1,3;2,4]; c = cat (3, a, 2*a);
for i = c
  i
endfor
```

In the above case, the multi-dimensional matrix c is reshaped to a two-dimensional matrix as `reshape (c, rows (c), prod (size (c)(2:end)))` and then the same behavior as a loop over a two dimensional matrix is produced.

Although it is possible to rewrite all `for` loops as `while` loops, the Octave language has both statements because often a `for` loop is both less work to type and more natural to think of. Counting the number of iterations is very common in loops and it can be easier to think of this counting as part of looping rather than as something to do inside the loop.

## 10.5.1 Looping Over Structure Elements

A special form of the `for` statement allows you to loop over all the elements of a structure:

```
for [ val, key ] = expression
  body
endfor
```

In this form of the `for` statement, the value of *expression* must be a structure. If it is, *key* and *val* are set to the name of the element and the corresponding value in turn, until there are no more elements. For example:

```
x.a = 1
x.b = [1, 2; 3, 4]
x.c = "string"
for [val, key] = x
  key
  val
endfor
```

        ⊣ key = a
        ⊣ val = 1
        ⊣ key = b
        ⊣ val =
        ⊣
        ⊣   1  2
        ⊣   3  4
        ⊣
        ⊣ key = c
        ⊣ val = string

The elements are not accessed in any particular order. If you need to cycle through the list in a particular way, you will have to use the function `fieldnames` and sort the list yourself.

The *key* variable may also be omitted. If it is, the brackets are also optional. This is useful for cycling through the values of all the structure elements when the names of the elements do not need to be known.

## 10.6 The break Statement

The `break` statement jumps out of the innermost `while`, `do-until`, or `for` loop that encloses it. The `break` statement may only be used within the body of a loop. The following example finds the smallest divisor of a given integer, and also identifies prime numbers:

```
num = 103;
div = 2;
while (div*div <= num)
  if (rem (num, div) == 0)
    break;
  endif
  div++;
endwhile
if (rem (num, div) == 0)
  printf ("Smallest divisor of %d is %d\n", num, div)
else
  printf ("%d is prime\n", num);
endif
```

When the remainder is zero in the first `while` statement, Octave immediately *breaks out* of the loop. This means that Octave proceeds immediately to the statement following the loop and continues processing. (This is very different from the `exit` statement which stops the entire Octave program.)

Here is another program equivalent to the previous one. It illustrates how the *condition* of a `while` statement could just as well be replaced with a `break` inside an `if`:

```
num = 103;
div = 2;
while (1)
  if (rem (num, div) == 0)
    printf ("Smallest divisor of %d is %d\n", num, div);
    break;
  endif
  div++;
  if (div*div > num)
    printf ("%d is prime\n", num);
    break;
  endif
endwhile
```

## 10.7 The continue Statement

The `continue` statement, like `break`, is used only inside `while`, `do-until`, or `for` loops. It skips over the rest of the loop body, causing the next cycle around the loop to begin immediately. Contrast this with `break`, which jumps out of the loop altogether. Here is an example:

```
# print elements of a vector of random
# integers that are even.

# first, create a row vector of 10 random
# integers with values between 0 and 100:

vec = round (rand (1, 10) * 100);

# print what we're interested in:

for x = vec
  if (rem (x, 2) != 0)
    continue;
  endif
  printf ("%d\n", x);
endfor
```

If one of the elements of *vec* is an odd number, this example skips the print statement for that element, and continues back to the first statement in the loop.

This is not a practical example of the `continue` statement, but it should give you a clear understanding of how it works. Normally, one would probably write the loop like this:

```
for x = vec
  if (rem (x, 2) == 0)
    printf ("%d\n", x);
  endif
endfor
```

## 10.8 The unwind_protect Statement

Octave supports a limited form of exception handling modeled after the unwind-protect form of Lisp.

The general form of an `unwind_protect` block looks like this:

```
unwind_protect
  body
unwind_protect_cleanup
  cleanup
end_unwind_protect
```

where *body* and *cleanup* are both optional and may contain any Octave expressions or commands. The statements in *cleanup* are guaranteed to be executed regardless of how control exits *body*.

This is useful to protect temporary changes to global variables from possible errors. For example, the following code will always restore the original value of the global variable `frobnosticate` even if an error occurs in the first part of the `unwind_protect` block.

```
save_frobnosticate = frobnosticate;
unwind_protect
  frobnosticate = true;
  ...
unwind_protect_cleanup
  frobnosticate = save_frobnosticate;
end_unwind_protect
```

Without `unwind_protect`, the value of *frobnosticate* would not be restored if an error occurs while evaluating the first part of the `unwind_protect` block because evaluation would stop at the point of the error and the statement to restore the value would not be executed.

In addition to unwind_protect, Octave supports another form of exception handling, the `try` block.

## 10.9 The try Statement

The original form of a `try` block looks like this:

```
try
  body
catch
  cleanup
end_try_catch
```

where *body* and *cleanup* are both optional and may contain any Octave expressions or commands. The statements in *cleanup* are only executed if an error occurs in *body*.

No warnings or error messages are printed while *body* is executing. If an error does occur during the execution of *body*, *cleanup* can use the functions `lasterr` or `lasterror` to access the text of the message that would have been printed, as well as its identifier. The alternative form,

```
try
  body
catch err
  cleanup
end_try_catch
```

will automatically store the output of `lasterror` in the structure *err*. See Chapter 12 [Errors and Warnings], page 205, for more information about the `lasterr` and `lasterror` functions.

## 10.10 Continuation Lines

In the Octave language, most statements end with a newline character and you must tell Octave to ignore the newline character in order to continue a statement from one line to the next. Lines that end with the characters ... are joined with the following line before they are divided into tokens by Octave's parser. For example, the lines

```
x = long_variable_name ...
    + longer_variable_name ...
    - 42
```

form a single statement.

Any text between the continuation marker and the newline character is ignored. For example, the statement

```
x = long_variable_name ...    # comment one
    + longer_variable_name ...comment two
    - 42                      # last comment
```

is equivalent to the one shown above.

Inside double-quoted string constants, the character \ has to be used as continuation marker. The \ must appear at the end of the line just before the newline character:

```
s = "This text starts in the first line \
and is continued in the second line."
```

Input that occurs inside parentheses can be continued to the next line without having to use a continuation marker. For example, it is possible to write statements like

```
if (fine_dining_destination == on_a_boat
    || fine_dining_destination == on_a_train)
  seuss (i, will, not, eat, them, sam, i, am, i,
         will, not, eat, green, eggs, and, ham);
endif
```

without having to add to the clutter with continuation markers.

# 11 Functions and Scripts

Complicated Octave programs can often be simplified by defining functions. Functions can be defined directly on the command line during interactive Octave sessions, or in external files, and can be called just like built-in functions.

## 11.1 Introduction to Function and Script Files

There are seven different things covered in this section.

1. Typing in a function at the command prompt.
2. Storing a group of commands in a file — called a script file.
3. Storing a function in a file—called a function file.
4. Subfunctions in function files.
5. Multiple functions in one script file.
6. Private functions.
7. Nested functions.

Both function files and script files end with an extension of .m, for MATLAB compatibility. If you want more than one independent functions in a file, it must be a script file (see Section 11.10 [Script Files], page 198), and to use these functions you must execute the script file before you can use the functions that are in the script file.

## 11.2 Defining Functions

In its simplest form, the definition of a function named *name* looks like this:

```
function name
  body
endfunction
```

A valid function name is like a valid variable name: a sequence of letters, digits and underscores, not starting with a digit. Functions share the same pool of names as variables.

The function *body* consists of Octave statements. It is the most important part of the definition, because it says what the function should actually *do*.

For example, here is a function that, when executed, will ring the bell on your terminal (assuming that it is possible to do so):

```
function wakeup
  printf ("\a");
endfunction
```

The `printf` statement (see Chapter 14 [Input and Output], page 231) simply tells Octave to print the string "\a". The special character '\a' stands for the alert character (ASCII 7). See Chapter 5 [Strings], page 67.

Once this function is defined, you can ask Octave to evaluate it by typing the name of the function.

Normally, you will want to pass some information to the functions you define. The syntax for passing parameters to a function in Octave is

```
function name (arg-list)
  body
endfunction
```

where *arg-list* is a comma-separated list of the function's arguments. When the function is called, the argument names are used to hold the argument values given in the call. The list of arguments may be empty, in which case this form is equivalent to the one shown above.

To print a message along with ringing the bell, you might modify the `wakeup` to look like this:

```
function wakeup (message)
  printf ("\a%s\n", message);
endfunction
```

Calling this function using a statement like this

```
wakeup ("Rise and shine!");
```

will cause Octave to ring your terminal's bell and print the message 'Rise and shine!', followed by a newline character (the '\n' in the first argument to the `printf` statement).

In most cases, you will also want to get some information back from the functions you define. Here is the syntax for writing a function that returns a single value:

```
function ret-var = name (arg-list)
  body
endfunction
```

The symbol *ret-var* is the name of the variable that will hold the value to be returned by the function. This variable must be defined before the end of the function body in order for the function to return a value.

Variables used in the body of a function are local to the function. Variables named in *arg-list* and *ret-var* are also local to the function. See Section 7.1 [Global Variables], page 124, for information about how to access global variables inside a function.

For example, here is a function that computes the average of the elements of a vector:

```
function retval = avg (v)
  retval = sum (v) / length (v);
endfunction
```

If we had written `avg` like this instead,

```
function retval = avg (v)
  if (isvector (v))
    retval = sum (v) / length (v);
  endif
endfunction
```

and then called the function with a matrix instead of a vector as the argument, Octave would have printed an error message like this:

```
error: value on right hand side of assignment is undefined
```

because the body of the `if` statement was never executed, and `retval` was never defined. To prevent obscure errors like this, it is a good idea to always make sure that the return variables will always have values, and to produce meaningful error messages when problems are encountered. For example, `avg` could have been written like this:

# Chapter 11: Functions and Scripts

```
function retval = avg (v)
  retval = 0;
  if (isvector (v))
    retval = sum (v) / length (v);
  else
    error ("avg: expecting vector argument");
  endif
endfunction
```

There is still one additional problem with this function. What if it is called without an argument? Without additional error checking, Octave will probably print an error message that won't really help you track down the source of the error. To allow you to catch errors like this, Octave provides each function with an automatic variable called `nargin`. Each time a function is called, `nargin` is automatically initialized to the number of arguments that have actually been passed to the function. For example, we might rewrite the `avg` function like this:

```
function retval = avg (v)
  retval = 0;
  if (nargin != 1)
    usage ("avg (vector)");
  endif
  if (isvector (v))
    retval = sum (v) / length (v);
  else
    error ("avg: expecting vector argument");
  endif
endfunction
```

Although Octave does not automatically report an error if you call a function with more arguments than expected, doing so probably indicates that something is wrong. Octave also does not automatically report an error if a function is called with too few arguments, but any attempt to use a variable that has not been given a value will result in an error. To avoid such problems and to provide useful messages, we check for both possibilities and issue our own error message.

**nargin ()** [Built-in Function]
**nargin (*fcn*)** [Built-in Function]
Report the number of input arguments to a function.

Called from within a function, return the number of arguments passed to the function. At the top level, return the number of command line arguments passed to Octave.

If called with the optional argument *fcn*—a function name or handle— return the declared number of arguments that the function can accept.

If the last argument to *fcn* is *varargin* the returned value is negative. For example, the function **union** for sets is declared as

```
function [y, ia, ib] = union (a, b, varargin)
```

and

```
nargin ("union")
    ⇒ -3
```

Programming Note: **nargin** does not work on built-in functions.

**See also:** [nargout], page 176, [narginchk], page 177, [varargin], page 182, [inputname], page 174.

**inputname** (*n*)　　　　　　　　　　　　　　　　　　　　　　　　　　　　　　　　　　[Function File]

Return the name of the *n*-th argument to the calling function.

If the argument is not a simple variable name, return an empty string. **inputname** may only be used within a function body, not at the command line.

**See also:** [nargin], page 173, [nthargout], page 175.

*val* = **silent_functions** ()　　　　　　　　　　　　　　　　　　　　　　　　　[Built-in Function]
*old_val* = **silent_functions** (*new_val*)　　　　　　　　　　　　　　　　　　[Built-in Function]
**silent_functions** (*new_val*, "*local*")　　　　　　　　　　　　　　　　　　[Built-in Function]

Query or set the internal variable that controls whether internal output from a function is suppressed.

If this option is disabled, Octave will display the results produced by evaluating expressions within a function body that are not terminated with a semicolon.

When called from inside a function with the "**local**" option, the variable is changed locally for the function and any subroutines it calls. The original variable value is restored when exiting the function.

## 11.3 Multiple Return Values

Unlike many other computer languages, Octave allows you to define functions that return more than one value. The syntax for defining functions that return multiple values is

```
function [ret-list] = name (arg-list)
  body
endfunction
```

where *name*, *arg-list*, and *body* have the same meaning as before, and *ret-list* is a comma-separated list of variable names that will hold the values returned from the function. The list of return values must have at least one element. If *ret-list* has only one element, this form of the **function** statement is equivalent to the form described in the previous section.

Here is an example of a function that returns two values, the maximum element of a vector and the index of its first occurrence in the vector.

## Chapter 11: Functions and Scripts

```
function [max, idx] = vmax (v)
  idx = 1;
  max = v (idx);
  for i = 2:length (v)
    if (v (i) > max)
      max = v (i);
      idx = i;
    endif
  endfor
endfunction
```

In this particular case, the two values could have been returned as elements of a single array, but that is not always possible or convenient. The values to be returned may not have compatible dimensions, and it is often desirable to give the individual return values distinct names.

It is possible to use the **nthargout** function to obtain only some of the return values or several at once in a cell array. See Section 3.1.5 [Cell Array Objects], page 44.

**nthargout** (*n*, *func*, ...)                                                                                [Function File]
**nthargout** (*n*, *ntot*, *func*, ...)                                                        [Function File]

Return the *n*th output argument of the function specified by the function handle or string *func*.

Any additional arguments are passed directly to *func*. The total number of arguments to call *func* with can be passed in *ntot*; by default *ntot* is *n*. The input *n* can also be a vector of indices of the output, in which case the output will be a cell array of the requested output arguments.

The intended use **nthargout** is to avoid intermediate variables. For example, when finding the indices of the maximum entry of a matrix, the following two compositions of nthargout

```
m = magic (5);
cell2mat (nthargout ([1, 2], @ind2sub, size (m),
                    nthargout (2, @max, m(:))))
    ⇒ 5    3
```

are completely equivalent to the following lines:

```
m = magic (5);
[~, idx] = max (M(:));
[i, j] = ind2sub (size (m), idx);
[i, j]
    ⇒ 5    3
```

It can also be helpful to have all output arguments in a single cell in the following manner:

```
USV = nthargout ([1:3], @svd, hilb (5));
```

**See also:** [nargin], page 173, [nargout], page 176, [varargin], page 182, [varargout], page 182, [isargout], page 184.

In addition to setting `nargin` each time a function is called, Octave also automatically initializes `nargout` to the number of values that are expected to be returned. This allows you to write functions that behave differently depending on the number of values that the user of the function has requested. The implicit assignment to the built-in variable `ans` does not figure in the count of output arguments, so the value of `nargout` may be zero.

The `svd` and `lu` functions are examples of built-in functions that behave differently depending on the value of `nargout`.

It is possible to write functions that only set some return values. For example, calling the function

```
function [x, y, z] = f ()
  x = 1;
  z = 2;
endfunction
```

as

```
[a, b, c] = f ()
```

produces:

```
a = 1

b = [](0x0)

c = 2
```

along with a warning.

**nargout ()**  [Built-in Function]
**nargout (*fcn*)**  [Built-in Function]
Report the number of output arguments from a function.

Called from within a function, return the number of values the caller expects to receive. At the top level, `nargout` with no argument is undefined and will produce an error.

If called with the optional argument *fcn*—a function name or handle—return the number of declared output values that the function can produce.

If the final output argument is *varargout* the returned value is negative.

For example,

```
f ()
```

will cause `nargout` to return 0 inside the function `f` and

```
[s, t] = f ()
```

will cause `nargout` to return 2 inside the function `f`.

In the second usage,

```
nargout (@histc) % or nargout ("histc")
```

will return 2, because `histc` has two outputs, whereas

```
nargout (@imread)
```

will return -2, because `imread` has two outputs and the second is *varargout*.

# Chapter 11: Functions and Scripts

Programming Note. `nargout` does not work for built-in functions and returns -1 for all anonymous functions.

**See also:** [nargin], page 173, [varargout], page 182, [isargout], page 184, [nthargout], page 175.

It is good practice at the head of a function to verify that it has been called correctly. In Octave the following idiom is seen frequently

```
if (nargin < min_#_inputs || nargin > max_#_inputs)
  print_usage ();
endif
```

which stops the function execution and prints a message about the correct way to call the function whenever the number of inputs is wrong.

For compatibility with MATLAB, `narginchk` and `nargoutchk` are available which provide similar error checking.

`narginchk (`*minargs*, *maxargs*`)` [Function File]

Check for correct number of input arguments.

Generate an error message if the number of arguments in the calling function is outside the range *minargs* and *maxargs*. Otherwise, do nothing.

Both *minargs* and *maxargs* must be scalar numeric values. Zero, Inf, and negative values are all allowed, and *minargs* and *maxargs* may be equal.

Note that this function evaluates `nargin` on the caller.

**See also:** [nargoutchk], page 177, [error], page 205, [nargout], page 176, [nargin], page 173.

`nargoutchk (`*minargs*, *maxargs*`)` [Function File]
*msgstr* = `nargoutchk (`*minargs*, *maxargs*, *nargs*`)` [Function File]
*msgstr* = `nargoutchk (`*minargs*, *maxargs*, *nargs*, "*string*"`)` [Function File]
*msgstruct* = `nargoutchk (`*minargs*, *maxargs*, *nargs*, "*struct*"`)` [Function File]

Check for correct number of output arguments.

In the first form, return an error if the number of arguments is not between *minargs* and *maxargs*. Otherwise, do nothing. Note that this function evaluates the value of `nargout` on the caller so its value must have not been tampered with.

Both *minargs* and *maxargs* must be numeric scalars. Zero, Inf, and negative are all valid, and they can have the same value.

For backwards compatibility, the other forms return an appropriate error message string (or structure) if the number of outputs requested is invalid.

This is useful for checking to that the number of output arguments supplied to a function is within an acceptable range.

**See also:** [narginchk], page 177, [error], page 205, [nargout], page 176, [nargin], page 173.

Besides the number of arguments, inputs can be checked for various properties. `validatestring` is used for string arguments and `validateattributes` for numeric arguments.

| | |
|---|---|
| `validstr = validatestring (str, strarray)` | [Function File] |
| `validstr = validatestring (str, strarray, funcname)` | [Function File] |
| `validstr = validatestring (str, strarray, funcname, varname)` | [Function File] |
| `validstr = validatestring (..., position)` | [Function File] |

Verify that *str* is an element, or substring of an element, in *strarray*.

When *str* is a character string to be tested, and *strarray* is a cellstr of valid values, then *validstr* will be the validated form of *str* where validation is defined as *str* being a member or substring of *validstr*. This is useful for both verifying and expanding short options, such as `"r"`, to their longer forms, such as `"red"`. If *str* is a substring of *validstr*, and there are multiple matches, the shortest match will be returned if all matches are substrings of each other. Otherwise, an error will be raised because the expansion of *str* is ambiguous. All comparisons are case insensitive.

The additional inputs *funcname*, *varname*, and *position* are optional and will make any generated validation error message more specific.

Examples:
```
validatestring ("r", {"red", "green", "blue"})
⇒ "red"

validatestring ("b", {"red", "green", "blue", "black"})
⇒ error: validatestring: multiple unique matches were found for 'b':
   blue, black
```

**See also:** [strcmp], page 76, [strcmpi], page 76, [validateattributes], page 178, [inputParser], page 180.

| | |
|---|---|
| `validateattributes (A, classes, attributes)` | [Function File] |
| `validateattributes (A, classes, attributes, arg_idx)` | [Function File] |
| `validateattributes (A, classes, attributes, func_name)` | [Function File] |
| `validateattributes (A, classes, attributes, func_name, arg_name)` | [Function File] |
| `validateattributes (A, classes, attributes, func_name, arg_name, arg_idx)` | [Function File] |

Check validity of input argument.

Confirms that the argument *A* is valid by belonging to one of *classes*, and holding all of the *attributes*. If it does not, an error is thrown, with a message formatted accordingly. The error message can be made further complete by the function name *fun_name*, the argument name *arg_name*, and its position in the input *arg_idx*.

*classes* must be a cell array of strings (an empty cell array is allowed) with the name of classes (remember that a class name is case sensitive). In addition to the class name, the following categories names are also valid:

`"float"`   Floating point value comprising classes `"double"` and `"single"`.

`"integer"`
           Integer value comprising classes (u)int8, (u)int16, (u)int32, (u)int64.

`"numeric"`
           Numeric value comprising either a floating point or integer value.

*attributes* must be a cell array with names of checks for *A*. Some of them require an additional value to be supplied right after the name (see details for each below).

"<="
: All values are less than or equal to the following value in *attributes*.

"<"
: All values are less than the following value in *attributes*.

">="
: All values are greater than or equal to the following value in *attributes*.

">"
: All values are greater than the following value in *attributes*.

"2d"
: A 2-dimensional matrix. Note that vectors and empty matrices have 2 dimensions, one of them being of length 1, or both length 0.

"3d"
: Has no more than 3 dimensions. A 2-dimensional matrix is a 3-D matrix whose 3rd dimension is of length 1.

"binary"
: All values are either 1 or 0.

"column"
: Values are arranged in a single column.

"decreasing"
: No value is *NaN*, and each is less than the preceding one.

"even"
: All values are even numbers.

"finite"
: All values are finite.

"increasing"
: No value is *NaN*, and each is greater than the preceding one.

"integer"
: All values are integer. This is different than using `isinteger` which only checks its an integer type. This checks that each value in *A* is an integer value, i.e., it has no decimal part.

"ncols"
: Has exactly as many columns as the next value in *attributes*.

"ndims"
: Has exactly as many dimensions as the next value in *attributes*.

"nondecreasing"
: No value is *NaN*, and each is greater than or equal to the preceding one.

"nonempty"
: It is not empty.

"nonincreasing"
: No value is *NaN*, and each is less than or equal to the preceding one.

"nonnan"
: No value is a `NaN`.

"non-negative"
: All values are non negative.

"nonsparse"
: It is not a sparse matrix.

"nonzero"
: No value is zero.

"nrows"    Has exactly as many rows as the next value in *attributes*.

"numel"    Has exactly as many elements as the next value in *attributes*.

"odd"      All values are odd numbers.

"positive"
           All values are positive.

"real"     It is a non-complex matrix.

"row"      Values are arranged in a single row.

"scalar"   It is a scalar.

"size"     Its size has length equal to the values of the next in *attributes*. The next value must is an array with the length for each dimension. To ignore the check for a certain dimension, the value of NaN can be used.

"square"   Is a square matrix.

"vector"   Values are arranged in a single vector (column or vector).

**See also:** [isa], page 39, [validatestring], page 177, [inputParser], page 180.

If none of the preceding functions is sufficient there is also the class `inputParser` which can perform extremely complex input checking for functions.

### *p* = inputParser ()    [Function File]
Create object *p* of the inputParser class.

This class is designed to allow easy parsing of function arguments. The class supports four types of arguments:

1. mandatory (see `addRequired`);
2. optional (see `addOptional`);
3. named (see `addParamValue`);
4. switch (see `addSwitch`).

After defining the function API with these methods, the supplied arguments can be parsed with the `parse` method and the parsing results accessed with the `Results` accessor.

### inputParser.Parameters    [Accessor method]
Return list of parameter names already defined.

### inputParser.Results    [Accessor method]
Return structure with argument names as fieldnames and corresponding values.

### inputParser.Unmatched    [Accessor method]
Return structure similar to `Results`, but for unmatched parameters. See the `KeepUnmatched` property.

### inputParser.UsingDefaults    [Accessor method]
Return cell array with the names of arguments that are using default values.

# Chapter 11: Functions and Scripts

inputParser.CaseSensitive = *boolean* [Class property]
    Set whether matching of argument names should be case sensitive. Defaults to false.

inputParser.FunctionName = *name* [Class property]
    Set function name to be used in error messages; Defaults to empty string.

inputParser.KeepUnmatched = *boolean* [Class property]
    Set whether an error should be given for non-defined arguments. Defaults to false. If set to true, the extra arguments can be accessed through Unmatched after the parse method. Note that since Switch and ParamValue arguments can be mixed, it is not possible to know the unmatched type. If argument is found unmatched it is assumed to be of the ParamValue type and it is expected to be followed by a value.

inputParser.StructExpand = *boolean* [Class property]
    Set whether a structure can be passed to the function instead of parameter/value pairs. Defaults to true. Not implemented yet.

    The following example shows how to use this class:

```
function check (varargin)
  p = inputParser ();                    # create object
  p.FunctionName = "check";              # set function name
  p.addRequired ("pack", @ischar);       # mandatory argument
  p.addOptional ("path", pwd(), @ischar); # optional argument

  ## create a function handle to anonymous functions for validators
  val_mat = @(x) isvector (x) && all (x <= 1) && all (x >= 0);
  p.addOptional ("mat", [0 0], val_mat);

  ## create two arguments of type "ParamValue"
  val_type = @(x) any (strcmp (x, {"linear", "quadratic"}));
  p.addParamValue ("type", "linear", val_type);
  val_verb = @(x) any (strcmp (x, {"low", "medium", "high"}));
  p.addParamValue ("tolerance", "low", val_verb);

  ## create a switch type of argument
  p.addSwitch ("verbose");

  p.parse (varargin{:});  # Run created parser on inputs

  ## the rest of the function can access inputs by using p.Results.
  ## for example, get the tolerance input with p.Results.tolerance
endfunction

check ("mech");              # valid, use defaults for other arguments
check ();                    # error, one argument is mandatory
check (1);                   # error, since ! ischar
check ("mech", "~/dev");     # valid, use defaults for other arguments
```

```
check ("mech", "~/dev", [0 1 0 0], "type", "linear");  # valid

## following is also valid.  Note how the Switch argument type can
## be mixed into or before the ParamValue argument type (but it
## must still appear after any Optional argument).
check ("mech", "~/dev", [0 1 0 0], "verbose", "tolerance", "high");

## following returns an error since not all optional arguments,
## 'path' and 'mat', were given before the named argument 'type'.
check ("mech", "~/dev", "type", "linear");
```

*Note 1*: A function can have any mixture of the four API types but they must appear in a specific order. `Required` arguments must be first and can be followed by any `Optional` arguments. Only the `ParamValue` and `Switch` arguments may be mixed together and they must appear at the end.

*Note 2*: If both `Optional` and `ParamValue` arguments are mixed in a function API then once a string Optional argument fails to validate it will be considered the end of the `Optional` arguments. The remaining arguments will be compared against any `ParamValue` or `Switch` arguments.

**See also:** [nargin], page 173, [validateattributes], page 178, [validatestring], page 177, [varargin], page 182.

## 11.4 Variable-length Argument Lists

Sometimes the number of input arguments is not known when the function is defined. As an example think of a function that returns the smallest of all its input arguments. For example:

```
a = smallest (1, 2, 3);
b = smallest (1, 2, 3, 4);
```

In this example both `a` and `b` would be 1. One way to write the `smallest` function is

```
function val = smallest (arg1, arg2, arg3, arg4, arg5)
  body
endfunction
```

and then use the value of `nargin` to determine which of the input arguments should be considered. The problem with this approach is that it can only handle a limited number of input arguments.

If the special parameter name `varargin` appears at the end of a function parameter list it indicates that the function takes a variable number of input arguments. Using `varargin` the function looks like this

```
function val = smallest (varargin)
  body
endfunction
```

In the function body the input arguments can be accessed through the variable `varargin`. This variable is a cell array containing all the input arguments. See Section 6.2 [Cell Arrays], page 112, for details on working with cell arrays. The `smallest` function can now be defined like this

Chapter 11: Functions and Scripts     183

```
function val = smallest (varargin)
  val = min ([varargin{:}]);
endfunction
```

This implementation handles any number of input arguments, but it's also a very simple solution to the problem.

A slightly more complex example of `varargin` is a function `print_arguments` that prints all input arguments. Such a function can be defined like this

```
function print_arguments (varargin)
  for i = 1:length (varargin)
    printf ("Input argument %d: ", i);
    disp (varargin{i});
  endfor
endfunction
```

This function produces output like this

```
print_arguments (1, "two", 3);
    ⊣ Input argument 1:  1
    ⊣ Input argument 2: two
    ⊣ Input argument 3:  3
```

[*reg*, *prop*] = parseparams (*params*)                           [Function File]
[*reg*, *var1*, ...] = parseparams (*params*, *name1*, *default1*,   [Function File]
        ...)

Return in *reg* the cell elements of *param* up to the first string element and in *prop* all remaining elements beginning with the first string element.

For example:

```
[reg, prop] = parseparams ({1, 2, "linewidth", 10})
reg =
{
  [1,1] = 1
  [1,2] = 2
}
prop =
{
  [1,1] = linewidth
  [1,2] = 10
}
```

The parseparams function may be used to separate regular numeric arguments from additional arguments given as property/value pairs of the *varargin* cell array.

In the second form of the call, available options are specified directly with their default values given as name-value pairs. If *params* do not form name-value pairs, or if an option occurs that does not match any of the available options, an error occurs.

When called from an m-file function, the error is prefixed with the name of the caller function.

The matching of options is case-insensitive.

**See also:** [varargin], page 182, [inputParser], page 180.

## 11.5 Ignoring Arguments

In the formal argument list, it is possible to use the dummy placeholder ~ instead of a name. This indicates that the corresponding argument value should be ignored and not stored to any variable.

```
function val = pick2nd (~, arg2)
  val = arg2;
endfunction
```

The value of `nargin` is not affected by using this declaration.

Return arguments can also be ignored using the same syntax. Functions may take advantage of ignored outputs to reduce the number of calculations performed. To do so, use the `isargout` function to query whether the output argument is wanted. For example:

```
function [out1, out2] = long_function (x, y, z)
  if (isargout (1))
    ## Long calculation
    ...
    out1 = result;
  endif
  ...
endfunction
```

`isargout (k)` [Built-in Function]

Within a function, return a logical value indicating whether the argument $k$ will be assigned to a variable on output.

If the result is false, the argument has been ignored during the function call through the use of the tilde (~) special output argument. Functions can use `isargout` to avoid performing unnecessary calculations for outputs which are unwanted.

If $k$ is outside the range `1:max (nargout)`, the function returns false. $k$ can also be an array, in which case the function works element-by-element and a logical array is returned. At the top level, `isargout` returns an error.

See also: [nargout], page 176, [varargout], page 182, [nthargout], page 175.

## 11.6 Variable-length Return Lists

It is possible to return a variable number of output arguments from a function using a syntax that's similar to the one used with the special `varargin` parameter name. To let a function return a variable number of output arguments the special output parameter name `varargout` is used. As with `varargin`, `varargout` is a cell array that will contain the requested output arguments.

As an example the following function sets the first output argument to 1, the second to 2, and so on.

```
function varargout = one_to_n ()
  for i = 1:nargout
    varargout{i} = i;
  endfor
endfunction
```

Chapter 11: Functions and Scripts 185

When called this function returns values like this

```
[a, b, c] = one_to_n ()
    ⇒ a =  1
    ⇒ b =  2
    ⇒ c =  3
```

If `varargin` (`varargout`) does not appear as the last element of the input (output) parameter list, then it is not special, and is handled the same as any other parameter name.

[r1, r2, ..., rn] = deal (a)                                       [Function File]
[r1, r2, ..., rn] = deal (a1, a2, ..., an)                         [Function File]
    Copy the input parameters into the corresponding output parameters.

If only a single input parameter is supplied, its value is copied to each of the outputs. For example,

```
[a, b, c] = deal (x, y, z);
```

is equivalent to

```
a = x;
b = y;
c = z;
```

and

```
[a, b, c] = deal (x);
```

is equivalent to

```
a = b = c = x;
```

Programming Note: `deal` is often used with comma separated lists derived from cell arrays or structures. This is unnecessary as the interpreter can perform the same action without the overhead of a function call. For example:

```
c = {[1 2], "Three", 4};
[x, y, z ] = c{:}
⇒
   x =

      1   2

   y = Three
   z =  4
```

**See also:** [cell2struct], page 120, [struct2cell], page 111, [repmat], page 417.

## 11.7 Returning from a Function

The body of a user-defined function can contain a `return` statement. This statement returns control to the rest of the Octave program. It looks like this:

    `return`

Unlike the `return` statement in C, Octave's `return` statement cannot be used to return a value from a function. Instead, you must assign values to the list of return variables that are part of the `function` statement. The `return` statement simply makes it easier to exit a function from a deeply nested loop or conditional statement.

Here is an example of a function that checks to see if any elements of a vector are nonzero.

```
function retval = any_nonzero (v)
  retval = 0;
  for i = 1:length (v)
    if (v (i) != 0)
      retval = 1;
      return;
    endif
  endfor
  printf ("no nonzero elements found\n");
endfunction
```

Note that this function could not have been written using the `break` statement to exit the loop once a nonzero value is found without adding extra logic to avoid printing the message if the vector does contain a nonzero element.

`return`                                                                    [Keyword]

When Octave encounters the keyword `return` inside a function or script, it returns control to the caller immediately. At the top level, the return statement is ignored. A `return` statement is assumed at the end of every function definition.

## 11.8 Default Arguments

Since Octave supports variable number of input arguments, it is very useful to assign default values to some input arguments. When an input argument is declared in the argument list it is possible to assign a default value to the argument like this

```
function name (arg1 = val1, ...)
  body
endfunction
```

If no value is assigned to *arg1* by the user, it will have the value *val1*.

As an example, the following function implements a variant of the classic "Hello, World" program.

```
function hello (who = "World")
  printf ("Hello, %s!\n", who);
endfunction
```

When called without an input argument the function prints the following

```
hello ();
    ⊣ Hello, World!
```

and when it's called with an input argument it prints the following

```
hello ("Beautiful World of Free Software");
    ⊣ Hello, Beautiful World of Free Software!
```

Sometimes it is useful to explicitly tell Octave to use the default value of an input argument. This can be done writing a ':' as the value of the input argument when calling the function.

```
hello (:);
     ⊣ Hello, World!
```

## 11.9 Function Files

Except for simple one-shot programs, it is not practical to have to define all the functions you need each time you need them. Instead, you will normally want to save them in a file so that you can easily edit them, and save them for use at a later time.

Octave does not require you to load function definitions from files before using them. You simply need to put the function definitions in a place where Octave can find them.

When Octave encounters an identifier that is undefined, it first looks for variables or functions that are already compiled and currently listed in its symbol table. If it fails to find a definition there, it searches a list of directories (the *path*) for files ending in '.m' that have the same base name as the undefined identifier.[1] Once Octave finds a file with a name that matches, the contents of the file are read. If it defines a *single* function, it is compiled and executed. See Section 11.10 [Script Files], page 198, for more information about how you can define more than one function in a single file.

When Octave defines a function from a function file, it saves the full name of the file it read and the time stamp on the file. If the time stamp on the file changes, Octave may reload the file. When Octave is running interactively, time stamp checking normally happens at most once each time Octave prints the prompt. Searching for new function definitions also occurs if the current working directory changes.

Checking the time stamp allows you to edit the definition of a function while Octave is running, and automatically use the new function definition without having to restart your Octave session.

To avoid degrading performance unnecessarily by checking the time stamps on functions that are not likely to change, Octave assumes that function files in the directory tree '*octave-home*/share/octave/*version*/m' will not change, so it doesn't have to check their time stamps every time the functions defined in those files are used. This is normally a very good assumption and provides a significant improvement in performance for the function files that are distributed with Octave.

If you know that your own function files will not change while you are running Octave, you can improve performance by calling `ignore_function_time_stamp ("all")`, so that Octave will ignore the time stamps for all function files. Passing `"system"` to this function resets the default behavior.

**edit** *name*   [Command]
**edit** *field value*   [Command]
*value* = **edit** *get field*   [Command]

    Edit the named function, or change editor settings.

    If `edit` is called with the name of a file or function as its argument it will be opened in the text editor defined by `EDITOR`.

- If the function *name* is available in a file on your path and that file is modifiable, then it will be edited in place. If it is a system function, then it will first be

---

[1] The '.m' suffix was chosen for compatibility with MATLAB.

copied to the directory HOME (see below) and then edited. If no file is found, then the m-file variant, ending with ".m", will be considered. If still no file is found, then variants with a leading "@" and then with both a leading "@" and trailing ".m" will be considered.

- If *name* is the name of a function defined in the interpreter but not in an m-file, then an m-file will be created in HOME to contain that function along with its current definition.

- If `name.cc` is specified, then it will search for `name.cc` in the path and try to modify it, otherwise it will create a new '.cc' file in the current directory. If *name* happens to be an m-file or interpreter defined function, then the text of that function will be inserted into the .cc file as a comment.

- If 'name.ext' is on your path then it will be edited, otherwise the editor will be started with 'name.ext' in the current directory as the filename. If 'name.ext' is not modifiable, it will be copied to HOME before editing.

  **Warning:** You may need to clear *name* before the new definition is available. If you are editing a .cc file, you will need to execute mkoctfile 'name.cc' before the definition will be available.

If edit is called with *field* and *value* variables, the value of the control field *field* will be set to *value*. If an output argument is requested and the first input argument is get then edit will return the value of the control field *field*. If the control field does not exist, edit will return a structure containing all fields and values. Thus, edit get all returns a complete control structure.

The following control fields are used:

'home'   This is the location of user local m-files. Be sure it is in your path. The default is '~/octave'.

'author'   This is the name to put after the "## Author:" field of new functions. By default it guesses from the gecos field of the password database.

'email'   This is the e-mail address to list after the name in the author field. By default it guesses <$LOGNAME@$HOSTNAME>, and if $HOSTNAME is not defined it uses uname -n. You probably want to override this. Be sure to use the format user@host.

'license'

    'gpl'   GNU General Public License (default).

    'bsd'   BSD-style license without advertising clause.

    'pd'   Public domain.

    '"text"'   Your own default copyright and license.

Unless you specify 'pd', edit will prepend the copyright statement with "Copyright (C) yyyy Function Author".

'mode'   This value determines whether the editor should be started in async mode (editor is started in the background and Octave continues) or sync mode (Octave waits until the editor exits). Set it to "sync" to start the editor in sync mode. The default is "async" (see [system], page 769).

Chapter 11: Functions and Scripts                                                189

'editinplace'
> Determines whether files should be edited in place, without regard to whether they are modifiable or not. The default is `false`.

**mfilename ()**                                                  [Built-in Function]
**mfilename ("***fullpath***")**                                  [Built-in Function]
**mfilename ("***fullpathext***")**                               [Built-in Function]
> Return the name of the currently executing file.
>
> When called from outside an m-file return the empty string.
>
> Given the argument `"fullpath"`, include the directory part of the file name, but not the extension.
>
> Given the argument `"fullpathext"`, include the directory part of the file name and the extension.

*val* **= ignore_function_time_stamp ()**                         [Built-in Function]
*old_val* **= ignore_function_time_stamp (***new_val***)**        [Built-in Function]
> Query or set the internal variable that controls whether Octave checks the time stamp on files each time it looks up functions defined in function files.
>
> If the internal variable is set to `"system"`, Octave will not automatically recompile function files in subdirectories of '*octave-home*/lib/*version*' if they have changed since they were last compiled, but will recompile other function files in the search path if they change.
>
> If set to `"all"`, Octave will not recompile any function files unless their definitions are removed with `clear`.
>
> If set to `"none"`, Octave will always check time stamps on files to determine whether functions defined in function files need to recompiled.

### 11.9.1 Manipulating the Load Path

When a function is called, Octave searches a list of directories for a file that contains the function declaration. This list of directories is known as the load path. By default the load path contains a list of directories distributed with Octave plus the current working directory. To see your current load path call the `path` function without any input or output arguments.

It is possible to add or remove directories to or from the load path using `addpath` and `rmpath`. As an example, the following code adds '~/Octave' to the load path.

    addpath ("~/Octave")

After this the directory '~/Octave' will be searched for functions.

**addpath (***dir1***, ...)**                                     [Built-in Function]
**addpath (***dir1***, ..., ***option***)**                       [Built-in Function]
> Add named directories to the function search path.
>
> If *option* is `"-begin"` or 0 (the default), prepend the directory name to the current path. If *option* is `"-end"` or 1, append the directory name to the current path. Directories added to the path must exist.
>
> In addition to accepting individual directory arguments, lists of directory names separated by `pathsep` are also accepted. For example:

```
addpath ("dir1:/dir2:~/dir3")
```

**See also:** [path], page 190, [rmpath], page 190, [genpath], page 190, [pathdef], page 191, [savepath], page 190, [pathsep], page 191.

**genpath** (*dir*) [Built-in Function]
**genpath** (*dir*, *skip*, ...) [Built-in Function]

Return a path constructed from *dir* and all its subdirectories.

If additional string parameters are given, the resulting path will exclude directories with those names.

**rmpath** (*dir1*, ...) [Built-in Function]

Remove *dir1*, ... from the current function search path.

In addition to accepting individual directory arguments, lists of directory names separated by `pathsep` are also accepted. For example:

```
rmpath ("dir1:/dir2:~/dir3")
```

**See also:** [path], page 190, [addpath], page 189, [genpath], page 190, [pathdef], page 191, [savepath], page 190, [pathsep], page 191.

**savepath** () [Function File]
**savepath** (*file*) [Function File]
*status* = **savepath** (...) [Function File]

Save the unique portion of the current function search path that is not set during Octave's initialization process to *file*.

If *file* is omitted, Octave looks in the current directory for a project-specific '.octaverc' file in which to save the path information. If no such file is present then the user's configuration file '~/.octaverc' is used.

If successful, `savepath` returns 0.

The `savepath` function makes it simple to customize a user's configuration file to restore the working paths necessary for a particular instance of Octave. Assuming no filename is specified, Octave will automatically restore the saved directory paths from the appropriate '.octaverc' file when starting up. If a filename has been specified then the paths may be restored manually by calling `source` *file*.

**See also:** [path], page 190, [addpath], page 189, [rmpath], page 190, [genpath], page 190, [pathdef], page 191.

**path** (...) [Built-in Function]

Modify or display Octave's load path.

If *nargin* and *nargout* are zero, display the elements of Octave's load path in an easy to read format.

If *nargin* is zero and nargout is greater than zero, return the current load path.

If *nargin* is greater than zero, concatenate the arguments, separating them with `pathsep`. Set the internal search path to the result and return it.

No checks are made for duplicate elements.

**See also:** [addpath], page 189, [rmpath], page 190, [genpath], page 190, [pathdef], page 191, [savepath], page 190, [pathsep], page 191.

Chapter 11: Functions and Scripts 191

`val = pathdef ()` [Function File]

Return the default path for Octave.

The path information is extracted from one of four sources. The possible sources, in order of preference, are:

1. '.octaverc'
2. '~/.octaverc'
3. '<OCTAVE_HOME>/.../<version>/m/startup/octaverc'
4. Octave's path prior to changes by any octaverc file.

**See also:** [path], page 190, [addpath], page 189, [rmpath], page 190, [genpath], page 190, [savepath], page 190.

`val = pathsep ()` [Built-in Function]
`old_val = pathsep (new_val)` [Built-in Function]

Query or set the character used to separate directories in a path.

**See also:** [filesep], page 761.

`rehash ()` [Built-in Function]

Reinitialize Octave's load path directory cache.

`file_in_loadpath (file)` [Built-in Function]
`file_in_loadpath (file, "all")` [Built-in Function]

Return the absolute name of *file* if it can be found in the list of directories specified by **path**.

If no file is found, return an empty character string.

If the first argument is a cell array of strings, search each directory of the loadpath for element of the cell array and return the first that matches.

If the second optional argument `"all"` is supplied, return a cell array containing the list of all files that have the same name in the path. If no files are found, return an empty cell array.

**See also:** [file_in_path], page 760, [dir_in_loadpath], page 191, [path], page 190.

`restoredefaultpath (...)` [Built-in Function]

Restore Octave's path to its initial state at startup.

**See also:** [path], page 190, [addpath], page 189, [rmpath], page 190, [genpath], page 190, [pathdef], page 191, [savepath], page 190, [pathsep], page 191.

`command_line_path (...)` [Built-in Function]

Return the command line path variable.

**See also:** [path], page 190, [addpath], page 189, [rmpath], page 190, [genpath], page 190, [pathdef], page 191, [savepath], page 190, [pathsep], page 191.

`dir_in_loadpath (dir)` [Built-in Function]
`dir_in_loadpath (dir, "all")` [Built-in Function]

Return the full name of the path element matching *dir*.

The match is performed at the end of each path element. For example, if *dir* is "foo/bar", it matches the path element "/some/dir/foo/bar", but not "/some/dir/foo/bar/baz" "/some/dir/allfoo/bar".

If the optional second argument is supplied, return a cell array containing all name matches rather than just the first.

**See also:** [file_in_path], page 760, [file_in_loadpath], page 191, [path], page 190.

### 11.9.2 Subfunctions

A function file may contain secondary functions called *subfunctions*. These secondary functions are only visible to the other functions in the same function file. For example, a file 'f.m' containing

```
function f ()
  printf ("in f, calling g\n");
  g ()
endfunction
function g ()
  printf ("in g, calling h\n");
  h ()
endfunction
function h ()
  printf ("in h\n")
endfunction
```

defines a main function f and two subfunctions. The subfunctions g and h may only be called from the main function f or from the other subfunctions, but not from outside the file 'f.m'.

### 11.9.3 Private Functions

In many cases one function needs to access one or more helper functions. If the helper function is limited to the scope of a single function, then subfunctions as discussed above might be used. However, if a single helper function is used by more than one function, then this is no longer possible. In this case the helper functions might be placed in a subdirectory, called "private", of the directory in which the functions needing access to this helper function are found.

As a simple example, consider a function func1, that calls a helper function func2 to do much of the work. For example:

```
function y = func1 (x)
  y = func2 (x);
endfunction
```

Then if the path to func1 is <directory>/func1.m, and if func2 is found in the directory <directory>/private/func2.m, then func2 is only available for use of the functions, like func1, that are found in <directory>.

### 11.9.4 Nested Functions

Nested functions are similar to subfunctions in that only the main function is visible outside the file. However, they also allow for child functions to access the local variables in their parent function. This shared access mimics using a global variable to share information — but a global variable which is not visible to the rest of Octave. As a programming strategy, sharing data this way can create code which is difficult to maintain. It is recommended to use subfunctions in place of nested functions when possible.

As a simple example, consider a parent function `foo`, that calls a nested child function `bar`, with a shared variable $x$.

```
function y = foo ()
  x = 10;
  bar ();
  y = x;

  function bar ()
    x = 20;
  endfunction
endfunction

foo ()
  ⇒ 20
```

Notice that there is no special syntax for sharing $x$. This can lead to problems with accidental variable sharing between a parent function and its child. While normally variables are inherited, child function parameters and return values are local to the child function.

Now consider the function `foobar` that uses variables $x$ and $y$. `foobar` calls a nested function `foo` which takes $x$ as a parameter and returns $y$. `foo` then calls `bat` which does some computation.

```
function z = foobar ()
  x = 0;
  y = 0;
  z = foo (5);
  z += x + y;

  function y = foo (x)
    y = x + bat ();

    function z = bat ()
      z = x;
    endfunction
  endfunction
endfunction

foobar ()
  ⇒ 10
```

It is important to note that the $x$ and $y$ in `foobar` remain zero, as in `foo` they are a return value and parameter respectively. The $x$ in `bat` refers to the $x$ in `foo`.

Variable inheritance leads to a problem for `eval` and scripts. If a new variable is created in a parent function, it is not clear what should happen in nested child functions. For example, consider a parent function `foo` with a nested child function `bar`:

```
function y = foo (to_eval)
  bar ();
  eval (to_eval);

  function bar ()
    eval ("x = 100;");
    eval ("y = x;");
  endfunction
endfunction

foo ("x = 5;")
    ⇒ error: can not add variable "x" to a static workspace

foo ("y = 10;")
    ⇒ 10

foo ("")
    ⇒ 100
```

The parent function `foo` is unable to create a new variable x, but the child function `bar` was successful. Furthermore, even in an `eval` statement y in `bar` is the same y as in its parent function `foo`. The use of `eval` in conjunction with nested functions is best avoided.

As with subfunctions, only the first nested function in a file may be called from the outside. Inside a function the rules are more complicated. In general a nested function may call:

0. Globally visible functions
1. Any function that the nested function's parent can call
2. Sibling functions (functions that have the same parents)
3. Direct children

As a complex example consider a parent function `ex_top` with two child functions, `ex_a` and `ex_b`. In addition, `ex_a` has two more child functions, `ex_aa` and `ex_ab`. For example:

```
function ex_top ()
  ## Can call: ex_top, ex_a, and ex_b
  ## Can NOT call: ex_aa and ex_ab

  function ex_a ()
    ## Call call everything

    function ex_aa ()
      ## Can call everything
    endfunction

    function ex_ab ()
```

```
      ## Can call everything
    endfunction
  endfunction

  function ex_b ()
    ## Can call: ex_top, ex_a, and ex_b
    ## Can NOT call: ex_aa and ex_ab
  endfunction
endfunction
```

### 11.9.5 Overloading and Autoloading

Functions can be overloaded to work with different input arguments. For example, the operator '+' has been overloaded in Octave to work with single, double, uint8, int32, and many other arguments. The preferred way to overload functions is through classes and object oriented programming (see Section 34.4.1 [Function Overloading], page 727). Occasionally, however, one needs to undo user overloading and call the default function associated with a specific type. The `builtin` function exists for this purpose.

[...] = builtin (*f*, ...)          [Built-in Function]

> Call the base function *f* even if *f* is overloaded to another function for the given type signature.
>
> This is normally useful when doing object-oriented programming and there is a requirement to call one of Octave's base functions rather than the overloaded one of a new class.
>
> A trivial example which redefines the `sin` function to be the `cos` function shows how `builtin` works.
>
> ```
> sin (0)
>    ⇒ 0
> function y = sin (x), y = cos (x); endfunction
> sin (0)
>    ⇒ 1
> builtin ("sin", 0)
>    ⇒ 0
> ```

A single dynamically linked file might define several functions. However, as Octave searches for functions based on the functions filename, Octave needs a manner in which to find each of the functions in the dynamically linked file. On operating systems that support symbolic links, it is possible to create a symbolic link to the original file for each of the functions which it contains.

However, there is at least one well known operating system that doesn't support symbolic links. Making copies of the original file for each of the functions is undesirable as it increases the amount of disk space used by Octave. Instead Octave supplies the `autoload` function, that permits the user to define in which file a certain function will be found.

*autoload_map* = autoload ()          [Built-in Function]
autoload (*function*, *file*)          [Built-in Function]

`autoload (..., "remove")` [Built-in Function]
    Define *function* to autoload from *file*.

    The second argument, *file*, should be an absolute file name or a file name in the same directory as the function or script from which the autoload command was run. *file should not* depend on the Octave load path.

    Normally, calls to `autoload` appear in PKG_ADD script files that are evaluated when a directory is added to Octave's load path. To avoid having to hardcode directory names in *file*, if *file* is in the same directory as the PKG_ADD script then

```
autoload ("foo", "bar.oct");
```

will load the function `foo` from the file `bar.oct`. The above usage when `bar.oct` is not in the same directory, or usages such as

```
autoload ("foo", file_in_loadpath ("bar.oct"))
```

are strongly discouraged, as their behavior may be unpredictable.

    With no arguments, return a structure containing the current autoload map.

    If a third argument `"remove"` is given, the function is cleared and not loaded anymore during the current Octave session.

    **See also:** [PKG_ADD], page 806.

### 11.9.6 Function Locking

It is sometime desirable to lock a function into memory with the `mlock` function. This is typically used for dynamically linked functions in Oct-files or mex-files that contain some initialization, and it is desirable that calling `clear` does not remove this initialization.

As an example,

```
function my_function ()
  mlock ();
  ...
```

prevents `my_function` from being removed from memory after it is called, even if `clear` is called. It is possible to determine if a function is locked into memory with the `mislocked`, and to unlock a function with `munlock`, which the following illustrates.

```
my_function ();
mislocked ("my_function")
⇒ ans = 1
munlock ("my_function");
mislocked ("my_function")
⇒ ans = 0
```

A common use of `mlock` is to prevent persistent variables from being removed from memory, as the following example shows:

Chapter 11: Functions and Scripts                                          197

```
function count_calls ()
  mlock ();
  persistent calls = 0;
  printf ("'count_calls' has been called %d times\n",
          ++calls);
endfunction

count_calls ();
⊣ 'count_calls' has been called 1 times

clear count_calls
count_calls ();
⊣ 'count_calls' has been called 2 times
```

`mlock` might equally be used to prevent changes to a function from having effect in Octave, though a similar effect can be had with the `ignore_function_time_stamp` function.

`mlock ()`                                                    [Built-in Function]

> Lock the current function into memory so that it can't be cleared.
>
> **See also:** [munlock], page 197, [mislocked], page 197, [persistent], page 126.

`munlock ()`                                                  [Built-in Function]
`munlock (fcn)`                                               [Built-in Function]

> Unlock the named function *fcn*.
>
> If no function is named then unlock the current function.
>
> **See also:** [mlock], page 197, [mislocked], page 197, [persistent], page 126.

`mislocked ()`                                                [Built-in Function]
`mislocked (fcn)`                                             [Built-in Function]

> Return true if the named function *fcn* is locked.
>
> If no function is named then return true if the current function is locked.
>
> **See also:** [mlock], page 197, [munlock], page 197, [persistent], page 126.

### 11.9.7 Function Precedence

Given the numerous different ways that Octave can define a function, it is possible and even likely that multiple versions of a function, might be defined within a particular scope. The precedence of which function will be used within a particular scope is given by

1. Subfunction A subfunction with the required function name in the given scope.

2. Private function A function defined within a private directory of the directory which contains the current function.

3. Class constructor A function that constructs a user class as defined in chapter Chapter 34 [Object Oriented Programming], page 717.

4. Class method An overloaded function of a class as in chapter Chapter 34 [Object Oriented Programming], page 717.

5. Command-line Function A function that has been defined on the command-line.

6. Autoload function A function that is marked as autoloaded with See [autoload], page 195.
7. A Function on the Path A function that can be found on the users load-path. There can also be Oct-file, mex-file or m-file versions of this function and the precedence between these versions are in that order.
8. Built-in function A function that is a part of core Octave such as `numel`, `size`, etc.

## 11.10 Script Files

A script file is a file containing (almost) any sequence of Octave commands. It is read and evaluated just as if you had typed each command at the Octave prompt, and provides a convenient way to perform a sequence of commands that do not logically belong inside a function.

Unlike a function file, a script file must *not* begin with the keyword `function`. If it does, Octave will assume that it is a function file, and that it defines a single function that should be evaluated as soon as it is defined.

A script file also differs from a function file in that the variables named in a script file are not local variables, but are in the same scope as the other variables that are visible on the command line.

Even though a script file may not begin with the `function` keyword, it is possible to define more than one function in a single script file and load (but not execute) all of them at once. To do this, the first token in the file (ignoring comments and other white space) must be something other than `function`. If you have no other statements to evaluate, you can use a statement that has no effect, like this:

```
# Prevent Octave from thinking that this
# is a function file:

1;

# Define function one:

function one ()
    ...
```

To have Octave read and compile these functions into an internal form, you need to make sure that the file is in Octave's load path (accessible through the `path` function), then simply type the base name of the file that contains the commands. (Octave uses the same rules to search for script files as it does to search for function files.)

If the first token in a file (ignoring comments) is `function`, Octave will compile the function and try to execute it, printing a message warning about any non-whitespace characters that appear after the function definition.

Note that Octave does not try to look up the definition of any identifier until it needs to evaluate it. This means that Octave will compile the following statements if they appear in a script file, or are typed at the command line,

# Chapter 11: Functions and Scripts

```
# not a function file:
1;
function foo ()
  do_something ();
endfunction
function do_something ()
  do_something_else ();
endfunction
```

even though the function `do_something` is not defined before it is referenced in the function `foo`. This is not an error because Octave does not need to resolve all symbols that are referenced by a function until the function is actually evaluated.

Since Octave doesn't look for definitions until they are needed, the following code will always print 'bar = 3' whether it is typed directly on the command line, read from a script file, or is part of a function body, even if there is a function or script file called 'bar.m' in Octave's path.

```
eval ("bar = 3");
bar
```

Code like this appearing within a function body could fool Octave if definitions were resolved as the function was being compiled. It would be virtually impossible to make Octave clever enough to evaluate this code in a consistent fashion. The parser would have to be able to perform the call to `eval` at compile time, and that would be impossible unless all the references in the string to be evaluated could also be resolved, and requiring that would be too restrictive (the string might come from user input, or depend on things that are not known until the function is evaluated).

Although Octave normally executes commands from script files that have the name '`file.m`', you can use the function `source` to execute commands from any file.

**source** (*file*)                                                          [Built-in Function]

    Parse and execute the contents of *file*.

    This is equivalent to executing commands from a script file, but without requiring the file to be named '`file.m`'.

    **See also:** [run], page 156.

## 11.11 Function Handles, Anonymous Functions, Inline Functions

It can be very convenient store a function in a variable so that it can be passed to a different function. For example, a function that performs numerical minimization needs access to the function that should be minimized.

### 11.11.1 Function Handles

A function handle is a pointer to another function and is defined with the syntax

    `@function-name`

For example,

```
f = @sin;
```
creates a function handle called `f` that refers to the function `sin`.

Function handles are used to call other functions indirectly, or to pass a function as an argument to another function like `quad` or `fsolve`. For example:

```
f = @sin;
quad (f, 0, pi)
    ⇒ 2
```

You may use `feval` to call a function using function handle, or simply write the name of the function handle followed by an argument list. If there are no arguments, you must use an empty argument list '()'. For example:

```
f = @sin;
feval (f, pi/4)
    ⇒ 0.70711
f (pi/4)
    ⇒ 0.70711
```

**is_function_handle (*x*)** [Built-in Function]
Return true if *x* is a function handle.

**See also:** [isa], page 39, [typeinfo], page 39, [class], page 39, [functions], page 200.

**s = functions (*fcn_handle*)** [Built-in Function]
Return a structure containing information about the function handle *fcn_handle*.

The structure *s* always contains these three fields:

function  The function name. For an anonymous function (no name) this will be the actual function definition.

type      Type of the function.

    anonymous
        The function is anonymous.

    private   The function is private.

    overloaded
        The function overloads an existing function.

    simple    The function is a built-in or m-file function.

    subfunction
        The function is a subfunction within an m-file.

file      The m-file that will be called to perform the function. This field is empty for anonymous and built-in functions.

In addition, some function types may return more information in additional fields.

**Warning:** `functions` is provided for debugging purposes only. It's behavior may change in the future and programs should not depend on a particular output.

Chapter 11: Functions and Scripts                                                      201

func2str (*fcn_handle*)                                               [Built-in Function]
    Return a string containing the name of the function referenced by the function handle *fcn_handle*.

    **See also:** [str2func], page 201, [functions], page 200.

str2func (*fcn_name*)                                                 [Built-in Function]
str2func (*fcn_name*, "*global*")                                     [Built-in Function]
    Return a function handle constructed from the string *fcn_name*.

    If the optional `"global"` argument is passed, locally visible functions are ignored in the lookup.

    **See also:** [func2str], page 201, [inline], page 202.

## 11.11.2 Anonymous Functions

Anonymous functions are defined using the syntax

    `@(argument-list) expression`

Any variables that are not found in the argument list are inherited from the enclosing scope. Anonymous functions are useful for creating simple unnamed functions from expressions or for wrapping calls to other functions to adapt them for use by functions like **quad**. For example,

```
f = @(x) x.^2;
quad (f, 0, 10)
    ⇒ 333.33
```

creates a simple unnamed function from the expression `x.^2` and passes it to **quad**,

```
quad (@(x) sin (x), 0, pi)
    ⇒ 2
```

wraps another function, and

```
a = 1;
b = 2;
quad (@(x) betainc (x, a, b), 0, 0.4)
    ⇒ 0.13867
```

adapts a function with several parameters to the form required by **quad**. In this example, the values of *a* and *b* that are passed to **betainc** are inherited from the current environment.

    Note that for performance reasons it is better to use handles to existing Octave functions, rather than to define anonymous functions which wrap an existing function. The integration of **sin (x)** is 5X faster if the code is written as

```
quad (@sin, 0, pi)
```

rather than using the anonymous function `@(x) sin (x)`. There are many operators which have functional equivalents that may be better choices than an anonymous function. Instead of writing

```
f = @(x, y) x + y
```

this should be coded as

```
f = @plus
```

    See Section 34.4.2 [Operator Overloading], page 727, for a list of operators which also have a functional form.

### 11.11.3 Inline Functions

An inline function is created from a string containing the function body using the `inline` function. The following code defines the function $f(x) = x^2 + 2$.

```
f = inline ("x^2 + 2");
```

After this it is possible to evaluate $f$ at any $x$ by writing `f(x)`.

**Caution**: MATLAB has begun the process of deprecating inline functions. At some point in the future support will be dropped and eventually Octave will follow MATLAB and also remove inline functions. Use anonymous functions in all new code.

`inline (str)` [Built-in Function]
`inline (str, arg1, ...)` [Built-in Function]
`inline (str, n)` [Built-in Function]

Create an inline function from the character string *str*.

If called with a single argument, the arguments of the generated function are extracted from the function itself. The generated function arguments will then be in alphabetical order. It should be noted that i and j are ignored as arguments due to the ambiguity between their use as a variable or their use as an built-in constant. All arguments followed by a parenthesis are considered to be functions. If no arguments are found, a function taking a single argument named `x` will be created.

If the second and subsequent arguments are character strings, they are the names of the arguments of the function.

If the second argument is an integer *n*, the arguments are `"x"`, `"P1"`, ..., `"PN"`.

Programming Note: The use of `inline` is discouraged and it may be removed from a future version of Octave. The preferred way to create functions from strings is through the use of anonymous functions (see Section 11.11.2 [Anonymous Functions], page 201) or `str2func`.

**See also:** [argnames], page 202, [formula], page 202, [vectorize], page 488, [str2func], page 201.

`argnames (fun)` [Built-in Function]

Return a cell array of character strings containing the names of the arguments of the inline function *fun*.

**See also:** [inline], page 202, [formula], page 202, [vectorize], page 488.

`formula (fun)` [Built-in Function]

Return a character string representing the inline function *fun*.

Note that `char (fun)` is equivalent to `formula (fun)`.

**See also:** [char], page 71, [argnames], page 202, [inline], page 202, [vectorize], page 488.

`vars = symvar (str)` [Function File]

Identify the symbolic variable names in the string *str*.

Common constant names such as `i`, `j`, `pi`, `Inf` and Octave functions such as `sin` or `plot` are ignored.

Any names identified are returned in a cell array of strings. The array is empty if no variables were found.

Example:

```
symvar ("x^2 + y^2 == 4")
⇒ {
    [1,1] = x
    [2,1] = y
  }
```

## 11.12 Commands

Commands are a special class of functions that only accept string input arguments. A command can be called as an ordinary function, but it can also be called without the parentheses. For example,

```
my_command hello world
```

is equivalent to

```
my_command ("hello", "world")
```

The general form of a command call is

    *cmdname* `arg1 arg2 ...`

which translates directly to

    *cmdname* `("arg1", "arg2", ...)`

Any regular function can be used as a command if it accepts string input arguments. For example:

```
toupper lower_case_arg
    ⇒ ans = LOWER_CASE_ARG
```

One difficulty of commands occurs when one of the string input arguments is stored in a variable. Because Octave can't tell the difference between a variable name and an ordinary string, it is not possible to pass a variable as input to a command. In such a situation a command must be called as a function. For example:

```
strvar = "hello world";
toupper strvar
    ⇒ ans = STRVAR
toupper (strvar)
    ⇒ ans = HELLO WORLD
```

## 11.13 Organization of Functions Distributed with Octave

Many of Octave's standard functions are distributed as function files. They are loosely organized by topic, in subdirectories of '*octave-home*/lib/octave/*version*/m', to make it easier to find them.

The following is a list of all the function file subdirectories, and the types of functions you will find there.

'audio'    Functions for playing and recording sounds.

'deprecated'
: Out-of-date functions which will eventually be removed from Octave.

'elfun'
: Elementary functions, principally trigonometric.

'@ftp'
: Class functions for the FTP object.

'general'
: Miscellaneous matrix manipulations, like `flipud`, `rot90`, and `triu`, as well as other basic functions, like `ismatrix`, `narginchk`, etc.

'geometry'
: Functions related to Delaunay triangulation.

'help'
: Functions for Octave's built-in help system.

'image'
: Image processing tools. These functions require the X Window System.

'io'
: Input-output functions.

'linear-algebra'
: Functions for linear algebra.

'miscellaneous'
: Functions that don't really belong anywhere else.

'optimization'
: Functions related to minimization, optimization, and root finding.

'path'
: Functions to manage the directory path Octave uses to find functions.

'pkg'
: Package manager for installing external packages of functions in Octave.

'plot'
: Functions for displaying and printing two- and three-dimensional graphs.

'polynomial'
: Functions for manipulating polynomials.

'prefs'
: Functions implementing user-defined preferences.

'set'
: Functions for creating and manipulating sets of unique values.

'signal'
: Functions for signal processing applications.

'sparse'
: Functions for handling sparse matrices.

'specfun'
: Special functions such as `bessel` or `factor`.

'special-matrix'
: Functions that create special matrix forms such as Hilbert or Vandermonde matrices.

'startup'
: Octave's system-wide startup file.

'statistics'
: Statistical functions.

'strings'
: Miscellaneous string-handling functions.

'testfun'
: Functions for performing unit tests on other functions.

'time'
: Functions related to time and date processing.

# 12 Errors and Warnings

Octave includes several functions for printing error and warning messages. When you write functions that need to take special action when they encounter abnormal conditions, you should print the error messages using the functions described in this chapter.

Since many of Octave's functions use these functions, it is also useful to understand them, so that errors and warnings can be handled.

## 12.1 Handling Errors

An error is something that occurs when a program is in a state where it doesn't make sense to continue. An example is when a function is called with too few input arguments. In this situation the function should abort with an error message informing the user of the lacking input arguments.

Since an error can occur during the evaluation of a program, it is very convenient to be able to detect that an error occurred, so that the error can be fixed. This is possible with the `try` statement described in Section 10.9 [The try Statement], page 168.

### 12.1.1 Raising Errors

The most common use of errors is for checking input arguments to functions. The following example calls the `error` function if the function `f` is called without any input arguments.

```
function f (arg1)
  if (nargin == 0)
    error ("not enough input arguments");
  endif
endfunction
```

When the `error` function is called, it prints the given message and returns to the Octave prompt. This means that no code following a call to `error` will be executed.

error (*template*, ...)  [Built-in Function]
error (*id*, *template*, ...)  [Built-in Function]
    Display an error message and stop m-file execution.

    Format the optional arguments under the control of the template string *template* using the same rules as the `printf` family of functions (see Section 14.2.4 [Formatted Output], page 253) and print the resulting message on the `stderr` stream. The message is prefixed by the character string 'error: '.

    Calling `error` also sets Octave's internal error state such that control will return to the top level without evaluating any further commands. This is useful for aborting from functions or scripts.

    If the error message does not end with a newline character, Octave will print a traceback of all the function calls leading to the error. For example, given the following function definitions:

```
function f () g (); end
function g () h (); end
function h () nargin == 1 || error ("nargin != 1"); end
```

calling the function `f` will result in a list of messages that can help you to quickly locate the exact location of the error:

```
f ()
error: nargin != 1
error: called from:
error:    error at line -1, column -1
error:    h at line 1, column 27
error:    g at line 1, column 15
error:    f at line 1, column 15
```

If the error message ends in a newline character, Octave will print the message but will not display any traceback messages as it returns control to the top level. For example, modifying the error message in the previous example to end in a newline causes Octave to only print a single message:

```
function h () nargin == 1 || error ("nargin != 1\n"); end
f ()
error: nargin != 1
```

A null string ("") input to `error` will be ignored and the code will continue running as if the statement were a NOP. This is for compatibility with MATLAB. It also makes it possible to write code such as

```
err_msg = "";
if (CONDITION 1)
  err_msg = "CONDITION 1 found";
elseif (CONDITION2)
  err_msg = "CONDITION 2 found";
...
endif
error (err_msg);
```

which will only stop execution if an error has been found.

Implementation Note: For compatibility with MATLAB, escape sequences in *template* (e.g., "\n" => newline) are processed regardless of whether *template* has been defined with single quotes, as long as there are two or more input arguments. To disable escape sequence expansion use a second backslash before the sequence (e.g., "\\n") or use the `regexptranslate` function.

**See also:** [warning], page 212, [lasterror], page 208.

Since it is common to use errors when there is something wrong with the input to a function, Octave supports functions to simplify such code. When the `print_usage` function is called, it reads the help text of the function calling `print_usage`, and presents a useful error. If the help text is written in Texinfo it is possible to present an error message that only contains the function prototypes as described by the `@deftypefn` parts of the help text. When the help text isn't written in Texinfo, the error message contains the entire help message.

Consider the following function.

Chapter 12: Errors and Warnings                                                         207

```
## -*- texinfo -*-
## @deftypefn {Function File} f (@var{arg1})
## Function help text goes here...
## @end deftypefn
function f (arg1)
  if (nargin == 0)
    print_usage ();
  endif
endfunction
```

When it is called with no input arguments it produces the following error.

```
f ()
```

    ⊣  error: Invalid call to f.  Correct usage is:
    ⊣
    ⊣   -- Function File: f (ARG1)
    ⊣
    ⊣
    ⊣  Additional help for built-in functions and operators is
    ⊣  available in the online version of the manual.  Use the command
    ⊣  'doc <topic>' to search the manual index.
    ⊣
    ⊣  Help and information about Octave is also available on the WWW
    ⊣  at http://www.octave.org and via the help@octave.org
    ⊣  mailing list.

**print_usage ()**                                                          [Function File]
**print_usage (*name*)**                                                    [Function File]
    Print the usage message for the function *name*.

    When called with no input arguments the `print_usage` function displays the usage message of the currently executing function.

    **See also:** [help], page 20.

**beep ()**                                                                 [Function File]
    Produce a beep from the speaker (or visual bell).

    This function sends the alarm character `"\a"` to the terminal. Depending on the user's configuration this may produce an audible beep, a visual bell, or nothing at all.

    **See also:** [puts], page 252, [fputs], page 251, [printf], page 253, [fprintf], page 253.

*val* = **beep_on_error ()**                                                [Built-in Function]
*old_val* = **beep_on_error (*new_val*)**                                   [Built-in Function]
**beep_on_error (*new_val*, "*local*")**                                    [Built-in Function]
    Query or set the internal variable that controls whether Octave will try to ring the terminal bell before printing an error message.

    When called from inside a function with the `"local"` option, the variable is changed locally for the function and any subroutines it calls. The original variable value is restored when exiting the function.

### 12.1.2 Catching Errors

When an error occurs, it can be detected and handled using the `try` statement as described in Section 10.9 [The try Statement], page 168. As an example, the following piece of code counts the number of errors that occurs during a `for` loop.

```
number_of_errors = 0;
for n = 1:100
  try
    ...
  catch
    number_of_errors++;
  end_try_catch
endfor
```

The above example treats all errors the same. In many situations it can however be necessary to discriminate between errors, and take different actions depending on the error. The `lasterror` function returns a structure containing information about the last error that occurred. As an example, the code above could be changed to count the number of errors related to the '*' operator.

```
number_of_errors = 0;
for n = 1:100
  try
    ...
  catch
    msg = lasterror.message;
    if (strfind (msg, "operator *"))
      number_of_errors++;
    endif
  end_try_catch
endfor
```

Alternatively, the output of the `lasterror` function can be found in a variable indicated immediately after the `catch` keyword, as in the example below showing how to redirect an error as a warning:

```
try
  ...
catch err
  warning(err.identifier, err.message);
  ...
end_try_catch
```

*lasterr* = lasterror ()  [Built-in Function]
lasterror (*err*)  [Built-in Function]
lasterror ("*reset*")  [Built-in Function]

> Query or set the last error message structure.
>
> When called without arguments, return a structure containing the last error message and other information related to this error. The elements of the structure are:
>
> message   The text of the last error message

identifier
: The message identifier of this error message

stack
: A structure containing information on where the message occurred. This may be an empty structure if the information cannot be obtained. The fields of the structure are:

    file
    : The name of the file where the error occurred

    name
    : The name of function in which the error occurred

    line
    : The line number at which the error occurred

    column
    : An optional field with the column number at which the error occurred

The last error structure may be set by passing a scalar structure, *err*, as input. Any fields of *err* that match those above are set while any unspecified fields are initialized with default values.

If `lasterror` is called with the argument `"reset"`, all fields are set to their default values.

**See also:** [lasterr], page 209, [error], page 205, [lastwarn], page 213.

[*msg*, *msgid*] = lasterr ()  [Built-in Function]
lasterr (*msg*)  [Built-in Function]
lasterr (*msg*, *msgid*)  [Built-in Function]
: Query or set the last error message.

When called without input arguments, return the last error message and message identifier.

With one argument, set the last error message to *msg*.

With two arguments, also set the last message identifier.

**See also:** [lasterror], page 208, [error], page 205, [lastwarn], page 213.

It is also possible to assign an identification string to an error. If an error has such an ID the user can catch this error as will be shown in the next example. To assign an ID to an error, simply call `error` with two string arguments, where the first is the identification string, and the second is the actual error. Note that error IDs are in the format `"NAMESPACE:ERROR-NAME"`. The namespace `"Octave"` is used for Octave's own errors. Any other string is available as a namespace for user's own errors.

The next example counts indexing errors. The errors are caught using the field identifier of the structure returned by the function `lasterror`.

```
      number_of_errors = 0;
      for n = 1:100
        try
          ...
        catch
          id = lasterror.identifier;
          if (strcmp (id, "Octave:invalid-indexing"))
            number_of_errors++;
          endif
        end_try_catch
      endfor
```

The functions distributed with Octave can issue one of the following errors.

`Octave:invalid-context`
:   Indicates the error was generated by an operation that cannot be executed in the scope from which it was called. For example, the function `print_usage ()` when called from the Octave prompt raises this error.

`Octave:invalid-input-arg`
:   Indicates that a function was called with invalid input arguments.

`Octave:invalid-fun-call`
:   Indicates that a function was called in an incorrect way, e.g., wrong number of input arguments.

`Octave:invalid-indexing`
:   Indicates that a data-type was indexed incorrectly, e.g., real-value index for arrays, nonexistent field of a structure.

`Octave:bad-alloc`
:   Indicates that memory couldn't be allocated.

`Octave:undefined-function`
:   Indicates a call to a function that is not defined. The function may exist but Octave is unable to find it in the search path.

When an error has been handled it is possible to raise it again. This can be useful when an error needs to be detected, but the program should still abort. This is possible using the `rethrow` function. The previous example can now be changed to count the number of errors related to the '*' operator, but still abort if another kind of error occurs.

Chapter 12: Errors and Warnings 211

```
number_of_errors = 0;
for n = 1:100
  try
    ...
  catch
    msg = lasterror.message;
    if (strfind (msg, "operator *"))
      number_of_errors++;
    else
      rethrow (lasterror);
    endif
  end_try_catch
endfor
```

**rethrow (*err*)** [Built-in Function]

Reissue a previous error as defined by *err*.

*err* is a structure that must contain at least the "`message`" and "`identifier`" fields. *err* can also contain a field "`stack`" that gives information on the assumed location of the error. Typically *err* is returned from `lasterror`.

See also: [lasterror], page 208, [lasterr], page 209, [error], page 205.

*err* = **errno ()** [Built-in Function]
*err* = **errno (*val*)** [Built-in Function]
*err* = **errno (*name*)** [Built-in Function]

Return the current value of the system-dependent variable errno, set its value to *val* and return the previous value, or return the named error code given *name* as a character string, or -1 if *name* is not found.

See also: [errno_list], page 211.

**errno_list ()** [Built-in Function]

Return a structure containing the system-dependent errno values.

See also: [errno], page 211.

## 12.1.3 Recovering From Errors

Octave provides several ways of recovering from errors. There are `try/catch` blocks, `unwind_protect/unwind_protect_cleanup` blocks, and finally the `onCleanup` command.

The `onCleanup` command associates an ordinary Octave variable (the trigger) with an arbitrary function (the action). Whenever the Octave variable ceases to exist—whether due to a function return, an error, or simply because the variable has been removed with `clear`—then the assigned function is executed.

The function can do anything necessary for cleanup such as closing open file handles, printing an error message, or restoring global variables to their initial values. The last example is a very convenient idiom for Octave code. For example:

```
function rand42
  old_state = rand ("state");
  restore_state = onCleanup (@() rand ("state", old_state));
  rand ("state", 42);
  ...
endfunction   # rand generator state restored by onCleanup
```

*obj* = onCleanup (*function*)  [Built-in Function]

Create a special object that executes a given function upon destruction.

If the object is copied to multiple variables (or cell or struct array elements) or returned from a function, *function* will be executed after clearing the last copy of the object. Note that if multiple local onCleanup variables are created, the order in which they are called is unspecified. For similar functionality See Section 10.8 [The unwind_protect Statement], page 168.

## 12.2 Handling Warnings

Like an error, a warning is issued when something unexpected happens. Unlike an error, a warning doesn't abort the currently running program. A simple example of a warning is when a number is divided by zero. In this case Octave will issue a warning and assign the value Inf to the result.

```
a = 1/0
     ⊣ warning: division by zero
     ⇒ a = Inf
```

### 12.2.1 Issuing Warnings

It is possible to issue warnings from any code using the `warning` function. In its most simple form, the `warning` function takes a string describing the warning as its input argument. As an example, the following code controls if the variable 'a' is non-negative, and if not issues a warning and sets 'a' to zero.

```
a = -1;
if (a < 0)
  warning ("'a' must be non-negative.  Setting 'a' to zero.");
  a = 0;
endif
     ⊣ 'a' must be non-negative.  Setting 'a' to zero.
```

Since warnings aren't fatal to a running program, it is not possible to catch a warning using the `try` statement or something similar. It is however possible to access the last warning as a string using the `lastwarn` function.

It is also possible to assign an identification string to a warning. If a warning has such an ID the user can enable and disable this warning as will be described in the next section. To assign an ID to a warning, simply call `warning` with two string arguments, where the first is the identification string, and the second is the actual warning. Note that warning IDs are in the format `"NAMESPACE:WARNING-NAME"`. The namespace `"Octave"` is used for Octave's own warnings. Any other string is available as a namespace for user's own warnings.

Chapter 12: Errors and Warnings                                          213

warning (*template*, ...)                                    [Built-in Function]
warning (*id*, *template*, ...)                              [Built-in Function]
warning ("*on*", *id*)                                       [Built-in Function]
warning ("*off*", *id*)                                      [Built-in Function]
warning ("*query*", *id*)                                    [Built-in Function]
warning ("*error*", *id*)                                    [Built-in Function]
warning (*state*, "*backtrace*")                             [Built-in Function]
warning (*state*, *id*, "*local*")                           [Built-in Function]

Display a warning message or control the behavior of Octave's warning system.

Format the optional arguments under the control of the template string *template* using the same rules as the `printf` family of functions (see Section 14.2.4 [Formatted Output], page 253) and print the resulting message on the `stderr` stream. The message is prefixed by the character string 'warning: '. You should use this function when you want to notify the user of an unusual condition, but only when it makes sense for your program to go on.

The optional message identifier allows users to enable or disable warnings tagged by *id*. A message identifier is of the form "NAMESPACE:WARNING-NAME". Octave's own warnings use the "Octave" namespace (see [XREFwarning_ids], page 214). The special identifier "all" may be used to set the state of all warnings.

If the first argument is "on" or "off", set the state of a particular warning using the identifier *id*. If the first argument is "query", query the state of this warning instead. If the identifier is omitted, a value of "all" is assumed. If you set the state of a warning to "error", the warning named by *id* is handled as if it were an error instead. So, for example, the following handles all warnings as errors:

    warning ("error");

If the state is "on" or "off" and the third argument is "backtrace", then a stack trace is printed along with the warning message when warnings occur inside function calls. This option is enabled by default.

If the state is "on", "off", or "error" and the third argument is "local", then the warning state will be set temporarily, until the end of the current function. Changes to warning states that are set locally affect the current function and all functions called from the current scope. The previous warning state is restored on return from the current function. The "local" option is ignored if used in the top-level workspace.

Implementation Note: For compatibility with MATLAB, escape sequences in *template* (e.g., "\n" => newline) are processed regardless of whether *template* has been defined with single quotes, as long as there are two or more input arguments. To disable escape sequence expansion use a second backslash before the sequence (e.g., "\\n") or use the `regexptranslate` function.

See also: [warning_ids], page 214, [lastwarn], page 213, [error], page 205.

[*msg*, *msgid*] = lastwarn ()                               [Built-in Function]
lastwarn (*msg*)                                             [Built-in Function]
lastwarn (*msg*, *msgid*)                                    [Built-in Function]

Query or set the last warning message.

When called without input arguments, return the last warning message and message identifier.

With one argument, set the last warning message to *msg*.

With two arguments, also set the last message identifier.

**See also:** [warning], page 212, [lasterror], page 208, [lasterr], page 209.

The functions distributed with Octave can issue one of the following warnings.

`Octave:abbreviated-property-match`
By default, the `Octave:abbreviated-property-match` warning is enabled.

`Octave:array-to-scalar`
If the `Octave:array-to-scalar` warning is enabled, Octave will warn when an implicit conversion from an array to a scalar value is attempted. By default, the `Octave:array-to-scalar` warning is disabled.

`Octave:array-to-vector`
If the `Octave:array-to-vector` warning is enabled, Octave will warn when an implicit conversion from an array to a vector value is attempted. By default, the `Octave:array-to-vector` warning is disabled.

`Octave:assign-as-truth-value`
If the `Octave:assign-as-truth-value` warning is enabled, a warning is issued for statements like

```
if (s = t)
   ...
```

since such statements are not common, and it is likely that the intent was to write

```
if (s == t)
   ...
```

instead.

There are times when it is useful to write code that contains assignments within the condition of a `while` or `if` statement. For example, statements like

```
while (c = getc ())
   ...
```

are common in C programming.

It is possible to avoid all warnings about such statements by disabling the `Octave:assign-as-truth-value` warning, but that may also let real errors like

```
if (x = 1)   # intended to test (x == 1)!
   ...
```

slip by.

In such cases, it is possible suppress errors for specific statements by writing them with an extra set of parentheses. For example, writing the previous example as

Chapter 12: Errors and Warnings                                                                                     215

```
          while ((c = getc ()))
              ...
```

will prevent the warning from being printed for this statement, while allowing Octave to warn about other assignments used in conditional contexts.

By default, the `Octave:assign-as-truth-value` warning is enabled.

`Octave:associativity-change`

If the `Octave:associativity-change` warning is enabled, Octave will warn about possible changes in the meaning of some code due to changes in associativity for some operators. Associativity changes have typically been made for MATLAB compatibility. By default, the `Octave:associativity-change` warning is enabled.

`Octave:autoload-relative-file-name`

If the `Octave:autoload-relative-file-name` is enabled, Octave will warn when parsing autoload() function calls with relative paths to function files. This usually happens when using autoload() calls in PKG_ADD files, when the PKG_ADD file is not in the same directory as the .oct file referred to by the autoload() command. By default, the `Octave:autoload-relative-file-name` warning is enabled.

`Octave:built-in-variable-assignment`

By default, the `Octave:built-in-variable-assignment` warning is enabled.

`Octave:deprecated-keyword`

If the `Octave:deprecated-keyword` warning is enabled, a warning is issued when Octave encounters a keyword that is obsolete and scheduled for removal from Octave. By default, the `Octave:deprecated-keyword` warning is enabled.

`Octave:divide-by-zero`

If the `Octave:divide-by-zero` warning is enabled, a warning is issued when Octave encounters a division by zero. By default, the `Octave:divide-by-zero` warning is enabled.

`Octave:fopen-file-in-path`

By default, the `Octave:fopen-file-in-path` warning is enabled.

`Octave:function-name-clash`

If the `Octave:function-name-clash` warning is enabled, a warning is issued when Octave finds that the name of a function defined in a function file differs from the name of the file. (If the names disagree, the name declared inside the file is ignored.) By default, the `Octave:function-name-clash` warning is enabled.

`Octave:future-time-stamp`

If the `Octave:future-time-stamp` warning is enabled, Octave will print a warning if it finds a function file with a time stamp that is in the future. By default, the `Octave:future-time-stamp` warning is enabled.

`Octave:glyph-render`

By default, the `Octave:glyph-render` warning is enabled.

`Octave:imag-to-real`

> If the `Octave:imag-to-real` warning is enabled, a warning is printed for implicit conversions of complex numbers to real numbers. By default, the `Octave:imag-to-real` warning is disabled.

`Octave:language-extension`

> Print warnings when using features that are unique to the Octave language and that may still be missing in MATLAB. By default, the `Octave:language-extension` warning is disabled. The '`--traditional`' or '`--braindead`' startup options for Octave may also be of use, see Section 2.1.1 [Command Line Options], page 15.

`Octave:load-file-in-path`

> By default, the `Octave:load-file-in-path` warning is enabled.

`Octave:logical-conversion`

> By default, the `Octave:logical-conversion` warning is enabled.

`Octave:md5sum-file-in-path`

> By default, the `Octave:md5sum-file-in-path` warning is enabled.

`Octave:missing-glyph`

> By default, the `Octave:missing-glyph` warning is enabled.

`Octave:missing-semicolon`

> If the `Octave:missing-semicolon` warning is enabled, Octave will warn when statements in function definitions don't end in semicolons. By default the `Octave:missing-semicolon` warning is disabled.

`Octave:mixed-string-concat`

> If the `Octave:mixed-string-concat` warning is enabled, print a warning when concatenating a mixture of double and single quoted strings. By default, the `Octave:mixed-string-concat` warning is disabled.

`Octave:neg-dim-as-zero`

> If the `Octave:neg-dim-as-zero` warning is enabled, print a warning for expressions like
>
> ```
> eye (-1)
> ```
>
> By default, the `Octave:neg-dim-as-zero` warning is disabled.

`Octave:nested-functions-coerced`

> By default, the `Octave:nested-functions-coerced` warning is enabled.

`Octave:noninteger-range-as-index`

> By default, the `Octave:noninteger-range-as-index` warning is enabled.

`Octave:num-to-str`

> If the `Octave:num-to-str` warning is enable, a warning is printed for implicit conversions of numbers to their ASCII character equivalents when strings are constructed using a mixture of strings and numbers in matrix notation. For example,
>
> ```
> [ "f", 111, 111 ]
> ```
> $\Rightarrow$ `"foo"`

elicits a warning if the `Octave:num-to-str` warning is enabled. By default, the `Octave:num-to-str` warning is enabled.

`Octave:possible-matlab-short-circuit-operator`

If the `Octave:possible-matlab-short-circuit-operator` warning is enabled, Octave will warn about using the not short circuiting operators `&` and `|` inside `if` or `while` conditions. They normally never short circuit, but MATLAB always short circuits if any logical operators are used in a condition. You can turn on the option

`do_braindead_shortcircuit_evaluation (1)`

if you would like to enable this short-circuit evaluation in Octave. Note that the `&&` and `||` operators always short circuit in both Octave and MATLAB, so it's only necessary to enable MATLAB-style short-circuiting if it's too arduous to modify existing code that relies on this behavior. By default, the `Octave:possible-matlab-short-circuit-operator` warning is enabled.

`Octave:precedence-change`

If the `Octave:precedence-change` warning is enabled, Octave will warn about possible changes in the meaning of some code due to changes in precedence for some operators. Precedence changes have typically been made for MATLAB compatibility. By default, the `Octave:precedence-change` warning is enabled.

`Octave:recursive-path-search`

By default, the `Octave:recursive-path-search` warning is enabled.

`Octave:remove-init-dir`

The `path` function changes the search path that Octave uses to find functions. It is possible to set the path to a value which excludes Octave's own built-in functions. If the `Octave:remove-init-dir` warning is enabled then Octave will warn when the `path` function has been used in a way that may render Octave unworkable. By default, the `Octave:remove-init-dir` warning is enabled.

`Octave:reload-forces-clear`

If several functions have been loaded from the same file, Octave must clear all the functions before any one of them can be reloaded. If the `Octave:reload-forces-clear` warning is enabled, Octave will warn you when this happens, and print a list of the additional functions that it is forced to clear. By default, the `Octave:reload-forces-clear` warning is enabled.

`Octave:resize-on-range-error`

If the `Octave:resize-on-range-error` warning is enabled, print a warning when a matrix is resized by an indexed assignment with indices outside the current bounds. By default, the ## `Octave:resize-on-range-error` warning is disabled.

`Octave:separator-insert`

Print warning if commas or semicolons might be inserted automatically in literal matrices. By default, the `Octave:separator-insert` warning is disabled.

`Octave:shadowed-function`

By default, the `Octave:shadowed-function` warning is enabled.

`Octave:single-quote-string`
> Print warning if a single quote character is used to introduce a string constant. By default, the `Octave:single-quote-string` warning is disabled.

`Octave:nearly-singular-matrix`
`Octave:singular-matrix`
> By default, the `Octave:nearly-singular-matrix` and `Octave:singular-matrix` warnings are enabled.

`Octave:sqrtm:SingularMatrix`
> By default, the `Octave:sqrtm:SingularMatrix` warning is enabled.

`Octave:str-to-num`
> If the `Octave:str-to-num` warning is enabled, a warning is printed for implicit conversions of strings to their numeric ASCII equivalents. For example,
>
> ```
> "abc" + 0
>      ⇒ 97 98 99
> ```
>
> elicits a warning if the `Octave:str-to-num` warning is enabled. By default, the `Octave:str-to-num` warning is disabled.

`Octave:undefined-return-values`
> If the `Octave:undefined-return-values` warning is disabled, print a warning if a function does not define all the values in the return list which are expected. By default, the `Octave:undefined-return-values` warning is enabled.

`Octave:variable-switch-label`
> If the `Octave:variable-switch-label` warning is enabled, Octave will print a warning if a switch label is not a constant or constant expression. By default, the `Octave:variable-switch-label` warning is disabled.

### 12.2.2 Enabling and Disabling Warnings

The `warning` function also allows you to control which warnings are actually printed to the screen. If the `warning` function is called with a string argument that is either `"on"` or `"off"` all warnings will be enabled or disabled.

It is also possible to enable and disable individual warnings through their string identifications. The following code will issue a warning

```
warning ("example:non-negative-variable",
         "'a' must be non-negative.  Setting 'a' to zero.");
```

while the following won't issue a warning

```
warning ("off", "example:non-negative-variable");
warning ("example:non-negative-variable",
         "'a' must be non-negative.  Setting 'a' to zero.");
```

# 13 Debugging

Octave includes a built-in debugger to aid in the development of scripts. This can be used to interrupt the execution of an Octave script at a certain point, or when certain conditions are met. Once execution has stopped, and debug mode is entered, the symbol table at the point where execution has stopped can be examined and modified to check for errors.

The normal command-line editing and history functions are available in debug mode.

## 13.1 Entering Debug Mode

There are two basic means of interrupting the execution of an Octave script. These are breakpoints (see Section 13.3 [Breakpoints], page 220), discussed in the next section, and interruption based on some condition.

Octave supports three means to stop execution based on the values set in the functions `debug_on_interrupt`, `debug_on_warning`, and `debug_on_error`.

*val* = debug_on_interrupt ()  [Built-in Function]
*old_val* = debug_on_interrupt (*new_val*)  [Built-in Function]
debug_on_interrupt (*new_val*, "*local*")  [Built-in Function]

> Query or set the internal variable that controls whether Octave will try to enter debugging mode when it receives an interrupt signal (typically generated with C-c).
>
> If a second interrupt signal is received before reaching the debugging mode, a normal interrupt will occur.
>
> When called from inside a function with the "local" option, the variable is changed locally for the function and any subroutines it calls. The original variable value is restored when exiting the function.
>
> **See also:** [debug_on_error], page 219, [debug_on_warning], page 219.

*val* = debug_on_warning ()  [Built-in Function]
*old_val* = debug_on_warning (*new_val*)  [Built-in Function]
debug_on_warning (*new_val*, "*local*")  [Built-in Function]

> Query or set the internal variable that controls whether Octave will try to enter the debugger when a warning is encountered.
>
> When called from inside a function with the "local" option, the variable is changed locally for the function and any subroutines it calls. The original variable value is restored when exiting the function.
>
> **See also:** [debug_on_error], page 219, [debug_on_interrupt], page 219.

*val* = debug_on_error ()  [Built-in Function]
*old_val* = debug_on_error (*new_val*)  [Built-in Function]
debug_on_error (*new_val*, "*local*")  [Built-in Function]

> Query or set the internal variable that controls whether Octave will try to enter the debugger when an error is encountered.
>
> This will also inhibit printing of the normal traceback message (you will only see the top-level error message).

When called from inside a function with the `"local"` option, the variable is changed locally for the function and any subroutines it calls. The original variable value is restored when exiting the function.

**See also:** [debug_on_warning], page 219, [debug_on_interrupt], page 219.

## 13.2 Leaving Debug Mode

Use either `dbcont` or `return` to leave the debug mode and continue the normal execution of the script.

`dbcont` [Command]

Leave command-line debugging mode and continue code execution normally.

**See also:** [dbstep], page 224, [dbquit], page 220.

To quit debug mode and return directly to the prompt without executing any additional code use `dbquit`.

`dbquit` [Command]

Quit debugging mode immediately without further code execution and return to the Octave prompt.

**See also:** [dbcont], page 220, [dbstep], page 224.

Finally, typing `exit` or `quit` at the debug prompt will result in Octave terminating normally.

## 13.3 Breakpoints

Breakpoints can be set in any m-file function by using the `dbstop` function.

| | |
|---|---:|
| `dbstop` *func* | [Command] |
| `dbstop` *func line* | [Command] |
| `dbstop` *func line1 line2* ... | [Command] |
| `dbstop` *line* ... | [Command] |
| *rline* = `dbstop ("`*func*`")` | [Built-in Function] |
| *rline* = `dbstop ("`*func*`",` *line*`)` | [Built-in Function] |
| *rline* = `dbstop ("`*func*`",` *line1, line2,* ...`)` | [Built-in Function] |
| `dbstop ("`*func*`", [`*line1,* ...`])` | [Built-in Function] |
| `dbstop (`*line,* ...`)` | [Built-in Function] |

Set a breakpoint at line number *line* in function *func*.

Arguments are

*func*  Function name as a string variable. When already in debug mode this argument can be omitted and the current function will be used.

*line*  Line number where the breakpoint should be set. Multiple lines may be given as separate arguments or as a vector.

When called with a single argument *func*, the breakpoint is set at the first executable line in the named function.

Chapter 13: Debugging 221

The optional output *rline* is the real line number where the breakpoint was set. This can differ from the specified line if the line is not executable. For example, if a breakpoint attempted on a blank line then Octave will set the real breakpoint at the next executable line.

**See also:** [dbclear], page 221, [dbstatus], page 221, [dbstep], page 224, [debug_on_error], page 219, [debug_on_warning], page 219, [debug_on_interrupt], page 219.

Breakpoints in class methods are also supported (e.g., `dbstop ("@class/method")`). However, breakpoints cannot be set in built-in functions (e.g., `sin`, etc.) or dynamically loaded functions (i.e., oct-files).

To set a breakpoint immediately upon entering a function use line number 1, or omit the line number entirely and just give the function name. When setting the breakpoint Octave will ignore the leading comment block, and the breakpoint will be set on the first executable statement in the function. For example:

```
dbstop ("asind", 1)
    ⇒ 29
```

Note that the return value of 29 means that the breakpoint was effectively set to line 29. The status of breakpoints in a function can be queried with `dbstatus`.

dbstatus ()  [Built-in Function]
*brk_list* = dbstatus ()  [Built-in Function]
*brk_list* = dbstatus ("*func*")  [Built-in Function]

Report the location of active breakpoints.

When called with no input or output arguments, print the list of all functions with breakpoints and the line numbers where those breakpoints are set.

If a function name *func* is specified then only report breakpoints for the named function.

The optional return argument *brk_list* is a struct array with the following fields.

name    The name of the function with a breakpoint.

file    The name of the m-file where the function code is located.

line    A line number, or vector of line numbers, with a breakpoint.

Note: When `dbstatus` is called from the debug prompt within a function, the list of breakpoints is automatically trimmed to the breakpoints in the current function.

**See also:** [dbclear], page 221, [dbwhere], page 223.

Reusing the previous example, `dbstatus ("asind")` will return 29. The breakpoints listed can then be cleared with the `dbclear` function.

dbclear *func*  [Command]
dbclear *func line*  [Command]
dbclear *func line1 line2* ...  [Command]
dbclear *line* ...  [Command]
dbclear *all*  [Command]

`dbclear ("`*func*`")` [Built-in Function]
`dbclear ("`*func*`", `*line*`)` [Built-in Function]
`dbclear ("`*func*`", `*line1*`, `*line2*`, ...)` [Built-in Function]
`dbclear ("`*func*`", [`*line1*`, ...])` [Built-in Function]
`dbclear (`*line*`, ...)` [Built-in Function]
`dbclear ("`*all*`")` [Built-in Function]

> Delete a breakpoint at line number *line* in the function *func*.
>
> Arguments are
>
> > *func*     Function name as a string variable. When already in debug mode this argument can be omitted and the current function will be used.
> >
> > *line*     Line number from which to remove a breakpoint. Multiple lines may be given as separate arguments or as a vector.
>
> When called without a line number specification all breakpoints in the named function are cleared.
>
> If the requested line is not a breakpoint no action is performed.
>
> The special keyword `"all"` will clear all breakpoints from all files.
>
> **See also:** [dbstop], page 220, [dbstatus], page 221, [dbwhere], page 223.

A breakpoint may also be set in a subfunction. For example, if a file contains the functions

```
function y = func1 (x)
  y = func2 (x);
endfunction
function y = func2 (x)
  y = x + 1;
endfunction
```

then a breakpoint can be set at the start of the subfunction directly with

```
dbstop (["func1", filemarker(), "func2"])
 ⇒ 5
```

Note that `filemarker` returns the character that marks subfunctions from the file containing them. Unless the default has been changed this character is '>'. Thus, a quicker and more normal way to set the breakpoint would be

```
dbstop func1>func2
```

Another simple way of setting a breakpoint in an Octave script is the use of the `keyboard` function.

`keyboard ()` [Built-in Function]
`keyboard ("`*prompt*`")` [Built-in Function]

> Stop m-file execution and enter debug mode.
>
> When the `keyboard` function is executed, Octave prints a prompt and waits for user input. The input strings are then evaluated and the results are printed. This makes it possible to examine the values of variables within a function, and to assign new values if necessary. To leave the prompt and return to normal execution type '`return`' or '`dbcont`'. The `keyboard` function does not return an exit status.

Chapter 13: Debugging 223

If `keyboard` is invoked without arguments, a default prompt of 'debug> ' is used.

**See also:** [dbstop], page 220, [dbcont], page 220, [dbquit], page 220.

The `keyboard` function is placed in a script at the point where the user desires that the execution be stopped. It automatically sets the running script into the debug mode.

## 13.4 Debug Mode

There are three additional support functions that allow the user to find out where in the execution of a script Octave entered the debug mode, and to print the code in the script surrounding the point where Octave entered debug mode.

`dbwhere` [Command]

In debugging mode, report the current file and line number where execution is stopped.

**See also:** [dbstatus], page 221, [dbcont], page 220, [dbstep], page 224, [dbup], page 225.

`dbtype` [Command]
`dbtype` *lineno* [Command]
`dbtype` *startl:endl* [Command]
`dbtype` *startl:end* [Command]
`dbtype` *func* [Command]
`dbtype` *func lineno* [Command]
`dbtype` *func startl:endl* [Command]
`dbtype` *func startl:end* [Command]

Display a script file with line numbers.

When called with no arguments in debugging mode, display the script file currently being debugged.

An optional range specification can be used to list only a portion of the file. The special keyword `"end"` is a valid line number specification for the last line of the file.

When called with the name of a function, list that script file with line numbers.

**See also:** [dbwhere], page 223, [dbstatus], page 221, [dbstop], page 220.

`dblist` [Command]
`dblist` *n* [Command]

In debugging mode, list *n* lines of the function being debugged centered around the current line to be executed.

If unspecified *n* defaults to 10 (+/- 5 lines)

**See also:** [dbwhere], page 223, [dbtype], page 223.

You may also use `isdebugmode` to determine whether the debugger is currently active.

`isdebugmode ()` [Built-in Function]

Return true if in debugging mode, otherwise false.

**See also:** [dbwhere], page 223, [dbstack], page 224, [dbstatus], page 221.

Debug mode also allows single line stepping through a function using the command `dbstep`.

`dbstep` [Command]
`dbstep` *n* [Command]
`dbstep` *in* [Command]
`dbstep` *out* [Command]
`dbnext` ... [Command]

> In debugging mode, execute the next *n* lines of code.
>
> If *n* is omitted, execute the next single line of code. If the next line of code is itself defined in terms of an m-file remain in the existing function.
>
> Using `dbstep in` will cause execution of the next line to step into any m-files defined on the next line.
>
> Using `dbstep out` will cause execution to continue until the current function returns.
>
> `dbnext` is an alias for `dbstep`.
>
> **See also:** [dbcont], page 220, [dbquit], page 220.

When in debug mode the RETURN key will execute the last entered command. This is useful, for example, after hitting a breakpoint and entering `dbstep` once. After that, one can advance line by line through the code with only a single key stroke.

## 13.5 Call Stack

The function being debugged may be the leaf node of a series of function calls. After examining values in the current subroutine it may turn out that the problem occurred in earlier pieces of code. Use `dbup` and `dbdown` to move up and down through the series of function calls to locate where variables first took on the wrong values. `dbstack` shows the entire series of function calls and at what level debugging is currently taking place.

`dbstack` [Command]
`dbstack` *n* [Command]
`dbstack` *-completenames* [Command]
[*stack, idx*] = dbstack (...) [Built-in Function]

> Display or return current debugging function stack information.
>
> With optional argument *n*, omit the *n* innermost stack frames.
>
> Although accepted, the argument *-completenames* is silently ignored. Octave always returns absolute file names.
>
> The arguments *n* and *-completenames* can be both specified in any order.
>
> The optional return argument *stack* is a struct array with the following fields:
>
> file     The name of the m-file where the function code is located.
>
> name    The name of the function with a breakpoint.
>
> line     The line number of an active breakpoint.
>
> column   The column number of the line where the breakpoint begins.
>
> scope    Undocumented.

Chapter 13: Debugging                                                                 225

   context      Undocumented.

   The return argument *idx* specifies which element of the *stack* struct array is currently active.

   **See also:** [dbup], page 225, [dbdown], page 225, [dbwhere], page 223, [dbstatus], page 221.

dbup                                                                            [Command]
dbup *n*                                                                        [Command]
   In debugging mode, move up the execution stack *n* frames.

   If *n* is omitted, move up one frame.

   **See also:** [dbstack], page 224, [dbdown], page 225.

dbdown                                                                          [Command]
dbdown *n*                                                                      [Command]
   In debugging mode, move down the execution stack *n* frames.

   If *n* is omitted, move down one frame.

   **See also:** [dbstack], page 224, [dbup], page 225.

## 13.6 Profiling

Octave supports profiling of code execution on a per-function level. If profiling is enabled, each call to a function (supporting built-ins, operators, functions in oct- and mex-files, user-defined functions in Octave code and anonymous functions) is recorded while running Octave code. After that, this data can aid in analyzing the code behavior, and is in particular helpful for finding "hot spots" in the code which use up a lot of computation time and are the best targets to spend optimization efforts on.

The main command for profiling is `profile`, which can be used to start or stop the profiler and also to query collected data afterwards. The data is returned in an Octave data structure which can then be examined or further processed by other routines or tools.

profile *on*                                                                    [Command]
profile *off*                                                                   [Command]
profile *resume*                                                                [Command]
profile *clear*                                                                 [Command]
S = profile ("*status*")                                                     [Function File]
T = profile ("*info*")                                                       [Function File]
   Control the built-in profiler.

   profile on
           Start the profiler, clearing all previously collected data if there is any.

   profile off
           Stop profiling. The collected data can later be retrieved and examined with T = `profile ("info")`.

   profile clear
           Clear all collected profiler data.

`profile resume`
>    Restart profiling without clearing the old data. All newly collected statistics are added to the existing ones.

`S = profile ("status")`
>    Return a structure with information about the current status of the profiler. At the moment, the only field is `ProfilerStatus` which is either `"on"` or `"off"`.

`T = profile ("info")`
>    Return the collected profiling statistics in the structure $T$. The flat profile is returned in the field `FunctionTable` which is an array of structures, each entry corresponding to a function which was called and for which profiling statistics are present. In addition, the field `Hierarchical` contains the hierarchical call tree. Each node has an index into the `FunctionTable` identifying the function it corresponds to as well as data fields for number of calls and time spent at this level in the call tree.
>
>    **See also:** [profshow], page 226, [profexplore], page 226.

An easy way to get an overview over the collected data is `profshow`. This function takes the profiler data returned by `profile` as input and prints a flat profile, for instance:

```
   Function Attr     Time (s)        Calls
-----------------------------------------------
     >myfib    R        2.195        13529
   binary <=            0.061        13529
   binary -             0.050        13528
   binary +             0.026         6764
```

This shows that most of the run time was spent executing the function 'myfib', and some minor proportion evaluating the listed binary operators. Furthermore, it is shown how often the function was called and the profiler also records that it is recursive.

`profshow (data)` [Function File]
`profshow (data, n)` [Function File]
`profshow ()` [Function File]
`profshow (n)` [Function File]
>    Display flat per-function profiler results.
>
>    Print out profiler data (execution time, number of calls) for the most critical $n$ functions. The results are sorted in descending order by the total time spent in each function. If $n$ is unspecified it defaults to 20.
>
>    The input *data* is the structure returned by `profile ("info")`. If unspecified, `profshow` will use the current profile dataset.
>
>    The attribute column displays 'R' for recursive functions, and is blank for all other function types.
>
>    **See also:** [profexplore], page 226, [profile], page 225.

`profexplore ()` [Function File]
`profexplore (data)` [Function File]
>    Interactively explore hierarchical profiler output.

Assuming *data* is the structure with profile data returned by `profile ("info")`, this command opens an interactive prompt that can be used to explore the call-tree. Type `help` to get a list of possible commands. If *data* is omitted, `profile ("info")` is called and used in its place.

**See also:** [profile], page 225, [profshow], page 226.

## 13.7 Profiler Example

Below, we will give a short example of a profiler session. See Section 13.6 [Profiling], page 225, for the documentation of the profiler functions in detail. Consider the code:

```
global N A;

N = 300;
A = rand (N, N);

function xt = timesteps (steps, x0, expM)
  global N;

  if (steps == 0)
    xt = NA (N, 0);
  else
    xt = NA (N, steps);
    x1 = expM * x0;
    xt(:, 1) = x1;
    xt(:, 2 : end) = timesteps (steps - 1, x1, expM);
  endif
endfunction

function foo ()
  global N A;

  initial = @(x) sin (x);
  x0 = (initial (linspace (0, 2 * pi, N)))';

  expA = expm (A);
  xt = timesteps (100, x0, expA);
endfunction

function fib = bar (N)
  if (N <= 2)
    fib = 1;
  else
    fib = bar (N - 1) + bar (N - 2);
  endif
endfunction
```

If we execute the two main functions, we get:

```
tic; foo; toc;
⇒ Elapsed time is 2.37338 seconds.

tic; bar (20); toc;
⇒ Elapsed time is 2.04952 seconds.
```

But this does not give much information about where this time is spent; for instance, whether the single call to `expm` is more expensive or the recursive time-stepping itself. To get a more detailed picture, we can use the profiler.

```
profile on;
foo;
profile off;

data = profile ("info");
profshow (data, 10);
```

This prints a table like:

| #  | Function   | Attr | Time (s) | Calls |
|----|------------|------|----------|-------|
| 7  | expm       |      | 1.034    | 1     |
| 3  | binary *   |      | 0.823    | 117   |
| 41 | binary \   |      | 0.188    | 1     |
| 38 | binary ^   |      | 0.126    | 2     |
| 43 | timesteps  | R    | 0.111    | 101   |
| 44 | NA         |      | 0.029    | 101   |
| 39 | binary +   |      | 0.024    | 8     |
| 34 | norm       |      | 0.011    | 1     |
| 40 | binary -   |      | 0.004    | 101   |
| 33 | balance    |      | 0.003    | 1     |

The entries are the individual functions which have been executed (only the 10 most important ones), together with some information for each of them. The entries like 'binary *' denote operators, while other entries are ordinary functions. They include both built-ins like `expm` and our own routines (for instance `timesteps`). From this profile, we can immediately deduce that `expm` uses up the largest proportion of the processing time, even though it is only called once. The second expensive operation is the matrix-vector product in the routine `timesteps`.[1]

Timing, however, is not the only information available from the profile. The attribute column shows us that `timesteps` calls itself recursively. This may not be that remarkable in this example (since it's clear anyway), but could be helpful in a more complex setting. As to the question of why is there a 'binary \' in the output, we can easily shed some light on that too. Note that `data` is a structure array (Section 6.1.2 [Structure Arrays], page 103) which contains the field `FunctionTable`. This stores the raw data for the profile shown. The number in the first column of the table gives the index under which the shown function can be found there. Looking up `data.FunctionTable(41)` gives:

---

[1] We only know it is the binary multiplication operator, but fortunately this operator appears only at one place in the code and thus we know which occurrence takes so much time. If there were multiple places, we would have to use the hierarchical profile to find out the exact place which uses up the time which is not covered in this example.

# Chapter 13: Debugging

```
    scalar structure containing the fields:

      FunctionName = binary \
      TotalTime =  0.18765
      NumCalls =  1
      IsRecursive = 0
      Parents =  7
      Children = [](1x0)
```

Here we see the information from the table again, but have additional fields `Parents` and `Children`. Those are both arrays, which contain the indices of functions which have directly called the function in question (which is entry 7, `expm`, in this case) or been called by it (no functions). Hence, the backslash operator has been used internally by `expm`.

Now let's take a look at `bar`. For this, we start a fresh profiling session (`profile on` does this; the old data is removed before the profiler is restarted):

```
profile on;
bar (20);
profile off;

profshow (profile ("info"));
```

This gives:

```
      #          Function Attr      Time (s)         Calls
     ---------------------------------------------------------
      1               bar    R        2.091         13529
      2         binary <=             0.062         13529
      3         binary  -             0.042         13528
      4         binary  +             0.023          6764
      5           profile             0.000             1
      8             false             0.000             1
      6            nargin             0.000             1
      7         binary !=             0.000             1
      9 __profiler_enable__           0.000             1
```

Unsurprisingly, `bar` is also recursive. It has been called 13,529 times in the course of recursively calculating the Fibonacci number in a suboptimal way, and most of the time was spent in `bar` itself.

Finally, let's say we want to profile the execution of both `foo` and `bar` together. Since we already have the run-time data collected for `bar`, we can restart the profiler without clearing the existing data and collect the missing statistics about `foo`. This is done by:

```
profile resume;
foo;
profile off;

profshow (profile ("info"), 10);
```

As you can see in the table below, now we have both profiles mixed together.

|  # | Function   | Attr | Time (s) | Calls |
|---:|-----------:|:----:|---------:|------:|
|  1 | bar        | R    | 2.091    | 13529 |
| 16 | expm       |      | 1.122    | 1     |
| 12 | binary *   |      | 0.798    | 117   |
| 46 | binary \   |      | 0.185    | 1     |
| 45 | binary ^   |      | 0.124    | 2     |
| 48 | timesteps  | R    | 0.115    | 101   |
|  2 | binary <=  |      | 0.062    | 13529 |
|  3 | binary -   |      | 0.045    | 13629 |
|  4 | binary +   |      | 0.041    | 6772  |
| 49 | NA         |      | 0.036    | 101   |

# 14 Input and Output

Octave supports several ways of reading and writing data to or from the prompt or a file. The simplest functions for data Input and Output (I/O) are easy to use, but only provide limited control of how data is processed. For more control, a set of functions modeled after the C standard library are also provided by Octave.

## 14.1 Basic Input and Output

### 14.1.1 Terminal Output

Since Octave normally prints the value of an expression as soon as it has been evaluated, the simplest of all I/O functions is a simple expression. For example, the following expression will display the value of 'pi'

```
pi
    ⊣ pi = 3.1416
```

This works well as long as it is acceptable to have the name of the variable (or 'ans') printed along with the value. To print the value of a variable without printing its name, use the function `disp`.

The `format` command offers some control over the way Octave prints values with `disp` and through the normal echoing mechanism.

**disp (*x*)**  [Built-in Function]
   Display the value of *x*.

   For example:

```
disp ("The value of pi is:"), disp (pi)

    ⊣ the value of pi is:
    ⊣ 3.1416
```

   Note that the output from `disp` always ends with a newline.

   If an output value is requested, `disp` prints nothing and returns the formatted output in a string.

   **See also:** [fdisp], page 242.

**list_in_columns (*arg*, *width*, *prefix*)**  [Built-in Function]
   Return a string containing the elements of *arg* listed in columns with an overall maximum width of *width* and optional prefix *prefix*.

   The argument *arg* must be a cell array of character strings or a character array.

   If *width* is not specified or is an empty matrix, or less than or equal to zero, the width of the terminal screen is used. Newline characters are used to break the lines in the output string. For example:

```
list_in_columns ({"abc", "def", "ghijkl", "mnop", "qrs", "tuv"}, 20)
    ⇒  abc     mnop
       def     qrs
       ghijkl  tuv

whos ans
    ⇒
    Variables in the current scope:

    Attr Name        Size                     Bytes  Class
    ==== ====        ====                     =====  =====
         ans         1x37                        37  char

    Total is 37 elements using 37 bytes
```

See also: [terminal_size], page 232.

**terminal_size ()**   *[Built-in Function]*

Return a two-element row vector containing the current size of the terminal window in characters (rows and columns).

See also: [list_in_columns], page 231.

**format**   *[Command]*
**format** *options*   *[Command]*

Reset or specify the format of the output produced by `disp` and Octave's normal echoing mechanism.

This command only affects the display of numbers but not how they are stored or computed. To change the internal representation from the default double use one of the conversion functions such as `single`, `uint8`, `int64`, etc.

By default, Octave displays 5 significant digits in a human readable form (option 'short' paired with 'loose' format for matrices). If `format` is invoked without any options, this default format is restored.

Valid formats for floating point numbers are listed in the following table.

- **short**  
  Fixed point format with 5 significant figures in a field that is a maximum of 10 characters wide. (default).

  If Octave is unable to format a matrix so that columns line up on the decimal point and all numbers fit within the maximum field width then it switches to an exponential 'e' format.

- **long**  
  Fixed point format with 15 significant figures in a field that is a maximum of 20 characters wide.

  As with the 'short' format, Octave will switch to an exponential 'e' format if it is unable to format a matrix properly using the current format.

- **short e**
- **long e**  
  Exponential format. The number to be represented is split between a mantissa and an exponent (power of 10). The mantissa has 5 significant digits in the short format and 15 digits in the long format. For example, with the 'short e' format, `pi` is displayed as `3.1416e+00`.

# Chapter 14: Input and Output

short E
long E
Identical to 'short e' or 'long e' but displays an uppercase 'E' to indicate the exponent. For example, with the 'long E' format, pi is displayed as 3.14159265358979E+00.

short g
long g
Optimally choose between fixed point and exponential format based on the magnitude of the number. For example, with the 'short g' format, pi .^ [2; 4; 8; 16; 32] is displayed as

    ans =

       9.8696
      97.409
    9488.5
    9.0032e+07
    8.1058e+15

short eng
long eng
Identical to 'short e' or 'long e' but displays the value using an engineering format, where the exponent is divisible by 3. For example, with the 'short eng' format, 10 * pi is displayed as 31.4159e+00.

long G
short G
Identical to 'short g' or 'long g' but displays an uppercase 'E' to indicate the exponent.

free
none
Print output in free format, without trying to line up columns of matrices on the decimal point. This also causes complex numbers to be formatted as numeric pairs like this '(0.60419, 0.60709)' instead of like this '0.60419 + 0.60709i'.

The following formats affect all numeric output (floating point and integer types).

"+"
"+" *chars*
plus
plus *chars*

Print a '+' symbol for matrix elements greater than zero, a '-' symbol for elements less than zero and a space for zero matrix elements. This format can be very useful for examining the structure of a large sparse matrix.

The optional argument *chars* specifies a list of 3 characters to use for printing values greater than zero, less than zero and equal to zero. For example, with the '"+" "+-."' format, [1, 0, -1; -1, 0, 1] is displayed as

    ans =

    +.-
    -.+

bank
Print in a fixed format with two digits to the right of the decimal point.

`native-hex`
: Print the hexadecimal representation of numbers as they are stored in memory. For example, on a workstation which stores 8 byte real values in IEEE format with the least significant byte first, the value of `pi` when printed in `native-hex` format is `400921fb54442d18`.

`hex`
: The same as `native-hex`, but always print the most significant byte first.

`native-bit`
: Print the bit representation of numbers as stored in memory. For example, the value of `pi` is

    0100000000001001001000011111111011
    0101010001000100001011010001100

    (shown here in two 32 bit sections for typesetting purposes) when printed in native-bit format on a workstation which stores 8 byte real values in IEEE format with the least significant byte first.

`bit`
: The same as `native-bit`, but always print the most significant bits first.

`rat`
: Print a rational approximation, i.e., values are approximated as the ratio of small integers. For example, with the 'rat' format, `pi` is displayed as `355/113`.

The following two options affect the display of all matrices.

`compact`
: Remove blank lines around column number labels and between matrices producing more compact output with more data per page.

`loose`
: Insert blank lines above and below column number labels and between matrices to produce a more readable output with less data per page. (default).

**See also:** [fixed_point_format], page 51, [output_max_field_width], page 49, [output_precision], page 50, [split_long_rows], page 50, [print_empty_dimensions], page 51, [rats], page 456.

### 14.1.1.1 Paging Screen Output

When running interactively, Octave normally sends any output intended for your terminal that is more than one screen long to a paging program, such as `less` or `more`. This avoids the problem of having a large volume of output stream by before you can read it. With `less` (and some versions of `more`) you can also scan forward and backward, and search for specific items.

Normally, no output is displayed by the pager until just before Octave is ready to print the top level prompt, or read from the standard input (for example, by using the `fscanf` or `scanf` functions). This means that there may be some delay before any output appears on your screen if you have asked Octave to perform a significant amount of work with a single command statement. The function `fflush` may be used to force output to be sent to the pager (or any other stream) immediately.

You can select the program to run as the pager using the `PAGER` function, and you can turn paging off by using the function `more`.

Chapter 14: Input and Output 235

**more** [Command]
**more** *on* [Command]
**more** *off* [Command]
> Turn output pagination on or off.
>
> Without an argument, `more` toggles the current state.
>
> The current state can be determined via `page_screen_output`.
>
> **See also:** [page_screen_output], page 235, [page_output_immediately], page 236, [PAGER], page 235, [PAGER_FLAGS], page 235.

*val* = PAGER () [Built-in Function]
*old_val* = PAGER (*new_val*) [Built-in Function]
PAGER (*new_val*, "*local*") [Built-in Function]
> Query or set the internal variable that specifies the program to use to display terminal output on your system.
>
> The default value is normally `"less"`, `"more"`, or `"pg"`, depending on what programs are installed on your system. See Appendix G [Installation], page 889.
>
> When called from inside a function with the `"local"` option, the variable is changed locally for the function and any subroutines it calls. The original variable value is restored when exiting the function.
>
> **See also:** [PAGER_FLAGS], page 235, [page_output_immediately], page 236, [more], page 234, [page_screen_output], page 235.

*val* = PAGER_FLAGS () [Built-in Function]
*old_val* = PAGER_FLAGS (*new_val*) [Built-in Function]
PAGER_FLAGS (*new_val*, "*local*") [Built-in Function]
> Query or set the internal variable that specifies the options to pass to the pager.
>
> When called from inside a function with the `"local"` option, the variable is changed locally for the function and any subroutines it calls. The original variable value is restored when exiting the function.
>
> **See also:** [PAGER], page 235, [more], page 234, [page_screen_output], page 235, [page_output_immediately], page 236.

*val* = page_screen_output () [Built-in Function]
*old_val* = page_screen_output (*new_val*) [Built-in Function]
page_screen_output (*new_val*, "*local*") [Built-in Function]
> Query or set the internal variable that controls whether output intended for the terminal window that is longer than one page is sent through a pager.
>
> This allows you to view one screenful at a time. Some pagers (such as `less`—see Appendix G [Installation], page 889) are also capable of moving backward on the output.
>
> When called from inside a function with the `"local"` option, the variable is changed locally for the function and any subroutines it calls. The original variable value is restored when exiting the function.
>
> **See also:** [more], page 234, [page_output_immediately], page 236, [PAGER], page 235, [PAGER_FLAGS], page 235.

`val = page_output_immediately ()` [Built-in Function]
`old_val = page_output_immediately (new_val)` [Built-in Function]
`page_output_immediately (new_val, "local")` [Built-in Function]

    Query or set the internal variable that controls whether Octave sends output to the pager as soon as it is available.

    Otherwise, Octave buffers its output and waits until just before the prompt is printed to flush it to the pager.

    When called from inside a function with the `"local"` option, the variable is changed locally for the function and any subroutines it calls. The original variable value is restored when exiting the function.

    **See also:** [page_screen_output], page 235, [more], page 234, [PAGER], page 235, [PAGER_FLAGS], page 235.

`fflush (fid)` [Built-in Function]

    Flush output to file descriptor *fid*.

    `fflush` returns 0 on success and an OS dependent error value (−1 on Unix) on error.

    Programming Note: Flushing is useful for ensuring that all pending output makes it to the screen before some other event occurs. For example, it is always a good idea to flush the standard output stream before calling `input`.

    **See also:** [fopen], page 250, [fclose], page 251.

## 14.1.2 Terminal Input

Octave has three functions that make it easy to prompt users for input. The `input` and `menu` functions are normally used for managing an interactive dialog with a user, and the `keyboard` function is normally used for doing simple debugging.

`ans = input (prompt)` [Built-in Function]
`ans = input (prompt, "s")` [Built-in Function]

    Print *prompt* and wait for user input.

    For example,

        `input ("Pick a number, any number! ")`

prints the prompt

        Pick a number, any number!

and waits for the user to enter a value. The string entered by the user is evaluated as an expression, so it may be a literal constant, a variable name, or any other valid Octave code.

    The number of return arguments, their size, and their class depend on the expression entered.

    If you are only interested in getting a literal string value, you can call `input` with the character string `"s"` as the second argument. This tells Octave to return the string entered by the user directly, without evaluating it first.

    Because there may be output waiting to be displayed by the pager, it is a good idea to always call `fflush (stdout)` before calling `input`. This will ensure that all pending output is written to the screen before your prompt.

Chapter 14: Input and Output 237

**See also:** [yes_or_no], page 237, [kbhit], page 237, [pause], page 749, [menu], page 237, [listdlg], page 794.

*choice* = menu (*title*, *opt1*, ...) [Function File]
*choice* = menu (*title*, {*opt1*, ...}) [Function File]
 Display a menu with heading *title* and options *opt1*, ..., and wait for user input.

 If the GUI is running, or Java is available, the menu is displayed graphically using `listdlg`. Otherwise, the title and menu options are printed on the console.

 *title* is a string and the options may be input as individual strings or as a cell array of strings.

 The return value *choice* is the number of the option selected by the user counting from 1.

 This function is useful for interactive programs. There is no limit to the number of options that may be passed in, but it may be confusing to present more than will fit easily on one screen.

 **See also:** [input], page 236, [listdlg], page 794.

*ans* = yes_or_no ("*prompt*") [Built-in Function]
 Ask the user a yes-or-no question.

 Return logical true if the answer is yes or false if the answer is no.

 Takes one argument, *prompt*, which is the string to display when asking the question. *prompt* should end in a space; `yes-or-no` adds the string '(yes or no) ' to it. The user must confirm the answer with RET and can edit it until it has been confirmed.

 **See also:** [input], page 236.

For `input`, the normal command line history and editing functions are available at the prompt.

Octave also has a function that makes it possible to get a single character from the keyboard without requiring the user to type a carriage return.

kbhit () [Built-in Function]
kbhit (*1*) [Built-in Function]
 Read a single keystroke from the keyboard.

 If called with an argument, don't wait for a keypress.

 For example,

        x = kbhit ();

 will set x to the next character typed at the keyboard as soon as it is typed.

        x = kbhit (1);

 is identical to the above example, but doesn't wait for a keypress, returning the empty string if no key is available.

 **See also:** [input], page 236, [pause], page 749.

### 14.1.3 Simple File I/O

The `save` and `load` commands allow data to be written to and read from disk files in various formats. The default format of files written by the `save` command can be controlled using the functions `save_default_options` and `save_precision`.

As an example the following code creates a 3-by-3 matrix and saves it to the file 'myfile.mat'.

```
A = [ 1:3; 4:6; 7:9 ];
save myfile.mat A
```

Once one or more variables have been saved to a file, they can be read into memory using the `load` command.

```
load myfile.mat
A
    ⊣ A =
    ⊣
    ⊣    1   2   3
    ⊣    4   5   6
    ⊣    7   8   9
```

| | |
|---|---|
| `save` *file* | [Command] |
| `save` *options file* | [Command] |
| `save` *options file v1 v2 ...* | [Command] |
| `save` *options file -struct STRUCT f1 f2 ...* | [Command] |
| `save` `"-"` *v1 v2 ...* | [Command] |
| *s* = `save` (`"-"` *v1 v2 ...*) | [Built-in Function] |

Save the named variables *v1*, *v2*, ..., in the file *file*.

The special filename '-' may be used to return the content of the variables as a string. If no variable names are listed, Octave saves all the variables in the current scope. Otherwise, full variable names or pattern syntax can be used to specify the variables to save. If the '-struct' modifier is used, fields *f1 f2* ... of the scalar structure *STRUCT* are saved as if they were variables with corresponding names. Valid options for the `save` command are listed in the following table. Options that modify the output format override the format specified by `save_default_options`.

If save is invoked using the functional form

```
save ("-option1", ..., "file", "v1", ...)
```

then the *options*, *file*, and variable name arguments (*v1*, ...) must be specified as character strings.

If called with a filename of `"-"`, write the output to stdout if nargout is 0, otherwise return the output in a character string.

`-append`   Append to the destination instead of overwriting.

`-ascii`    Save a single matrix in a text file without header or any other information.

`-binary`   Save the data in Octave's binary data format.

## Chapter 14: Input and Output

-float-binary
: Save the data in Octave's binary data format but only using single precision. Only use this format if you know that all the values to be saved can be represented in single precision.

-hdf5
: Save the data in HDF5 format. (HDF5 is a free, portable binary format developed by the National Center for Supercomputing Applications at the University of Illinois.) This format is only available if Octave was built with a link to the HDF5 libraries.

-float-hdf5
: Save the data in HDF5 format but only using single precision. Only use this format if you know that all the values to be saved can be represented in single precision.

-V7
-v7
-7
-mat7-binary
: Save the data in MATLAB's v7 binary data format.

-V6
-v6
-6
-mat
-mat-binary
: Save the data in MATLAB's v6 binary data format.

-V4
-v4
-4
-mat4-binary
: Save the data in the binary format written by MATLAB version 4.

-text
: Save the data in Octave's text data format. (default).

-zip
-z
: Use the gzip algorithm to compress the file. This works equally on files that are compressed with gzip outside of octave, and gzip can equally be used to convert the files for backward compatibility. This option is only available if Octave was built with a link to the zlib libraries.

The list of variables to save may use wildcard patterns containing the following special characters:

?
: Match any single character.

*
: Match zero or more characters.

[ list ]
: Match the list of characters specified by *list*. If the first character is ! or ^, match all characters except those specified by *list*. For example, the pattern [a-zA-Z] will match all lower and uppercase alphabetic characters.

Wildcards may also be used in the field name specifications when using the '-struct' modifier (but not in the struct name itself).

Except when using the MATLAB binary data file format or the '-ascii' format, saving global variables also saves the global status of the variable. If the variable is restored at a later time using 'load', it will be restored as a global variable.

The command

```
save -binary data a b*
```

saves the variable 'a' and all variables beginning with 'b' to the file 'data' in Octave's binary format.

**See also:** [load], page 241, [save_default_options], page 240, [save_header_format_string], page 240, [dlmread], page 244, [csvread], page 244, [fread], page 262.

There are three functions that modify the behavior of save.

*val* = save_default_options ()     [Built-in Function]
*old_val* = save_default_options (*new_val*)     [Built-in Function]
save_default_options (*new_val*, "*local*")     [Built-in Function]

Query or set the internal variable that specifies the default options for the save command, and defines the default format.

Typical values include "-ascii", "-text -zip". The default value is '-text'.

When called from inside a function with the "local" option, the variable is changed locally for the function and any subroutines it calls. The original variable value is restored when exiting the function.

**See also:** [save], page 238.

*val* = save_precision ()     [Built-in Function]
*old_val* = save_precision (*new_val*)     [Built-in Function]
save_precision (*new_val*, "*local*")     [Built-in Function]

Query or set the internal variable that specifies the number of digits to keep when saving data in text format.

When called from inside a function with the "local" option, the variable is changed locally for the function and any subroutines it calls. The original variable value is restored when exiting the function.

*val* = save_header_format_string ()     [Built-in Function]
*old_val* = save_header_format_string (*new_val*)     [Built-in Function]
save_header_format_string (*new_val*, "*local*")     [Built-in Function]

Query or set the internal variable that specifies the format string used for the comment line written at the beginning of text-format data files saved by Octave.

The format string is passed to strftime and should begin with the character '#' and contain no newline characters. If the value of save_header_format_string is the empty string, the header comment is omitted from text-format data files. The default value is

```
"# Created by Octave VERSION, %a %b %d %H:%M:%S %Y %Z <USER@HOST>"
```

Chapter 14: Input and Output                                                        241

When called from inside a function with the "local" option, the variable is changed
locally for the function and any subroutines it calls. The original variable value is
restored when exiting the function.

**See also:** [strftime], page 745, [save], page 238.

load *file*                                                                  [Command]
load *options file*                                                          [Command]
load *options file v1 v2 ...*                                                [Command]
S = load ("*options*", "*file*", "*v1*", "*v2*", ...)                        [Command]
load *file options*                                                          [Command]
load *file options v1 v2 ...*                                                [Command]
S = load ("*file*", "*options*", "*v1*", "*v2*", ...)                        [Command]

Load the named variables *v1*, *v2*, ..., from the file *file*.

If no variables are specified then all variables found in the file will be loaded. As with
save, the list of variables to extract can be full names or use a pattern syntax. The
format of the file is automatically detected but may be overridden by supplying the
appropriate option.

If load is invoked using the functional form

```
load ("-option1", ..., "file", "v1", ...)
```

then the *options*, *file*, and variable name arguments (*v1*, ...) must be specified as
character strings.

If a variable that is not marked as global is loaded from a file when a global symbol
with the same name already exists, it is loaded in the global symbol table. Also, if
a variable is marked as global in a file and a local symbol exists, the local symbol is
moved to the global symbol table and given the value from the file.

If invoked with a single output argument, Octave returns data instead of inserting variables in the symbol table. If the data file contains only numbers (TAB- or
space-delimited columns), a matrix of values is returned. Otherwise, load returns a
structure with members corresponding to the names of the variables in the file.

The load command can read data stored in Octave's text and binary formats,
and MATLAB's binary format. If compiled with zlib support, it can also load
gzip-compressed files. It will automatically detect the type of file and do conversion
from different floating point formats (currently only IEEE big and little endian,
though other formats may be added in the future).

Valid options for load are listed in the following table.

-force     This option is accepted for backward compatibility but is ignored. Octave
           now overwrites variables currently in memory with those of the same name
           found in the file.

-ascii     Force Octave to assume the file contains columns of numbers in text
           format without any header or other information. Data in the file will be
           loaded as a single numeric matrix with the name of the variable derived
           from the name of the file.

-binary    Force Octave to assume the file is in Octave's binary format.

-hdf5      Force Octave to assume the file is in HDF5 format. (HDF5 is a free, portable binary format developed by the National Center for Supercomputing Applications at the University of Illinois.) Note that Octave can read HDF5 files not created by itself, but may skip some datasets in formats that it cannot support. This format is only available if Octave was built with a link to the HDF5 libraries.

-import    This option is accepted for backward compatibility but is ignored. Octave can now support multi-dimensional HDF data and automatically modifies variable names if they are invalid Octave identifiers.

-mat
-mat-binary
-6
-v6
-7
-v7        Force Octave to assume the file is in MATLAB's version 6 or 7 binary format.

-mat4-binary
-4
-v4
-V4        Force Octave to assume the file is in the binary format written by MATLAB version 4.

-text      Force Octave to assume the file is in Octave's text format.

**See also:** [save], page 238, [dlmwrite], page 243, [csvwrite], page 244, [fwrite], page 265.

*str* = fileread (*filename*)                                    [Function File]
Read the contents of *filename* and return it as a string.

**See also:** [fread], page 262, [textread], page 244, [sscanf], page 260.

native_float_format ()                                           [Built-in Function]
Return the native floating point format as a string.

It is possible to write data to a file in a similar way to the disp function for writing data to the screen. The fdisp works just like disp except its first argument is a file pointer as created by fopen. As an example, the following code writes to data 'myfile.txt'.

    fid = fopen ("myfile.txt", "w");
    fdisp (fid, "3/8 is ");
    fdisp (fid, 3/8);
    fclose (fid);

See Section 14.2.1 [Opening and Closing Files], page 249, for details on how to use fopen and fclose.

fdisp (*fid*, *x*)                                               [Built-in Function]
Display the value of *x* on the stream *fid*.

For example:

Chapter 14: Input and Output                                                    243

```
          fdisp (stdout, "The value of pi is:"), fdisp (stdout, pi)
```

⊣ the value of pi is:
⊣ 3.1416

Note that the output from `fdisp` always ends with a newline.

**See also:** [disp], page 231.

Octave can also read and write matrices text files such as comma separated lists.

**dlmwrite** (*file*, *M*)                                              [Function File]
**dlmwrite** (*file*, *M*, *delim*, *r*, *c*)                           [Function File]
**dlmwrite** (*file*, *M*, *key*, *val* ...)                            [Function File]
**dlmwrite** (*file*, *M*, "-append", ...)                              [Function File]
**dlmwrite** (*fid*, ...)                                               [Function File]

Write the matrix $M$ to the named file using delimiters.

*file* should be a file name or writable file ID given by `fopen`.

The parameter *delim* specifies the delimiter to use to separate values on a row.

The value of $r$ specifies the number of delimiter-only lines to add to the start of the file.

The value of $c$ specifies the number of delimiters to prepend to each line of data.

If the argument "-append" is given, append to the end of *file*.

In addition, the following keyword value pairs may appear at the end of the argument list:

"append"   Either "on" or "off". See "-append" above.

"delimiter"
           See *delim* above.

"newline"
           The character(s) to use to separate each row. Three special cases exist for this option. "unix" is changed into "\n", "pc" is changed into "\r\n", and "mac" is changed into "\r". Any other value is used directly as the newline separator.

"roffset"
           See $r$ above.

"coffset"
           See $c$ above.

"precision"
           The precision to use when writing the file. It can either be a format string (as used by fprintf) or a number of significant digits.

```
          dlmwrite ("file.csv", reshape (1:16, 4, 4));
          dlmwrite ("file.tex", a, "delimiter", "&", "newline", "\n")
```

**See also:** [dlmread], page 244, [csvread], page 244, [csvwrite], page 244.

`data = dlmread (file)` [Built-in Function]
`data = dlmread (file, sep)` [Built-in Function]
`data = dlmread (file, sep, r0, c0)` [Built-in Function]
`data = dlmread (file, sep, range)` [Built-in Function]
`data = dlmread (..., "emptyvalue", EMPTYVAL)` [Built-in Function]

Read the matrix *data* from a text file which uses the delimiter *sep* between data values.

If *sep* is not defined the separator between fields is determined from the file itself.

Given two scalar arguments *r0* and *c0*, these define the starting row and column of the data to be read. These values are indexed from zero, such that the first row corresponds to an index of zero.

The *range* parameter may be a 4-element vector containing the upper left and lower right corner `[R0,C0,R1,C1]` where the lowest index value is zero. Alternatively, a spreadsheet style range such as `"A2..Q15"` or `"T1:AA5"` can be used. The lowest alphabetical index 'A' refers to the first column. The lowest row index is 1.

*file* should be a file name or file id given by `fopen`. In the latter case, the file is read until end of file is reached.

The `"emptyvalue"` option may be used to specify the value used to fill empty fields. The default is zero.

See also: [csvread], page 244, [textscan], page 246, [textread], page 244, [dlmwrite], page 243.

`csvwrite (filename, x)` [Function File]
`csvwrite (filename, x, dlm_opts)` [Function File]

Write the matrix *x* to the file *filename* in comma-separated-value format.

This function is equivalent to

   `dlmwrite (filename, x, ",", ...)`

See also: [csvread], page 244, [dlmwrite], page 243, [dlmread], page 244.

`x = csvread (filename)` [Function File]
`x = csvread (filename, dlm_opts)` [Function File]

Read the comma-separated-value file *filename* into the matrix *x*.

This function is equivalent to

   `x = dlmread (filename, "," , ...)`

See also: [csvwrite], page 244, [dlmread], page 244, [dlmwrite], page 243.

Formatted data from can be read from, or written to, text files as well.

`[a, ...] = textread (filename)` [Function File]
`[a, ...] = textread (filename, format)` [Function File]
`[a, ...] = textread (filename, format, n)` [Function File]
`[a, ...] = textread (filename, format, prop1, value1, ...)` [Function File]
`[a, ...] = textread (filename, format, n, prop1, value1, ...)` [Function File]

Read data from a text file.

# Chapter 14: Input and Output

The file *filename* is read and parsed according to *format*. The function behaves like **strread** except it works by parsing a file instead of a string. See the documentation of **strread** for details.

In addition to the options supported by **strread**, this function supports two more:

- "headerlines": The first *value* number of lines of *filename* are skipped.
- "endofline": Specify a single character or "\r\n". If no value is given, it will be inferred from the file. If set to "" (empty string) EOLs are ignored as delimiters.

The optional input n (format repeat count) specifies the number of times the format string is to be used or the number of lines to be read, whichever happens first while reading. The former is equivalent to requesting that the data output vectors should be of length N. Note that when reading files with format strings referring to multiple lines, n should rather be the number of lines to be read than the number of format string uses.

If the format string is empty (not just omitted) and the file contains only numeric data (excluding headerlines), textread will return a rectangular matrix with the number of columns matching the number of numeric fields on the first data line of the file. Empty fields are returned as zero values.

Examples:

```
Assume a data file like:
1 a 2 b
3 c 4 d
5 e
[a, b] = textread (f, "%f %s")
returns two columns of data, one with doubles, the other a
cellstr array:
a = [1; 2; 3; 4; 5]
b = {"a"; "b"; "c"; "d"; "e"}
[a, b] = textread (f, "%f %s", 3)
(read data into two culumns, try to use the format string
three times)
returns
a = [1; 2; 3]
b = {"a"; "b"; "c"}

With a data file like:
1
a
2
b

[a, b] = textread (f, "%f %s", 2)
returns a = 1 and b = {"a"}; i.e., the format string is used
only once because the format string refers to 2 lines of the
data file. To obtain 2x1 data output columns, specify N = 4
```

(number of data lines containing all requested data) rather
than 2.

**See also:** [strread], page 84, [load], page 241, [dlmread], page 244, [fscanf], page 259,
[textscan], page 246.

`C = textscan (fid, format)`   [Function File]
`C = textscan (fid, format, n)`   [Function File]
`C = textscan (fid, format, param, value, ...)`   [Function File]
`C = textscan (fid, format, n, param, value, ...)`   [Function File]
`C = textscan (str, ...)`   [Function File]
`[C, position] = textscan (fid, ...)`   [Function File]

Read data from a text file or string.

The string *str* or file associated with *fid* is read from and parsed according to *format*. The function behaves like `strread` except it can also read from file instead of a string. See the documentation of `strread` for details.

In addition to the options supported by `strread`, this function supports a few more:

- `"collectoutput"`: A value of 1 or true instructs textscan to concatenate consecutive columns of the same class in the output cell array. A value of 0 or false (default) leaves output in distinct columns.

- `"endofline"`: Specify `"\r"`, `"\n"` or `"\r\n"` (for CR, LF, or CRLF). If no value is given, it will be inferred from the file. If set to `""` (empty string) EOLs are ignored as delimiters and added to whitespace.

- `"headerlines"`: The first *value* number of lines of *fid* are skipped.

- `"returnonerror"`: If set to numerical 1 or true (default), return normally when read errors have been encountered. If set to 0 or false, return an error and no data. As the string or file is read by columns rather than by rows, and because textscan is fairly forgiving as regards read errors, setting this option may have little or no actual effect.

When reading from a character string, optional input argument *n* specifies the number of times *format* should be used (i.e., to limit the amount of data read). When reading from file, *n* specifies the number of data lines to read; in this sense it differs slightly from the format repeat count in strread.

The output *C* is a cell array whose second dimension is determined by the number of format specifiers.

The second output, *position*, provides the position, in characters, from the beginning of the file.

If the format string is empty (not: omitted) and the file contains only numeric data (excluding headerlines), textscan will return data in a number of columns matching the number of numeric fields on the first data line of the file.

**See also:** [dlmread], page 244, [fscanf], page 259, [load], page 241, [strread], page 84, [textread], page 244.

The `importdata` function has the ability to work with a wide variety of data.

Chapter 14: Input and Output    247

`A = importdata (`*fname*`)`   [Function File]
`A = importdata (`*fname, delimiter*`)`   [Function File]
`A = importdata (`*fname, delimiter, header_rows*`)`   [Function File]
`[A, `*delimiter*`] = importdata (...)`   [Function File]
`[A, `*delimiter, header_rows*`] = importdata (...)`   [Function File]

Import data from the file *fname*.

Input parameters:

- *fname* The name of the file containing data.
- *delimiter* The character separating columns of data. Use `\t` for tab. (Only valid for ASCII files)
- *header_rows* The number of header rows before the data begins. (Only valid for ASCII files)

Different file types are supported:

- ASCII table

  Import ASCII table using the specified number of header rows and the specified delimiter.

- Image file
- MATLAB file
- Spreadsheet files (depending on external software)
- WAV file

**See also:** [textscan], page 246, [dlmread], page 244, [csvread], page 244, [load], page 241.

### 14.1.3.1 Saving Data on Unexpected Exits

If Octave for some reason exits unexpectedly it will by default save the variables available in the workspace to a file in the current directory. By default this file is named 'octave-workspace' and can be loaded into memory with the `load` command. While the default behavior most often is reasonable it can be changed through the following functions.

`val = crash_dumps_octave_core ()`   [Built-in Function]
`old_val = crash_dumps_octave_core (`*new_val*`)`   [Built-in Function]
`crash_dumps_octave_core (`*new_val*`, "`*local*`")`   [Built-in Function]

Query or set the internal variable that controls whether Octave tries to save all current variables to the file 'octave-workspace' if it crashes or receives a hangup, terminate or similar signal.

When called from inside a function with the `"local"` option, the variable is changed locally for the function and any subroutines it calls. The original variable value is restored when exiting the function.

**See also:** [octave_core_file_limit], page 248, [octave_core_file_name], page 248, [octave_core_file_options], page 248.

`val = sighup_dumps_octave_core ()`   [Built-in Function]
`old_val = sighup_dumps_octave_core (`*new_val*`)`   [Built-in Function]

`sighup_dumps_octave_core (new_val, "local")`        [Built-in Function]

    Query or set the internal variable that controls whether Octave tries to save all current variables to the file 'octave-workspace' if it receives a hangup signal.

    When called from inside a function with the `"local"` option, the variable is changed locally for the function and any subroutines it calls. The original variable value is restored when exiting the function.

`val = sigterm_dumps_octave_core ()`        [Built-in Function]
`old_val = sigterm_dumps_octave_core (new_val)`        [Built-in Function]
`sigterm_dumps_octave_core (new_val, "local")`        [Built-in Function]

    Query or set the internal variable that controls whether Octave tries to save all current variables to the file 'octave-workspace' if it receives a terminate signal.

    When called from inside a function with the `"local"` option, the variable is changed locally for the function and any subroutines it calls. The original variable value is restored when exiting the function.

`val = octave_core_file_options ()`        [Built-in Function]
`old_val = octave_core_file_options (new_val)`        [Built-in Function]
`octave_core_file_options (new_val, "local")`        [Built-in Function]

    Query or set the internal variable that specifies the options used for saving the workspace data if Octave aborts.

    The value of `octave_core_file_options` should follow the same format as the options for the `save` function. The default value is Octave's binary format.

    When called from inside a function with the `"local"` option, the variable is changed locally for the function and any subroutines it calls. The original variable value is restored when exiting the function.

    **See also:** [crash_dumps_octave_core], page 247, [octave_core_file_name], page 248, [octave_core_file_limit], page 248.

`val = octave_core_file_limit ()`        [Built-in Function]
`old_val = octave_core_file_limit (new_val)`        [Built-in Function]
`octave_core_file_limit (new_val, "local")`        [Built-in Function]

    Query or set the internal variable that specifies the maximum amount of memory (in kilobytes) of the top-level workspace that Octave will attempt to save when writing data to the crash dump file (the name of the file is specified by *octave_core_file_name*).

    If *octave_core_file_options* flags specify a binary format, then *octave_core_file_limit* will be approximately the maximum size of the file. If a text file format is used, then the file could be much larger than the limit. The default value is -1 (unlimited)

    When called from inside a function with the `"local"` option, the variable is changed locally for the function and any subroutines it calls. The original variable value is restored when exiting the function.

    **See also:** [crash_dumps_octave_core], page 247, [octave_core_file_name], page 248, [octave_core_file_options], page 248.

`val = octave_core_file_name ()`        [Built-in Function]
`old_val = octave_core_file_name (new_val)`        [Built-in Function]

Chapter 14: Input and Output       249

octave_core_file_name (*new_val*, "*local*")          [Built-in Function]
: Query or set the internal variable that specifies the name of the file used for saving data from the top-level workspace if Octave aborts.

    The default value is "octave-workspace"

    When called from inside a function with the "local" option, the variable is changed locally for the function and any subroutines it calls. The original variable value is restored when exiting the function.

    **See also:** [crash_dumps_octave_core], page 247, [octave_core_file_name], page 248, [octave_core_file_options], page 248.

## 14.2 C-Style I/O Functions

Octave's C-style input and output functions provide most of the functionality of the C programming language's standard I/O library. The argument lists for some of the input functions are slightly different, however, because Octave has no way of passing arguments by reference.

In the following, *file* refers to a file name and `fid` refers to an integer file number, as returned by `fopen`.

There are three files that are always available. Although these files can be accessed using their corresponding numeric file ids, you should always use the symbolic names given in the table below, since it will make your programs easier to understand.

stdin ()                                              [Built-in Function]
: Return the numeric value corresponding to the standard input stream.

    When Octave is used interactively, stdin is filtered through the command line editing functions.

    **See also:** [stdout], page 249, [stderr], page 249.

stdout ()                                             [Built-in Function]
: Return the numeric value corresponding to the standard output stream.

    Data written to the standard output is normally filtered through the pager.

    **See also:** [stdin], page 249, [stderr], page 249.

stderr ()                                             [Built-in Function]
: Return the numeric value corresponding to the standard error stream.

    Even if paging is turned on, the standard error is not sent to the pager. It is useful for error messages and prompts.

    **See also:** [stdin], page 249, [stdout], page 249.

### 14.2.1 Opening and Closing Files

When reading data from a file it must be opened for reading first, and likewise when writing to a file. The `fopen` function returns a pointer to an open file that is ready to be read or written. Once all data has been read from or written to the opened file it should be closed. The `fclose` function does this. The following code illustrates the basic pattern for writing to a file, but a very similar pattern is used when reading a file.

```
filename = "myfile.txt";
fid = fopen (filename, "w");
# Do the actual I/O here...
fclose (fid);
```

*fid* = fopen (*name*)     [Built-in Function]
*fid* = fopen (*name*, *mode*)     [Built-in Function]
*fid* = fopen (*name*, *mode*, *arch*)     [Built-in Function]
[*fid*, *msg*] = fopen (...)     [Built-in Function]
*fid_list* = fopen ("*all*")     [Built-in Function]
[*file*, *mode*, *arch*] = fopen (*fid*)     [Built-in Function]

Open a file for low-level I/O or query open files and file descriptors.

The first form of the `fopen` function opens the named file with the specified mode (read-write, read-only, etc.) and architecture interpretation (IEEE big endian, IEEE little endian, etc.), and returns an integer value that may be used to refer to the file later. If an error occurs, *fid* is set to $-1$ and *msg* contains the corresponding system error message. The *mode* is a one or two character string that specifies whether the file is to be opened for reading, writing, or both.

The second form of the `fopen` function returns a vector of file ids corresponding to all the currently open files, excluding the `stdin`, `stdout`, and `stderr` streams.

The third form of the `fopen` function returns information about the open file given its file id.

For example,

```
myfile = fopen ("splat.dat", "r", "ieee-le");
```

opens the file 'splat.dat' for reading. If necessary, binary numeric values will be read assuming they are stored in IEEE format with the least significant bit first, and then converted to the native representation.

Opening a file that is already open simply opens it again and returns a separate file id. It is not an error to open a file several times, though writing to the same file through several different file ids may produce unexpected results.

The possible values '`mode`' may have are

'r' (default)
: Open a file for reading.

'w'
: Open a file for writing. The previous contents are discarded.

'a'
: Open or create a file for writing at the end of the file.

'r+'
: Open an existing file for reading and writing.

'w+'
: Open a file for reading or writing. The previous contents are discarded.

'a+'
: Open or create a file for reading or writing at the end of the file.

Append a "t" to the mode string to open the file in text mode or a "b" to open in binary mode. On Windows and Macintosh systems, text mode reading and writing automatically converts linefeeds to the appropriate line end character for the system (carriage-return linefeed on Windows, carriage-return on Macintosh). The default when no mode is specified is binary mode.

Chapter 14: Input and Output                                              251

Additionally, you may append a "z" to the mode string to open a gzipped file for reading or writing. For this to be successful, you must also open the file in binary mode.

The parameter *arch* is a string specifying the default data format for the file. Valid values for *arch* are:

'native (default)'
: The format of the current machine.

'ieee-be'   IEEE big endian format.

'ieee-le'   IEEE little endian format.

however, conversions are currently only supported for 'native' 'ieee-be', and 'ieee-le' formats.

When opening a new file that does not yet exist, permissions will be set to 0666 - *umask*.

**See also:** [fclose], page 251, [fgets], page 252, [fgetl], page 252, [fscanf], page 259, [fread], page 262, [fputs], page 251, [fdisp], page 242, [fprintf], page 253, [fwrite], page 265, [fskipl], page 253, [fseek], page 268, [frewind], page 268, [ftell], page 268, [feof], page 267, [ferror], page 267, [fclear], page 267, [fflush], page 236, [freport], page 267, [umask], page 756.

fclose (*fid*)                                              [Built-in Function]
fclose ("*all*")                                            [Built-in Function]
status = fclose ("*all*")                                   [Built-in Function]
: Close the file specified by the file descriptor *fid*.

If successful, `fclose` returns 0, otherwise, it returns -1. The second form of the `fclose` call closes all open files except `stdout`, `stderr`, and `stdin`.

Programming Note: When using "all" the file descriptors associated with gnuplot will also be closed. This will prevent further plotting with gnuplot until Octave is closed and restarted.

**See also:** [fopen], page 250, [fflush], page 236, [freport], page 267.

is_valid_file_id (*fid*)                                         [Function File]
: Return true if *fid* refers to an open file.

**See also:** [freport], page 267, [fopen], page 250.

## 14.2.2 Simple Output

Once a file has been opened for writing a string can be written to the file using the `fputs` function. The following example shows how to write the string 'Free Software is needed for Free Science' to the file 'free.txt'.

```
filename = "free.txt";
fid = fopen (filename, "w");
fputs (fid, "Free Software is needed for Free Science");
fclose (fid);
```

`fputs (`*`fid`*`, `*`string`*`)` [Built-in Function]
`status = fputs (`*`fid`*`, `*`string`*`)` [Built-in Function]

    Write the string *string* to the file with file descriptor *fid*.

    The string is written to the file with no additional formatting. Use `fdisp` instead to automatically append a newline character appropriate for the local machine.

    Return a non-negative number on success or EOF on error.

    **See also:** [fdisp], page 242, [fprintf], page 253, [fwrite], page 265, [fopen], page 250.

A function much similar to `fputs` is available for writing data to the screen. The `puts` function works just like `fputs` except it doesn't take a file pointer as its input.

`puts (`*`string`*`)` [Built-in Function]
`status = puts (`*`string`*`)` [Built-in Function]

    Write a string to the standard output with no formatting.

    The string is written verbatim to the standard output. Use `disp` to automatically append a newline character appropriate for the local machine.

    Return a non-negative number on success and EOF on error.

    **See also:** [fputs], page 251, [disp], page 231.

### 14.2.3 Line-Oriented Input

To read from a file it must be opened for reading using `fopen`. Then a line can be read from the file using `fgetl` as the following code illustrates

```
fid = fopen ("free.txt");
txt = fgetl (fid)
      ⊣ Free Software is needed for Free Science
fclose (fid);
```

This of course assumes that the file 'free.txt' exists and contains the line 'Free Software is needed for Free Science'.

`str = fgetl (`*`fid`*`)` [Built-in Function]
`str = fgetl (`*`fid`*`, `*`len`*`)` [Built-in Function]

    Read characters from a file, stopping after a newline, or EOF, or *len* characters have been read.

    The characters read, excluding the possible trailing newline, are returned as a string.

    If *len* is omitted, `fgetl` reads until the next newline character.

    If there are no more characters to read, `fgetl` returns $-1$.

    To read a line and return the terminating newline see `fgets`.

    **See also:** [fgets], page 252, [fscanf], page 259, [fread], page 262, [fopen], page 250.

`str = fgets (`*`fid`*`)` [Built-in Function]
`str = fgets (`*`fid`*`, `*`len`*`)` [Built-in Function]

    Read characters from a file, stopping after a newline, or EOF, or *len* characters have been read.

    The characters read, including the possible trailing newline, are returned as a string.

# Chapter 14: Input and Output

If *len* is omitted, `fgets` reads until the next newline character.

If there are no more characters to read, `fgets` returns −1.

To read a line and discard the terminating newline see `fgetl`.

**See also:** [fputs], page 251, [fgetl], page 252, [fscanf], page 259, [fread], page 262, [fopen], page 250.

*nlines* = fskipl (*fid*)                                                                      [Built-in Function]
*nlines* = fskipl (*fid*, *count*)                                                     [Built-in Function]
*nlines* = fskipl (*fid*, *Inf*)                                                        [Built-in Function]

Read and skip *count* lines from the file specified by the file descriptor *fid*.

`fskipl` discards characters until an end-of-line is encountered exactly *count*-times, or until the end-of-file marker is found.

If *count* is omitted, it defaults to 1. *count* may also be `Inf`, in which case lines are skipped until the end of the file. This form is suitable for counting the number of lines in a file.

Returns the number of lines skipped (end-of-line sequences encountered).

**See also:** [fgetl], page 252, [fgets], page 252, [fscanf], page 259, [fopen], page 250.

## 14.2.4 Formatted Output

This section describes how to call `printf` and related functions.

The following functions are available for formatted output. They are modeled after the C language functions of the same name, but they interpret the format template differently in order to improve the performance of printing vector and matrix values.

Implementation Note: For compatibility with MATLAB, escape sequences in the template string (e.g., `"\n"` => newline) are expanded even when the template string is defined with single quotes.

printf (*template*, ...)                                                           [Built-in Function]

Print optional arguments under the control of the template string *template* to the stream `stdout` and return the number of characters printed.

See the Formatted Output section of the GNU Octave manual for a complete description of the syntax of the template string.

Implementation Note: For compatibility with MATLAB, escape sequences in the template string (e.g., `"\n"` => newline) are expanded even when the template string is defined with single quotes.

**See also:** [fprintf], page 253, [sprintf], page 254, [scanf], page 259.

fprintf (*fid*, *template*, ...)                                             [Built-in Function]
fprintf (*template*, ...)                                                      [Built-in Function]
*numbytes* = fprintf (...)                                              [Built-in Function]

This function is equivalent to `printf`, except that the output is written to the file descriptor *fid* instead of `stdout`.

If *fid* is omitted, the output is written to `stdout` making the function exactly equivalent to `printf`.

The optional output returns the number of bytes written to the file.

Implementation Note: For compatibility with MATLAB, escape sequences in the template string (e.g., `"\n"` => newline) are expanded even when the template string is defined with single quotes.

**See also:** [fputs], page 251, [fdisp], page 242, [fwrite], page 265, [fscanf], page 259, [printf], page 253, [sprintf], page 254, [fopen], page 250.

**sprintf** (*template*, ...)                                                                                             [Built-in Function]

This is like `printf`, except that the output is returned as a string.

Unlike the C library function, which requires you to provide a suitably sized string as an argument, Octave's `sprintf` function returns the string, automatically sized to hold all of the items converted.

Implementation Note: For compatibility with MATLAB, escape sequences in the template string (e.g., `"\n"` => newline) are expanded even when the template string is defined with single quotes.

**See also:** [printf], page 253, [fprintf], page 253, [sscanf], page 260.

The `printf` function can be used to print any number of arguments. The template string argument you supply in a call provides information not only about the number of additional arguments, but also about their types and what style should be used for printing them.

Ordinary characters in the template string are simply written to the output stream as-is, while *conversion specifications* introduced by a '%' character in the template cause subsequent arguments to be formatted and written to the output stream. For example,

```
pct = 37;
filename = "foo.txt";
printf ("Processed %d%% of '%s'.\nPlease be patient.\n",
        pct, filename);
```

produces output like

```
Processed 37% of 'foo.txt'.
Please be patient.
```

This example shows the use of the '%d' conversion to specify that a scalar argument should be printed in decimal notation, the '%s' conversion to specify printing of a string argument, and the '%%' conversion to print a literal '%' character.

There are also conversions for printing an integer argument as an unsigned value in octal, decimal, or hexadecimal radix ('%o', '%u', or '%x', respectively); or as a character value ('%c').

Floating-point numbers can be printed in normal, fixed-point notation using the '%f' conversion or in exponential notation using the '%e' conversion. The '%g' conversion uses either '%e' or '%f' format, depending on what is more appropriate for the magnitude of the particular number.

You can control formatting more precisely by writing *modifiers* between the '%' and the character that indicates which conversion to apply. These slightly alter the ordinary behavior of the conversion. For example, most conversion specifications permit you to specify a minimum field width and a flag indicating whether you want the result left- or right-justified within the field.

The specific flags and modifiers that are permitted and their interpretation vary depending on the particular conversion. They're all described in more detail in the following sections.

### 14.2.5 Output Conversion for Matrices

When given a matrix value, Octave's formatted output functions cycle through the format template until all the values in the matrix have been printed. For example:

```
printf ("%4.2f %10.2e %8.4g\n", hilb (3));
```

```
⊣ 1.00    5.00e-01    0.3333
⊣ 0.50    3.33e-01    0.25
⊣ 0.33    2.50e-01    0.2
```

If more than one value is to be printed in a single call, the output functions do not return to the beginning of the format template when moving on from one value to the next. This can lead to confusing output if the number of elements in the matrices are not exact multiples of the number of conversions in the format template. For example:

```
printf ("%4.2f %10.2e %8.4g\n", [1, 2], [3, 4]);
```

```
⊣ 1.00    2.00e+00           3
⊣ 4.00
```

If this is not what you want, use a series of calls instead of just one.

### 14.2.6 Output Conversion Syntax

This section provides details about the precise syntax of conversion specifications that can appear in a `printf` template string.

Characters in the template string that are not part of a conversion specification are printed as-is to the output stream.

The conversion specifications in a `printf` template string have the general form:

`% flags width [ . precision ] type conversion`

For example, in the conversion specifier '`%-10.8ld`', the '`-`' is a flag, '`10`' specifies the field width, the precision is '`8`', the letter '`l`' is a type modifier, and '`d`' specifies the conversion style. (This particular type specifier says to print a numeric argument in decimal notation, with a minimum of 8 digits left-justified in a field at least 10 characters wide.)

In more detail, output conversion specifications consist of an initial '`%`' character followed in sequence by:

- Zero or more *flag characters* that modify the normal behavior of the conversion specification.
- An optional decimal integer specifying the *minimum field width*. If the normal conversion produces fewer characters than this, the field is padded with spaces to the specified width. This is a *minimum* value; if the normal conversion produces more characters than this, the field is *not* truncated. Normally, the output is right-justified within the field.

You can also specify a field width of '*'. This means that the next argument in the argument list (before the actual value to be printed) is used as the field width. The value is rounded to the nearest integer. If the value is negative, this means to set the '-' flag (see below) and to use the absolute value as the field width.

- An optional *precision* to specify the number of digits to be written for the numeric conversions. If the precision is specified, it consists of a period ('.') followed optionally by a decimal integer (which defaults to zero if omitted).

  You can also specify a precision of '*'. This means that the next argument in the argument list (before the actual value to be printed) is used as the precision. The value must be an integer, and is ignored if it is negative.

- An optional *type modifier character*. This character is ignored by Octave's `printf` function, but is recognized to provide compatibility with the C language `printf`.

- A character that specifies the conversion to be applied.

The exact options that are permitted and how they are interpreted vary between the different conversion specifiers. See the descriptions of the individual conversions for information about the particular options that they use.

### 14.2.7 Table of Output Conversions

Here is a table summarizing what all the different conversions do:

'%d', '%i'   Print an integer as a signed decimal number. See Section 14.2.8 [Integer Conversions], page 257, for details. '%d' and '%i' are synonymous for output, but are different when used with `scanf` for input (see Section 14.2.13 [Table of Input Conversions], page 261).

'%o'   Print an integer as an unsigned octal number. See Section 14.2.8 [Integer Conversions], page 257, for details.

'%u'   Print an integer as an unsigned decimal number. See Section 14.2.8 [Integer Conversions], page 257, for details.

'%x', '%X'   Print an integer as an unsigned hexadecimal number. '%x' uses lowercase letters and '%X' uses uppercase. See Section 14.2.8 [Integer Conversions], page 257, for details.

'%f'   Print a floating-point number in normal (fixed-point) notation. See Section 14.2.9 [Floating-Point Conversions], page 258, for details.

'%e', '%E'   Print a floating-point number in exponential notation. '%e' uses lowercase letters and '%E' uses uppercase. See Section 14.2.9 [Floating-Point Conversions], page 258, for details.

'%g', '%G'   Print a floating-point number in either normal (fixed-point) or exponential notation, whichever is more appropriate for its magnitude. '%g' uses lowercase letters and '%G' uses uppercase. See Section 14.2.9 [Floating-Point Conversions], page 258, for details.

'%c'   Print a single character. See Section 14.2.10 [Other Output Conversions], page 258.

'%s'   Print a string. See Section 14.2.10 [Other Output Conversions], page 258.

Chapter 14: Input and Output 257

'%%'   Print a literal '%' character. See Section 14.2.10 [Other Output Conversions], page 258.

If the syntax of a conversion specification is invalid, unpredictable things will happen, so don't do this. In particular, MATLAB allows a bare percentage sign '%' with no subsequent conversion character. Octave will emit an error and stop if it sees such code. When the string variable to be processed cannot be guaranteed to be free of potential format codes it is better to use the two argument form of any of the `printf` functions and set the format string to `%s`. Alternatively, for code which is not required to be backwards-compatible with MATLAB the Octave function `puts` or `disp` can be used.

```
printf (strvar);         # Unsafe if strvar contains format codes
printf ("%s", strvar);   # Safe
puts (strvar);           # Safe
```

If there aren't enough function arguments provided to supply values for all the conversion specifications in the template string, or if the arguments are not of the correct types, the results are unpredictable. If you supply more arguments than conversion specifications, the extra argument values are simply ignored; this is sometimes useful.

## 14.2.8 Integer Conversions

This section describes the options for the '%d', '%i', '%o', '%u', '%x', and '%X' conversion specifications. These conversions print integers in various formats.

The '%d' and '%i' conversion specifications both print an numeric argument as a signed decimal number; while '%o', '%u', and '%x' print the argument as an unsigned octal, decimal, or hexadecimal number (respectively). The '%X' conversion specification is just like '%x' except that it uses the characters 'ABCDEF' as digits instead of 'abcdef'.

The following flags are meaningful:

'-'    Left-justify the result in the field (instead of the normal right-justification).

'+'    For the signed '%d' and '%i' conversions, print a plus sign if the value is positive.

' '    For the signed '%d' and '%i' conversions, if the result doesn't start with a plus or minus sign, prefix it with a space character instead. Since the '+' flag ensures that the result includes a sign, this flag is ignored if you supply both of them.

'#'    For the '%o' conversion, this forces the leading digit to be '0', as if by increasing the precision. For '%x' or '%X', this prefixes a leading '0x' or '0X' (respectively) to the result. This doesn't do anything useful for the '%d', '%i', or '%u' conversions.

'0'    Pad the field with zeros instead of spaces. The zeros are placed after any indication of sign or base. This flag is ignored if the '-' flag is also specified, or if a precision is specified.

If a precision is supplied, it specifies the minimum number of digits to appear; leading zeros are produced if necessary. If you don't specify a precision, the number is printed with as many digits as it needs. If you convert a value of zero with an explicit precision of zero, then no characters at all are produced.

### 14.2.9 Floating-Point Conversions

This section discusses the conversion specifications for floating-point numbers: the '%f', '%e', '%E', '%g', and '%G' conversions.

The '%f' conversion prints its argument in fixed-point notation, producing output of the form [-]ddd.ddd, where the number of digits following the decimal point is controlled by the precision you specify.

The '%e' conversion prints its argument in exponential notation, producing output of the form [-]d.ddde[+|-]dd. Again, the number of digits following the decimal point is controlled by the precision. The exponent always contains at least two digits. The '%E' conversion is similar but the exponent is marked with the letter 'E' instead of 'e'.

The '%g' and '%G' conversions print the argument in the style of '%e' or '%E' (respectively) if the exponent would be less than -4 or greater than or equal to the precision; otherwise they use the '%f' style. Trailing zeros are removed from the fractional portion of the result and a decimal-point character appears only if it is followed by a digit.

The following flags can be used to modify the behavior:

'-'   Left-justify the result in the field. Normally the result is right-justified.

'+'   Always include a plus or minus sign in the result.

' '   If the result doesn't start with a plus or minus sign, prefix it with a space instead. Since the '+' flag ensures that the result includes a sign, this flag is ignored if you supply both of them.

'#'   Specifies that the result should always include a decimal point, even if no digits follow it. For the '%g' and '%G' conversions, this also forces trailing zeros after the decimal point to be left in place where they would otherwise be removed.

'0'   Pad the field with zeros instead of spaces; the zeros are placed after any sign. This flag is ignored if the '-' flag is also specified.

The precision specifies how many digits follow the decimal-point character for the '%f', '%e', and '%E' conversions. For these conversions, the default precision is 6. If the precision is explicitly 0, this suppresses the decimal point character entirely. For the '%g' and '%G' conversions, the precision specifies how many significant digits to print. Significant digits are the first digit before the decimal point, and all the digits after it. If the precision is 0 or not specified for '%g' or '%G', it is treated like a value of 1. If the value being printed cannot be expressed precisely in the specified number of digits, the value is rounded to the nearest number that fits.

### 14.2.10 Other Output Conversions

This section describes miscellaneous conversions for `printf`.

The '%c' conversion prints a single character. The '-' flag can be used to specify left-justification in the field, but no other flags are defined, and no precision or type modifier can be given. For example:

    printf ("%c%c%c%c%c", "h", "e", "l", "l", "o");

prints 'hello'.

# Chapter 14: Input and Output

The '%s' conversion prints a string. The corresponding argument must be a string. A precision can be specified to indicate the maximum number of characters to write; otherwise characters in the string up to but not including the terminating null character are written to the output stream. The '-' flag can be used to specify left-justification in the field, but no other flags or type modifiers are defined for this conversion. For example:

```
printf ("%3s%-6s", "no", "where");
```

prints ' nowhere ' (note the leading and trailing spaces).

## 14.2.11 Formatted Input

Octave provides the `scanf`, `fscanf`, and `sscanf` functions to read formatted input. There are two forms of each of these functions. One can be used to extract vectors of data from a file, and the other is more 'C-like'.

[val, count, errmsg] = fscanf (*fid, template, size*)   [Built-in Function]
[v1, v2, ..., count, errmsg] = fscanf (*fid, template,*   [Built-in Function]
    "*C*")

> In the first form, read from *fid* according to *template*, returning the result in the matrix *val*.
>
> The optional argument *size* specifies the amount of data to read and may be one of
>
> Inf           Read as much as possible, returning a column vector.
>
> nr            Read up to *nr* elements, returning a column vector.
>
> [nr, Inf]     Read as much as possible, returning a matrix with *nr* rows. If the number of elements read is not an exact multiple of *nr*, the last column is padded with zeros.
>
> [nr, nc]      Read up to *nr * nc* elements, returning a matrix with *nr* rows. If the number of elements read is not an exact multiple of *nr*, the last column is padded with zeros.
>
> If *size* is omitted, a value of Inf is assumed.
>
> A string is returned if *template* specifies only character conversions.
>
> The number of items successfully read is returned in *count*.
>
> If an error occurs, *errmsg* contains a system-dependent error message.
>
> In the second form, read from *fid* according to *template*, with each conversion specifier in *template* corresponding to a single scalar return value. This form is more "C-like", and also compatible with previous versions of Octave. The number of successful conversions is returned in *count*
>
> See the Formatted Input section of the GNU Octave manual for a complete description of the syntax of the template string.
>
> **See also:** [fgets], page 252, [fgetl], page 252, [fread], page 262, [scanf], page 259, [sscanf], page 260, [fopen], page 250.

[val, count, errmsg] = scanf (*template, size*)   [Built-in Function]
[v1, v2, ..., count, errmsg]] = scanf (*template,* "*C*")   [Built-in Function]
> This is equivalent to calling `fscanf` with *fid* = `stdin`.

It is currently not useful to call `scanf` in interactive programs.

**See also:** [fscanf], page 259, [sscanf], page 260, [printf], page 253.

[*val*, *count*, *errmsg*, *pos*] = sscanf (*string*, *template*, *size*)     [Built-in Function]

[*v1*, *v2*, ..., *count*, *errmsg*] = sscanf (*string*, *template*, "*C*")     [Built-in Function]

This is like `fscanf`, except that the characters are taken from the string *string* instead of from a stream.

Reaching the end of the string is treated as an end-of-file condition. In addition to the values returned by `fscanf`, the index of the next character to be read is returned in *pos*.

**See also:** [fscanf], page 259, [scanf], page 259, [sprintf], page 254.

Calls to `scanf` are superficially similar to calls to `printf` in that arbitrary arguments are read under the control of a template string. While the syntax of the conversion specifications in the template is very similar to that for `printf`, the interpretation of the template is oriented more towards free-format input and simple pattern matching, rather than fixed-field formatting. For example, most `scanf` conversions skip over any amount of "white space" (including spaces, tabs, and newlines) in the input file, and there is no concept of precision for the numeric input conversions as there is for the corresponding output conversions. Ordinarily, non-whitespace characters in the template are expected to match characters in the input stream exactly.

When a *matching failure* occurs, `scanf` returns immediately, leaving the first non-matching character as the next character to be read from the stream, and `scanf` returns all the items that were successfully converted.

The formatted input functions are not used as frequently as the formatted output functions. Partly, this is because it takes some care to use them properly. Another reason is that it is difficult to recover from a matching error.

### 14.2.12 Input Conversion Syntax

A `scanf` template string is a string that contains ordinary multibyte characters interspersed with conversion specifications that start with '%'.

Any whitespace character in the template causes any number of whitespace characters in the input stream to be read and discarded. The whitespace characters that are matched need not be exactly the same whitespace characters that appear in the template string. For example, write ' , ' in the template to recognize a comma with optional whitespace before and after.

Other characters in the template string that are not part of conversion specifications must match characters in the input stream exactly; if this is not the case, a matching failure occurs.

The conversion specifications in a `scanf` template string have the general form:

`% flags width type conversion`

In more detail, an input conversion specification consists of an initial '%' character followed in sequence by:

Chapter 14: Input and Output 261

- An optional *flag character* '*', which says to ignore the text read for this specification. When `scanf` finds a conversion specification that uses this flag, it reads input as directed by the rest of the conversion specification, but it discards this input, does not return any value, and does not increment the count of successful assignments.

- An optional decimal integer that specifies the *maximum field width*. Reading of characters from the input stream stops either when this maximum is reached or when a non-matching character is found, whichever happens first. Most conversions discard initial whitespace characters, and these discarded characters don't count towards the maximum field width. Conversions that do not discard initial whitespace are explicitly documented.

- An optional type modifier character. This character is ignored by Octave's `scanf` function, but is recognized to provide compatibility with the C language `scanf`.

- A character that specifies the conversion to be applied.

The exact options that are permitted and how they are interpreted vary between the different conversion specifiers. See the descriptions of the individual conversions for information about the particular options that they allow.

## 14.2.13 Table of Input Conversions

Here is a table that summarizes the various conversion specifications:

'%d'  Matches an optionally signed integer written in decimal. See Section 14.2.14 [Numeric Input Conversions], page 262.

'%i'  Matches an optionally signed integer in any of the formats that the C language defines for specifying an integer constant. See Section 14.2.14 [Numeric Input Conversions], page 262.

'%o'  Matches an unsigned integer written in octal radix. See Section 14.2.14 [Numeric Input Conversions], page 262.

'%u'  Matches an unsigned integer written in decimal radix. See Section 14.2.14 [Numeric Input Conversions], page 262.

'%x', '%X'  Matches an unsigned integer written in hexadecimal radix. See Section 14.2.14 [Numeric Input Conversions], page 262.

'%e', '%f', '%g', '%E', '%G'
Matches an optionally signed floating-point number. See Section 14.2.14 [Numeric Input Conversions], page 262.

'%s'  Matches a string containing only non-whitespace characters. See Section 14.2.15 [String Input Conversions], page 262.

'%c'  Matches a string of one or more characters; the number of characters read is controlled by the maximum field width given for the conversion. See Section 14.2.15 [String Input Conversions], page 262.

'%%'  This matches a literal '%' character in the input stream. No corresponding argument is used.

If the syntax of a conversion specification is invalid, the behavior is undefined. If there aren't enough function arguments provided to supply addresses for all the conversion specifications in the template strings that perform assignments, or if the arguments are not of the correct types, the behavior is also undefined. On the other hand, extra arguments are simply ignored.

### 14.2.14 Numeric Input Conversions

This section describes the `scanf` conversions for reading numeric values.

The '`%d`' conversion matches an optionally signed integer in decimal radix.

The '`%i`' conversion matches an optionally signed integer in any of the formats that the C language defines for specifying an integer constant.

For example, any of the strings '`10`', '`0xa`', or '`012`' could be read in as integers under the '`%i`' conversion. Each of these specifies a number with decimal value 10.

The '`%o`', '`%u`', and '`%x`' conversions match unsigned integers in octal, decimal, and hexadecimal radices, respectively.

The '`%X`' conversion is identical to the '`%x`' conversion. They both permit either uppercase or lowercase letters to be used as digits.

Unlike the C language `scanf`, Octave ignores the '`h`', '`l`', and '`L`' modifiers.

### 14.2.15 String Input Conversions

This section describes the `scanf` input conversions for reading string and character values: '`%s`' and '`%c`'.

The '`%c`' conversion is the simplest: it matches a fixed number of characters, always. The maximum field with says how many characters to read; if you don't specify the maximum, the default is 1. This conversion does not skip over initial whitespace characters. It reads precisely the next *n* characters, and fails if it cannot get that many.

The '`%s`' conversion matches a string of non-whitespace characters. It skips and discards initial whitespace, but stops when it encounters more whitespace after having read something.

For example, reading the input:

    hello, world

with the conversion '`%10c`' produces `" hello, wo"`, but reading the same input with the conversion '`%10s`' produces `"hello,"`.

### 14.2.16 Binary I/O

Octave can read and write binary data using the functions `fread` and `fwrite`, which are patterned after the standard C functions with the same names. They are able to automatically swap the byte order of integer data and convert among the supported floating point formats as the data are read.

| | |
|---|---:|
| *val* = fread (*fid*) | [Built-in Function] |
| *val* = fread (*fid*, *size*) | [Built-in Function] |
| *val* = fread (*fid*, *size*, *precision*) | [Built-in Function] |
| *val* = fread (*fid*, *size*, *precision*, *skip*) | [Built-in Function] |

Chapter 14: Input and Output                                                           263

*val* = fread (*fid, size, precision, skip, arch*)                    [Built-in Function]
[*val, count*] = fread (...)                                          [Built-in Function]

> Read binary data from the file specified by the file descriptor *fid*.
>
> The optional argument *size* specifies the amount of data to read and may be one of
>
> Inf
> : Read as much as possible, returning a column vector.
>
> *nr*
> : Read up to *nr* elements, returning a column vector.
>
> [*nr*, Inf]
> : Read as much as possible, returning a matrix with *nr* rows. If the number of elements read is not an exact multiple of *nr*, the last column is padded with zeros.
>
> [*nr, nc*]
> : Read up to *nr* * *nc* elements, returning a matrix with *nr* rows. If the number of elements read is not an exact multiple of *nr*, the last column is padded with zeros.
>
> If *size* is omitted, a value of Inf is assumed.
>
> The optional argument *precision* is a string specifying the type of data to read and may be one of
>
> "schar"
> "signed char"
> : Signed character.
>
> "uchar"
> "unsigned char"
> : Unsigned character.
>
> "int8"
> "integer*1"
> : 8-bit signed integer.
>
> "int16"
> "integer*2"
> : 16-bit signed integer.
>
> "int32"
> "integer*4"
> : 32-bit signed integer.
>
> "int64"
> "integer*8"
> : 64-bit signed integer.
>
> "uint8"
> : 8-bit unsigned integer.
>
> "uint16"
> : 16-bit unsigned integer.
>
> "uint32"
> : 32-bit unsigned integer.
>
> "uint64"
> : 64-bit unsigned integer.
>
> "single"
> "float32"
> "real*4"
> : 32-bit floating point number.

`"double"`
`"float64"`
`"real*8"`   64-bit floating point number.

`"char"`
`"char*1"`   Single character.

`"short"`    Short integer (size is platform dependent).

`"int"`      Integer (size is platform dependent).

`"long"`     Long integer (size is platform dependent).

`"ushort"`
`"unsigned short"`
             Unsigned short integer (size is platform dependent).

`"uint"`
`"unsigned int"`
             Unsigned integer (size is platform dependent).

`"ulong"`
`"unsigned long"`
             Unsigned long integer (size is platform dependent).

`"float"`    Single precision floating point number (size is platform dependent).

The default precision is `"uchar"`.

The *precision* argument may also specify an optional repeat count. For example, '32*single' causes `fread` to read a block of 32 single precision floating point numbers. Reading in blocks is useful in combination with the *skip* argument.

The *precision* argument may also specify a type conversion. For example, 'int16=>int32' causes `fread` to read 16-bit integer values and return an array of 32-bit integer values. By default, `fread` returns a double precision array. The special form '*TYPE' is shorthand for 'TYPE=>TYPE'.

The conversion and repeat counts may be combined. For example, the specification '32*single=>single' causes `fread` to read blocks of single precision floating point values and return an array of single precision values instead of the default array of double precision values.

The optional argument *skip* specifies the number of bytes to skip after each element (or block of elements) is read. If it is not specified, a value of 0 is assumed. If the final block read is not complete, the final skip is omitted. For example,

        `fread (f, 10, "3*single=>single", 8)`

will omit the final 8-byte skip because the last read will not be a complete block of 3 values.

The optional argument *arch* is a string specifying the data format for the file. Valid values are

`"native"`   The format of the current machine.

`"ieee-be"`
             IEEE big endian.

"ieee-le"
: IEEE little endian.

The output argument *val* contains the data read from the file. The optional return value *count* contains the number of elements read.

**See also:** [fwrite], page 265, [fgets], page 252, [fgetl], page 252, [fscanf], page 259, [fopen], page 250.

fwrite (*fid, data*)                                     [Built-in Function]
fwrite (*fid, data, precision*)                          [Built-in Function]
fwrite (*fid, data, precision, skip*)                    [Built-in Function]
fwrite (*fid, data, precision, skip, arch*)              [Built-in Function]
count = fwrite (...)                                     [Built-in Function]
: Write data in binary form to the file specified by the file descriptor *fid*, returning the number of values *count* successfully written to the file.

The argument *data* is a matrix of values that are to be written to the file. The values are extracted in column-major order.

The remaining arguments *precision*, *skip*, and *arch* are optional, and are interpreted as described for `fread`.

The behavior of `fwrite` is undefined if the values in *data* are too large to fit in the specified precision.

**See also:** [fread], page 262, [fputs], page 251, [fprintf], page 253, [fopen], page 250.

### 14.2.17 Temporary Files

Sometimes one needs to write data to a file that is only temporary. This is most commonly used when an external program launched from within Octave needs to access data. When Octave exits all temporary files will be deleted, so this step need not be executed manually.

[*fid, name, msg*] = mkstemp ("*template*")              [Built-in Function]
[*fid, name, msg*] = mkstemp ("*template*", *delete*)    [Built-in Function]
: Return the file descriptor *fid* corresponding to a new temporary file with a unique name created from *template*.

The last six characters of *template* must be "XXXXXX" and these are replaced with a string that makes the filename unique. The file is then created with mode read/write and permissions that are system dependent (on GNU/Linux systems, the permissions will be 0600 for versions of glibc 2.0.7 and later). The file is opened in binary mode and with the O_EXCL flag.

If the optional argument *delete* is supplied and is true, the file will be deleted automatically when Octave exits.

If successful, *fid* is a valid file ID, *name* is the name of the file, and *msg* is an empty string. Otherwise, *fid* is -1, *name* is empty, and *msg* contains a system-dependent error message.

**See also:** [tempname], page 266, [tempdir], page 266, [P_tmpdir], page 266, [tmpfile], page 266, [fopen], page 250.

[fid, msg] = tmpfile ()                                    [Built-in Function]
> Return the file ID corresponding to a new temporary file with a unique name.
>
> The file is opened in binary read/write ("w+b") mode and will be deleted automatically when it is closed or when Octave exits.
>
> If successful, fid is a valid file ID and msg is an empty string. Otherwise, fid is -1 and msg contains a system-dependent error message.
>
> **See also:** [tempname], page 266, [mkstemp], page 265, [tempdir], page 266, [P_tmpdir], page 266.

fname = tempname ()                                        [Built-in Function]
fname = tempname (dir)                                     [Built-in Function]
fname = tempname (dir, prefix)                             [Built-in Function]
> Return a unique temporary file name as a string.
>
> If prefix is omitted, a value of "oct-" is used.
>
> If dir is also omitted, the default directory for temporary files (P_tmpdir) is used. If dir is provided, it must exist, otherwise the default directory for temporary files is used.
>
> Programming Note: Because the named file is not opened by **tempname**, it is possible, though relatively unlikely, that it will not be available by the time your program attempts to open it. If this is a concern, see **tmpfile**.
>
> **See also:** [mkstemp], page 265, [tempdir], page 266, [P_tmpdir], page 266, [tmpfile], page 266.

dir = tempdir ()                                           [Function File]
> Return the name of the host system's directory for temporary files.
>
> The directory name is taken first from the environment variable TMPDIR. If that does not exist the system default returned by P_tmpdir is used.
>
> **See also:** [P_tmpdir], page 266, [tempname], page 266, [mkstemp], page 265, [tmpfile], page 266.

P_tmpdir ()                                                [Built-in Function]
> Return the name of the host system's **default** directory for temporary files.
>
> Programming Note: The value returned by P_tmpdir is always the default location. This value may not agree with that returned from **tempdir** if the user has overridden the default with the TMPDIR environment variable.
>
> **See also:** [tempdir], page 266, [tempname], page 266, [mkstemp], page 265, [tmpfile], page 266.

### 14.2.18 End of File and Errors

Once a file has been opened its status can be acquired. As an example the `feof` functions determines if the end of the file has been reached. This can be very useful when reading small parts of a file at a time. The following example shows how to read one line at a time from a file until the end has been reached.

Chapter 14: Input and Output                                              267

```
filename = "myfile.txt";
fid = fopen (filename, "r");
while (! feof (fid) )
  text_line = fgetl (fid);
endwhile
fclose (fid);
```

Note that in some situations it is more efficient to read the entire contents of a file and then process it, than it is to read it line by line. This has the potential advantage of removing the loop in the above code.

*status* = feof (*fid*) [Built-in Function]

> Return 1 if an end-of-file condition has been encountered for the file specified by file descriptor *fid* and 0 otherwise.
>
> Note that feof will only return 1 if the end of the file has already been encountered, not if the next read operation will result in an end-of-file condition.
>
> **See also:** [fread], page 262, [frewind], page 268, [fseek], page 268, [fclear], page 267, [fopen], page 250.

*msg* = ferror (*fid*) [Built-in Function]
[*msg*, *err*] = ferror (*fid*) [Built-in Function]
[*dots*] = ferror (*fid*, "*clear*") [Built-in Function]

> Query the error status of the stream specified by file descriptor *fid*
>
> If an error condition exists then return a string *msg* describing the error. Otherwise, return an empty string "".
>
> The second input "clear" is optional. If supplied, the error state on the stream will be cleared.
>
> The optional second output is a numeric indication of the error status. *err* is 1 if an error condition has been encountered and 0 otherwise.
>
> Note that ferror indicates if an error has already occurred, not whether the next operation will result in an error condition.
>
> **See also:** [fclear], page 267, [fopen], page 250.

fclear (*fid*) [Built-in Function]

> Clear the stream state for the file specified by the file descriptor *fid*.
>
> **See also:** [ferror], page 267, [fopen], page 250.

freport () [Built-in Function]

> Print a list of which files have been opened, and whether they are open for reading, writing, or both.
>
> For example:

```
          freport ()
        -|  number  mode  arch      name
        -|  ------  ----  ----      ----
        -|    0      r    ieee-le   stdin
        -|    1      w    ieee-le   stdout
        -|    2      w    ieee-le   stderr
        -|    3      r    ieee-le   myfile
```

**See also:** [fopen], page 250, [fclose], page 251, [is_valid_file_id], page 251.

### 14.2.19 File Positioning

Three functions are available for setting and determining the position of the file pointer for a given file.

*pos* = ftell (*fid*)  [Built-in Function]

Return the position of the file pointer as the number of characters from the beginning of the file specified by file descriptor *fid*.

**See also:** [fseek], page 268, [frewind], page 268, [feof], page 267, [fopen], page 250.

fseek (*fid, offset*)  [Built-in Function]
fseek (*fid, offset, origin*)  [Built-in Function]
*status* = fseek (...)  [Built-in Function]

Set the file pointer to the location *offset* within the file *fid*.

The pointer is positioned *offset* characters from the *origin*, which may be one of the predefined variables SEEK_CUR (current position), SEEK_SET (beginning), or SEEK_END (end of file) or strings "cof", "bof" or "eof". If *origin* is omitted, SEEK_SET is assumed. *offset* may be positive, negative, or zero but not all combinations of *origin* and *offset* can be realized.

fseek returns 0 on success and -1 on error.

**See also:** [fskipl], page 253, [frewind], page 268, [ftell], page 268, [fopen], page 250.

SEEK_SET ()  [Built-in Function]
SEEK_CUR ()  [Built-in Function]
SEEK_END ()  [Built-in Function]

Return the numerical value to pass to fseek to perform one of the following actions:

SEEK_SET   Position file relative to the beginning.

SEEK_CUR   Position file relative to the current position.

SEEK_END   Position file relative to the end.

**See also:** [fseek], page 268.

frewind (*fid*)  [Built-in Function]
*status* = frewind (*fid*)  [Built-in Function]

Move the file pointer to the beginning of the file specified by file descriptor *fid*.

frewind returns 0 for success, and -1 if an error is encountered. It is equivalent to fseek (*fid*, 0, SEEK_SET).

**See also:** [fseek], page 268, [ftell], page 268, [fopen], page 250.

The following example stores the current file position in the variable `marker`, moves the pointer to the beginning of the file, reads four characters, and then returns to the original position.

```
marker = ftell (myfile);
frewind (myfile);
fourch = fgets (myfile, 4);
fseek (myfile, marker, SEEK_SET);
```

# 15 Plotting

## 15.1 Introduction to Plotting

Earlier versions of Octave provided plotting through the use of gnuplot. This capability is still available. But, a newer plotting capability is provided by access to OpenGL. Which plotting system is used is controlled by the `graphics_toolkit` function. See Section 15.4.7 [Graphics Toolkits], page 400.

The function call `graphics_toolkit ("fltk")` selects the FLTK/OpenGL system, and `graphics_toolkit ("gnuplot")` selects the gnuplot system. The two systems may be used selectively through the use of the `graphics_toolkit` property of the graphics handle for each figure. This is explained in Section 15.3 [Graphics Data Structures], page 352. **Caution:** The FLTK toolkit uses single precision variables internally which limits the maximum value that can be displayed to approximately $10^{38}$. If your data contains larger values you must use the gnuplot toolkit which supports values up to $10^{308}$.

## 15.2 High-Level Plotting

Octave provides simple means to create many different types of two- and three-dimensional plots using high-level functions.

If you need more detailed control, see Section 15.3 [Graphics Data Structures], page 352 and Section 15.4 [Advanced Plotting], page 384.

### 15.2.1 Two-Dimensional Plots

The `plot` function allows you to create simple x-y plots with linear axes. For example,

```
x = -10:0.1:10;
plot (x, sin (x));
```

displays a sine wave shown in Figure 15.1. On most systems, this command will open a separate plot window to display the graph.

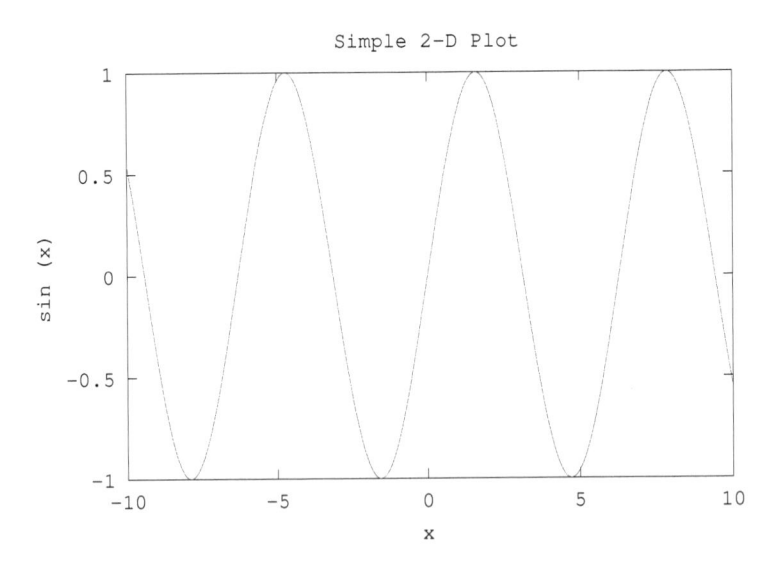

Figure 15.1: Simple Two-Dimensional Plot.

| | |
|---|---|
| `plot (y)` | [Function File] |
| `plot (x, y)` | [Function File] |
| `plot (x, y, fmt)` | [Function File] |
| `plot (..., property, value, ...)` | [Function File] |
| `plot (x1, y1, ..., xn, yn)` | [Function File] |
| `plot (hax, ...)` | [Function File] |
| `h = plot (...)` | [Function File] |

Produce 2-D plots.

Many different combinations of arguments are possible. The simplest form is

```
plot (y)
```

where the argument is taken as the set of y coordinates and the x coordinates are taken to be the range `1:numel (y)`.

If more than one argument is given, they are interpreted as

```
plot (y, property, value, ...)
```

or

```
plot (x, y, property, value, ...)
```

or

```
plot (x, y, fmt, ...)
```

and so on. Any number of argument sets may appear. The x and y values are interpreted as follows:

- If a single data argument is supplied, it is taken as the set of y coordinates and the x coordinates are taken to be the indices of the elements, starting with 1.
- If x and y are scalars, a single point is plotted.
- `squeeze()` is applied to arguments with more than two dimensions, but no more than two singleton dimensions.

# Chapter 15: Plotting

- If both arguments are vectors, the elements of y are plotted versus the elements of x.
- If x is a vector and y is a matrix, then the columns (or rows) of y are plotted versus x. (using whichever combination matches, with columns tried first.)
- If the x is a matrix and y is a vector, y is plotted versus the columns (or rows) of x. (using whichever combination matches, with columns tried first.)
- If both arguments are matrices, the columns of y are plotted versus the columns of x. In this case, both matrices must have the same number of rows and columns and no attempt is made to transpose the arguments to make the number of rows match.

Multiple property-value pairs may be specified, but they must appear in pairs. These arguments are applied to the line objects drawn by `plot`. Useful properties to modify are `"linestyle"`, `"linewidth"`, `"color"`, `"marker"`, `"markersize"`, `"markeredgecolor"`, `"markerfacecolor"`.

The *fmt* format argument can also be used to control the plot style. The format is composed of three parts: linestyle, markerstyle, color. When a markerstyle is specified, but no linestyle, only the markers are plotted. Similarly, if a linestyle is specified, but no markerstyle, then only lines are drawn. If both are specified then lines and markers will be plotted. If no *fmt* and no *property/value* pairs are given, then the default plot style is solid lines with no markers and the color determined by the `"colororder"` property of the current axes.

Format arguments:

linestyle

| | | |
|---|---|---|
| | '-'  | Use solid lines (default). |
| | '--' | Use dashed lines. |
| | ':'  | Use dotted lines. |
| | '-.' | Use dash-dotted lines. |

markerstyle

| | | |
|---|---|---|
| | '+' | crosshair |
| | 'o' | circle |
| | '*' | star |
| | '.' | point |
| | 'x' | cross |
| | 's' | square |
| | 'd' | diamond |
| | '^' | upward-facing triangle |
| | 'v' | downward-facing triangle |
| | '>' | right-facing triangle |
| | '<' | left-facing triangle |
| | 'p' | pentagram |
| | 'h' | hexagram |

color

| | | |
|---|---|---|
| | 'k' | blacK |
| | 'r' | Red |

| | |
|---|---|
| 'g' | Green |
| 'b' | Blue |
| 'm' | Magenta |
| 'c' | Cyan |
| 'w' | White |

`";key;"`   Here `"key"` is the label to use for the plot legend.

The *fmt* argument may also be used to assign legend keys. To do so, include the desired label between semicolons after the formatting sequence described above, e.g., `"+b;Key Title;"`. Note that the last semicolon is required and Octave will generate an error if it is left out.

Here are some plot examples:

```
plot (x, y, "or", x, y2, x, y3, "m", x, y4, "+")
```

This command will plot y with red circles, y2 with solid lines, y3 with solid magenta lines, and y4 with points displayed as '+'.

```
plot (b, "*", "markersize", 10)
```

This command will plot the data in the variable b, with points displayed as '*' and a marker size of 10.

```
t = 0:0.1:6.3;
plot (t, cos(t), "-;cos(t);", t, sin(t), "-b;sin(t);");
```

This will plot the cosine and sine functions and label them accordingly in the legend.

If the first argument *hax* is an axes handle, then plot into this axis, rather than the current axes returned by `gca`.

The optional return value *h* is a vector of graphics handles to the created line objects.

To save a plot, in one of several image formats such as PostScript or PNG, use the `print` command.

**See also:** [axis], page 299, [box], page 330, [grid], page 330, [hold], page 338, [legend], page 327, [title], page 327, [xlabel], page 329, [ylabel], page 329, [xlim], page 301, [ylim], page 301, [ezplot], page 302, [errorbar], page 288, [fplot], page 301, [line], page 355, [plot3], page 317, [polar], page 292, [loglog], page 276, [semilogx], page 275, [semilogy], page 275, [subplot], page 334.

The `plotyy` function may be used to create a plot with two independent y axes.

| | |
|---|---|
| `plotyy (x1, y1, x2, y2)` | [Function File] |
| `plotyy (..., fun)` | [Function File] |
| `plotyy (..., fun1, fun2)` | [Function File] |
| `plotyy (hax, ...)` | [Function File] |
| `[ax, h1, h2] = plotyy (...)` | [Function File] |

Plot two sets of data with independent y-axes and a common x-axis.

The arguments *x1* and *y1* define the arguments for the first plot and *x1* and *y2* for the second.

By default the arguments are evaluated with `feval (@plot, x, y)`. However the type of plot can be modified with the *fun* argument, in which case the plots are generated

# Chapter 15: Plotting

by `feval (fun, x, y)`. *fun* can be a function handle, an inline function, or a string of a function name.

The function to use for each of the plots can be independently defined with *fun1* and *fun2*.

If the first argument *hax* is an axes handle, then it defines the principal axis in which to plot the *x1* and *y1* data.

The return value *ax* is a vector with the axis handles of the two y-axes. *h1* and *h2* are handles to the objects generated by the plot commands.

```
x = 0:0.1:2*pi;
y1 = sin (x);
y2 = exp (x - 1);
ax = plotyy (x, y1, x - 1, y2, @plot, @semilogy);
xlabel ("X");
ylabel (ax(1), "Axis 1");
ylabel (ax(2), "Axis 2");
```

**See also:** [plot], page 272.

The functions `semilogx`, `semilogy`, and `loglog` are similar to the `plot` function, but produce plots in which one or both of the axes use log scales.

| | |
|---|---:|
| `semilogx (y)` | [Function File] |
| `semilogx (x, y)` | [Function File] |
| `semilogx (x, y, property, value, ...)` | [Function File] |
| `semilogx (x, y, fmt)` | [Function File] |
| `semilogx (hax, ...)` | [Function File] |
| `h = semilogx (...)` | [Function File] |

Produce a 2-D plot using a logarithmic scale for the x-axis.

See the documentation of `plot` for a description of the arguments that `semilogx` will accept.

If the first argument *hax* is an axes handle, then plot into this axis, rather than the current axes returned by `gca`.

The optional return value *h* is a graphics handle to the created plot.

**See also:** [plot], page 272, [semilogy], page 275, [loglog], page 276.

| | |
|---|---:|
| `semilogy (y)` | [Function File] |
| `semilogy (x, y)` | [Function File] |
| `semilogy (x, y, property, value, ...)` | [Function File] |
| `semilogy (x, y, fmt)` | [Function File] |
| `semilogy (h, ...)` | [Function File] |
| `h = semilogy (...)` | [Function File] |

Produce a 2-D plot using a logarithmic scale for the y-axis.

See the documentation of `plot` for a description of the arguments that `semilogy` will accept.

If the first argument *hax* is an axes handle, then plot into this axis, rather than the current axes returned by `gca`.

The optional return value h is a graphics handle to the created plot.

**See also:** [plot], page 272, [semilogx], page 275, [loglog], page 276.

loglog (*y*)        [Function File]
loglog (*x*, *y*)        [Function File]
loglog (*x*, *y*, *prop*, *value*, ...)        [Function File]
loglog (*x*, *y*, *fmt*)        [Function File]
loglog (*hax*, ...)        [Function File]
h = loglog (...)        [Function File]

Produce a 2-D plot using logarithmic scales for both axes.

See the documentation of `plot` for a description of the arguments that `loglog` will accept.

If the first argument *hax* is an axes handle, then plot into this axis, rather than the current axes returned by `gca`.

The optional return value h is a graphics handle to the created plot.

**See also:** [plot], page 272, [semilogx], page 275, [semilogy], page 275.

The functions `bar`, `barh`, `stairs`, and `stem` are useful for displaying discrete data. For example,

    `hist (randn (10000, 1), 30);`

produces the histogram of 10,000 normally distributed random numbers shown in Figure 15.2.

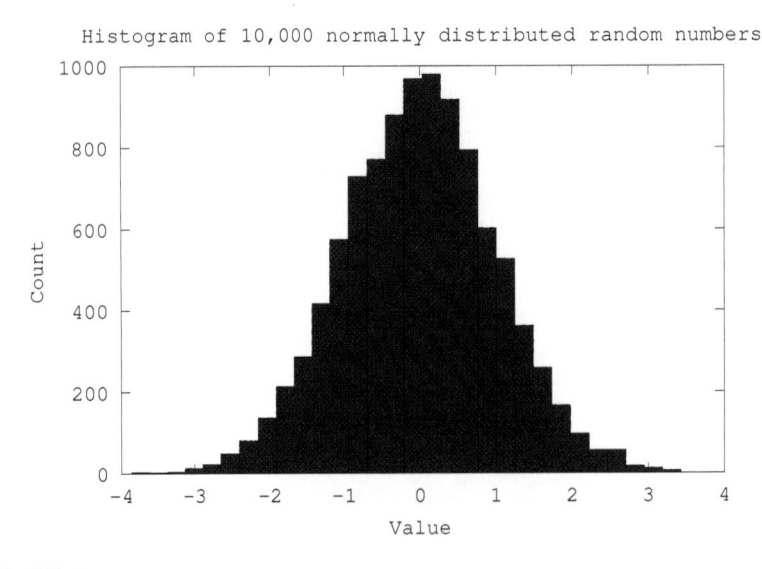

Figure 15.2: Histogram.

bar (*y*)        [Function File]
bar (*x*, *y*)        [Function File]
bar (..., *w*)        [Function File]
bar (..., *style*)        [Function File]
bar (..., *prop*, *val*, ...)        [Function File]

## Chapter 15: Plotting

bar (*hax*, ...)                                                                [Function File]
*h* = bar (..., *prop*, *val*, ...)                                  [Function File]

Produce a bar graph from two vectors of X-Y data.

If only one argument is given, *y*, it is taken as a vector of Y values and the X coordinates are the range `1:numel` (*y*).

The optional input *w* controls the width of the bars. A value of 1.0 will cause each bar to exactly touch any adjacent bars. The default width is 0.8.

If *y* is a matrix, then each column of *y* is taken to be a separate bar graph plotted on the same graph. By default the columns are plotted side-by-side. This behavior can be changed by the *style* argument which can take the following values:

"grouped" (default)
: Side-by-side bars with a gap between bars and centered over the X-coordinate.

"stacked"
: Bars are stacked so that each X value has a single bar composed of multiple segments.

"hist"
: Side-by-side bars with no gap between bars and centered over the X-coordinate.

"histc"
: Side-by-side bars with no gap between bars and left-aligned to the X-coordinate.

Optional property/value pairs are passed directly to the underlying patch objects.

If the first argument *hax* is an axes handle, then plot into this axis, rather than the current axes returned by `gca`.

The optional return value *h* is a vector of handles to the created "bar series" hggroups with one handle per column of the variable *y*. This series makes it possible to change a common element in one bar series object and have the change reflected in the other "bar series". For example,

```
h = bar (rand (5, 10));
set (h(1), "basevalue", 0.5);
```

changes the position on the base of all of the bar series.

The following example modifies the face and edge colors using property/value pairs.

```
bar (randn (1, 100), "facecolor", "r", "edgecolor", "b");
```

The color of the bars is taken from the figure's colormap, such that

```
bar (rand (10, 3));
colormap (summer (64));
```

will change the colors used for the bars. The color of bars can also be set manually using the "facecolor" property as shown below.

```
h = bar (rand (10, 3));
set (h(1), "facecolor", "r")
set (h(2), "facecolor", "g")
set (h(3), "facecolor", "b")
```

**See also:** [barh], page 278, [hist], page 278, [pie], page 293, [plot], page 272, [patch], page 355.

barh (*y*) [Function File]
barh (*x*, *y*) [Function File]
barh (..., *w*) [Function File]
barh (..., *style*) [Function File]
barh (..., *prop*, *val*, ...) [Function File]
barh (*hax*, ...) [Function File]
*h* = barh (..., *prop*, *val*, ...) [Function File]

Produce a horizontal bar graph from two vectors of X-Y data.

If only one argument is given, it is taken as a vector of Y values and the X coordinates are the range `1:numel (y)`.

The optional input *w* controls the width of the bars. A value of 1.0 will cause each bar to exactly touch any adjacent bars. The default width is 0.8.

If *y* is a matrix, then each column of *y* is taken to be a separate bar graph plotted on the same graph. By default the columns are plotted side-by-side. This behavior can be changed by the *style* argument which can take the following values:

"grouped" (default)
: Side-by-side bars with a gap between bars and centered over the Y-coordinate.

"stacked"
: Bars are stacked so that each Y value has a single bar composed of multiple segments.

"hist"
: Side-by-side bars with no gap between bars and centered over the Y-coordinate.

"histc"
: Side-by-side bars with no gap between bars and left-aligned to the Y-coordinate.

Optional property/value pairs are passed directly to the underlying patch objects.

If the first argument *hax* is an axes handle, then plot into this axis, rather than the current axes returned by `gca`.

The optional return value *h* is a graphics handle to the created bar series hggroup. For a description of the use of the bar series, see [bar], page 276.

**See also:** [bar], page 276, [hist], page 278, [pie], page 293, [plot], page 272, [patch], page 355.

hist (*y*) [Function File]
hist (*y*, *x*) [Function File]
hist (*y*, *nbins*) [Function File]
hist (*y*, *x*, *norm*) [Function File]
hist (..., *prop*, *val*, ...) [Function File]
hist (*hax*, ...) [Function File]
[*nn*, *xx*] = hist (...) [Function File]

Produce histogram counts or plots.

With one vector input argument, *y*, plot a histogram of the values with 10 bins. The range of the histogram bins is determined by the range of the data. With one matrix input argument, *y*, plot a histogram where each bin contains a bar per input column.

Chapter 15: Plotting 279

Given a second vector argument, *x*, use that as the centers of the bins, with the width of the bins determined from the adjacent values in the vector.

If scalar, the second argument, *nbins*, defines the number of bins.

If a third argument is provided, the histogram is normalized such that the sum of the bars is equal to *norm*.

Extreme values are lumped into the first and last bins.

The histogram's appearance may be modified by specifying property/value pairs. For example the face and edge color may be modified.

```
hist (randn (1, 100), 25, "facecolor", "r", "edgecolor", "b");
```

The histogram's colors also depend upon the current colormap.

```
hist (rand (10, 3));
colormap (summer ());
```

If the first argument *hax* is an axes handle, then plot into this axis, rather than the current axes returned by `gca`.

With two output arguments, produce the values *nn* (numbers of elements) and *xx* (bin centers) such that `bar (xx, nn)` will plot the histogram.

**See also:** [histc], page 604, [bar], page 276, [pie], page 293, [rose], page 285.

stemleaf (*x, caption*) [Function File]
stemleaf (*x, caption, stem_sz*) [Function File]
*plotstr* = stemleaf (...) [Function File]
    Compute and display a stem and leaf plot of the vector *x*.

The input *x* should be a vector of integers. Any non-integer values will be converted to integer by `x = fix (x)`. By default each element of *x* will be plotted with the last digit of the element as a leaf value and the remaining digits as the stem. For example, 123 will be plotted with the stem '12' and the leaf '3'. The second argument, *caption*, should be a character array which provides a description of the data. It is included as a heading for the output.

The optional input *stem_sz* sets the width of each stem. The stem width is determined by `10^(stem_sz + 1)`. The default stem width is 10.

The output of `stemleaf` is composed of two parts: a "Fenced Letter Display," followed by the stem-and-leaf plot itself. The Fenced Letter Display is described in *Exploratory Data Analysis*. Briefly, the entries are as shown:

```
          Fenced Letter Display
    #% nx|_____     nx = numel (x)
    M% mi|       md        |      mi median index, md median
    H% hi|hl             hu| hs   hi lower hinge index, hl,hu hinges,
    1     |x(1)        x(nx)|     hs h_spreadx(1), x(nx) first
            -------                and last data value.
          _____|step |_____    step 1.5*h_spread
         f|ifl            ifh|    inner fence, lower and higher
          |nfl            nfh|    no.\ of data points within fences
         F|ofl            ofh|    outer fence, lower and higher
          |nFl            nFh|    no.\ of data points outside outer
                                  fences
```

The stem-and-leaf plot shows on each line the stem value followed by the string made up of the leaf digits. If the *stem_sz* is not 1 the successive leaf values are separated by ",".

With no return argument, the plot is immediately displayed. If an output argument is provided, the plot is returned as an array of strings.

The leaf digits are not sorted. If sorted leaf values are desired, use `xs = sort (x)` before calling `stemleaf (xs)`.

The stem and leaf plot and associated displays are described in: Ch. 3, *Exploratory Data Analysis* by J. W. Tukey, Addison-Wesley, 1977.

See also: [hist], page 278, [printd], page 280.

**printd** (*obj*, *filename*)  [Function File]
*out_file* = **printd** (...)  [Function File]

Convert any object acceptable to `disp` into the format selected by the suffix of *filename*.

If the return argument *out_file* is given, the name of the created file is returned.

This function is intended to facilitate manipulation of the output of functions such as `stemleaf`.

See also: [stemleaf], page 279.

**stairs** (*y*)  [Function File]
**stairs** (*x*, *y*)  [Function File]
**stairs** (..., *style*)  [Function File]
**stairs** (..., *prop*, *val*, ...)  [Function File]
**stairs** (*hax*, ...)  [Function File]
*h* = **stairs** (...)  [Function File]
[*xstep*, *ystep*] = **stairs** (...)  [Function File]

Produce a stairstep plot.

The arguments *x* and *y* may be vectors or matrices. If only one argument is given, it is taken as a vector of Y values and the X coordinates are taken to be the indices of the elements.

The style to use for the plot can be defined with a line style *style* of the same format as the `plot` command.

Multiple property/value pairs may be specified, but they must appear in pairs.

If the first argument *hax* is an axis handle, then plot into this axis, rather than the current axis handle returned by `gca`.

If one output argument is requested, return a graphics handle to the created plot. If two output arguments are specified, the data are generated but not plotted. For example,

```
stairs (x, y);
```

and

```
[xs, ys] = stairs (x, y);
plot (xs, ys);
```

are equivalent.

**See also:** [bar], page 276, [hist], page 278, [plot], page 272, [stem], page 281.

stem (*y*) [Function File]
stem (*x*, *y*) [Function File]
stem (..., *linespec*) [Function File]
stem (..., "*filled*") [Function File]
stem (..., *prop*, *val*, ...) [Function File]
stem (*hax*, ...) [Function File]
h = stem (...) [Function File]

Plot a 2-D stem graph.

If only one argument is given, it is taken as the y-values and the x-coordinates are taken from the indices of the elements.

If *y* is a matrix, then each column of the matrix is plotted as a separate stem graph. In this case *x* can either be a vector, the same length as the number of rows in *y*, or it can be a matrix of the same size as *y*.

The default color is "b" (blue), the default line style is "-", and the default marker is "o". The line style can be altered by the `linespec` argument in the same manner as the `plot` command. If the "`filled`" argument is present the markers at the top of the stems will be filled in. For example,

```
x = 1:10;
y = 2*x;
stem (x, y, "r");
```

plots 10 stems with heights from 2 to 20 in red;

Optional property/value pairs may be specified to control the appearance of the plot.

If the first argument *hax* is an axes handle, then plot into this axis, rather than the current axes returned by `gca`.

The optional return value *h* is a handle to a "stem series" hggroup. The single hggroup handle has all of the graphical elements comprising the plot as its children; This allows the properties of multiple graphics objects to be changed by modifying just a single property of the "stem series" hggroup.

For example,

```
x = [0:10]';
y = [sin(x), cos(x)]
h = stem (x, y);
set (h(2), "color", "g");
set (h(1), "basevalue", -1)
```

changes the color of the second "stem series" and moves the base line of the first.

Stem Series Properties

linestyle    The linestyle of the stem. (Default: `"-"`)

linewidth    The width of the stem. (Default: 0.5)

color        The color of the stem, and if not separately specified, the marker. (Default: `"b"` [blue])

marker       The marker symbol to use at the top of each stem. (Default: `"o"`)

markeredgecolor
             The edge color of the marker. (Default: `"color"` property)

markerfacecolor
             The color to use for "filling" the marker. (Default: `"none"` [unfilled])

markersize
             The size of the marker. (Default: 6)

baseline     The handle of the line object which implements the baseline. Use `set` with the returned handle to change graphic properties of the baseline.

basevalue    The y-value where the baseline is drawn. (Default: 0)

**See also:** [stem3], page 282, [bar], page 276, [hist], page 278, [plot], page 272, [stairs], page 280.

stem3 (*x, y, z*)                                              [Function File]
stem3 (..., *linespec*)                                        [Function File]
stem3 (..., "*filled*")                                        [Function File]
stem3 (..., *prop, val*, ...)                                  [Function File]
stem3 (*hax*, ...)                                             [Function File]
h = stem3 (...)                                                [Function File]

Plot a 3-D stem graph.

Stems are drawn from the height z to the location in the x-y plane determined by x and y. The default color is `"b"` (blue), the default line style is `"-"`, and the default marker is `"o"`.

The line style can be altered by the `linespec` argument in the same manner as the `plot` command. If the `"filled"` argument is present the markers at the top of the stems will be filled in.

Optional property/value pairs may be specified to control the appearance of the plot.

If the first argument *hax* is an axes handle, then plot into this axis, rather than the current axes returned by `gca`.

Chapter 15: Plotting                                                                     283

The optional return value *h* is a handle to the "stem series" hggroup containing the line and marker objects used for the plot. See [stem], page 281, for a description of the "stem series" object.

Example:

```
theta = 0:0.2:6;
stem3 (cos (theta), sin (theta), theta);
```

plots 31 stems with heights from 0 to 6 lying on a circle.

Implementation Note: Color definitions with RGB-triples are not valid.

**See also:** [stem], page 281, [bar], page 276, [hist], page 278, [plot], page 272.

scatter (*x, y*)                                                        [Function File]
scatter (*x, y, s*)                                                     [Function File]
scatter (*x, y, s, c*)                                                  [Function File]
scatter (..., *style*)                                                  [Function File]
scatter (..., "*filled*")                                               [Function File]
scatter (..., *prop, val*, ...)                                         [Function File]
scatter (*hax*, ...)                                                    [Function File]
*h* = scatter (...)                                                     [Function File]

Draw a 2-D scatter plot.

A marker is plotted at each point defined by the coordinates in the vectors *x* and *y*.

The size of the markers is determined by *s*, which can be a scalar or a vector of the same length as *x* and *y*. If *s* is not given, or is an empty matrix, then a default value of 8 points is used.

The color of the markers is determined by *c*, which can be a string defining a fixed color; a 3-element vector giving the red, green, and blue components of the color; a vector of the same length as *x* that gives a scaled index into the current colormap; or an Nx3 matrix defining the RGB color of each marker individually.

The marker to use can be changed with the *style* argument, that is a string defining a marker in the same manner as the `plot` command. If no marker is specified it defaults to "o" or circles. If the argument "`filled`" is given then the markers are filled.

Additional property/value pairs are passed directly to the underlying patch object.

If the first argument *hax* is an axes handle, then plot into this axis, rather than the current axes returned by `gca`.

The optional return value *h* is a graphics handle to the created patch object.

Example:

```
x = randn (100, 1);
y = randn (100, 1);
scatter (x, y, [], sqrt (x.^2 + y.^2));
```

**See also:** [scatter3], page 320, [patch], page 355, [plot], page 272.

plotmatrix (*x, y*)                                                     [Function File]
plotmatrix (*x*)                                                        [Function File]
plotmatrix (..., *style*)                                               [Function File]
plotmatrix (*hax*, ...)                                                 [Function File]

`[h, ax, bigax, p, pax] = plotmatrix (…)`  [Function File]

Scatter plot of the columns of one matrix against another.

Given the arguments *x* and *y* that have a matching number of rows, `plotmatrix` plots a set of axes corresponding to

`plot (x(:, i), y(:, j))`

When called with a single argument *x* this is equivalent to

`plotmatrix (x, x)`

except that the diagonal of the set of axes will be replaced with the histogram `hist (x(:, i))`.

The marker to use can be changed with the *style* argument, that is a string defining a marker in the same manner as the `plot` command.

If the first argument *hax* is an axes handle, then plot into this axis, rather than the current axes returned by `gca`.

The optional return value *h* provides handles to the individual graphics objects in the scatter plots, whereas *ax* returns the handles to the scatter plot axis objects.

*bigax* is a hidden axis object that surrounds the other axes, such that the commands `xlabel`, `title`, etc., will be associated with this hidden axis.

Finally, *p* returns the graphics objects associated with the histogram and *pax* the corresponding axes objects.

Example:

`plotmatrix (randn (100, 3), "g+")`

See also: [scatter], page 283, [plot], page 272.

`pareto (y)`  [Function File]
`pareto (y, x)`  [Function File]
`pareto (hax, …)`  [Function File]
`h = pareto (…)`  [Function File]

Draw a Pareto chart.

A Pareto chart is a bar graph that arranges information in such a way that priorities for process improvement can be established; It organizes and displays information to show the relative importance of data. The chart is similar to the histogram or bar chart, except that the bars are arranged in decreasing magnitude from left to right along the x-axis.

The fundamental idea (Pareto principle) behind the use of Pareto diagrams is that the majority of an effect is due to a small subset of the causes. For quality improvement, the first few contributing causes (leftmost bars as presented on the diagram) to a problem usually account for the majority of the result. Thus, targeting these "major causes" for elimination results in the most cost-effective improvement scheme.

Typically only the magnitude data *y* is present in which case *x* is taken to be the range `1 : length (y)`. If *x* is given it may be a string array, a cell array of strings, or a numerical vector.

If the first argument *hax* is an axes handle, then plot into this axis, rather than the current axes returned by `gca`.

Chapter 15: Plotting 285

The optional return value *h* is a 2-element vector with a graphics handle for the created bar plot and a second handle for the created line plot.

An example of the use of `pareto` is

```
Cheese = {"Cheddar", "Swiss", "Camembert", ...
          "Munster", "Stilton", "Blue"};
Sold = [105, 30, 70, 10, 15, 20];
pareto (Sold, Cheese);
```

**See also:** [bar], page 276, [barh], page 278, [hist], page 278, [pie], page 293, [plot], page 272.

rose (*th*)     [Function File]
rose (*th*, *nbins*)     [Function File]
rose (*th*, *bins*)     [Function File]
rose (*hax*, ...)     [Function File]
*h* = rose (...)     [Function File]
[*thout rout*] = rose (...)     [Function File]

Plot an angular histogram.

With one vector argument, *th*, plot the histogram with 20 angular bins. If *th* is a matrix then each column of *th* produces a separate histogram.

If *nbins* is given and is a scalar, then the histogram is produced with *nbin* bins. If *bins* is a vector, then the center of each bin is defined by the values of *bins* and the number of bins is given by the number of elements in *bins*.

If the first argument *hax* is an axes handle, then plot into this axis, rather than the current axes returned by `gca`.

The optional return value *h* is a vector of graphics handles to the line objects representing each histogram.

If two output arguments are requested then no plot is made and the polar vectors necessary to plot the histogram are returned instead.

```
[th, r] = rose ([2*randn(1e5,1), pi + 2*randn(1e5,1)]);
polar (th, r);
```

**See also:** [hist], page 278, [polar], page 292.

The `contour`, `contourf` and `contourc` functions produce two-dimensional contour plots from three-dimensional data.

contour (*z*)     [Function File]
contour (*z*, *vn*)     [Function File]
contour (*x*, *y*, *z*)     [Function File]
contour (*x*, *y*, *z*, *vn*)     [Function File]
contour (..., *style*)     [Function File]
contour (*hax*, ...)     [Function File]
[*c*, *h*] = contour (...)     [Function File]

Create a 2-D contour plot.

Plot level curves (contour lines) of the matrix *z*, using the contour matrix *c* computed by `contourc` from the same arguments; see the latter for their interpretation.

The appearance of contour lines can be defined with a line style *style* in the same manner as `plot`. Only line style and color are used; Any markers defined by *style* are ignored.

If the first argument *hax* is an axes handle, then plot into this axis, rather than the current axes returned by `gca`.

The optional output *c* contains the contour levels in `contourc` format.

The optional return value *h* is a graphics handle to the hggroup comprising the contour lines.

Example:

```
x = 0:2;
y = x;
z = x' * y;
contour (x, y, z, 2:3)
```

**See also:** [ezcontour], page 303, [contourc], page 287, [contourf], page 286, [contour3], page 287, [clabel], page 329, [meshc], page 307, [surfc], page 309, [caxis], page 300, [colormap], page 698, [plot], page 272.

| | |
|---|---|
| contourf (*z*) | [Function File] |
| contourf (*z*, *vn*) | [Function File] |
| contourf (*x*, *y*, *z*) | [Function File] |
| contourf (*x*, *y*, *z*, *vn*) | [Function File] |
| contourf (…, *style*) | [Function File] |
| contourf (*hax*, …) | [Function File] |
| [*c*, *h*] = contourf (…) | [Function File] |

Create a 2-D contour plot with filled intervals.

Plot level curves (contour lines) of the matrix *z* and fill the region between lines with colors from the current colormap.

The level curves are taken from the contour matrix *c* computed by `contourc` for the same arguments; see the latter for their interpretation.

The appearance of contour lines can be defined with a line style *style* in the same manner as `plot`. Only line style and color are used; Any markers defined by *style* are ignored.

If the first argument *hax* is an axes handle, then plot into this axis, rather than the current axes returned by `gca`.

The optional output *c* contains the contour levels in `contourc` format.

The optional return value *h* is a graphics handle to the hggroup comprising the contour lines.

The following example plots filled contours of the `peaks` function.

```
[x, y, z] = peaks (50);
contourf (x, y, z, -7:9)
```

**See also:** [ezcontourf], page 303, [contour], page 285, [contourc], page 287, [contour3], page 287, [clabel], page 329, [meshc], page 307, [surfc], page 309, [caxis], page 300, [colormap], page 698, [plot], page 272.

Chapter 15: Plotting                                                                 287

[c, lev] = contourc (z)                                                   [Function File]
[c, lev] = contourc (z, vn)                                               [Function File]
[c, lev] = contourc (x, y, z)                                             [Function File]
[c, lev] = contourc (x, y, z, vn)                                         [Function File]
 Compute contour lines (isolines of constant Z value).

 The matrix z contains height values above the rectangular grid determined by x and y. If only a single input z is provided then x is taken to be `1:rows (z)` and y is taken to be `1:columns (z)`.

 The optional input vn is either a scalar denoting the number of contour lines to compute or a vector containing the Z values where lines will be computed. When vn is a vector the number of contour lines is `numel (vn)`. However, to compute a single contour line at a given value use vn = [val, val]. If vn is omitted it defaults to 10.

 The return value c is a 2x$n$ matrix containing the contour lines in the following format

```
c = [lev1, x1, x2, ..., levn, x1, x2, ...
     len1, y1, y2, ..., lenn, y1, y2, ...]
```

in which contour line $n$ has a level (height) of lev$n$ and length of len$n$.

 The optional return value lev is a vector with the Z values of the contour levels.

 Example:
```
x = 0:2;
y = x;
z = x' * y;
contourc (x, y, z, 2:3)
    ⇒   2.0000   2.0000   1.0000   3.0000   1.5000   2.0000
        2.0000   1.0000   2.0000   2.0000   2.0000   1.5000
```

 **See also:** [contour], page 285, [contourf], page 286, [contour3], page 287, [clabel], page 329.

contour3 (z)                                                              [Function File]
contour3 (z, vn)                                                          [Function File]
contour3 (x, y, z)                                                        [Function File]
contour3 (x, y, z, vn)                                                    [Function File]
contour3 (..., style)                                                     [Function File]
contour3 (hax, ...)                                                       [Function File]
[c, h] = contour3 (...)                                                   [Function File]
 Create a 3-D contour plot.

 `contour3` plots level curves (contour lines) of the matrix z at a Z level corresponding to each contour. This is in contrast to `contour` which plots all of the contour lines at the same Z level and produces a 2-D plot.

 The level curves are taken from the contour matrix c computed by `contourc` for the same arguments; see the latter for their interpretation.

 The appearance of contour lines can be defined with a line style *style* in the same manner as `plot`. Only line style and color are used; Any markers defined by *style* are ignored.

 If the first argument hax is an axes handle, then plot into this axis, rather than the current axes returned by `gca`.

The optional output *c* are the contour levels in `contourc` format.

The optional return value *h* is a graphics handle to the hggroup comprising the contour lines.

Example:
```
contour3 (peaks (19));
colormap cool;
hold on;
surf (peaks (19), "facecolor", "none", "edgecolor", "black");
```

**See also:** [contour], page 285, [contourc], page 287, [contourf], page 286, [clabel], page 329, [meshc], page 307, [surfc], page 309, [caxis], page 300, [colormap], page 698, [plot], page 272.

The `errorbar`, `semilogxerr`, `semilogyerr`, and `loglogerr` functions produce plots with error bar markers. For example,
```
x = 0:0.1:10;
y = sin (x);
yp =  0.1 .* randn (size (x));
ym = -0.1 .* randn (size (x));
errorbar (x, sin (x), ym, yp);
```
produces the figure shown in Figure 15.3.

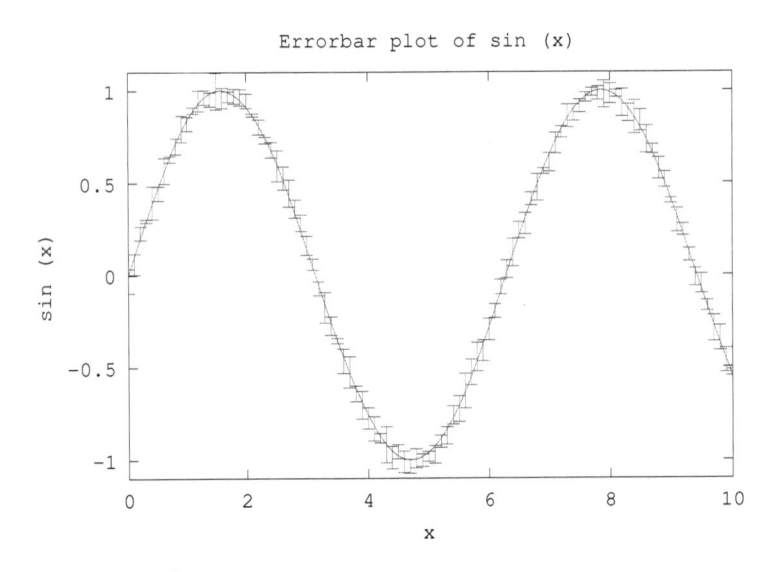

Figure 15.3: Errorbar plot.

| | |
|---|---|
| errorbar (*y*, *ey*) | [Function File] |
| errorbar (*y*, ..., *fmt*) | [Function File] |
| errorbar (*x*, *y*, *ey*) | [Function File] |
| errorbar (*x*, *y*, *err*, *fmt*) | [Function File] |
| errorbar (*x*, *y*, *lerr*, *uerr*, *fmt*) | [Function File] |
| errorbar (*x*, *y*, *ex*, *ey*, *fmt*) | [Function File] |
| errorbar (*x*, *y*, *lx*, *ux*, *ly*, *uy*, *fmt*) | [Function File] |

Chapter 15: Plotting                                                                 289

**errorbar** (*x1*, *y1*, ..., *fmt*, *xn*, *yn*, ...)                    [Function File]
**errorbar** (*hax*, ...)                                                [Function File]
*h* = **errorbar** (...)                                                 [Function File]

Create a 2-D plot with errorbars.

Many different combinations of arguments are possible. The simplest form is

    **errorbar** (*y*, *ey*)

where the first argument is taken as the set of *y* coordinates, the second argument *ey* are the errors around the *y* values, and the *x* coordinates are taken to be the indices of the elements (`1:numel (y)`).

The general form of the function is

    **errorbar** (*x*, *y*, *err1*, ..., *fmt*, ...)

After the *x* and *y* arguments there can be 1, 2, or 4 parameters specifying the error values depending on the nature of the error values and the plot format *fmt*.

*err* (scalar)
: When the error is a scalar all points share the same error value. The errorbars are symmetric and are drawn from *data-err* to *data+err*. The *fmt* argument determines whether *err* is in the x-direction, y-direction (default), or both.

*err* (vector or matrix)
: Each data point has a particular error value. The errorbars are symmetric and are drawn from *data*(n)-*err*(n) to *data*(n)+*err*(n).

*lerr*, *uerr* (scalar)
: The errors have a single low-side value and a single upper-side value. The errorbars are not symmetric and are drawn from *data-lerr* to *data+uerr*.

*lerr*, *uerr* (vector or matrix)
: Each data point has a low-side error and an upper-side error. The errorbars are not symmetric and are drawn from *data*(n)-*lerr*(n) to *data*(n)+*uerr*(n).

Any number of data sets (*x1*,*y1*, *x2*,*y2*, ...) may appear as long as they are separated by a format string *fmt*.

If *y* is a matrix, *x* and the error parameters must also be matrices having the same dimensions. The columns of *y* are plotted versus the corresponding columns of *x* and errorbars are taken from the corresponding columns of the error parameters.

If *fmt* is missing, the yerrorbars (`"~"`) plot style is assumed.

If the *fmt* argument is supplied then it is interpreted, as in normal plots, to specify the line style, marker, and color. In addition, *fmt* may include an errorbar style which **must precede** the ordinary format codes. The following errorbar styles are supported:

'~'        Set yerrorbars plot style (default).

'>'        Set xerrorbars plot style.

'~>'       Set xyerrorbars plot style.

'#~'       Set yboxes plot style.

'#'         Set xboxes plot style.

'#~>'       Set xyboxes plot style.

If the first argument *hax* is an axes handle, then plot into this axis, rather than the current axes returned by `gca`.

The optional return value *h* is a handle to the hggroup object representing the data plot and errorbars.

Note: For compatibility with MATLAB a line is drawn through all data points. However, most scientific errorbar plots are a scatter plot of points with errorbars. To accomplish this, add a marker style to the *fmt* argument such as ".". Alternatively, remove the line by modifying the returned graphic handle with `set (h, "linestyle", "none")`.

Examples:

        errorbar (x, y, ex, ">.r")

produces an xerrorbar plot of y versus x with x errorbars drawn from x-ex to x+ex. The marker "." is used so no connecting line is drawn and the errorbars appear in red.

        errorbar (x, y1, ey, "~",
                  x, y2, ly, uy)

produces yerrorbar plots with y1 and y2 versus x. Errorbars for y1 are drawn from y1-ey to y1+ey, errorbars for y2 from y2-ly to y2+uy.

        errorbar (x, y, lx, ux,
                  ly, uy, "~>")

produces an xyerrorbar plot of y versus x in which x errorbars are drawn from x-lx to x+ux and y errorbars from y-ly to y+uy.

**See also:** [semilogxerr], page 290, [semilogyerr], page 291, [loglogerr], page 291, [plot], page 272.

---

`semilogxerr (y, ey)`                               [Function File]
`semilogxerr (y, ..., fmt)`                         [Function File]
`semilogxerr (x, y, ey)`                            [Function File]
`semilogxerr (x, y, err, fmt)`                      [Function File]
`semilogxerr (x, y, lerr, uerr, fmt)`               [Function File]
`semilogxerr (x, y, ex, ey, fmt)`                   [Function File]
`semilogxerr (x, y, lx, ux, ly, uy, fmt)`           [Function File]
`semilogxerr (x1, y1, ..., fmt, xn, yn, ...)`       [Function File]
`semilogxerr (hax, ...)`                            [Function File]
`h = semilogxerr (...)`                             [Function File]

Produce 2-D plots using a logarithmic scale for the x-axis and errorbars at each data point.

Many different combinations of arguments are possible. The most common form is

        `semilogxerr (x, y, ey, fmt)`

which produces a semi-logarithmic plot of y versus x with errors in the y-scale defined by *ey* and the plot format defined by *fmt*. See [errorbar], page 288, for available formats and additional information.

Chapter 15: Plotting

If the first argument *hax* is an axes handle, then plot into this axis, rather than the current axes returned by `gca`.

**See also:** [errorbar], page 288, [semilogyerr], page 291, [loglogerr], page 291.

| | |
|---|---:|
| `semilogyerr (y, ey)` | [Function File] |
| `semilogyerr (y, ..., fmt)` | [Function File] |
| `semilogyerr (x, y, ey)` | [Function File] |
| `semilogyerr (x, y, err, fmt)` | [Function File] |
| `semilogyerr (x, y, lerr, uerr, fmt)` | [Function File] |
| `semilogyerr (x, y, ex, ey, fmt)` | [Function File] |
| `semilogyerr (x, y, lx, ux, ly, uy, fmt)` | [Function File] |
| `semilogyerr (x1, y1, ..., fmt, xn, yn, ...)` | [Function File] |
| `semilogyerr (hax, ...)` | [Function File] |
| `h = semilogyerr (...)` | [Function File] |

Produce 2-D plots using a logarithmic scale for the y-axis and errorbars at each data point.

Many different combinations of arguments are possible. The most common form is

    `semilogyerr (x, y, ey, fmt)`

which produces a semi-logarithmic plot of *y* versus *x* with errors in the *y*-scale defined by *ey* and the plot format defined by *fmt*. See [errorbar], page 288, for available formats and additional information.

If the first argument *hax* is an axes handle, then plot into this axis, rather than the current axes returned by `gca`.

**See also:** [errorbar], page 288, [semilogxerr], page 290, [loglogerr], page 291.

| | |
|---|---:|
| `loglogerr (y, ey)` | [Function File] |
| `loglogerr (y, ..., fmt)` | [Function File] |
| `loglogerr (x, y, ey)` | [Function File] |
| `loglogerr (x, y, err, fmt)` | [Function File] |
| `loglogerr (x, y, lerr, uerr, fmt)` | [Function File] |
| `loglogerr (x, y, ex, ey, fmt)` | [Function File] |
| `loglogerr (x, y, lx, ux, ly, uy, fmt)` | [Function File] |
| `loglogerr (x1, y1, ..., fmt, xn, yn, ...)` | [Function File] |
| `loglogerr (hax, ...)` | [Function File] |
| `h = loglogerr (...)` | [Function File] |

Produce 2-D plots on a double logarithm axis with errorbars.

Many different combinations of arguments are possible. The most common form is

    `loglogerr (x, y, ey, fmt)`

which produces a double logarithm plot of *y* versus *x* with errors in the *y*-scale defined by *ey* and the plot format defined by *fmt*. See [errorbar], page 288, for available formats and additional information.

If the first argument *hax* is an axes handle, then plot into this axis, rather than the current axes returned by `gca`.

**See also:** [errorbar], page 288, [semilogxerr], page 290, [semilogyerr], page 291.

Finally, the `polar` function allows you to easily plot data in polar coordinates. However, the display coordinates remain rectangular and linear. For example,

```
polar (0:0.1:10*pi, 0:0.1:10*pi);
```

produces the spiral plot shown in Figure 15.4.

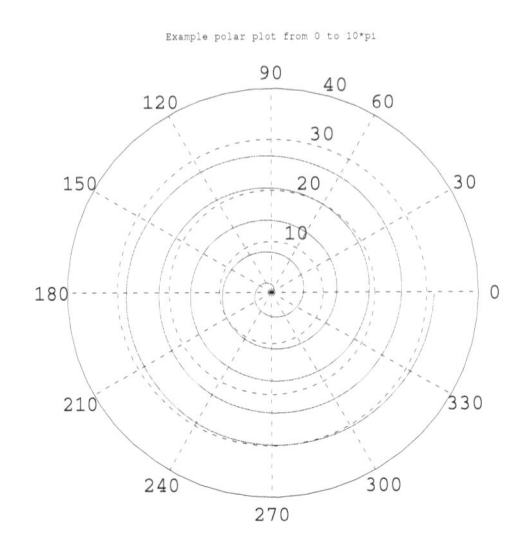

Figure 15.4: Polar plot.

| | |
|---|---|
| polar (*theta*, *rho*) | [Function File] |
| polar (*theta*, *rho*, *fmt*) | [Function File] |
| polar (*cplx*) | [Function File] |
| polar (*cplx*, *fmt*) | [Function File] |
| polar (*hax*, ...) | [Function File] |
| h = polar (...) | [Function File] |

Create a 2-D plot from polar coordinates *theta* and *rho*.

If a single complex input *cplx* is given then the real part is used for *theta* and the imaginary part is used for *rho*.

The optional argument *fmt* specifies the line format in the same way as `plot`.

If the first argument *hax* is an axes handle, then plot into this axis, rather than the current axes returned by `gca`.

The optional return value *h* is a graphics handle to the created plot.

Implementation Note: The polar axis is drawn using line and text objects encapsulated in an hggroup. The hggroup properties are linked to the original axes object such that altering an appearance property, for example `fontname`, will update the polar axis. Two new properties are added to the original axes–`rtick`, `ttick`–which replace `xtick`, `ytick`. The first is a list of tick locations in the radial (rho) direction; The second is a list of tick locations in the angular (theta) direction specified in degrees, i.e., in the range 0–359.

**See also:** [rose], page 285, [compass], page 295, [plot], page 272.

Chapter 15: Plotting    293

pie (*x*)    [Function File]
pie (..., *explode*)    [Function File]
pie (..., *labels*)    [Function File]
pie (*hax*, ...);    [Function File]
h = pie (...);    [Function File]

Plot a 2-D pie chart.

When called with a single vector argument, produce a pie chart of the elements in *x*. The size of the ith slice is the percentage that the element *xi* represents of the total sum of *x*: pct = x(i) / sum (x).

The optional input *explode* is a vector of the same length as *x* that, if nonzero, "explodes" the slice from the pie chart.

The optional input *labels* is a cell array of strings of the same length as *x* specifying the label for each slice.

If the first argument *hax* is an axes handle, then plot into this axis, rather than the current axes returned by gca.

The optional return value *h* is a list of handles to the patch and text objects generating the plot.

Note: If sum (x) $\leq$ 1 then the elements of *x* are interpreted as percentages directly and are not normalized by sum (x). Furthermore, if the sum is less than 1 then there will be a missing slice in the pie plot to represent the missing, unspecified percentage.

**See also:** [pie3], page 293, [bar], page 276, [hist], page 278, [rose], page 285.

pie3 (*x*)    [Function File]
pie3 (..., *explode*)    [Function File]
pie3 (..., *labels*)    [Function File]
pie3 (*hax*, ...);    [Function File]
h = pie3 (...);    [Function File]

Plot a 3-D pie chart.

Called with a single vector argument, produces a 3-D pie chart of the elements in *x*. The size of the ith slice is the percentage that the element *xi* represents of the total sum of *x*: pct = x(i) / sum (x).

The optional input *explode* is a vector of the same length as *x* that, if nonzero, "explodes" the slice from the pie chart.

The optional input *labels* is a cell array of strings of the same length as *x* specifying the label for each slice.

If the first argument *hax* is an axes handle, then plot into this axis, rather than the current axes returned by gca.

The optional return value *h* is a list of graphics handles to the patch, surface, and text objects generating the plot.

Note: If sum (x) $\leq$ 1 then the elements of *x* are interpreted as percentages directly and are not normalized by sum (x). Furthermore, if the sum is less than 1 then there will be a missing slice in the pie plot to represent the missing, unspecified percentage.

**See also:** [pie], page 293, [bar], page 276, [hist], page 278, [rose], page 285.

| | |
|---|---|
| quiver (u, v) | [Function File] |
| quiver (x, y, u, v) | [Function File] |
| quiver (..., s) | [Function File] |
| quiver (..., style) | [Function File] |
| quiver (..., "filled") | [Function File] |
| quiver (hax, ...) | [Function File] |
| h = quiver (...) | [Function File] |

Plot a 2-D vector field with arrows.

Plot the (u, v) components of a vector field in an (x, y) meshgrid. If the grid is uniform then x and y can be specified as vectors.

If x and y are undefined they are assumed to be (1:m, 1:n) where [m, n] = size (u).

The variable s is a scalar defining a scaling factor to use for the arrows of the field relative to the mesh spacing. A value of 0 disables all scaling. The default value is 0.9.

The style to use for the plot can be defined with a line style *style* of the same format as the `plot` command. If a marker is specified then markers at the grid points of the vectors are drawn rather than arrows. If the argument `"filled"` is given then the markers are filled.

If the first argument *hax* is an axes handle, then plot into this axis, rather than the current axes returned by `gca`.

The optional return value *h* is a graphics handle to a quiver object. A quiver object regroups the components of the quiver plot (body, arrow, and marker), and allows them to be changed together.

Example:

```
[x, y] = meshgrid (1:2:20);
h = quiver (x, y, sin (2*pi*x/10), sin (2*pi*y/10));
set (h, "maxheadsize", 0.33);
```

**See also:** [quiver3], page 294, [compass], page 295, [feather], page 295, [plot], page 272.

| | |
|---|---|
| quiver3 (u, v, w) | [Function File] |
| quiver3 (x, y, z, u, v, w) | [Function File] |
| quiver3 (..., s) | [Function File] |
| quiver3 (..., style) | [Function File] |
| quiver3 (..., "filled") | [Function File] |
| quiver3 (hax, ...) | [Function File] |
| h = quiver3 (...) | [Function File] |

Plot a 3-D vector field with arrows.

Plot the (u, v, w) components of a vector field in an (x, y, z) meshgrid. If the grid is uniform then x, y, and z can be specified as vectors.

If x, y, and z are undefined they are assumed to be (1:m, 1:n, 1:p) where [m, n] = size (u) and p = max (size (w)).

The variable *s* is a scalar defining a scaling factor to use for the arrows of the field relative to the mesh spacing. A value of 0 disables all scaling. The default value is 0.9.

The style to use for the plot can be defined with a line style *style* of the same format as the `plot` command. If a marker is specified then markers at the grid points of the vectors are drawn rather than arrows. If the argument `"filled"` is given then the markers are filled.

If the first argument *hax* is an axes handle, then plot into this axis, rather than the current axes returned by `gca`.

The optional return value *h* is a graphics handle to a quiver object. A quiver object regroups the components of the quiver plot (body, arrow, and marker), and allows them to be changed together.

```
[x, y, z] = peaks (25);
surf (x, y, z);
hold on;
[u, v, w] = surfnorm (x, y, z / 10);
h = quiver3 (x, y, z, u, v, w);
set (h, "maxheadsize", 0.33);
```

**See also:** [quiver], page 294, [compass], page 295, [feather], page 295, [plot], page 272.

compass (*u*, *v*)     [Function File]
compass (*z*)     [Function File]
compass (..., *style*)     [Function File]
compass (*hax*, ...)     [Function File]
*h* = compass (...)     [Function File]

Plot the (*u*, *v*) components of a vector field emanating from the origin of a polar plot.

The arrow representing each vector has one end at the origin and the tip at [*u*(i), *v*(i)]. If a single complex argument *z* is given, then `u = real (z)` and `v = imag (z)`.

The style to use for the plot can be defined with a line style *style* of the same format as the `plot` command.

If the first argument *hax* is an axes handle, then plot into this axis, rather than the current axes returned by `gca`.

The optional return value *h* is a vector of graphics handles to the line objects representing the drawn vectors.

```
a = toeplitz ([1;randn(9,1)], [1,randn(1,9)]);
compass (eig (a));
```

**See also:** [polar], page 292, [feather], page 295, [quiver], page 294, [rose], page 285, [plot], page 272.

feather (*u*, *v*)     [Function File]
feather (*z*)     [Function File]
feather (..., *style*)     [Function File]
feather (*hax*, ...)     [Function File]

`h = feather (…)`     [Function File]

Plot the (u, v) components of a vector field emanating from equidistant points on the x-axis.

If a single complex argument z is given, then `u = real (z)` and `v = imag (z)`.

The style to use for the plot can be defined with a line style *style* of the same format as the `plot` command.

If the first argument *hax* is an axes handle, then plot into this axis, rather than the current axes returned by `gca`.

The optional return value *h* is a vector of graphics handles to the line objects representing the drawn vectors.

```
phi = [0 : 15 : 360] * pi/180;
feather (sin (phi), cos (phi));
```

See also: [plot], page 272, [quiver], page 294, [compass], page 295.

`pcolor (x, y, c)`     [Function File]
`pcolor (c)`     [Function File]
`pcolor (hax, …)`     [Function File]
`h = pcolor (…)`     [Function File]

Produce a 2-D density plot.

A `pcolor` plot draws rectangles with colors from the matrix *c* over the two-dimensional region represented by the matrices *x* and *y*. *x* and *y* are the coordinates of the mesh's vertices and are typically the output of `meshgrid`. If *x* and *y* are vectors, then a typical vertex is $(x(j), y(i), c(i,j))$. Thus, columns of *c* correspond to different *x* values and rows of *c* correspond to different *y* values.

The values in *c* are scaled to span the range of the current colormap. Limits may be placed on the color axis by the command `caxis`, or by setting the `clim` property of the parent axis.

The face color of each cell of the mesh is determined by interpolating the values of *c* for each of the cell's vertices; Contrast this with `imagesc` which renders one cell for each element of *c*.

`shading` modifies an attribute determining the manner by which the face color of each cell is interpolated from the values of *c*, and the visibility of the cells' edges. By default the attribute is `"faceted"`, which renders a single color for each cell's face with the edge visible.

If the first argument *hax* is an axes handle, then plot into this axis, rather than the current axes returned by `gca`.

The optional return value *h* is a graphics handle to the created surface object.

See also: [caxis], page 300, [shading], page 319, [meshgrid], page 316, [contour], page 285, [imagesc], page 696.

`area (y)`     [Function File]
`area (x, y)`     [Function File]
`area (…, lvl)`     [Function File]
`area (…, prop, val, …)`     [Function File]

Chapter 15: Plotting 297

area (*hax*, ...)  [Function File]
h = area (...)  [Function File]

    Area plot of the columns of *y*.

    This plot shows the contributions of each column value to the row sum. It is functionally similar to plot (*x*, cumsum (*y*, 2)), except that the area under the curve is shaded.

    If the *x* argument is omitted it defaults to 1:rows (*y*). A value *lvl* can be defined that determines where the base level of the shading under the curve should be defined. The default level is 0.

    Additional property/value pairs are passed directly to the underlying patch object.

    If the first argument *hax* is an axes handle, then plot into this axis, rather than the current axes returned by gca.

    The optional return value *h* is a graphics handle to the hggroup object comprising the area patch objects. The "BaseValue" property of the hggroup can be used to adjust the level where shading begins.

    Example: Verify identity $\sin^2 + \cos^2 = 1$

```
t = linspace (0, 2*pi, 100)';
y = [sin(t).^2, cos(t).^2];
area (t, y);
legend ("sin^2", "cos^2", "location", "NorthEastOutside");
```

    **See also:** [plot], page 272, [patch], page 355.

fill (*x*, *y*, *c*)  [Function File]
fill (*x1*, *y1*, *c1*, *x2*, *y2*, *c2*)  [Function File]
fill (..., *prop*, *val*)  [Function File]
fill (*hax*, ...)  [Function File]
h = fill (...)  [Function File]

    Create one or more filled 2-D polygons.

    The inputs *x* and *y* are the coordinates of the polygon vertices. If the inputs are matrices then the rows represent different vertices and each column produces a different polygon. fill will close any open polygons before plotting.

    The input *c* determines the color of the polygon. The simplest form is a single color specification such as a plot format or an RGB-triple. In this case the polygon(s) will have one unique color. If *c* is a vector or matrix then the color data is first scaled using caxis and then indexed into the current colormap. A row vector will color each polygon (a column from matrices *x* and *y*) with a single computed color. A matrix *c* of the same size as *x* and *y* will compute the color of each vertex and then interpolate the face color between the vertices.

    Multiple property/value pairs for the underlying patch object may be specified, but they must appear in pairs.

    If the first argument *hax* is an axes handle, then plot into this axis, rather than the current axes returned by gca.

    The optional return value *h* is a vector of graphics handles to the created patch objects.

    Example: red square

```
vertices = [0 0
            1 0
            1 1
            0 1];
fill (vertices(:,1), vertices(:,2), "r");
axis ([-0.5 1.5, -0.5 1.5])
axis equal
```

See also: [patch], page 355, [caxis], page 300, [colormap], page 698.

comet (*y*)    [Function File]
comet (*x*, *y*)    [Function File]
comet (*x*, *y*, *p*)    [Function File]
comet (*hax*, ...)    [Function File]

Produce a simple comet style animation along the trajectory provided by the input coordinate vectors (*x*, *y*).

If *x* is not specified it defaults to the indices of *y*.

The speed of the comet may be controlled by *p*, which represents the time each point is displayed before moving to the next one. The default for *p* is 0.1 seconds.

If the first argument *hax* is an axes handle, then plot into this axis, rather than the current axes returned by **gca**.

See also: [comet3], page 298.

comet3 (*z*)    [Function File]
comet3 (*x*, *y*, *z*)    [Function File]
comet3 (*x*, *y*, *z*, *p*)    [Function File]
comet3 (*hax*, ...)    [Function File]

Produce a simple comet style animation along the trajectory provided by the input coordinate vectors (*x*, *y*, *z*).

If only *z* is specified then *x*, *y* default to the indices of *z*.

The speed of the comet may be controlled by *p*, which represents the time each point is displayed before moving to the next one. The default for *p* is 0.1 seconds.

If the first argument *hax* is an axes handle, then plot into this axis, rather than the current axes returned by **gca**.

See also: [comet], page 298.

[*x*, *map*] = frame2im (*f*)    [Function File]

Convert movie frame to indexed image.

A movie frame is simply a struct with the fields "**cdata**" and "**colormap**".

Support for N-dimensional images or movies is given when *f* is a struct array. In such cases, *x* will be a MxNx1xK or MxNx3xK for indexed and RGB movies respectively, with each frame concatenated along the 4th dimension.

See also: [im2frame], page 299.

Chapter 15: Plotting 299

im2frame (*rgb*) [Function File]
im2frame (*x*, *map*) [Function File]
    Convert image to movie frame.

    A movie frame is simply a struct with the fields "cdata" and "colormap".

    Support for N-dimensional images is given when each image projection, matrix sizes of MxN and MxNx3 for RGB images, is concatenated along the fourth dimension. In such cases, the returned value is a struct array.

    **See also:** [frame2im], page 298.

### 15.2.1.1 Axis Configuration

The axis function may be used to change the axis limits of an existing plot and various other axis properties, such as the aspect ratio and the appearance of tic marks.

axis () [Function File]
axis ([*x_lo x_hi*]) [Function File]
axis ([*x_lo x_hi y_lo y_hi*]) [Function File]
axis ([*x_lo x_hi y_lo y_hi z_lo z_hi*]) [Function File]
axis (*option*) [Function File]
axis (..., *option*) [Function File]
axis (*hax*, ...) [Function File]
*limits* = axis () [Function File]
    Set axis limits and appearance.

    The argument *limits* should be a 2-, 4-, or 6-element vector. The first and second elements specify the lower and upper limits for the x-axis. The third and fourth specify the limits for the y-axis, and the fifth and sixth specify the limits for the z-axis. The special values -Inf and Inf may be used to indicate that the limit should automatically be computed based on the data in the axis.

    Without any arguments, axis turns autoscaling on.

    With one output argument, *limits* = axis returns the current axis limits.

    The vector argument specifying limits is optional, and additional string arguments may be used to specify various axis properties. For example,

        axis ([1, 2, 3, 4], "square");

forces a square aspect ratio, and

        axis ("tic", "labely");

turns tic marks on for all axes and tic mark labels on for the y-axis only.

    The following options control the aspect ratio of the axes.

    "square"    Force a square aspect ratio.

    "equal"     Force x distance to equal y-distance.

    "normal"    Restore default aspect ratio.

    The following options control the way axis limits are interpreted.

    "auto"      Set the specified axes to have nice limits around the data or all if no axes are specified.

`"manual"`   Fix the current axes limits.

`"tight"`   Fix axes to the limits of the data.

`"image"`   Equivalent to `"tight"` and `"equal"`.

The following options affect the appearance of tic marks.

`"on"`   Turn tic marks and labels on for all axes.

`"off"`   Turn tic marks off for all axes.

`"tic[xyz]"`
: Turn tic marks on for all axes, or turn them on for the specified axes and off for the remainder.

`"label[xyz]"`
: Turn tic labels on for all axes, or turn them on for the specified axes and off for the remainder.

`"nolabel"`
: Turn tic labels off for all axes.

Note, if there are no tic marks for an axis, there can be no labels.

The following options affect the direction of increasing values on the axes.

`"ij"`   Reverse y-axis, so lower values are nearer the top.

`"xy"`   Restore y-axis, so higher values are nearer the top.

If the first argument *hax* is an axes handle, then operate on this axes rather than the current axes returned by `gca`.

**See also:** [xlim], page 301, [ylim], page 301, [zlim], page 301, [daspect], page 321, [pbaspect], page 322, [box], page 330, [grid], page 330.

Similarly the axis limits of the colormap can be changed with the caxis function.

`caxis ([`*cmin cmax*`])`                                    [Function File]
`caxis ("`*auto*`")`                                          [Function File]
`caxis ("`*manual*`")`                                        [Function File]
`caxis (`*hax*`, ...)`                                        [Function File]
*limits* `= caxis ()`                                         [Function File]

Query or set color axis limits for plots.

The limits argument should be a 2-element vector specifying the lower and upper limits to assign to the first and last value in the colormap. Data values outside this range are clamped to the first and last colormap entries.

If the `"auto"` option is given then automatic colormap limits are applied. The automatic algorithm sets *cmin* to the minimum data value and *cmax* to the maximum data value. If `"manual"` is specified then the `"climmode"` property is set to `"manual"` and the numeric values in the `"clim"` property are used for limits.

If the first argument *hax* is an axes handle, then operate on this axis rather than the current axes returned by `gca`.

Called without arguments the current color axis limits are returned.

**See also:** [colormap], page 698.

Chapter 15: Plotting                                                                301

The xlim, ylim, and zlim functions may be used to get or set individual axis limits. Each has the same form.

*xlimits* = xlim ()                                                        [Function File]
*xmode* = xlim ("*mode*")                                                  [Function File]
xlim ([*x_lo x_hi*])                                                       [Function File]
xlim ("*auto*")                                                            [Function File]
xlim ("*manual*")                                                          [Function File]
xlim (*hax*, ...)                                                          [Function File]

Query or set the limits of the x-axis for the current plot.

Called without arguments xlim returns the x-axis limits of the current plot.

With the input query "mode", return the current x-limit calculation mode which is either "auto" or "manual".

If passed a 2-element vector [*x_lo x_hi*], the limits of the x-axis are set to these values and the mode is set to "manual".

The current plotting mode can be changed by using either "auto" or "manual" as the argument.

If the first argument *hax* is an axes handle, then operate on this axis rather than the current axes returned by gca.

See also: [ylim], page 301, [zlim], page 301, [axis], page 299, [set], page 359, [get], page 359, [gca], page 358.

### 15.2.1.2 Two-dimensional Function Plotting

Octave can plot a function from a function handle, inline function, or string defining the function without the user needing to explicitly create the data to be plotted. The function fplot also generates two-dimensional plots with linear axes using a function name and limits for the range of the x-coordinate instead of the x and y data. For example,

```
fplot (@sin, [-10, 10], 201);
```

produces a plot that is equivalent to the one above, but also includes a legend displaying the name of the plotted function.

fplot (*fn, limits*)                                                       [Function File]
fplot (..., *tol*)                                                         [Function File]
fplot (..., *n*)                                                           [Function File]
fplot (..., *fmt*)                                                         [Function File]
[*x, y*] = fplot (...)                                                     [Function File]

Plot a function *fn* within the range defined by *limits*.

*fn* is a function handle, inline function, or string containing the name of the function to evaluate.

The limits of the plot are of the form [*xlo, xhi*] or [*xlo, xhi, ylo, yhi*].

The next three arguments are all optional and any number of them may be given in any order.

*tol* is the relative tolerance to use for the plot and defaults to 2e-3 (.2%).

$n$ is the minimum number of points to use. When $n$ is specified, the maximum stepsize will be `xhi - xlo / n`. More than $n$ points may still be used in order to meet the relative tolerance requirement.

The *fmt* argument specifies the linestyle to be used by the plot command.

If the first argument *hax* is an axes handle, then plot into this axis, rather than the current axes returned by `gca`.

With no output arguments the results are immediately plotted. With two output arguments the 2-D plot data is returned. The data can subsequently be plotted manually with `plot (x, y)`.

Example:
```
fplot (@cos, [0, 2*pi])
fplot ("[cos(x), sin(x)]", [0, 2*pi])
```

Programming Notes:

`fplot` works best with continuous functions. Functions with discontinuities are unlikely to plot well. This restriction may be removed in the future.

`fplot` requires that the function accept and return a vector argument. Consider this when writing user-defined functions and use `.*`, `./`, etc. See the function `vectorize` for potentially converting inline or anonymous functions to vectorized versions.

**See also:** [ezplot], page 302, [plot], page 272, [vectorize], page 488.

Other functions that can create two-dimensional plots directly from a function include `ezplot`, `ezcontour`, `ezcontourf` and `ezpolar`.

| | |
|---|---|
| `ezplot (f)` | [Function File] |
| `ezplot (f2v)` | [Function File] |
| `ezplot (fx, fy)` | [Function File] |
| `ezplot (..., dom)` | [Function File] |
| `ezplot (..., n)` | [Function File] |
| `ezplot (hax, ...)` | [Function File] |
| `h = ezplot (...)` | [Function File] |

Plot the 2-D curve defined by the function f.

The function $f$ may be a string, inline function, or function handle and can have either one or two variables. If $f$ has one variable, then the function is plotted over the domain `-2*pi < x < 2*pi` with 500 points.

If *f2v* is a function of two variables then the implicit function $f(x,y) = 0$ is calculated over the meshed domain `-2*pi <= x | y <= 2*pi` with 60 points in each dimension.

For example:
```
ezplot (@(x, y) x.^2 - y.^2 - 1)
```

If two functions are passed as inputs then the parametric function
```
x = fx (t)
y = fy (t)
```
is plotted over the domain `-2*pi <= t <= 2*pi` with 500 points.

Chapter 15: Plotting    303

If *dom* is a two element vector, it represents the minimum and maximum values of both x and y, or t for a parametric plot. If *dom* is a four element vector, then the minimum and maximum values are [xmin xmax ymin ymax].

n is a scalar defining the number of points to use in plotting the function.

If the first argument *hax* is an axes handle, then plot into this axis, rather than the current axes returned by gca.

The optional return value *h* is a vector of graphics handles to the created line objects.

**See also:** [plot], page 272, [ezplot3], page 322, [ezpolar], page 304, [ezcontour], page 303, [ezcontourf], page 303, [ezmesh], page 322, [ezmeshc], page 323, [ezsurf], page 324, [ezsurfc], page 325.

ezcontour (*f*)                            [Function File]
ezcontour (..., *dom*)                     [Function File]
ezcontour (..., *n*)                       [Function File]
ezcontour (*hax*, ...)                     [Function File]
h = ezcontour (...)                        [Function File]

Plot the contour lines of a function.

f is a string, inline function, or function handle with two arguments defining the function. By default the plot is over the meshed domain -2*pi <= x | y <= 2*pi with 60 points in each dimension.

If *dom* is a two element vector, it represents the minimum and maximum values of both x and y. If *dom* is a four element vector, then the minimum and maximum values are [xmin xmax ymin ymax].

n is a scalar defining the number of points to use in each dimension.

If the first argument *hax* is an axes handle, then plot into this axis, rather than the current axes returned by gca.

The optional return value *h* is a graphics handle to the created plot.

Example:

```
f = @(x,y) sqrt (abs (x .* y)) ./ (1 + x.^2 + y.^2);
ezcontour (f, [-3, 3]);
```

**See also:** [contour], page 285, [ezcontourf], page 303, [ezplot], page 302, [ezmeshc], page 323, [ezsurfc], page 325.

ezcontourf (*f*)                           [Function File]
ezcontourf (..., *dom*)                    [Function File]
ezcontourf (..., *n*)                      [Function File]
ezcontourf (*hax*, ...)                    [Function File]
h = ezcontourf (...)                       [Function File]

Plot the filled contour lines of a function.

f is a string, inline function, or function handle with two arguments defining the function. By default the plot is over the meshed domain -2*pi <= x | y <= 2*pi with 60 points in each dimension.

If *dom* is a two element vector, it represents the minimum and maximum values of both x and y. If *dom* is a four element vector, then the minimum and maximum values are [xmin xmax ymin ymax].

*n* is a scalar defining the number of points to use in each dimension.

If the first argument *hax* is an axes handle, then plot into this axis, rather than the current axes returned by gca.

The optional return value *h* is a graphics handle to the created plot.

Example:

```
f = @(x,y) sqrt (abs (x .* y)) ./ (1 + x.^2 + y.^2);
ezcontourf (f, [-3, 3]);
```

**See also:** [contourf], page 286, [ezcontour], page 303, [ezplot], page 302, [ezmeshc], page 323, [ezsurfc], page 325.

| | |
|---|---|
| ezpolar (*f*) | [Function File] |
| ezpolar (..., *dom*) | [Function File] |
| ezpolar (..., *n*) | [Function File] |
| ezpolar (*hax*, ...) | [Function File] |
| *h* = ezpolar (...) | [Function File] |

Plot a 2-D function in polar coordinates.

The function *f* is a string, inline function, or function handle with a single argument. The expected form of the function is rho = f(theta). By default the plot is over the domain 0 <= theta <= 2*pi with 500 points.

If *dom* is a two element vector, it represents the minimum and maximum values of *theta*.

*n* is a scalar defining the number of points to use in plotting the function.

If the first argument *hax* is an axes handle, then plot into this axis, rather than the current axes returned by gca.

The optional return value *h* is a graphics handle to the created plot.

Example:

```
ezpolar (@(t) sin (5/4 * t), [0, 8*pi]);
```

**See also:** [polar], page 292, [ezplot], page 302.

### 15.2.1.3 Two-dimensional Geometric Shapes

| | |
|---|---|
| rectangle () | [Function File] |
| rectangle (..., "*Position*", *pos*) | [Function File] |
| rectangle (..., "*Curvature*", *curv*) | [Function File] |
| rectangle (..., "*EdgeColor*", *ec*) | [Function File] |
| rectangle (..., "*FaceColor*", *fc*) | [Function File] |
| rectangle (*hax*, ...) | [Function File] |
| *h* = rectangle (...) | [Function File] |

Draw a rectangular patch defined by *pos* and *curv*.

The variable *pos*(1:2) defines the lower left-hand corner of the patch and *pos*(3:4) defines its width and height. By default, the value of *pos* is [0, 0, 1, 1].

The variable *curv* defines the curvature of the sides of the rectangle and may be a scalar or two-element vector with values between 0 and 1. A value of 0 represents no curvature of the side, whereas a value of 1 means that the side is entirely curved

Chapter 15: Plotting 305

into the arc of a circle. If *curv* is a two-element vector, then the first element is the curvature along the x-axis of the patch and the second along y-axis.

If *curv* is a scalar, it represents the curvature of the shorter of the two sides of the rectangle and the curvature of the other side is defined by

```
min (pos(1:2)) / max (pos(1:2)) * curv
```

Additional property/value pairs are passed to the underlying patch command.

If the first argument *hax* is an axes handle, then plot into this axis, rather than the current axes returned by `gca`.

The optional return value *h* is a graphics handle to the created rectangle object.

**See also:** [patch], page 355, [line], page 355, [cylinder], page 325, [ellipsoid], page 326, [sphere], page 326.

### 15.2.2 Three-Dimensional Plots

The function `mesh` produces mesh surface plots. For example,

```
tx = ty = linspace (-8, 8, 41)';
[xx, yy] = meshgrid (tx, ty);
r = sqrt (xx .^ 2 + yy .^ 2) + eps;
tz = sin (r) ./ r;
mesh (tx, ty, tz);
```

produces the familiar "sombrero" plot shown in Figure 15.5. Note the use of the function `meshgrid` to create matrices of X and Y coordinates to use for plotting the Z data. The `ndgrid` function is similar to `meshgrid`, but works for N-dimensional matrices.

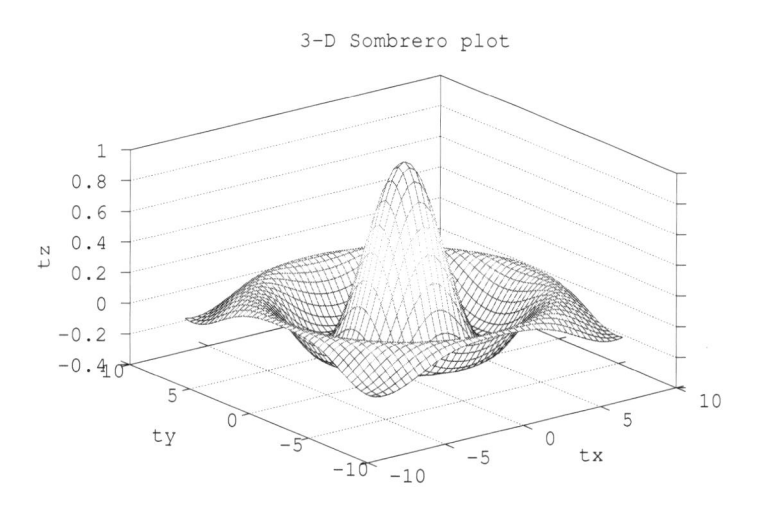

Figure 15.5: Mesh plot.

The `meshc` function is similar to `mesh`, but also produces a plot of contours for the surface.

The `plot3` function displays arbitrary three-dimensional data, without requiring it to form a surface. For example,

```
t = 0:0.1:10*pi;
r = linspace (0, 1, numel (t));
z = linspace (0, 1, numel (t));
plot3 (r.*sin(t), r.*cos(t), z);
```
displays the spiral in three dimensions shown in Figure 15.6.

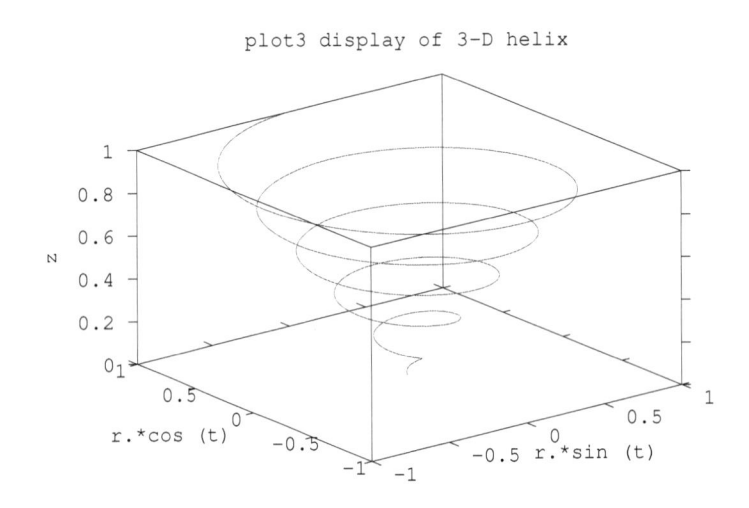

Figure 15.6: Three-dimensional spiral.

Finally, the `view` function changes the viewpoint for three-dimensional plots.

| | |
|---|---|
| mesh (x, y, z) | [Function File] |
| mesh (z) | [Function File] |
| mesh (..., c) | [Function File] |
| mesh (..., prop, val, ...) | [Function File] |
| mesh (hax, ...) | [Function File] |
| h = mesh (...) | [Function File] |

Plot a 3-D wireframe mesh.

The wireframe mesh is plotted using rectangles. The vertices of the rectangles [x, y] are typically the output of `meshgrid`. over a 2-D rectangular region in the x-y plane. z determines the height above the plane of each vertex. If only a single z matrix is given, then it is plotted over the meshgrid x = 1:columns (z), y = 1:rows (z). Thus, columns of z correspond to different x values and rows of z correspond to different y values.

The color of the mesh is computed by linearly scaling the z values to fit the range of the current colormap. Use `caxis` and/or change the colormap to control the appearance.

Optionally, the color of the mesh can be specified independently of z by supplying a color matrix, c.

Any property/value pairs are passed directly to the underlying surface object.

If the first argument hax is an axes handle, then plot into this axis, rather than the current axes returned by `gca`.

The optional return value h is a graphics handle to the created surface object.

Chapter 15: Plotting	307

> **See also:** [ezmesh], page 322, [meshc], page 307, [meshz], page 307, [trimesh], page 663, [contour], page 285, [surf], page 308, [surface], page 356, [meshgrid], page 316, [hidden], page 308, [shading], page 319, [colormap], page 698, [caxis], page 300.

meshc (*x, y, z*)        [Function File]
meshc (*z*)        [Function File]
meshc (..., *c*)        [Function File]
meshc (..., *prop, val,* ...)        [Function File]
meshc (*hax,* ...)        [Function File]
h = meshc (...)        [Function File]

> Plot a 3-D wireframe mesh with underlying contour lines.
>
> The wireframe mesh is plotted using rectangles. The vertices of the rectangles [*x, y*] are typically the output of `meshgrid`. over a 2-D rectangular region in the x-y plane. *z* determines the height above the plane of each vertex. If only a single *z* matrix is given, then it is plotted over the meshgrid x = 1:columns (*z*), y = 1:rows (*z*). Thus, columns of *z* correspond to different *x* values and rows of *z* correspond to different *y* values.
>
> The color of the mesh is computed by linearly scaling the *z* values to fit the range of the current colormap. Use `caxis` and/or change the colormap to control the appearance.
>
> Optionally the color of the mesh can be specified independently of *z* by supplying a color matrix, *c*.
>
> Any property/value pairs are passed directly to the underlying surface object.
>
> If the first argument *hax* is an axes handle, then plot into this axis, rather than the current axes returned by `gca`.
>
> The optional return value *h* is a 2-element vector with a graphics handle to the created surface object and to the created contour plot.
>
> **See also:** [ezmeshc], page 323, [mesh], page 306, [meshz], page 307, [contour], page 285, [surfc], page 309, [surface], page 356, [meshgrid], page 316, [hidden], page 308, [shading], page 319, [colormap], page 698, [caxis], page 300.

meshz (*x, y, z*)        [Function File]
meshz (*z*)        [Function File]
meshz (..., *c*)        [Function File]
meshz (..., *prop, val,* ...)        [Function File]
meshz (*hax,* ...)        [Function File]
h = meshz (...)        [Function File]

> Plot a 3-D wireframe mesh with a surrounding curtain.
>
> The wireframe mesh is plotted using rectangles. The vertices of the rectangles [*x, y*] are typically the output of `meshgrid`. over a 2-D rectangular region in the x-y plane. *z* determines the height above the plane of each vertex. If only a single *z* matrix is given, then it is plotted over the meshgrid x = 0:columns (*z*) - 1, y = 0:rows (*z*) - 1. Thus, columns of *z* correspond to different *x* values and rows of *z* correspond to different *y* values.
>
> The color of the mesh is computed by linearly scaling the *z* values to fit the range of the current colormap. Use `caxis` and/or change the colormap to control the appearance.

Optionally the color of the mesh can be specified independently of z by supplying a color matrix, c.

Any property/value pairs are passed directly to the underlying surface object.

If the first argument hax is an axes handle, then plot into this axis, rather than the current axes returned by `gca`.

The optional return value h is a graphics handle to the created surface object.

**See also:** [mesh], page 306, [meshc], page 307, [contour], page 285, [surf], page 308, [surface], page 356, [waterfall], page 320, [meshgrid], page 316, [hidden], page 308, [shading], page 319, [colormap], page 698, [caxis], page 300.

`hidden`                                    [Command]
`hidden` *on*                               [Command]
`hidden` *off*                              [Command]
*mode* = `hidden` (…)                       [Function File]

Control mesh hidden line removal.

When called with no argument the hidden line removal state is toggled.

When called with one of the modes `"on"` or `"off"` the state is set accordingly.

The optional output argument *mode* is the current state.

Hidden Line Removal determines what graphic objects behind a mesh plot are visible. The default is for the mesh to be opaque and lines behind the mesh are not visible. If hidden line removal is turned off then objects behind the mesh can be seen through the faces (openings) of the mesh, although the mesh grid lines are still opaque.

**See also:** [mesh], page 306, [meshc], page 307, [meshz], page 307, [ezmesh], page 322, [ezmeshc], page 323, [trimesh], page 663, [waterfall], page 320.

`surf (x, y, z)`                            [Function File]
`surf (z)`                                  [Function File]
`surf (…, c)`                               [Function File]
`surf (…, prop, val, …)`                    [Function File]
`surf (hax, …)`                             [Function File]
`h = surf (…)`                              [Function File]

Plot a 3-D surface mesh.

The surface mesh is plotted using shaded rectangles. The vertices of the rectangles [x, y] are typically the output of `meshgrid`. over a 2-D rectangular region in the x-y plane. z determines the height above the plane of each vertex. If only a single z matrix is given, then it is plotted over the meshgrid `x = 1:columns (z), y = 1:rows (z)`. Thus, columns of z correspond to different x values and rows of z correspond to different y values.

The color of the surface is computed by linearly scaling the z values to fit the range of the current colormap. Use `caxis` and/or change the colormap to control the appearance.

Optionally, the color of the surface can be specified independently of z by supplying a color matrix, c.

Any property/value pairs are passed directly to the underlying surface object.

Chapter 15: Plotting

If the first argument *hax* is an axes handle, then plot into this axis, rather than the current axes returned by `gca`.

The optional return value *h* is a graphics handle to the created surface object.

Note: The exact appearance of the surface can be controlled with the `shading` command or by using `set` to control surface object properties.

**See also:** [ezsurf], page 324, [surfc], page 309, [surfl], page 309, [surfnorm], page 310, [trisurf], page 664, [contour], page 285, [mesh], page 306, [surface], page 356, [meshgrid], page 316, [hidden], page 308, [shading], page 319, [colormap], page 698, [caxis], page 300.

| | |
|---|---|
| surfc (*x, y, z*) | [Function File] |
| surfc (*z*) | [Function File] |
| surfc (..., *c*) | [Function File] |
| surfc (..., *prop, val,* ...) | [Function File] |
| surfc (*hax,* ...) | [Function File] |
| h = surfc (...) | [Function File] |

Plot a 3-D surface mesh with underlying contour lines.

The surface mesh is plotted using shaded rectangles. The vertices of the rectangles [*x, y*] are typically the output of `meshgrid`. over a 2-D rectangular region in the x-y plane. *z* determines the height above the plane of each vertex. If only a single *z* matrix is given, then it is plotted over the meshgrid $x$ = 1:columns (*z*), $y$ = 1:rows (*z*). Thus, columns of *z* correspond to different *x* values and rows of *z* correspond to different *y* values.

The color of the surface is computed by linearly scaling the *z* values to fit the range of the current colormap. Use `caxis` and/or change the colormap to control the appearance.

Optionally, the color of the surface can be specified independently of *z* by supplying a color matrix, *c*.

Any property/value pairs are passed directly to the underlying surface object.

If the first argument *hax* is an axes handle, then plot into this axis, rather than the current axes returned by `gca`.

The optional return value *h* is a graphics handle to the created surface object.

Note: The exact appearance of the surface can be controlled with the `shading` command or by using `set` to control surface object properties.

**See also:** [ezsurfc], page 325, [surf], page 308, [surfl], page 309, [surfnorm], page 310, [trisurf], page 664, [contour], page 285, [mesh], page 306, [surface], page 356, [meshgrid], page 316, [hidden], page 308, [shading], page 319, [colormap], page 698, [caxis], page 300.

| | |
|---|---|
| surfl (*z*) | [Function File] |
| surfl (*x, y, z*) | [Function File] |
| surfl (..., *lsrc*) | [Function File] |
| surfl (*x, y, z, lsrc, P*) | [Function File] |
| surfl (..., "*cdata*") | [Function File] |
| surfl (..., "*light*") | [Function File] |

`surfl (hax, ...)` [Function File]
`h = surfl (...)` [Function File]

Plot a 3-D surface using shading based on various lighting models.

The surface mesh is plotted using shaded rectangles. The vertices of the rectangles [x, y] are typically the output of `meshgrid`. over a 2-D rectangular region in the x-y plane. z determines the height above the plane of each vertex. If only a single z matrix is given, then it is plotted over the meshgrid `x = 1:columns (z)`, `y = 1:rows (z)`. Thus, columns of z correspond to different x values and rows of z correspond to different y values.

The default lighting mode `"cdata"`, changes the cdata property of the surface object to give the impression of a lighted surface. **Warning:** The alternative mode `"light"` mode which creates a light object to illuminate the surface is not implemented (yet).

The light source location can be specified using *lsrc*. It can be given as a 2-element vector [azimuth, elevation] in degrees, or as a 3-element vector [lx, ly, lz]. The default value is rotated 45 degrees counterclockwise to the current view.

The material properties of the surface can specified using a 4-element vector $P = [AM\ D\ SP\ exp]$ which defaults to $p = [0.55\ 0.6\ 0.4\ 10]$.

`"AM"` strength of ambient light
`"D"` strength of diffuse reflection
`"SP"` strength of specular reflection
`"EXP"` specular exponent

If the first argument *hax* is an axes handle, then plot into this axis, rather than the current axes returned by `gca`.

The optional return value *h* is a graphics handle to the created surface object.

Example:

```
colormap (bone (64));
surfl (peaks);
shading interp;
```

See also: [diffuse], page 316, [specular], page 316, [surf], page 308, [shading], page 319, [colormap], page 698, [caxis], page 300.

`surfnorm (x, y, z)` [Function File]
`surfnorm (z)` [Function File]
`surfnorm (..., prop, val, ...)` [Function File]
`surfnorm (hax, ...)` [Function File]
`[nx, ny, nz] = surfnorm (...)` [Function File]

Find the vectors normal to a meshgridded surface.

If x and y are vectors, then a typical vertex is $(x(j), y(i), z(i,j))$. Thus, columns of z correspond to different x values and rows of z correspond to different y values. If only a single input z is given then x is taken to be `1:rows (z)` and y is `1:columns (z)`.

If no return arguments are requested, a surface plot with the normal vectors to the surface is plotted.

Any property/value input pairs are assigned to the surface object.

If the first argument *hax* is an axes handle, then plot into this axis, rather than the current axes returned by `gca`.

If output arguments are requested then the components of the normal vectors are returned in *nx*, *ny*, and *nz* and no plot is made.

An example of the use of `surfnorm` is

```
surfnorm (peaks (25));
```

Algorithm: The normal vectors are calculated by taking the cross product of the diagonals of each of the quadrilaterals in the meshgrid to find the normal vectors of the centers of these quadrilaterals. The four nearest normal vectors to the meshgrid points are then averaged to obtain the normal to the surface at the meshgridded points.

**See also:** [isonormals], page 312, [quiver3], page 294, [surf], page 308, [meshgrid], page 316.

| | |
|---|---|
| `[fv] = isosurface (val, iso)` | [Function File] |
| `[fv] = isosurface (x, y, z, val, iso)` | [Function File] |
| `[fv] = isosurface (..., "noshare", "verbose")` | [Function File] |
| `[fvc] = isosurface (..., col)` | [Function File] |
| `[f, v] = isosurface (x, y, z, val, iso)` | [Function File] |
| `[f, v, c] = isosurface (x, y, z, val, iso, col)` | [Function File] |
| `isosurface (x, y, z, val, iso, col, opt)` | [Function File] |

Calculate isosurface of 3-D data.

If called with one output argument and the first input argument *val* is a three-dimensional array that contains the data of an isosurface geometry and the second input argument *iso* keeps the isovalue as a scalar value then return a structure array *fv* that contains the fields *Faces* and *Vertices* at computed points `[x, y, z] = meshgrid (1:l, 1:m, 1:n)`. The output argument *fv* can directly be taken as an input argument for the `patch` function.

If called with further input arguments *x*, *y* and *z* which are three-dimensional arrays with the same size than *val* then the volume data is taken at those given points.

The string input argument `"noshare"` is only for compatibility and has no effect. If given the string input argument `"verbose"` then print messages to the command line interface about the current progress.

If called with the input argument *col* which is a three-dimensional array of the same size than *val* then take those values for the interpolation of coloring the isosurface geometry. Add the field *FaceVertexCData* to the structure array *fv*.

If called with two or three output arguments then return the information about the faces *f*, vertices *v* and color data *c* as separate arrays instead of a single structure array.

If called with no output argument then directly process the isosurface geometry with the `patch` command.

For example,

```
[x, y, z] = meshgrid (1:5, 1:5, 1:5);
val = rand (5, 5, 5);
isosurface (x, y, z, val, .5);
```

will directly draw a random isosurface geometry in a graphics window. Another example for an isosurface geometry with different additional coloring

```
N = 15;    # Increase number of vertices in each direction
iso = .4;  # Change isovalue to .1 to display a sphere
lin = linspace (0, 2, N);
[x, y, z] = meshgrid (lin, lin, lin);
c = abs ((x-.5).^2 + (y-.5).^2 + (z-.5).^2);
figure (); # Open another figure window

subplot (2,2,1); view (-38, 20);
[f, v] = isosurface (x, y, z, c, iso);
p = patch ("Faces", f, "Vertices", v, "EdgeColor", "none");
set (gca, "PlotBoxAspectRatioMode", "manual", ...
          "PlotBoxAspectRatio", [1 1 1]);
# set (p, "FaceColor", "green", "FaceLighting", "phong");
# light ("Position", [1 1 5]); # Available with the JHandles package

subplot (2,2,2); view (-38, 20);
p = patch ("Faces", f, "Vertices", v, "EdgeColor", "blue");
set (gca, "PlotBoxAspectRatioMode", "manual", ...
          "PlotBoxAspectRatio", [1 1 1]);
# set (p, "FaceColor", "none", "FaceLighting", "phong");
# light ("Position", [1 1 5]);

subplot (2,2,3); view (-38, 20);
[f, v, c] = isosurface (x, y, z, c, iso, y);
p = patch ("Faces", f, "Vertices", v, "FaceVertexCData", c, ...
           "FaceColor", "interp", "EdgeColor", "none");
set (gca, "PlotBoxAspectRatioMode", "manual", ...
          "PlotBoxAspectRatio", [1 1 1]);
# set (p, "FaceLighting", "phong");
# light ("Position", [1 1 5]);

subplot (2,2,4); view (-38, 20);
p = patch ("Faces", f, "Vertices", v, "FaceVertexCData", c, ...
           "FaceColor", "interp", "EdgeColor", "blue");
set (gca, "PlotBoxAspectRatioMode", "manual", ...
          "PlotBoxAspectRatio", [1 1 1]);
# set (p, "FaceLighting", "phong");
# light ("Position", [1 1 5]);
```

**See also:** [isonormals], page 312, [isocolors], page 313.

| | |
|---|---|
| [n] = isonormals (*val*, *v*) | [Function File] |
| [n] = isonormals (*val*, *p*) | [Function File] |
| [n] = isonormals (*x*, *y*, *z*, *val*, *v*) | [Function File] |
| [n] = isonormals (*x*, *y*, *z*, *val*, *p*) | [Function File] |
| [n] = isonormals (..., "*negate*") | [Function File] |
| isonormals (..., *p*) | [Function File] |

Calculate normals to an isosurface.

If called with one output argument and the first input argument *val* is a three-dimensional array that contains the data for an isosurface geometry and the second input argument *v* keeps the vertices of an isosurface then return the normals *n* in form of a matrix with the same size than *v* at computed points [x, y, z] =

## Chapter 15: Plotting

meshgrid (1:l, 1:m, 1:n). The output argument n can be taken to manually set *VertexNormals* of a patch.

If called with further input arguments x, y and z which are three-dimensional arrays with the same size than *val* then the volume data is taken at those given points. Instead of the vertices data v a patch handle p can be passed to this function.

If given the string input argument "negate" as last input argument then compute the reverse vector normals of an isosurface geometry.

If no output argument is given then directly redraw the patch that is given by the patch handle p.

For example:

```
function [] = isofinish (p)
  set (gca, "PlotBoxAspectRatioMode", "manual", ...
       "PlotBoxAspectRatio", [1 1 1]);
  set (p, "VertexNormals", -get (p,"VertexNormals")); # Revert normals
  set (p, "FaceColor", "interp");
  ## set (p, "FaceLighting", "phong");
  ## light ("Position", [1 1 5]); # Available with JHandles
endfunction

N = 15;    # Increase number of vertices in each direction
iso = .4;  # Change isovalue to .1 to display a sphere
lin = linspace (0, 2, N);
[x, y, z] = meshgrid (lin, lin, lin);
c = abs ((x-.5).^2 + (y-.5).^2 + (z-.5).^2);
figure (); # Open another figure window

subplot (2,2,1); view (-38, 20);
[f, v, cdat] = isosurface (x, y, z, c, iso, y);
p = patch ("Faces", f, "Vertices", v, "FaceVertexCData", cdat, ...
           "FaceColor", "interp", "EdgeColor", "none");
isofinish (p);  # Call user function isofinish

subplot (2,2,2); view (-38, 20);
p = patch ("Faces", f, "Vertices", v, "FaceVertexCData", cdat, ...
           "FaceColor", "interp", "EdgeColor", "none");
isonormals (x, y, z, c, p); # Directly modify patch
isofinish (p);

subplot (2,2,3); view (-38, 20);
p = patch ("Faces", f, "Vertices", v, "FaceVertexCData", cdat, ...
           "FaceColor", "interp", "EdgeColor", "none");
n = isonormals (x, y, z, c, v); # Compute normals of isosurface
set (p, "VertexNormals", n);    # Manually set vertex normals
isofinish (p);

subplot (2,2,4); view (-38, 20);
p = patch ("Faces", f, "Vertices", v, "FaceVertexCData", cdat, ...
           "FaceColor", "interp", "EdgeColor", "none");
isonormals (x, y, z, c, v, "negate"); # Use reverse directly
isofinish (p);
```

**See also:** [isosurface], page 311, [isocolors], page 313.

[*cd*] = isocolors (*c*, *v*)                                                                          [Function File]
[*cd*] = isocolors (*x*, *y*, *z*, *c*, *v*)                                                 [Function File]

[cd] = isocolors (x, y, z, r, g, b, v)                                [Function File]
[cd] = isocolors (r, g, b, v)                                         [Function File]
[cd] = isocolors (..., p)                                             [Function File]
isocolors (...)                                                       [Function File]

Compute isosurface colors.

If called with one output argument and the first input argument $c$ is a three-dimensional array that contains color values and the second input argument $v$ keeps the vertices of a geometry then return a matrix $cd$ with color data information for the geometry at computed points [x, y, z] = meshgrid (1:1, 1:m, 1:n). The output argument $cd$ can be taken to manually set FaceVertexCData of a patch.

If called with further input arguments $x$, $y$ and $z$ which are three–dimensional arrays of the same size than $c$ then the color data is taken at those given points. Instead of the color data $c$ this function can also be called with RGB values $r$, $g$, $b$. If input argumnets $x$, $y$, $z$ are not given then again meshgrid computed values are taken.

Optionally, the patch handle $p$ can be given as the last input argument to all variations of function calls instead of the vertices data $v$. Finally, if no output argument is given then directly change the colors of a patch that is given by the patch handle $p$.

For example:

```
function [] = isofinish (p)
  set (gca, "PlotBoxAspectRatioMode", "manual", ...
            "PlotBoxAspectRatio", [1 1 1]);
  set (p, "FaceColor", "interp");
  ## set (p, "FaceLighting", "flat");
  ## light ("Position", [1 1 5]);  # Available with JHandles
endfunction

N = 15;    # Increase number of vertices in each direction
iso = .4;  # Change isovalue to .1 to display a sphere
lin = linspace (0, 2, N);
[x, y, z] = meshgrid (lin, lin, lin);
c = abs ((x-.5).^2 + (y-.5).^2 + (z-.5).^2);
figure (); # Open another figure window

subplot (2,2,1); view (-38, 20);
[f, v] = isosurface (x, y, z, c, iso);
p = patch ("Faces", f, "Vertices", v, "EdgeColor", "none");
cdat = rand (size (c));           # Compute random patch color data
isocolors (x, y, z, cdat, p); # Directly set colors of patch
isofinish (p);                    # Call user function isofinish

subplot (2,2,2); view (-38, 20);
p = patch ("Faces", f, "Vertices", v, "EdgeColor", "none");
[r, g, b] = meshgrid (lin, 2-lin, 2-lin);
cdat = isocolors (x, y, z, c, v); # Compute color data vertices
set (p, "FaceVertexCData", cdat); # Set color data manually
isofinish (p);
```

# Chapter 15: Plotting

```
subplot (2,2,3); view (-38, 20);
p = patch ("Faces", f, "Vertices", v, "EdgeColor", "none");
cdat = isocolors (r, g, b, c, p); # Compute color data patch
set (p, "FaceVertexCData", cdat); # Set color data manually
isofinish (p);

subplot (2,2,4); view (-38, 20);
p = patch ("Faces", f, "Vertices", v, "EdgeColor", "none");
r = g = b = repmat ([1:N] / N, [N, 1, N]); # Black to white
cdat = isocolors (x, y, z, r, g, b, v);
set (p, "FaceVertexCData", cdat);
isofinish (p);
```

See also: [isosurface], page 311, [isonormals], page 312.

shrinkfaces (*p*, *sf*)     [Function File]
*nfv* = shrinkfaces (*p*, *sf*)     [Function File]
*nfv* = shrinkfaces (*fv*, *sf*)     [Function File]
*nfv* = shrinkfaces (*f*, *v*, *sf*)     [Function File]
[*nf*, *nv*] = shrinkfaces (...)     [Function File]

Reduce the size of faces in a patch by the shrink factor *sf*.

The patch object can be specified by a graphics handle (*p*), a patch structure (*fv*) with the fields `"faces"` and `"vertices"`, or as two separate matrices (*f*, *v*) of faces and vertices.

The shrink factor *sf* is a positive number specifying the percentage of the original area the new face will occupy. If no factor is given the default is 0.3 (a reduction to 30% of the original size). A factor greater than 1.0 will result in the expansion of faces.

Given a patch handle as the first input argument and no output parameters, perform the shrinking of the patch faces in place and redraw the patch.

If called with one output argument, return a structure with fields `"faces"`, `"vertices"`, and `"facevertexcdata"` containing the data after shrinking. This structure can be used directly as an input argument to the `patch` function.

**Caution::** Performing the shrink operation on faces which are not convex can lead to undesirable results.

Example: a triangulated 3/4 circle and the corresponding shrunken version.

```
[phi r] = meshgrid (linspace (0, 1.5*pi, 16), linspace (1, 2, 4));
tri = delaunay (phi(:), r(:));
v = [r(:).*sin(phi(:)) r(:).*cos(phi(:))];
clf ()
p = patch ("Faces", tri, "Vertices", v, "FaceColor", "none");
fv = shrinkfaces (p);
patch (fv)
axis equal
grid on
```

**See also:** [patch], page 355.

**diffuse** (*sx*, *sy*, *sz*, *lv*)  [Function File]

Calculate the diffuse reflection strength of a surface defined by the normal vector elements *sx*, *sy*, *sz*.

The light source location vector *lv* can be given as a 2-element vector [azimuth, elevation] in degrees or as a 3-element vector [x, y, z].

**See also:** [specular], page 316, [surfl], page 309.

**specular** (*sx*, *sy*, *sz*, *lv*, *vv*)  [Function File]
**specular** (*sx*, *sy*, *sz*, *lv*, *vv*, *se*)  [Function File]

Calculate the specular reflection strength of a surface defined by the normal vector elements *sx*, *sy*, *sz* using Phong's approximation.

The light source location and viewer location vectors are specified using parameters *lv* and *vv* respectively. The location vectors can given as 2-element vectors [azimuth, elevation] in degrees or as 3-element vectors [x, y, z].

An optional sixth argument specifies the specular exponent (spread) *se*. If not given, *se* defaults to 10.

**See also:** [diffuse], page 316, [surfl], page 309.

[*xx*, *yy*] = **meshgrid** (*x*, *y*)  [Function File]
[*xx*, *yy*, *zz*] = **meshgrid** (*x*, *y*, *z*)  [Function File]
[*xx*, *yy*] = **meshgrid** (*x*)  [Function File]
[*xx*, *yy*, *zz*] = **meshgrid** (*x*)  [Function File]

Given vectors of *x* and *y* coordinates, return matrices *xx* and *yy* corresponding to a full 2-D grid.

The rows of *xx* are copies of *x*, and the columns of *yy* are copies of *y*. If *y* is omitted, then it is assumed to be the same as *x*.

If the optional *z* input is given, or *zz* is requested, then the output will be a full 3-D grid.

`meshgrid` is most frequently used to produce input for a 2-D or 3-D function that will be plotted. The following example creates a surface plot of the "sombrero" function.

```
f = @(x,y) sin (sqrt (x.^2 + y.^2)) ./ sqrt (x.^2 + y.^2);
range = linspace (-8, 8, 41);
[X, Y] = meshgrid (range, range);
Z = f (X, Y);
surf (X, Y, Z);
```

Programming Note: `meshgrid` is restricted to 2-D or 3-D grid generation. The `ndgrid` function will generate 1-D through N-D grids. However, the functions are not completely equivalent. If *x* is a vector of length M and *y* is a vector of length N, then `meshgrid` will produce an output grid which is NxM. `ndgrid` will produce an output which is MxN (transpose) for the same input. Some core functions expect `meshgrid` input and others expect `ndgrid` input. Check the documentation for the function in question to determine the proper input format.

**See also:** [ndgrid], page 317, [mesh], page 306, [contour], page 285, [surf], page 308.

Chapter 15: Plotting 317

[y1, y2, ..., yn] = ndgrid (x1, x2, ..., xn)  [Function File]
[y1, y2, ..., yn] = ndgrid (x)  [Function File]

Given n vectors $x1, \ldots, xn$, ndgrid returns n arrays of dimension n.

The elements of the i-th output argument contains the elements of the vector $xi$ repeated over all dimensions different from the i-th dimension. Calling ndgrid with only one input argument $x$ is equivalent to calling ndgrid with all n input arguments equal to $x$:

$[y1, y2, \ldots, yn] = \text{ndgrid}(x, \ldots, x)$

Programming Note: ndgrid is very similar to the function meshgrid except that the first two dimensions are transposed in comparison to meshgrid. Some core functions expect meshgrid input and others expect ndgrid input. Check the documentation for the function in question to determine the proper input format.

**See also:** [meshgrid], page 316.

plot3 (x, y, z)  [Function File]
plot3 (x, y, z, prop, value, ...)  [Function File]
plot3 (x, y, z, fmt)  [Function File]
plot3 (x, cplx)  [Function File]
plot3 (cplx)  [Function File]
plot3 (hax, ...)  [Function File]
h = plot3 (...)  [Function File]

Produce 3-D plots.

Many different combinations of arguments are possible. The simplest form is

plot3 (x, y, z)

in which the arguments are taken to be the vertices of the points to be plotted in three dimensions. If all arguments are vectors of the same length, then a single continuous line is drawn. If all arguments are matrices, then each column of is treated as a separate line. No attempt is made to transpose the arguments to make the number of rows match.

If only two arguments are given, as

plot3 (x, cplx)

the real and imaginary parts of the second argument are used as the y and z coordinates, respectively.

If only one argument is given, as

plot3 (cplx)

the real and imaginary parts of the argument are used as the y and z values, and they are plotted versus their index.

Arguments may also be given in groups of three as

plot3 (x1, y1, z1, x2, y2, z2, ...)

in which each set of three arguments is treated as a separate line or set of lines in three dimensions.

To plot multiple one- or two-argument groups, separate each group with an empty format string, as

```
plot3 (x1, c1, "", c2, "", …)
```

Multiple property-value pairs may be specified which will affect the line objects drawn by `plot3`. If the *fmt* argument is supplied it will format the line objects in the same manner as `plot`.

If the first argument *hax* is an axes handle, then plot into this axis, rather than the current axes returned by `gca`.

The optional return value *h* is a graphics handle to the created plot.

Example:

```
z = [0:0.05:5];
plot3 (cos (2*pi*z), sin (2*pi*z), z, ";helix;");
plot3 (z, exp (2i*pi*z), ";complex sinusoid;");
```

**See also:** [ezplot3], page 322, [plot], page 272.

| | |
|---|---|
| `view (`*azimuth*`, `*elevation*`)` | [Function File] |
| `view ([`*azimuth elevation*`])` | [Function File] |
| `view ([`*x y z*`])` | [Function File] |
| `view (2)` | [Function File] |
| `view (3)` | [Function File] |
| `view (`*hax*`, …)` | [Function File] |
| `[`*azimuth, elevation*`] = view ()` | [Function File] |

Query or set the viewpoint for the current axes.

The parameters *azimuth* and *elevation* can be given as two arguments or as 2-element vector. The viewpoint can also be specified with Cartesian coordinates *x*, *y*, and *z*.

The call `view (2)` sets the viewpoint to *azimuth* = 0 and *elevation* = 90, which is the default for 2-D graphs.

The call `view (3)` sets the viewpoint to *azimuth* = -37.5 and *elevation* = 30, which is the default for 3-D graphs.

If the first argument *hax* is an axes handle, then operate on this axis rather than the current axes returned by `gca`.

If no inputs are given, return the current *azimuth* and *elevation*.

| | |
|---|---|
| `slice (`*x, y, z, v, sx, sy, sz*`)` | [Function File] |
| `slice (`*x, y, z, v, xi, yi, zi*`)` | [Function File] |
| `slice (`*v, sx, sy, sz*`)` | [Function File] |
| `slice (`*v, xi, yi, zi*`)` | [Function File] |
| `slice (…, `*method*`)` | [Function File] |
| `slice (`*hax*`, …)` | [Function File] |
| `h = slice (…)` | [Function File] |

Plot slices of 3-D data/scalar fields.

Each element of the 3-dimensional array *v* represents a scalar value at a location given by the parameters *x*, *y*, and *z*. The parameters *x*, *x*, and *z* are either 3-dimensional arrays of the same size as the array *v* in the `"meshgrid"` format or vectors. The parameters *xi*, etc. respect a similar format to *x*, etc., and they represent the points at which the array *vi* is interpolated using interp3. The vectors *sx*, *sy*, and *sz* contain points of orthogonal slices of the respective axes.

# Chapter 15: Plotting

If *x*, *y*, *z* are omitted, they are assumed to be x = 1:size (v, 2), y = 1:size (v, 1) and z = 1:size (v, 3).

*method* is one of:

"nearest"
: Return the nearest neighbor.

"linear"
: Linear interpolation from nearest neighbors.

"cubic"
: Cubic interpolation from four nearest neighbors (not implemented yet).

"spline"
: Cubic spline interpolation—smooth first and second derivatives throughout the curve.

The default method is "linear".

If the first argument *hax* is an axes handle, then plot into this axis, rather than the current axes returned by gca.

The optional return value *h* is a graphics handle to the created surface object.

Examples:
```
[x, y, z] = meshgrid (linspace (-8, 8, 32));
v = sin (sqrt (x.^2 + y.^2 + z.^2)) ./ (sqrt (x.^2 + y.^2 + z.^2));
slice (x, y, z, v, [], 0, []);

[xi, yi] = meshgrid (linspace (-7, 7));
zi = xi + yi;
slice (x, y, z, v, xi, yi, zi);
```

**See also:** [interp3], page 658, [surface], page 356, [pcolor], page 296.

---

ribbon (*y*)                                                                                                          [Function File]
ribbon (*x*, *y*)                                                                        [Function File]
ribbon (*x*, *y*, *width*)                                                  [Function File]
ribbon (*hax*, ...)                                                  [Function File]
*h* = ribbon (...)                                                 [Function File]

Draw a ribbon plot for the columns of *y* vs. *x*.

The optional parameter *width* specifies the width of a single ribbon (default is 0.75). If *x* is omitted, a vector containing the row numbers is assumed (1:rows (Y)).

If the first argument *hax* is an axes handle, then plot into this axis, rather than the current axes returned by gca.

The optional return value *h* is a vector of graphics handles to the surface objects representing each ribbon.

**See also:** [surface], page 356, [waterfall], page 320.

---

shading (*type*)                                                          [Function File]
shading (*hax*, *type*)                                               [Function File]

Set the shading of patch or surface graphic objects.

Valid arguments for *type* are

"flat"
: Single colored patches with invisible edges.

"faceted"
: Single colored patches with visible edges.

"interp"
: Color between patch vertices are interpolated and the patch edges are invisible.

If the first argument *hax* is an axes handle, then plot into this axis, rather than the current axes returned by `gca`.

**See also:** [fill], page 297, [mesh], page 306, [patch], page 355, [pcolor], page 296, [surf], page 308, [surface], page 356, [hidden], page 308.

scatter3 (*x, y, z*)         [Function File]
scatter3 (*x, y, z, s*)         [Function File]
scatter3 (*x, y, z, s, c*)         [Function File]
scatter3 (..., *style*)         [Function File]
scatter3 (..., "filled")         [Function File]
scatter3 (..., *prop, val*)         [Function File]
scatter3 (*hax*, ...)         [Function File]
h = scatter3 (...)         [Function File]

Draw a 3-D scatter plot.

A marker is plotted at each point defined by the coordinates in the vectors *x, y,* and *z*.

The size of the markers is determined by *s*, which can be a scalar or a vector of the same length as *x, y,* and *z*. If *s* is not given, or is an empty matrix, then a default value of 8 points is used.

The color of the markers is determined by *c*, which can be a string defining a fixed color; a 3-element vector giving the red, green, and blue components of the color; a vector of the same length as *x* that gives a scaled index into the current colormap; or an Nx3 matrix defining the RGB color of each marker individually.

The marker to use can be changed with the *style* argument, that is a string defining a marker in the same manner as the `plot` command. If no marker is specified it defaults to "o" or circles. If the argument "filled" is given then the markers are filled.

Additional property/value pairs are passed directly to the underlying patch object.

If the first argument *hax* is an axes handle, then plot into this axis, rather than the current axes returned by `gca`.

The optional return value *h* is a graphics handle to the hggroup object representing the points.

```
[x, y, z] = peaks (20);
scatter3 (x(:), y(:), z(:), [], z(:));
```

**See also:** [scatter], page 283, [patch], page 355, [plot], page 272.

waterfall (*x, y, z*)         [Function File]
waterfall (*z*)         [Function File]
waterfall (..., *c*)         [Function File]
waterfall (..., *prop, val*, ...)         [Function File]
waterfall (*hax*, ...)         [Function File]

Chapter 15: Plotting 321

*h* = waterfall (...)  [Function File]

Plot a 3-D waterfall plot.

A waterfall plot is similar to a `meshz` plot except only mesh lines for the rows of *z* (x-values) are shown.

The wireframe mesh is plotted using rectangles. The vertices of the rectangles [*x*, *y*] are typically the output of `meshgrid`. over a 2-D rectangular region in the x-y plane. *z* determines the height above the plane of each vertex. If only a single *z* matrix is given, then it is plotted over the meshgrid *x* = 1:columns (*z*), *y* = 1:rows (*z*). Thus, columns of *z* correspond to different *x* values and rows of *z* correspond to different *y* values.

The color of the mesh is computed by linearly scaling the *z* values to fit the range of the current colormap. Use `caxis` and/or change the colormap to control the appearance.

Optionally the color of the mesh can be specified independently of *z* by supplying a color matrix, *c*.

Any property/value pairs are passed directly to the underlying surface object.

If the first argument *hax* is an axes handle, then plot into this axis, rather than the current axes returned by `gca`.

The optional return value *h* is a graphics handle to the created surface object.

**See also:** [meshz], page 307, [mesh], page 306, [meshc], page 307, [contour], page 285, [surf], page 308, [surface], page 356, [ribbon], page 319, [meshgrid], page 316, [hidden], page 308, [shading], page 319, [colormap], page 698, [caxis], page 300.

### 15.2.2.1 Aspect Ratio

For three-dimensional plots the aspect ratio can be set for data with `daspect` and for the plot box with `pbaspect`. See Section 15.2.1.1 [Axis Configuration], page 299, for controlling the x-, y-, and z-limits for plotting.

*data_aspect_ratio* = daspect ()  [Function File]
daspect (*data_aspect_ratio*)  [Function File]
daspect (*mode*)  [Function File]
*data_aspect_ratio_mode* = daspect ("*mode*")  [Function File]
daspect (*hax*, ...)  [Function File]

Query or set the data aspect ratio of the current axes.

The aspect ratio is a normalized 3-element vector representing the span of the x, y, and z-axis limits.

daspect (*mode*)

Set the data aspect ratio mode of the current axes. *mode* is either "auto" or "manual".

daspect ("mode")

Return the data aspect ratio mode of the current axes.

daspect (*hax*, ...)

Operate on the axes in handle *hax* instead of the current axes.

**See also:** [axis], page 299, [pbaspect], page 322, [xlim], page 301, [ylim], page 301, [zlim], page 301.

`plot_box_aspect_ratio = pbaspect ( )`                    [Function File]
`pbaspect (plot_box_aspect_ratio)`                        [Function File]
`pbaspect (mode)`                                         [Function File]
`plot_box_aspect_ratio_mode = pbaspect ("mode")`          [Function File]
`pbaspect (hax, ...)`                                     [Function File]

Query or set the plot box aspect ratio of the current axes.

The aspect ratio is a normalized 3-element vector representing the rendered lengths of the x, y, and z axes.

`pbaspect(mode)`

Set the plot box aspect ratio mode of the current axes. *mode* is either "auto" or "manual".

`pbaspect ("mode")`

Return the plot box aspect ratio mode of the current axes.

`pbaspect (hax, ...)`

Operate on the axes in handle *hax* instead of the current axes.

**See also:** [axis], page 299, [daspect], page 321, [xlim], page 301, [ylim], page 301, [zlim], page 301.

### 15.2.2.2 Three-dimensional Function Plotting

`ezplot3 (fx, fy, fz)`                                    [Function File]
`ezplot3 (..., dom)`                                      [Function File]
`ezplot3 (..., n)`                                        [Function File]
`ezplot3 (hax, ...)`                                      [Function File]
`h = ezplot3 (...)`                                       [Function File]

Plot a parametrically defined curve in three dimensions.

*fx*, *fy*, and *fz* are strings, inline functions, or function handles with one argument defining the function. By default the plot is over the domain `0 <= t <= 2*pi` with 500 points.

If *dom* is a two element vector, it represents the minimum and maximum values of *t*.

*n* is a scalar defining the number of points to use in plotting the function.

If the first argument *hax* is an axes handle, then plot into this axis, rather than the current axes returned by `gca`.

The optional return value *h* is a graphics handle to the created plot.

```
fx = @(t) cos (t);
fy = @(t) sin (t);
fz = @(t) t;
ezplot3 (fx, fy, fz, [0, 10*pi], 100);
```

**See also:** [plot3], page 317, [ezplot], page 302, [ezmesh], page 322, [ezsurf], page 324.

`ezmesh (f)`                                              [Function File]
`ezmesh (fx, fy, fz)`                                     [Function File]
`ezmesh (..., dom)`                                       [Function File]
`ezmesh (..., n)`                                         [Function File]

Chapter 15: Plotting

| | |
|---|---:|
| ezmesh (..., "*circ*") | [Function File] |
| ezmesh (*hax*, ...) | [Function File] |
| h = ezmesh (...) | [Function File] |

Plot the mesh defined by a function.

*f* is a string, inline function, or function handle with two arguments defining the function. By default the plot is over the meshed domain -2*pi <= x | y <= 2*pi with 60 points in each dimension.

If three functions are passed, then plot the parametrically defined function [fx (s, t), fy (s, t), fz (s, t)].

If *dom* is a two element vector, it represents the minimum and maximum values of both *x* and *y*. If *dom* is a four element vector, then the minimum and maximum values are [xmin xmax ymin ymax].

*n* is a scalar defining the number of points to use in each dimension.

If the argument "circ" is given, then the function is plotted over a disk centered on the middle of the domain *dom*.

If the first argument *hax* is an axes handle, then plot into this axis, rather than the current axes returned by gca.

The optional return value *h* is a graphics handle to the created surface object.

Example 1: 2-argument function

```
f = @(x,y) sqrt (abs (x .* y)) ./ (1 + x.^2 + y.^2);
ezmesh (f, [-3, 3]);
```

Example 2: parametrically defined function

```
fx = @(s,t) cos (s) .* cos (t);
fy = @(s,t) sin (s) .* cos (t);
fz = @(s,t) sin (t);
ezmesh (fx, fy, fz, [-pi, pi, -pi/2, pi/2], 20);
```

**See also:** [mesh], page 306, [ezmeshc], page 323, [ezplot], page 302, [ezsurf], page 324, [ezsurfc], page 325, [hidden], page 308.

| | |
|---|---:|
| ezmeshc (*f*) | [Function File] |
| ezmeshc (*fx, fy, fz*) | [Function File] |
| ezmeshc (..., *dom*) | [Function File] |
| ezmeshc (..., *n*) | [Function File] |
| ezmeshc (..., "*circ*") | [Function File] |
| ezmeshc (*hax*, ...) | [Function File] |
| h = ezmeshc (...) | [Function File] |

Plot the mesh and contour lines defined by a function.

*f* is a string, inline function, or function handle with two arguments defining the function. By default the plot is over the meshed domain -2*pi <= x | y <= 2*pi with 60 points in each dimension.

If three functions are passed, then plot the parametrically defined function [fx (s, t), fy (s, t), fz (s, t)].

If *dom* is a two element vector, it represents the minimum and maximum values of both *x* and *y*. If *dom* is a four element vector, then the minimum and maximum values are [xmin xmax ymin ymax].

*n* is a scalar defining the number of points to use in each dimension.

If the argument "circ" is given, then the function is plotted over a disk centered on the middle of the domain *dom*.

If the first argument *hax* is an axes handle, then plot into this axis, rather than the current axes returned by gca.

The optional return value *h* is a 2-element vector with a graphics handle for the created mesh plot and a second handle for the created contour plot.

Example: 2-argument function
```
f = @(x,y) sqrt (abs (x .* y)) ./ (1 + x.^2 + y.^2);
ezmeshc (f, [-3, 3]);
```

**See also:** [meshc], page 307, [ezmesh], page 322, [ezplot], page 302, [ezsurf], page 324, [ezsurfc], page 325, [hidden], page 308.

ezsurf (*f*)   [Function File]
ezsurf (*fx, fy, fz*)   [Function File]
ezsurf (..., *dom*)   [Function File]
ezsurf (..., *n*)   [Function File]
ezsurf (..., "*circ*")   [Function File]
ezsurf (*hax*, ...)   [Function File]
*h* = ezsurf (...)   [Function File]

Plot the surface defined by a function.

*f* is a string, inline function, or function handle with two arguments defining the function. By default the plot is over the meshed domain -2*pi <= x | y <= 2*pi with 60 points in each dimension.

If three functions are passed, then plot the parametrically defined function [*fx (s, t), fy (s, t), fz (s, t)*].

If *dom* is a two element vector, it represents the minimum and maximum values of both *x* and *y*. If *dom* is a four element vector, then the minimum and maximum values are [xmin xmax ymin ymax].

*n* is a scalar defining the number of points to use in each dimension.

If the argument "circ" is given, then the function is plotted over a disk centered on the middle of the domain *dom*.

If the first argument *hax* is an axes handle, then plot into this axis, rather than the current axes returned by gca.

The optional return value *h* is a graphics handle to the created surface object.

Example 1: 2-argument function
```
f = @(x,y) sqrt (abs (x .* y)) ./ (1 + x.^2 + y.^2);
ezsurf (f, [-3, 3]);
```

Example 2: parametrically defined function

Chapter 15: Plotting 325

```
fx = @(s,t) cos (s) .* cos (t);
fy = @(s,t) sin (s) .* cos (t);
fz = @(s,t) sin (t);
ezsurf (fx, fy, fz, [-pi, pi, -pi/2, pi/2], 20);
```

**See also:** [surf], page 308, [ezsurfc], page 325, [ezplot], page 302, [ezmesh], page 322, [ezmeshc], page 323, [shading], page 319.

| | |
|---|---|
| ezsurfc (*f*) | [Function File] |
| ezsurfc (*fx, fy, fz*) | [Function File] |
| ezsurfc (..., *dom*) | [Function File] |
| ezsurfc (..., *n*) | [Function File] |
| ezsurfc (..., "*circ*") | [Function File] |
| ezsurfc (*hax*, ...) | [Function File] |
| h = ezsurfc (...) | [Function File] |

Plot the surface and contour lines defined by a function.

*f* is a string, inline function, or function handle with two arguments defining the function. By default the plot is over the meshed domain -2*pi <= x | y <= 2*pi with 60 points in each dimension.

If three functions are passed, then plot the parametrically defined function [*fx* (*s, t*), *fy* (*s, t*), *fz* (*s, t*)].

If *dom* is a two element vector, it represents the minimum and maximum values of both *x* and *y*. If *dom* is a four element vector, then the minimum and maximum values are [xmin xmax ymin ymax].

*n* is a scalar defining the number of points to use in each dimension.

If the argument "circ" is given, then the function is plotted over a disk centered on the middle of the domain *dom*.

If the first argument *hax* is an axes handle, then plot into this axis, rather than the current axes returned by gca.

The optional return value *h* is a 2-element vector with a graphics handle for the created surface plot and a second handle for the created contour plot.

Example:

```
f = @(x,y) sqrt (abs (x .* y)) ./ (1 + x.^2 + y.^2);
ezsurfc (f, [-3, 3]);
```

**See also:** [surfc], page 309, [ezsurf], page 324, [ezplot], page 302, [ezmesh], page 322, [ezmeshc], page 323, [shading], page 319.

### 15.2.2.3 Three-dimensional Geometric Shapes

| | |
|---|---|
| cylinder | [Command] |
| cylinder (*r*) | [Function File] |
| cylinder (*r, n*) | [Function File] |
| cylinder (*hax*, ...) | [Function File] |
| [x, y, z] = cylinder (...) | [Function File] |

Plot a 3-D unit cylinder.

The optional input *r* is a vector specifying the radius along the unit z-axis. The default is [1 1] indicating radius 1 at Z == 0 and at Z == 1.

The optional input *n* determines the number of faces around the circumference of the cylinder. The default value is 20.

If the first argument *hax* is an axes handle, then plot into this axis, rather than the current axes returned by `gca`.

If outputs are requested `cylinder` returns three matrices in `meshgrid` format, such that `surf (x, y, z)` generates a unit cylinder.

Example:

```
[x, y, z] = cylinder (10:-1:0, 50);
surf (x, y, z);
title ("a cone");
```

See also: [ellipsoid], page 326, [rectangle], page 304, [sphere], page 326.

**sphere ()** [Function File]
**sphere (*n*)** [Function File]
**sphere (*hax*, ...)** [Function File]
**[*x*, *y*, *z*] = sphere (...)** [Function File]

Plot a 3-D unit sphere.

The optional input *n* determines the number of faces around the circumference of the sphere. The default value is 20.

If the first argument *hax* is an axes handle, then plot into this axis, rather than the current axes returned by `gca`.

If outputs are requested `sphere` returns three matrices in `meshgrid` format such that `surf (x, y, z)` generates a unit sphere.

Example:

```
[x, y, z] = sphere (40);
surf (3*x, 3*y, 3*z);
axis equal;
title ("sphere of radius 3");
```

See also: [cylinder], page 325, [ellipsoid], page 326, [rectangle], page 304.

**ellipsoid (*xc*, *yc*, *zc*, *xr*, *yr*, *zr*, *n*)** [Function File]
**ellipsoid (..., *n*)** [Function File]
**ellipsoid (*hax*, ...)** [Function File]
**[*x*, *y*, *z*] = ellipsoid (...)** [Function File]

Plot a 3-D ellipsoid.

The inputs *xc*, *yc*, *zc* specify the center of the ellipsoid. The inputs *xr*, *yr*, *zr* specify the semi-major axis lengths.

The optional input *n* determines the number of faces around the circumference of the cylinder. The default value is 20.

If the first argument *hax* is an axes handle, then plot into this axis, rather than the current axes returned by `gca`.

Chapter 15: Plotting                                                              327

If outputs are requested `ellipsoid` returns three matrices in `meshgrid` format, such that surf (x, y, z) generates the ellipsoid.

**See also:** [cylinder], page 325, [rectangle], page 304, [sphere], page 326.

### 15.2.3 Plot Annotations

You can add titles, axis labels, legends, and arbitrary text to an existing plot. For example:

```
x = -10:0.1:10;
plot (x, sin (x));
title ("sin(x) for x = -10:0.1:10");
xlabel ("x");
ylabel ("sin (x)");
text (pi, 0.7, "arbitrary text");
legend ("sin (x)");
```

The functions `grid` and `box` may also be used to add grid and border lines to the plot. By default, the grid is off and the border lines are on.

Finally, arrows, text and rectangular or elliptic boxes can be added to highlight parts of a plot using the `annotation` function. Those objects are drawn in an invisible axes, on top of every other axes.

title (*string*)                                                         [Function File]
title (*string*, *prop*, *val*, ...)                                     [Function File]
title (*hax*, ...)                                                       [Function File]
h = title (...)                                                          [Function File]

Specify the string used as a title for the current axis.

An optional list of *property/value* pairs can be used to change the appearance of the created title text object.

If the first argument *hax* is an axes handle, then plot into this axis, rather than the current axes returned by `gca`.

The optional return value *h* is a graphics handle to the created text object.

**See also:** [xlabel], page 329, [ylabel], page 329, [zlabel], page 329, [text], page 329.

legend (*str1*, *str2*, ...)                                             [Function File]
legend (*matstr*)                                                        [Function File]
legend (*cellstr*)                                                       [Function File]
legend (..., "*location*", *pos*)                                        [Function File]
legend (..., "*orientation*", *orient*)                                  [Function File]
legend (*hax*, ...)                                                      [Function File]
legend (*hobjs*, ...)                                                    [Function File]
legend (*hax*, *hobjs*, ...)                                             [Function File]
legend ("*option*")                                                      [Function File]
[*hleg*, *hleg_obj*, *hplot*, *labels*] = legend (...)                   [Function File]

Display a legend for the current axes using the specified strings as labels.

Legend entries may be specified as individual character string arguments, a character array, or a cell array of character strings.

If the first argument *hax* is an axes handle, then plot into this axis, rather than the current axes returned by `gca`. If the handles, *hobjs*, are not specified then the legend's strings will be associated with the axes' descendants. `legend` works on line graphs, bar graphs, etc. A plot must exist before legend is called.

The optional parameter *pos* specifies the location of the legend as follows:

| pos | location of the legend |
|---|---|
| north | center top |
| south | center bottom |
| east | right center |
| west | left center |
| northeast | right top (default) |
| northwest | left top |
| southeast | right bottom |
| southwest | left bottom |
| outside | can be appended to any location string |

The optional parameter *orient* determines if the key elements are placed vertically or horizontally. The allowed values are `"vertical"` (default) or `"horizontal"`.

The following customizations are available using *option*:

`"show"`     Show legend on the plot

`"hide"`     Hide legend on the plot

`"toggle"`     Toggles between `"hide"` and `"show"`

`"boxon"`     Show a box around legend (default)

`"boxoff"`     Hide the box around legend

`"right"`     Place label text to the right of the keys (default)

`"left"`     Place label text to the left of the keys

`"off"`     Delete the legend object

The optional output values are

*hleg*     The graphics handle of the legend object.

*hleg_obj*     Graphics handles to the text and line objects which make up the legend.

*hplot*     Graphics handles to the plot objects which were used in making the legend.

*labels*     A cell array of strings of the labels in the legend.

The legend label text is either provided in the call to `legend` or is taken from the DisplayName property of graphics objects. If no labels or DisplayNames are available, then the label text is simply `"data1"`, `"data2"`, ..., `"dataN"`.

Implementation Note: A legend is implemented as an additional axes object of the current figure with the `"tag"` set to `"legend"`. Properties of the legend object may be manipulated directly by using `set`.

Chapter 15: Plotting                                                                 329

text (*x*, *y*, *string*)                                                       [Function File]
text (*x*, *y*, *z*, *string*)                                                  [Function File]
text (..., *prop*, *val*, ...)                                                  [Function File]
h = text (...)                                                                  [Function File]
> Create a text object with text *string* at position x, y, (z) on the current axes.
>
> Multiple locations can be specified if x, y, (z) are vectors. Multiple strings can be specified with a character matrix or a cell array of strings.
>
> Optional property/value pairs may be used to control the appearance of the text.
>
> The optional return value h is a vector of graphics handles to the created text objects.
>
> **See also:** [gtext], page 350, [title], page 327, [xlabel], page 329, [ylabel], page 329, [zlabel], page 329.

See Section 15.3.3.5 [Text Properties], page 374 for the properties that you can set.

xlabel (*string*)                                                               [Function File]
xlabel (*string*, *property*, *val*, ...)                                       [Function File]
xlabel (*hax*, ...)                                                             [Function File]
h = xlabel (...)                                                                [Function File]
> Specify the string used to label the x-axis of the current axis.
>
> An optional list of *property/value* pairs can be used to change the properties of the created text label.
>
> If the first argument *hax* is an axes handle, then operate on this axis rather than the current axes returned by `gca`.
>
> The optional return value h is a graphics handle to the created text object.
>
> **See also:** [ylabel], page 329, [zlabel], page 329, [datetick], page 754, [title], page 327, [text], page 329.

clabel (*c*, *h*)                                                               [Function File]
clabel (*c*, *h*, *v*)                                                          [Function File]
clabel (*c*, *h*, "*manual*")                                                   [Function File]
clabel (*c*)                                                                    [Function File]
clabel (..., *prop*, *val*, ...)                                                [Function File]
h = clabel (...)                                                                [Function File]
> Add labels to the contours of a contour plot.
>
> The contour levels are specified by the contour matrix c which is returned by `contour`, `contourc`, `contourf`, and `contour3`. Contour labels are rotated to match the local line orientation and centered on the line. The position of labels along the contour line is chosen randomly.
>
> If the argument h is a handle to a contour group object, then label this plot rather than the one in the current axes returned by `gca`.
>
> By default, all contours are labeled. However, the contours to label can be specified by the vector v. If the "manual" argument is given then the contours to label can be selected with the mouse.

Additional property/value pairs that are valid properties of text objects can be given and are passed to the underlying text objects. Moreover, the contour group property "LabelSpacing" is available which determines the spacing between labels on a contour to be specified. The default is 144 points, or 2 inches.

The optional return value h is a vector of graphics handles to the text objects representing each label. The "userdata" property of the text objects contains the numerical value of the contour label.

An example of the use of clabel is

```
[c, h] = contour (peaks (), -4 : 6);
clabel (c, h, -4:2:6, "fontsize", 12);
```

See also: [contour], page 285, [contourf], page 286, [contour3], page 287, [meshc], page 307, [surfc], page 309, [text], page 329.

box [Command]
box *on* [Command]
box *off* [Command]
box (*hax*, ...) [Function File]

Control display of the axis border.

The argument may be either "on" or "off". If it is omitted, the current box state is toggled.

If the first argument *hax* is an axes handle, then operate on this axis rather than the current axes returned by gca.

See also: [axis], page 299, [grid], page 330.

grid [Command]
grid *on* [Command]
grid *off* [Command]
grid *minor* [Command]
grid *minor on* [Command]
grid *minor off* [Command]
grid (*hax*, ...) [Function File]

Control the display of plot grid lines.

The function state input may be either "on" or "off". If it is omitted, the current grid state is toggled.

When the first argument is "minor" all subsequent commands modify the minor grid rather than the major grid.

If the first argument *hax* is an axes handle, then operate on this axis rather than the current axes returned by gca.

To control the grid lines for an individual axis use the set function. For example:

```
set (gca, "ygrid", "on");
```

See also: [axis], page 299, [box], page 330.

colorbar [Command]
colorbar (*loc*) [Function File]

Chapter 15: Plotting                                                              331

colorbar (*delete_option*)                                                [Function File]
colorbar (*hcb*, ...)                                                     [Function File]
colorbar (*hax*, ...)                                                     [Function File]
colorbar (..., "*peer*", *hax*, ...)                                      [Function File]
colorbar (..., "*location*", *loc*, ...)                                  [Function File]
colorbar (..., *prop*, *val*, ...)                                        [Function File]
h = colorbar (...)                                                        [Function File]

> Add a colorbar to the current axes.
>
> A colorbar displays the current colormap along with numerical rulings so that the color scale can be interpreted.
>
> The optional input *loc* determines the location of the colorbar. Valid values for *loc* are
>
> "EastOutside"
> : Place the colorbar outside the plot to the right. This is the default.
>
> "East"    Place the colorbar inside the plot to the right.
>
> "WestOutside"
> : Place the colorbar outside the plot to the left.
>
> "West"    Place the colorbar inside the plot to the left.
>
> "NorthOutside"
> : Place the colorbar above the plot.
>
> "North"   Place the colorbar at the top of the plot.
>
> "SouthOutside"
> : Place the colorbar under the plot.
>
> "South"   Place the colorbar at the bottom of the plot.
>
> To remove a colorbar from a plot use any one of the following keywords for the *delete_option*: "delete", "hide", "off".
>
> If the argument "peer" is given, then the following argument is treated as the axes handle in which to add the colorbar. Alternatively, If the first argument *hax* is an axes handle, then the colorbar is added to this axis, rather than the current axes returned by gca.
>
> If the first argument *hcb* is a handle to a colorbar object, then operate on this colorbar directly.
>
> Additional property/value pairs are passed directly to the underlying axes object.
>
> The optional return value *h* is a graphics handle to the created colorbar object.
>
> Implementation Note: A colorbar is created as an additional axes to the current figure with the "tag" property set to "colorbar". The created axes object has the extra property "location" which controls the positioning of the colorbar.
>
> **See also:** [colormap], page 698.

annotation (*type*)                                                       [Function File]
annotation ("*line*", *x*, *y*)                                           [Function File]

`annotation ("`*arrow*`", x, y)` [Function File]
`annotation ("`*doublearrow*`", x, y)` [Function File]
`annotation ("`*textarrow*`", x, y)` [Function File]
`annotation ("`*textbox*`", pos)` [Function File]
`annotation ("`*rectangle*`", pos)` [Function File]
`annotation ("`*ellipse*`", pos)` [Function File]
`annotation (..., prop, val)` [Function File]
`annotation (hf, ...)` [Function File]
`h = annotation (...)` [Function File]

Draw annotations to emphasize parts of a figure.

You may build a default annotation by specifying only the *type* of the annotation.

Otherwise you can select the type of annotation and then set its position using either *x* and *y* coordinates for line-based annotations or a position vector *pos* for others. In either case, coordinates are interpreted using the `"units"` property of the annotation object. The default is `"normalized"`, which means the lower left hand corner of the figure has coordinates '`[0 0]`' and the upper right hand corner '`[1 1]`'.

If the first argument *hf* is a figure handle, then plot into this figure, rather than the current figure returned by `gcf`.

Further arguments can be provided in the form of *prop/val* pairs to customize the annotation appearance.

The optional return value *h* is a graphics handle to the created annotation object. This can be used with the `set` function to customize an existing annotation object.

All annotation objects share two properties:

- `"units"`: the units in which coordinates are interpreted.
  Its value may be one of `"centimeters"` | `"characters"` | `"inches"` | `"{normalized}"` | `"pixels"` | `"points"`.
- `"position"`: a four-element vector [x0 y0 width height].
  The vector specifies the coordinates (x0,y0) of the origin of the annotation object, its width, and its height. The width and height may be negative, depending on the orientation of the object.

Valid annotation types and their specific properties are described below:

`"line"`     Constructs a line. *x* and *y* must be two-element vectors specifying the x and y coordinates of the two ends of the line.

The line can be customized using `"linewidth"`, `"linestyle"`, and `"color"` properties the same way as for `line` objects.

`"arrow"`     Construct an arrow. The second point in vectors *x* and *y* specifies the arrowhead coordinates.

Besides line properties, the arrowhead can be customized using `"headlength"`, `"headwidth"`, and `"headstyle"` properties. Supported values for `"headstyle"` property are: [`"diamond"` | `"ellipse"` | `"plain"` | `"rectangle"` | `"vback1"` | `"{vback2}"` | `"vback3"`]

`"doublearrow"`

Construct a double arrow. Vectors *x* and *y* specify the arrowhead coordinates.

Chapter 15: Plotting                                                            333

> The line and the arrowhead can be customized as for arrow annotations, but some property names are duplicated: `"head1length"`/`"head2length"`, `"head1width"`/`"head2width"`, etc. The index 1 marks the properties of the arrowhead at the first point in *x* and *y* coordinates.

`"textarrow"`
> Construct an arrow with a text label at the opposite end from the arrowhead.
>
> The line and the arrowhead can be customized as for arrow annotations, and the text can be customized using the same properties as `text` graphics objects. Note, however, that some text property names are prefixed with "text" to distinguish them from arrow properties: `"textbackgroundcolor"`, `"textcolor"`, `"textedgecolor"`, `"textlinewidth"`, `"textmargin"`, `"textrotation"`.

`"textbox"`
> Construct a box with text inside. *pos* specifies the `"position"` property of the annotation.
>
> You may use `"backgroundcolor"`, `"edgecolor"`, `"linestyle"`, and `"linewidth"` properties to customize the box background color and edge appearance. A limited set of `text` objects properties are also available; Besides `"font..."` properties, you may also use `"horizontalalignment"` and `"verticalalignment"` to position the text inside the box.
>
> Finally, the `"fitboxtotext"` property controls the actual extent of the box. If `"on"` (the default) the box limits are fitted to the text extent.

`"rectangle"`
> Construct a rectangle. *pos* specifies the `"position"` property of the annotation.
>
> You may use `"facecolor"`, `"color"`, `"linestyle"`, and `"linewidth"` properties to customize the rectangle background color and edge appearance.

`"ellipse"`
> Construct an ellipse. *pos* specifies the `"position"` property of the annotation.
>
> See `"rectangle"` annotations for customization.

**See also:** [xlabel], page 329, [ylabel], page 329, [zlabel], page 329, [title], page 327, [text], page 329, [gtext], page 350, [legend], page 327, [colorbar], page 330.

## 15.2.4 Multiple Plots on One Page

Octave can display more than one plot in a single figure. The simplest way to do this is to use the `subplot` function to divide the plot area into a series of subplot windows that are indexed by an integer. For example,

```
subplot (2, 1, 1)
fplot (@sin, [-10, 10]);
subplot (2, 1, 2)
fplot (@cos, [-10, 10]);
```

creates a figure with two separate axes, one displaying a sine wave and the other a cosine wave. The first call to subplot divides the figure into two plotting areas (two rows and one column) and makes the first plot area active. The grid of plot areas created by `subplot` is numbered in column-major order (top to bottom, left to right).

**subplot** (*rows*, *cols*, *index*)  [Function File]
**subplot** (*rcn*)  [Function File]
**subplot** (*hax*)  [Function File]
**subplot** (..., "*align*")  [Function File]
**subplot** (..., "*replace*")  [Function File]
**subplot** (..., "*position*", *pos*)  [Function File]
**subplot** (..., *prop*, *val*, ...)  [Function File]
*hax* = **subplot** (...)  [Function File]

Set up a plot grid with *rows* by *cols* subwindows and set the current axes for plotting (`gca`) to the location given by *index*.

If only one numeric argument is supplied, then it must be a three digit value specifying the number of rows in digit 1, the number of columns in digit 2, and the plot index in digit 3.

The plot index runs row-wise; First, all columns in a row are numbered and then the next row is filled.

For example, a plot with 2x3 grid will have plot indices running as follows:

| 1 | 2 | 3 |
|---|---|---|
| 4 | 5 | 6 |

*index* may also be a vector. In this case, the new axis will enclose the grid locations specified. The first demo illustrates this:

```
demo ("subplot", 1)
```

The index of the subplot to make active may also be specified by its axes handle, *hax*, returned from a previous `subplot` command.

If the option `"align"` is given then the plot boxes of the subwindows will align, but this may leave no room for axis tick marks or labels.

If the option `"replace"` is given then the subplot axis will be reset, rather than just switching the current axis for plotting to the requested subplot.

The `"position"` property can be used to exactly position the subplot axes within the current figure. The option *pos* is a 4-element vector [x, y, width, height] that determines the location and size of the axes. The values in *pos* are normalized in the range [0,1].

Any property/value pairs are passed directly to the underlying axes object.

# Chapter 15: Plotting

If the output *hax* is requested, subplot returns the axis handle for the subplot. This is useful for modifying the properties of a subplot using `set`.

**See also:** [axes], page 355, [plot], page 272, [gca], page 358, [set], page 359.

### 15.2.5 Multiple Plot Windows

You can open multiple plot windows using the `figure` function. For example,

```
figure (1);
fplot (@sin, [-10, 10]);
figure (2);
fplot (@cos, [-10, 10]);
```

creates two figures, with the first displaying a sine wave and the second a cosine wave. Figure numbers must be positive integers.

| | |
|---|---|
| `figure` | [Command] |
| `figure n` | [Command] |
| `figure (n)` | [Function File] |
| `figure (..., "property", value, ...)` | [Function File] |
| `h = figure (...)` | [Function File] |

Create a new figure window for plotting.

If no arguments are specified, a new figure with the next available number is created.

If called with an integer *n*, and no such numbered figure exists, then a new figure with the specified number is created. If the figure already exists then it is made visible and becomes the current figure for plotting.

Multiple property-value pairs may be specified for the figure object, but they must appear in pairs.

The optional return value *h* is a graphics handle to the created figure object.

**See also:** [axes], page 355, [gcf], page 357, [clf], page 339, [close], page 340.

### 15.2.6 Manipulation of Plot Objects

| | |
|---|---|
| `pan` | [Command] |
| `pan` *on* | [Command] |
| `pan` *off* | [Command] |
| `pan` *xon* | [Command] |
| `pan` *yon* | [Command] |
| `pan (hfig, option)` | [Function File] |

Control the interactive panning mode of a figure in the GUI.

Given the option `"on"` or `"off"`, set the interactive pan mode on or off.

With no arguments, toggle the current pan mode on or off.

Given the option `"xon"` or `"yon"`, enable pan mode for the x or y axis only.

If the first argument *hfig* is a figure, then operate on the given figure rather than the current figure as returned by `gcf`.

**See also:** [rotate3d], page 336, [zoom], page 336.

`rotate (h, dir, alpha)` [Function File]
`rotate (..., origin)` [Function File]
> Rotate the plot object *h* through *alpha* degrees around the line with direction *dir* and origin *origin*.
>
> The default value of *origin* is the center of the axes object that is the parent of *h*.
>
> If *h* is a vector of handles, they must all have the same parent axes object.
>
> Graphics objects that may be rotated are lines, surfaces, patches, and images.

`rotate3d` [Command]
`rotate3d` *on* [Command]
`rotate3d` *off* [Command]
`rotate3d (hfig, option)` [Function File]
> Control the interactive 3-D rotation mode of a figure in the GUI.
>
> Given the option `"on"` or `"off"`, set the interactive rotate mode on or off.
>
> With no arguments, toggle the current rotate mode on or off.
>
> If the first argument *hfig* is a figure, then operate on the given figure rather than the current figure as returned by `gcf`.
>
> **See also:** [pan], page 335, [zoom], page 336.

`zoom` [Command]
`zoom (factor)` [Command]
`zoom` *on* [Command]
`zoom` *off* [Command]
`zoom` *xon* [Command]
`zoom` *yon* [Command]
`zoom` *out* [Command]
`zoom` *reset* [Command]
`zoom (hfig, option)` [Command]
> Zoom the current axes object or control the interactive zoom mode of a figure in the GUI.
>
> Given a numeric argument greater than zero, zoom by the given factor. If the zoom factor is greater than one, zoom in on the plot. If the factor is less than one, zoom out. If the zoom factor is a two- or three-element vector, then the elements specify the zoom factors for the x, y, and z axes respectively.
>
> Given the option `"on"` or `"off"`, set the interactive zoom mode on or off.
>
> With no arguments, toggle the current zoom mode on or off.
>
> Given the option `"xon"` or `"yon"`, enable zoom mode for the x or y-axis only.
>
> Given the option `"out"`, zoom to the initial zoom setting.
>
> Given the option `"reset"`, store the current zoom setting so that `zoom out` will return to this zoom level.
>
> If the first argument *hfig* is a figure, then operate on the given figure rather than the current figure as returned by `gcf`.
>
> **See also:** [pan], page 335, [rotate3d], page 336.

Chapter 15: Plotting                                                          337

## 15.2.7 Manipulation of Plot Windows

By default, Octave refreshes the plot window when a prompt is printed, or when waiting for input. The `drawnow` function is used to cause a plot window to be updated.

drawnow ()                                                          [Built-in Function]
drawnow ("*expose*")                                                [Built-in Function]
drawnow (*term, file, mono, debug_file*)                            [Built-in Function]
    Update figure windows and their children.

    The event queue is flushed and any callbacks generated are executed.

    With the optional argument "`expose`", only graphic objects are updated and no other events or callbacks are processed.

    The third calling form of `drawnow` is for debugging and is undocumented.

    **See also:** [refresh], page 337.

Only figures that are modified will be updated. The `refresh` function can also be used to cause an update of the current figure, even if it is not modified.

refresh ()                                                          [Function File]
refresh (*h*)                                                       [Function File]
    Refresh a figure, forcing it to be redrawn.

    When called without an argument the current figure is redrawn. Otherwise, the figure with graphic handle *h* is redrawn.

    **See also:** [drawnow], page 337.

Normally, high-level plot functions like `plot` or `mesh` call `newplot` to initialize the state of the current axes so that the next plot is drawn in a blank window with default property settings. To have two plots superimposed over one another, use the `hold` function. For example,

```
hold on;
x = -10:0.1:10;
plot (x, sin (x));
plot (x, cos (x));
hold off;
```

displays sine and cosine waves on the same axes. If the hold state is off, consecutive plotting commands like this will only display the last plot.

newplot ()                                                          [Function File]
newplot (*hfig*)                                                    [Function File]
newplot (*hax*)                                                     [Function File]
*hax* = newplot (...)                                               [Function File]
    Prepare graphics engine to produce a new plot.

    This function is called at the beginning of all high-level plotting functions. It is not normally required in user programs. `newplot` queries the "`NextPlot`" field of the current figure and axis to determine what to do.

    **Figure NextPlot**          **Action**

| | |
|---|---|
| `"new"` | Create a new figure and make it the current figure. |
| `"add"` (default) | Add new graphic objects to the current figure. |
| `"replacechildren"` | Delete child objects whose HandleVisibility is set to `"on"`. Set NextPlot property to `"add"`. This typically clears a figure, but leaves in place hidden objects such as menubars. This is equivalent to `clf`. |
| `"replace"` | Delete all child objects of the figure and reset all figure properties to their defaults. However, the following four properties are not reset: Position, Units, PaperPosition, PaperUnits. This is equivalent to `clf reset`. |

| Axis NextPlot | Action |
|---|---|
| `"add"` | Add new graphic objects to the current axes. This is equivalent to `hold on`. |
| `"replacechildren"` | Delete child objects whose HandleVisibility is set to `"on"`, but leave axis properties unmodified. This typically clears a plot, but preserves special settings such as log scaling for axes. This is equivalent to `cla`. |
| `"replace"` (default) | Delete all child objects of the axis and reset all axis properties to their defaults. However, the following properties are not reset: Position, Units. This is equivalent to `cla reset`. |

If the optional input *hfig* or *hax* is given then prepare the specified figure or axes rather than the current figure and axes.

The optional return value *hax* is a graphics handle to the created axes object (not figure).

**Caution:** Calling `newplot` may change the current figure and current axis.

**hold**                                                                                                                                                                                                                                                                      [Command]
**hold** *on*                       [Command]
**hold** *off*                       [Command]
**hold** *all*                       [Command]
**hold** (*hax*, ...)                  [Function File]

Toggle or set the `"hold"` state of the plotting engine which determines whether new graphic objects are added to the plot or replace the existing objects.

    `hold on`     Retain plot data and settings so that subsequent plot commands are displayed on a single graph.

    `hold all`     Retain plot line color, line style, data, and settings so that subsequent plot commands are displayed on a single graph with the next line color and style.

    `hold off`     Restore default graphics settings which clear the graph and reset axis properties before each new plot command. (default).

# Chapter 15: Plotting

> hold        Toggle the current hold state.
>
> When given the additional argument *hax*, the hold state is modified for this axis rather than the current axes returned by `gca`.
>
> To query the current hold state use the `ishold` function.
>
> **See also:** [ishold], page 339, [cla], page 339, [clf], page 339, [newplot], page 337.

ishold                                                    [Command]
ishold (*hax*)                              [Function File]
ishold (*hfig*)                           [Function File]

> Return true if the next plot will be added to the current plot, or false if the plot device will be cleared before drawing the next plot.
>
> If the first argument is an axes handle *hax* or figure handle *hfig* then operate on this plot rather than the current one.
>
> **See also:** [hold], page 338, [newplot], page 337.

To clear the current figure, call the `clf` function. To clear the current axis, call the `cla` function. To bring the current figure to the top of the window stack, call the `shg` function. To delete a graphics object, call `delete` on its index. To close the figure window, call the `close` function.

clf                                                          [Command]
clf *reset*                                           [Command]
clf (*hfig*)                                       [Function File]
clf (*hfig*, "*reset*")                 [Function File]
h = clf (...)                           [Function File]

> Clear the current figure window.
>
> `clf` operates by deleting child graphics objects with visible handles (HandleVisibility = "on").
>
> If the optional argument "reset" is specified, delete all child objects including those with hidden handles and reset all figure properties to their defaults. However, the following properties are not reset: Position, Units, PaperPosition, PaperUnits.
>
> If the first argument *hfig* is a figure handle, then operate on this figure rather than the current figure returned by `gcf`.
>
> The optional return value *h* is the graphics handle of the figure window that was cleared.
>
> **See also:** [cla], page 339, [close], page 340, [delete], page 340, [reset], page 384.

cla                                                        [Command]
cla *reset*                                        [Command]
cla (*hax*)                                      [Function File]
cla (*hax*, "*reset*")               [Function File]

> Clear the current axes.
>
> `cla` operates by deleting child graphic objects with visible handles (HandleVisibility = "on").

If the optional argument `"reset"` is specified, delete all child objects including those with hidden handles and reset all axis properties to their defaults. However, the following properties are not reset: Position, Units.

If the first argument *hax* is an axes handle, then operate on this axis rather than the current axes returned by `gca`.

**See also:** [clf], page 339, [delete], page 340, [reset], page 384.

**shg** [Command]

Show the graph window.

Currently, this is the same as executing `drawnow`.

**See also:** [drawnow], page 337, [figure], page 335.

**delete (*file*)** [Function File]
**delete (*file1*, *file2*, ...)** [Function File]
**delete (*handle*)** [Function File]

Delete the named file or graphics handle.

*file* may contain globbing patterns such as '*'. Multiple files to be deleted may be specified in the same function call.

*handle* may be a scalar or vector of graphic handles to delete.

Programming Note: Deleting graphics objects is the proper way to remove features from a plot without clearing the entire figure.

**See also:** [clf], page 339, [cla], page 339, [unlink], page 755, [rmdir], page 756.

**close** [Command]
**close (*h*)** [Command]
**close *h*** [Command]
**close** all [Command]
**close** all hidden [Command]
**close** all force [Command]

Close figure window(s).

When called with no arguments, close the current figure. This is equivalent to `close (gcf)`. If the input *h* is a graphic handle, or vector of graphics handles, then close each figure in *h*.

If the argument `"all"` is given then all figures with visible handles (HandleVisibility = `"on"`) are closed.

If the argument `"all hidden"` is given then all figures, including hidden ones, are closed.

If the argument `"all force"` is given then all figures are closed even when `"closerequestfcn"` has been altered to prevent closing the window.

Implementation Note: `close` operates by calling the function specified by the `"closerequestfcn"` property for each figure. By default, the function `closereq` is used. It is possible that the function invoked will delay or abort removing the figure. To remove a figure without executing any callback functions use `delete`. When writing a callback function to close a window do not use `close` to avoid recursion.

**See also:** [closereq], page 341, [delete], page 340.

`closereq ()` [Function File]
> Close the current figure and delete all graphics objects associated with it.
>
> By default, the `"closerequestfcn"` property of a new plot figure points to this function.
>
> **See also:** [close], page 340, [delete], page 340.

### 15.2.8 Use of the `interpreter` Property

All text objects—such as titles, labels, legends, and text—include the property `"interpreter"` that determines the manner in which special control sequences in the text are rendered.

The interpreter property can take three values: `"none"`, `"tex"`, `"latex"`. If the interpreter is set to `"none"` then no special rendering occurs—the displayed text is a verbatim copy of the specified text. Currently, the `"latex"` interpreter is not implemented and is equivalent to `"none"`.

The `"tex"` option implements a subset of TeX functionality when rendering text. This allows the insertion of special glyphs such as Greek characters or mathematical symbols. The special characters are inserted with a code following a backslash (\) character, as shown in Table 15.1.

Note that for on-screen display the interpreter property is honored by all graphics toolkits. However for printing, **only** the `"gnuplot"` toolkit renders TeX instructions.

Besides special glyphs, the formatting of text can be changed within the string by using the codes

| | |
|---|---|
| \bf | Bold font |
| \it | Italic font |
| \sl | Oblique Font |
| \rm | Normal font |

These codes may be used in conjunction with the { and } characters to limit the change to just a part of the string. For example,

```
xlabel ('{\bf H} = a {\bf V}')
```

where the character 'a' will not appear in a bold font. Note that to avoid having Octave interpret the backslash characters in the strings, the strings should be in single quotes.

It is also possible to change the fontname and size within the text

| | |
|---|---|
| \fontname{fontname} | Specify the font to use |
| \fontsize{size} | Specify the size of the font to use |

Finally, superscripting and subscripting can be controlled with the '^' and '_' characters. If the '^' or '_' is followed by a { character, then all of the block surrounded by the { } pair is super- or sub-scripted. Without the { } pair, only the character immediately following the '^' or '_' is super- or sub-scripted.

### Greek Lowercase Letters

| Code | Sym | Code | Sym | Code | Sym |
|---|---|---|---|---|---|
| \alpha | $\alpha$ | \beta | $\beta$ | \gamma | $\gamma$ |
| \delta | $\delta$ | \epsilon | $\epsilon$ | \zeta | $\zeta$ |
| \eta | $\eta$ | \theta | $\theta$ | \vartheta | $\vartheta$ |
| \iota | $\iota$ | \kappa | $\kappa$ | \lambda | $\lambda$ |
| \mu | $\mu$ | \nu | $\nu$ | \xi | $\xi$ |
| \o | $o$ | \pi | $\pi$ | \varpi | $\varpi$ |
| \rho | $\rho$ | \sigma | $\sigma$ | \varsigma | $\varsigma$ |
| \tau | $\tau$ | \upsilon | $\upsilon$ | \phi | $\phi$ |
| \chi | $\chi$ | \psi | $\psi$ | \omega | $\omega$ |

### Greek Uppercase Letters

| Code | Sym | Code | Sym | Code | Sym |
|---|---|---|---|---|---|
| \Gamma | $\Gamma$ | \Delta | $\Delta$ | \Theta | $\Theta$ |
| \Lambda | $\Lambda$ | \Xi | $\Xi$ | \Pi | $\Pi$ |
| \Sigma | $\Sigma$ | \Upsilon | $\Upsilon$ | \Phi | $\Phi$ |
| \Psi | $\Psi$ | \Omega | $\Omega$ | | |

### Misc Symbols Type Ord

| Code | Sym | Code | Sym | Code | Sym |
|---|---|---|---|---|---|
| \aleph | $\aleph$ | \wp | $\wp$ | \Re | $\Re$ |
| \Im | $\Im$ | \partial | $\partial$ | \infty | $\infty$ |
| \prime | $\prime$ | \nabla | $\nabla$ | \surd | $\surd$ |
| \angle | $\angle$ | \forall | $\forall$ | \exists | $\exists$ |
| \neg | $\neg$ | \clubsuit | $\clubsuit$ | \diamondsuit | $\diamondsuit$ |
| \heartsuit | $\heartsuit$ | \spadesuit | $\spadesuit$ | | |

### "Large" Operators

| Code | Sym | Code | Sym | Code | Sym |
|---|---|---|---|---|---|
| \int | $\int$ | | | | |

### Binary operators

| Code | Sym | Code | Sym | Code | Sym |
|---|---|---|---|---|---|
| \pm | $\pm$ | \cdot | $\cdot$ | \times | $\times$ |
| \ast | $\ast$ | \circ | $\circ$ | \bullet | $\bullet$ |
| \div | $\div$ | \cap | $\cap$ | \cup | $\cup$ |
| \vee | $\vee$ | \wedge | $\wedge$ | \oplus | $\oplus$ |
| \otimes | $\otimes$ | \oslash | $\oslash$ | | |

Table 15.1: Available special characters in TeX mode

# Chapter 15: Plotting

Relations

| Code | Sym | Code | Sym | Code | Sym |
|---|---|---|---|---|---|
| \leq | ≤ | \subset | ⊂ | \subseteq | ⊆ |
| \in | ∈ | \geq | ≥ | \supset | ⊃ |
| \supseteq | ⊇ | \ni | ∋ | \mid | \| |
| \equiv | ≡ | \sim | ∼ | \approx | ≈ |
| \cong | ≅ | \propto | ∝ | \perp | ⊥ |

Arrows

| Code | Sym | Code | Sym | Code | Sym |
|---|---|---|---|---|---|
| \leftarrow | ← | \Leftarrow | ⇐ | \rightarrow | → |
| \Rightarrow | ⇒ | \leftrightarrow | ↔ | \uparrow | ↑ |
| \downarrow | ↓ | | | | |

Openings and Closings

| Code | Sym | Code | Sym | Code | Sym |
|---|---|---|---|---|---|
| \lfloor | ⌊ | \langle | ⟨ | \lceil | ⌈ |
| \rfloor | ⌋ | \rangle | ⟩ | \rceil | ⌉ |

Alternate Names

| Code | Sym | Code | Sym | Code | Sym |
|---|---|---|---|---|---|
| \neq | ≠ | | | | |

Other (not in Appendix F Tables)

| Code | Sym | Code | Sym | Code | Sym |
|---|---|---|---|---|---|
| \ldots | ... | \O | ⊘ | \copyright | © |
| \deg | ° | | | | |

Table 15.1: Available special characters in TEX mode (cont.)

A complete example showing the capabilities of the extended text is

```
x = 0:0.01:3;
plot (x, erf (x));
hold on;
plot (x,x,"r");
axis ([0, 3, 0, 1]);
text (0.65, 0.6175, strcat ('\leftarrow x = {2/\surd\pi',
' {\fontsize{16}\int_{\fontsize{8}0}^{\fontsize{8}x}}',
' e^{-t^2} dt} = 0.6175'))
```

The result of which can be seen in Figure 15.7

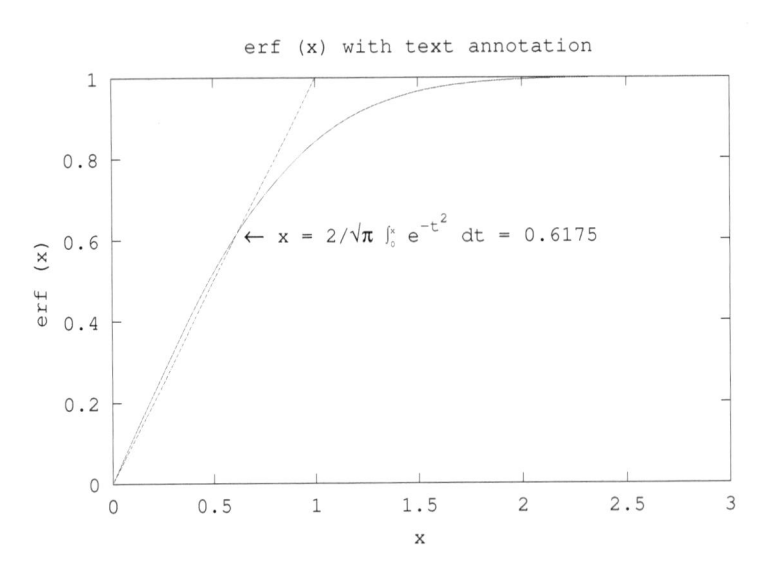

Figure 15.7: Example of inclusion of text with the TeX interpreter

### 15.2.9 Printing and Saving Plots

The `print` command allows you to send plots to you printer and to save plots in a variety of formats. For example,

    `print -dpsc`

prints the current figure to a color PostScript printer. And,

    `print -deps foo.eps`

saves the current figure to an encapsulated PostScript file called 'foo.eps'.

The different graphic toolkits have different print capabilities. In particular, the OpenGL based toolkits such as `fltk` do not support the `"interpreter"` property of text objects. This means special symbols drawn with the `"tex"` interpreter will appear correctly on-screen but will be rendered with interpreter `"none"` when printing. Switch graphics toolkits for printing if this is a concern.

| | |
|---|---:|
| `print ()` | [Function File] |
| `print (options)` | [Function File] |
| `print (filename, options)` | [Function File] |
| `print (h, filename, options)` | [Function File] |

    Print a plot, or save it to a file.

    Both output formatted for printing (PDF and PostScript), and many bitmapped and vector image formats are supported.

    *filename* defines the name of the output file. If the file name has no suffix, one is inferred from the specified device and appended to the file name. If no filename is specified, the output is sent to the printer.

    *h* specifies the handle of the figure to print. If no handle is specified the current figure is used.

    For output to a printer, PostScript file, or PDF file, the paper size is specified by the figure's `papersize` property. The location and size of the image on the page

# Chapter 15: Plotting

are specified by the figure's `paperposition` property. The orientation of the page is specified by the figure's `paperorientation` property.

The width and height of images are specified by the figure's `paperpositon(3:4)` property values.

The `print` command supports many *options*:

-f*h*  Specify the handle, *h*, of the figure to be printed. The default is the current figure.

-P*printer*
Set the *printer* name to which the plot is sent if no *filename* is specified.

-G*ghostscript_command*
Specify the command for calling Ghostscript. For Unix and Windows the defaults are `"gs"` and `"gswin32c"`, respectively.

-color
-mono  Color or monochrome output.

-solid
-dashed  Force all lines to be solid or dashed, respectively.

-portrait
-landscape
Specify the orientation of the plot for printed output. For non-printed output the aspect ratio of the output corresponds to the plot area defined by the `"paperposition"` property in the orientation specified. This option is equivalent to changing the figure's `"paperorientation"` property.

-TextAlphaBits=*n*
-GraphicsAlphaBits=*n*
Octave is able to produce output for various printers, bitmaps, and vector formats by using Ghostscript. For bitmap and printer output anti-aliasing is applied using Ghostscript's TextAlphaBits and GraphicsAlphaBits options. The default number of bits for each is 4. Allowed values for *N* are 1, 2, or 4.

-d*device*  The available output format is specified by the option *device*, and is one of:

    ps
    ps2
    psc
    psc2    PostScript (level 1 and 2, mono and color). The FLTK graphics toolkit generates PostScript level 3.0.

    eps
    eps2
    epsc
    epsc2    Encapsulated PostScript (level 1 and 2, mono and color). The FLTK graphic toolkit generates PostScript level 3.0.

pslatex
epslatex
pdflatex
pslatexstandalone
epslatexstandalone
pdflatexstandalone

    Generate a LaTeX file 'filename.tex' for the text portions of a plot and a file 'filename.(ps|eps|pdf)' for the remaining graphics. The graphics file suffix .ps|eps|pdf is determined by the specified device type. The LaTeX file produced by the 'standalone' option can be processed directly by LaTeX. The file generated without the 'standalone' option is intended to be included from another LaTeX document. In either case, the LaTeX file contains an \includegraphics command so that the generated graphics file is automatically included when the LaTeX file is processed. The text that is written to the LaTeX file contains the strings **exactly** as they were specified in the plot. If any special characters of the TeX mode interpreter were used, the file must be edited before LaTeX processing. Specifically, the special characters must be enclosed with dollar signs ($ ... $), and other characters that are recognized by LaTeX may also need editing (.e.g., braces). The 'pdflatex' device, and any of the 'standalone' formats, are not available with the Gnuplot toolkit.

tikz
    Generate a LaTeX file using PGF/TikZ. For the FLTK toolkit the result is PGF.

ill
aifm
    Adobe Illustrator (Obsolete for Gnuplot versions > 4.2)

cdr
corel
    CorelDraw

dxf
    AutoCAD

emf
meta
    Microsoft Enhanced Metafile

fig
    XFig. For the Gnuplot graphics toolkit, the additional options '-textspecial' or '-textnormal' can be used to control whether the special flag should be set for the text in the figure. (default is '-textnormal')

hpgl
    HP plotter language

mf
    Metafont

png
    Portable network graphics

jpg
jpeg
    JPEG image

## Chapter 15: Plotting

        gif        GIF image (only available for the Gnuplot graphics toolkit)

        pbm        PBMplus

        svg        Scalable vector graphics

        pdf        Portable document format

If the device is omitted, it is inferred from the file extension, or if there is no filename it is sent to the printer as PostScript.

**-d***ghostscript_device*

Additional devices are supported by Ghostscript. Some examples are:

        pdfwrite    Produces pdf output from eps

        ljet2p      HP LaserJet IIP

        pcx24b      24-bit color PCX file format

        ppm         Portable Pixel Map file format

For a complete list, type `system ("gs -h")` to see what formats and devices are available.

When Ghostscript output is sent to a printer the size is determined by the figure's `"papersize"` property. When the output is sent to a file the size is determined by the plot box defined by the figure's `"paperposition"` property.

**-append**    Append PostScript or PDF output to a pre-existing file of the same type.

**-r***NUM*     Resolution of bitmaps in pixels per inch. For both metafiles and SVG the default is the screen resolution; for other formats it is 150 dpi. To specify screen resolution, use "-r0".

**-loose**

**-tight**     Force a tight or loose bounding box for eps files. The default is loose.

**-***preview*  Add a preview to eps files. Supported formats are:

        **-interchange**

                Provide an interchange preview.

        **-metafile**

                Provide a metafile preview.

        **-pict**      Provide pict preview.

        **-tiff**      Provide a tiff preview.

**-S***xsize,ysize*

Plot size in pixels for EMF, GIF, JPEG, PBM, PNG, and SVG. For PS, EPS, PDF, and other vector formats the plot size is in points. This option is equivalent to changing the size of the plot box associated with the `"paperposition"` property. When using the command form of the print function you must quote the *xsize,ysize* option. For example, by writing "-S640,480".

      -F*fontname*
      -F*fontname*:*size*
      -F:*size*    Use *fontname* and/or *fontsize* for all text. *fontname* is ignored for some devices: dxf, fig, hpgl, etc.

The filename and options can be given in any order.

Example: Print to a file using the pdf device.

```
figure (1);
clf ();
surf (peaks);
print figure1.pdf
```

Example: Print to a file using jpg device.

```
clf ();
surf (peaks);
print -djpg figure2.jpg
```

Example: Print to printer named PS_printer using ps format.

```
clf ();
surf (peaks);
print -dpswrite -PPS_printer
```

See also: [saveas], page 348, [hgsave], page 349, [orient], page 348, [figure], page 335.

**saveas (*h, filename*)**                                                                   [Function File]
**saveas (*h, filename, fmt*)**                                                       [Function File]

Save graphic object *h* to the file *filename* in graphic format *fmt*.

*fmt* should be one of the following formats:

| | |
|---|---|
| ps | PostScript |
| eps | Encapsulated PostScript |
| jpg | JPEG Image |
| png | PNG Image |
| emf | Enhanced Meta File |
| pdf | Portable Document Format |

All device formats specified in `print` may also be used. If *fmt* is omitted it is extracted from the extension of *filename*. The default format is `"pdf"`.

```
clf ();
surf (peaks);
saveas (1, "figure1.png");
```

See also: [print], page 344, [hgsave], page 349, [orient], page 348.

**orient (*orientation*)**                                                                      [Function File]
**orient (*hfig, orientation*)**                                                       [Function File]
*orientation* = **orient ()**                                                                       [Function File]
*orientation* = **orient (*hfig*)**                                                        [Function File]

Query or set the print orientation for figure *hfig*.

Chapter 15: Plotting    349

Valid values for *orientation* are "portrait", "landscape", and "tall".

The "landscape" option changes the orientation so the plot width is larger than the plot height. The "paperposition" is also modified so that the plot fills the page, while leaving a 0.25 inch border.

The "tall" option sets the orientation to "portrait" and fills the page with the plot, while leaving a 0.25 inch border.

The "portrait" option (default) changes the orientation so the plot height is larger than the plot width. It also restores the default "paperposition" property.

When called with no arguments, return the current print orientation.

If the argument *hfig* is omitted, then operate on the current figure returned by gcf.

**See also:** [print], page 344, [saveas], page 348.

print and saveas are used when work on a plot has finished and the output must be in a publication-ready format. During intermediate stages it is often better to save the graphics object and all of its associated information so that changes—to colors, axis limits, marker styles, etc.—can be made easily from within Octave. The hgsave/hgload commands can be used to save and re-create a graphics object.

hgsave (*filename*)                                           [Function File]
hgsave (*h*, *filename*)                                      [Function File]
hgsave (*h*, *filename*, *fmt*)                               [Function File]

Save the graphics handle *h* to the file *filename* in the format *fmt*.

If unspecified, *h* is the current figure as returned by gcf.

When *filename* does not have an extension the default filename extension '.ofig' will be appended.

If present, *fmt* should be one of the following:

- '-binary', '-float-binary'
- '-hdf5', '-float-hdf5'
- '-V7', '-v7', -7, '-mat7-binary'
- '-V6', '-v6', -6, '-mat6-binary'
- '-text'
- '-zip', '-z'

When producing graphics for final publication use print or saveas. When it is important to be able to continue to edit a figure as an Octave object, use hgsave/hgload.

**See also:** [hgload], page 349, [hdl2struct], page 361, [saveas], page 348, [print], page 344.

h = hgload (*filename*)                                       [Function File]

Load the graphics object in *filename* into the graphics handle *h*.

If *filename* has no extension, Octave will try to find the file with and without the standard extension of '.ofig'.

**See also:** [hgsave], page 349, [struct2hdl], page 361.

### 15.2.10 Interacting with Plots

The user can select points on a plot with the `ginput` function or selection the position at which to place text on the plot with the `gtext` function using the mouse. Menus may also be created and populated with specific user commands via the `uimenu` function.

`[x, y, buttons] = ginput (n)` [Function File]
`[x, y, buttons] = ginput ()` [Function File]

Return the position and type of mouse button clicks and/or key strokes in the current figure window.

If n is defined, then capture n events before returning. When n is not defined `ginput` will loop until the return key RET is pressed.

The return values *x*, *y* are the coordinates where the mouse was clicked in the units of the current axes. The return value *button* is 1, 2, or 3 for the left, middle, or right button. If a key is pressed the ASCII value is returned in *button*.

Implementation Note: `ginput` is intenteded for 2-D plots. For 3-D plots see the *currentpoint* property of the current axes which can be transformed with knowledge of the current `view` into data units.

**See also:** [gtext], page 350, [waitforbuttonpress], page 350.

`waitforbuttonpress ()` [Function File]
`b = waitforbuttonpress ()` [Function File]

Wait for mouse click or key press over the current figure window.

The return value of *b* is 0 if a mouse button was pressed or 1 if a key was pressed.

**See also:** [waitfor], page 738, [ginput], page 350, [kbhit], page 237.

`gtext (s)` [Function File]
`gtext ({s1, s2, ...})` [Function File]
`gtext ({s1; s2; ...})` [Function File]
`gtext (..., prop, val, ...)` [Function File]
`h = gtext (...)` [Function File]

Place text on the current figure using the mouse.

The text is defined by the string *s*. If *s* is a cell string organized as a row vector then each string of the cell array is written to a separate line. If *s* is organized as a column vector then one string element of the cell array is placed for every mouse click.

Optional property/value pairs are passed directly to the underlying text objects.

The optional return value *h* is a graphics handle to the created text object(s).

**See also:** [ginput], page 350, [text], page 329.

`hui = uimenu (property, value, ...)` [Function File]
`hui = uimenu (h, property, value, ...)` [Function File]

Create a uimenu object and return a handle to it.

If *h* is omitted then a top-level menu for the current figure is created. If *h* is given then a submenu relative to *h* is created.

uimenu objects have the following specific properties:

Chapter 15: Plotting

"accelerator"
: A string containing the key combination together with CTRL to execute this menu entry (e.g., "x" for CTRL+x).

"callback"
: Is the function called when this menu entry is executed. It can be either a function string (e.g., "myfun"), a function handle (e.g., @myfun) or a cell array containing the function handle and arguments for the callback function (e.g., {@myfun, arg1, arg2}).

"checked"
: Can be set "on" or "off". Sets a mark at this menu entry.

"enable"
: Can be set "on" or "off". If disabled the menu entry cannot be selected and it is grayed out.

"foregroundcolor"
: A color value setting the text color for this menu entry.

"label"
: A string containing the label for this menu entry. A "&"-symbol can be used to mark the "accelerator" character (e.g., "E&xit")

"position"
: An scalar value containing the relative menu position. The entry with the lowest value is at the first position starting from left or top.

"separator"
: Can be set "on" or "off". If enabled it draws a separator line above the current position. It is ignored for top level entries.

Examples:
```
f = uimenu ("label", "&File", "accelerator", "f");
e = uimenu ("label", "&Edit", "accelerator", "e");
uimenu (f, "label", "Close", "accelerator", "q", ...
        "callback", "close (gcf)");
uimenu (e, "label", "Toggle &Grid", "accelerator", "g", ...
        "callback", "grid (gca)");
```

See also: [figure], page 335.

### 15.2.11 Test Plotting Functions

The functions **sombrero** and **peaks** provide a way to check that plotting is working. Typing either **sombrero** or **peaks** at the Octave prompt should display a three-dimensional plot.

sombrero ()  [Function File]
sombrero (*n*)  [Function File]
z = sombrero (...)  [Function File]
[x, y, z] = sombrero (...)  [Function File]

: Plot the familiar 3-D sombrero function.

The function plotted is

$$z = \frac{\sin(\sqrt{(x^2 + y^2)})}{\sqrt{(x^2 + y^2)}}$$

Called without a return argument, `sombrero` plots the surface of the above function over the meshgrid [-8,8] using `surf`.

If n is a scalar the plot is made with n grid lines. The default value for n is 41.

When called with output arguments, return the data for the function evaluated over the meshgrid. This can subsequently be plotted with `surf (x, y, z)`.

**See also:** [peaks], page 352, [meshgrid], page 316, [mesh], page 306, [surf], page 308.

peaks ()                                              [Function File]
peaks (*n*)                                           [Function File]
peaks (*x, y*)                                        [Function File]
z = peaks (...)                                       [Function File]
[x, y, z] = peaks (...)                               [Function File]

Plot a function with lots of local maxima and minima.

The function has the form

$$f(x,y) = 3(1-x)^2 e^{\left(-x^2-(y+1)^2\right)} - 10\left(\frac{x}{5} - x^3 - y^5\right) - \frac{1}{3}e^{\left(-(x+1)^2 - y^2\right)}$$

Called without a return argument, `peaks` plots the surface of the above function using `surf`.

If n is a scalar, `peaks` plots the value of the above function on an n-by-n mesh over the range [-3,3]. The default value for n is 49.

If n is a vector, then it represents the grid values over which to calculate the function. If x and y are specified then the function value is calculated over the specified grid of vertices.

When called with output arguments, return the data for the function evaluated over the meshgrid. This can subsequently be plotted with `surf (x, y, z)`.

**See also:** [sombrero], page 351, [meshgrid], page 316, [mesh], page 306, [surf], page 308.

## 15.3 Graphics Data Structures

### 15.3.1 Introduction to Graphics Structures

The graphics functions use pointers, which are of class graphics_handle, in order to address the data structures which control visual display. A graphics handle may point to any one of a number of different base object types and these objects are the graphics data structures themselves. The primitive graphic object types are: `figure`, `axes`, `line`, `text`, `patch`, `surface`, `text`, and `image`.

Each of these objects has a function by the same name, and, each of these functions returns a graphics handle pointing to an object of the corresponding type. In addition there are several functions which operate on properties of the graphics objects and which also return handles: the functions `plot` and `plot3` return a handle pointing to an object of type line, the function `subplot` returns a handle pointing to an object of type axes, the function `fill` returns a handle pointing to an object of type patch, the functions `area`, `bar`, `barh`, `contour`, `contourf`, `contour3`, `surf`, `mesh`, `surfc`, `meshc`, `errorbar`, `quiver`,

# Chapter 15: Plotting

`quiver3`, `scatter`, `scatter3`, `stair`, `stem`, `stem3` each return a handle to a complex data structure as documented in [Data Sources], page 392.

The graphics objects are arranged in a hierarchy:

1. The root is at 0. In other words, `get (0)` returns the properties of the root object.
2. Below the root are `figure` objects.
3. Below the `figure` objects are `axes` objects.
4. Below the `axes` objects are `line`, `text`, `patch`, `surface`, and `image` objects.

Graphics handles may be distinguished from function handles (see Section 11.11.1 [Function Handles], page 199) by means of the function `ishandle`. `ishandle` returns true if its argument is a handle of a graphics object. In addition, a figure or axes object may be tested using `isfigure` or `isaxes` respectively. The test functions return true only if the argument is both a handle and of the correct type (figure or axes).

The `whos` function can be used to show the object type of each currently defined graphics handle. (Note: this is not true today, but it is, I hope, considered an error in whos. It may be better to have whos just show graphics_handle as the class, and provide a new function which, given a graphics handle, returns its object type. This could generalize the ishandle() functions and, in fact, replace them.)

The `get` and `set` commands are used to obtain and set the values of properties of graphics objects. In addition, the `get` command may be used to obtain property names.

For example, the property `"type"` of the graphics object pointed to by the graphics handle h may be displayed by:

```
get (h, "type")
```

The properties and their current values are returned by `get (h)` where h is a handle of a graphics object. If only the names of the allowed properties are wanted they may be displayed by: `get (h, "")`.

Thus, for example:
```
h = figure ();
get (h, "type")
ans = figure
get (h, "");
error: get: ambiguous figure property name ; possible matches:

    __enhanced__            hittest               resize
    __graphics_toolkit__    integerhandle         resizefcn
    __guidata__             interruptible         selected
    __modified__            inverthardcopy        selectionhighlight
    __myhandle__            keypressfcn           selectiontype
    __plot_stream__         keyreleasefcn         tag
    alphamap                menubar               toolbar
    beingdeleted            mincolormap           type
    busyaction              name                  uicontextmenu
    buttondownfcn           nextplot              units
    children                numbertitle           userdata
    clipping                outerposition         visible
    closerequestfcn         paperorientation      windowbuttondownfcn
    color                   paperposition         windowbuttonmotionfcn
    colormap                paperpositionmode     windowbuttonupfcn
    createfcn               papersize             windowkeypressfcn
    currentaxes             papertype             windowkeyreleasefcn
```

| | | |
|---|---|---|
| currentcharacter | paperunits | windowscrollwheelfcn |
| currentobject | parent | windowstyle |
| currentpoint | pointer | wvisual |
| deletefcn | pointershapecdata | wvisualmode |
| dockcontrols | pointershapehotspot | xdisplay |
| doublebuffer | position | xvisual |
| filename | renderer | xvisualmode |
| handlevisibility | renderermode | |

The root figure has index 0. Its properties may be displayed by: `get (0, "")`.

The uses of `get` and `set` are further explained in [get], page 359, [set], page 359.

**`res = isprop (obj, "prop")`**　　　　　　　　　　　　　　　　　　　　　　　[Function File]

Return true if *prop* is a property of the object *obj*.

*obj* may also be an array of objects in which case *res* will be a logical array indicating whether each handle has the property *prop*.

For plotting, *obj* is a handle to a graphics object. Otherwise, *obj* should be an instance of a class.

**See also:** [get], page 359, [set], page 359, [ismethod], page 719, [isobject], page 718.

### 15.3.2 Graphics Objects

The hierarchy of graphics objects was explained above. See Section 15.3.1 [Introduction to Graphics Structures], page 352. Here the specific objects are described, and the properties contained in these objects are discussed. Keep in mind that graphics objects are always referenced by *handle*.

root figure　the top level of the hierarchy and the parent of all figure objects. The handle index of the root figure is 0.

figure　　A figure window.

axes　　　A set of axes. This object is a child of a figure object and may be a parent of line, text, image, patch, or surface objects.

line　　　A line in two or three dimensions.

text　　　Text annotations.

image　　A bitmap image.

patch　　A filled polygon, currently limited to two dimensions.

surface　A three-dimensional surface.

### 15.3.2.1 Creating Graphics Objects

You can create any graphics object primitive by calling the function of the same name as the object; In other words, `figure`, `axes`, `line`, `text`, `image`, `patch`, and `surface` functions. These fundamental graphic objects automatically become children of the current axes object as if `hold on` was in place. Seperately, axes will automatically become children of the current figure object and figures will become children of the root object 0.

If this auto-joining feature is not desired then it is important to call `newplot` first to prepare a new figure and axes for plotting. Alternatively, the easier way is to call a high-level graphics routine which will both create the plot and then populate it with low-level

Chapter 15: Plotting    355

graphics objects. Instead of calling `line`, use `plot`. Or use `surf` instead of `surface`. Or use `fill` instead of `patch`.

`axes ()`                                                  [Function File]
`axes (`*property*, *value*, ...`)`                         [Function File]
`axes (`*hax*`)`                                            [Function File]
`h = axes (`...`)`                                          [Function File]

> Create an axes object and return a handle to it, or set the current axes to *hax*.
>
> Called without any arguments, or with *property*/*value* pairs, construct a new axes. For accepted properties and corresponding values, see [set], page 359.
>
> Called with a single axes handle argument *hax*, the function makes *hax* the current axis. It also restacks the axes in the corresponding figure so that *hax* is the first entry in the list of children. This causes *hax* to be displayed on top of any other axes objects (Z-order stacking).
>
> **See also:** [gca], page 358, [set], page 359, [get], page 359.

`line ()`                                                   [Function File]
`line (`*x*, *y*`)`                                          [Function File]
`line (`*x*, *y*, *property*, *value*, ...`)`                [Function File]
`line (`*x*, *y*, *z*`)`                                     [Function File]
`line (`*x*, *y*, *z*, *property*, *value*, ...`)`           [Function File]
`line (`*property*, *value*, ...`)`                          [Function File]
`line (`*hax*, ...`)`                                        [Function File]
`h = line (`...`)`                                           [Function File]

> Create line object from *x* and *y* (and possibly *z*) and insert in the current axes.
>
> Multiple property-value pairs may be specified for the line object, but they must appear in pairs.
>
> If the first argument *hax* is an axes handle, then plot into this axis, rather than the current axes returned by `gca`.
>
> The optional return value *h* is a graphics handle (or vector of handles) to the line objects created.
>
> **See also:** [image], page 695, [patch], page 355, [rectangle], page 304, [surface], page 356, [text], page 329.

`patch ()`                                                  [Function File]
`patch (`*x*, *y*, *c*`)`                                    [Function File]
`patch (`*x*, *y*, *z*, *c*`)`                               [Function File]
`patch (`*fv*`)`                                             [Function File]
`patch ("`Faces`", `*faces*`, "`Vertices`", `*verts*`, ...`)`  [Function File]
`patch (`..., *prop*, *val*, ...`)`                          [Function File]
`patch (`*hax*, ...`)`                                       [Function File]
`h = patch (`...`)`                                          [Function File]

> Create patch object in the current axes with vertices at locations (*x*, *y*) and of color *c*.
>
> If the vertices are matrices of size MxN then each polygon patch has M vertices and a total of N polygons will be created. If some polygons do not have M vertices use NaN to represent "no vertex". If the *z* input is present then 3-D patches will be created.

The color argument c can take many forms. To create polygons which all share a single color use a string value (e.g., "r" for red), a scalar value which is scaled by `caxis` and indexed into the current colormap, or a 3-element RGB vector with the precise TrueColor.

If c is a vector of length N then the ith polygon will have a color determined by scaling entry $c(i)$ according to `caxis` and then indexing into the current colormap. More complicated coloring situations require directly manipulating patch property/value pairs.

Instead of specifying polygons by matrices x and y, it is possible to present a unique list of vertices and then a list of polygon faces created from those vertices. In this case the "Vertices" matrix will be an Nx2 (2-D patch) or Nx3 (3-D path). The MxN "Faces" matrix describes M polygons having N vertices—each row describes a single polygon and each column entry is an index into the "Vertices" matrix to identify a vertex. The patch object can be created by directly passing the property/value pairs "Vertices"/*verts*, "Faces"/*faces* as inputs.

A third input form is to create a structure *fv* with the fields "vertices", "faces", and optionally "facevertexcdata".

If the first argument *hax* is an axes handle, then plot into this axis, rather than the current axes returned by `gca`.

The optional return value *h* is a graphics handle to the created patch object.

Implementation Note: Patches are highly configurable objects. To truly customize them requires setting patch properties directly. Useful patch properties are: "cdata", "edgecolor", "facecolor", "faces", "facevertexcdata".

**See also:** [fill], page 297, [get], page 359, [set], page 359.

surface (*x, y, z, c*) [Function File]
surface (*x, y, z*) [Function File]
surface (*z, c*) [Function File]
surface (*z*) [Function File]
surface (..., *prop*, *val*, ...) [Function File]
surface (*hax*, ...) [Function File]
*h* = surface (...) [Function File]

Create a surface graphic object given matrices x and y from `meshgrid` and a matrix of values z corresponding to the x and y coordinates of the surface.

If x and y are vectors, then a typical vertex is $(x(j), y(i), z(i,j))$. Thus, columns of z correspond to different x values and rows of z correspond to different y values. If only a single input z is given then x is taken to be `1:rows (z)` and y is `1:columns (z)`.

Any property/value input pairs are assigned to the surface object.

If the first argument *hax* is an axes handle, then plot into this axis, rather than the current axes returned by `gca`.

The optional return value *h* is a graphics handle to the created surface object.

**See also:** [surf], page 308, [mesh], page 306, [patch], page 355, [line], page 355.

# Chapter 15: Plotting

## 15.3.2.2 Handle Functions

To determine whether a variable is a graphics object index, or an index to an axes or figure, use the functions `ishandle`, `isaxes`, and `isfigure`.

**ishandle (h)**  [Built-in Function]

  Return true if $h$ is a graphics handle and false otherwise.

  $h$ may also be a matrix of handles in which case a logical array is returned that is true where the elements of $h$ are graphics handles and false where they are not.

  **See also:** [isaxes], page 357, [isfigure], page 357.

**ishghandle (h)**  [Function File]

  Return true if $h$ is a graphics handle and false otherwise.

  This function is equivalent to `ishandle` and is provided for compatibility with MATLAB.

  **See also:** [ishandle], page 357.

**isaxes (h)**  [Function File]

  Return true if $h$ is an axes graphics handle and false otherwise.

  If $h$ is a matrix then return a logical array which is true where the elements of $h$ are axes graphics handles and false where they are not.

  **See also:** [isaxes], page 357, [ishandle], page 357.

**isfigure (h)**  [Function File]

  Return true if $h$ is a figure graphics handle and false otherwise.

  If $h$ is a matrix then return a logical array which is true where the elements of $h$ are figure graphics handles and false where they are not.

  **See also:** [isaxes], page 357, [ishandle], page 357.

The function `gcf` returns an index to the current figure object, or creates one if none exists. Similarly, `gca` returns the current axes object, or creates one (and its parent figure object) if none exists.

**h = gcf ()**  [Function File]

  Return a handle to the current figure.

  The current figure is the default target for graphics output. If multiple figures exist, `gcf` returns the last created figure or the last figure that was clicked on with the mouse.

  If a current figure does not exist, create one and return its handle. The handle may then be used to examine or set properties of the figure. For example,

      fplot (@sin, [-10, 10]);
      fig = gcf ();
      set (fig, "numbertitle", "off", "name", "sin plot")

  plots a sine wave, finds the handle of the current figure, and then renames the figure window to describe the contents.

  Note: To find the current figure without creating a new one if it does not exist, query the `"CurrentFigure"` property of the root graphics object.

```
get (0, "currentfigure");
```

**See also:** [gca], page 358, [gco], page 358, [gcbf], page 386, [gcbo], page 386, [get], page 359, [set], page 359.

**h = gca ()**  [Function File]

Return a handle to the current axis object.

The current axis is the default target for graphics output. In the case of a figure with multiple axes, `gca` returns the last created axes or the last axes that was clicked on with the mouse.

If no current axes object exists, create one and return its handle. The handle may then be used to examine or set properties of the axes. For example,

```
ax = gca ();
set (ax, "position", [0.5, 0.5, 0.5, 0.5]);
```

creates an empty axes object and then changes its location and size in the figure window.

Note: To find the current axis without creating a new axes object if it does not exist, query the `"CurrentAxes"` property of a figure.

```
get (gcf, "currentaxes");
```

**See also:** [gcf], page 357, [gco], page 358, [gcbf], page 386, [gcbo], page 386, [get], page 359, [set], page 359.

**h = gco ()**  [Function File]
**h = gco (*fig*)**  [Function File]

Return a handle to the current object of the current figure, or a handle to the current object of the figure with handle *fig*.

The current object of a figure is the object that was last clicked on. It is stored in the `"CurrentObject"` property of the target figure.

If the last mouse click did not occur on any child object of the figure, then the current object is the figure itself.

If no mouse click occurred in the target figure, this function returns an empty matrix.

Programming Note: The value returned by this function is not necessarily the same as the one returned by `gcbo` during callback execution. An executing callback can be interrupted by another callback and the current object may be changed.

**See also:** [gcbo], page 386, [gca], page 358, [gcf], page 357, [gcbf], page 386, [get], page 359, [set], page 359.

The `get` and `set` functions may be used to examine and set properties for graphics objects. For example,

Chapter 15: Plotting 359

```
        get (0)
     ⇒ ans =
        {
          type = root
          currentfigure = [](0x0)
          children = [](0x0)
          visible = on
          ...
        }
```

returns a structure containing all the properties of the root figure. As with all functions in Octave, the structure is returned by value, so modifying it will not modify the internal root figure plot object. To do that, you must use the `set` function. Also, note that in this case, the `currentfigure` property is empty, which indicates that there is no current figure window.

The `get` function may also be used to find the value of a single property. For example,

```
    get (gca (), "xlim")
       ⇒ [ 0 1 ]
```

returns the range of the x-axis for the current axes object in the current figure.

To set graphics object properties, use the set function. For example,

```
    set (gca (), "xlim", [-10, 10]);
```

sets the range of the x-axis for the current axes object in the current figure to '[-10, 10]'.

Default property values can also be queried if the `set` function is called without a value argument. When only one argument is given (a graphic handle) then a structure with defaults for all properties of the given object type is returned. For example,

```
    set (gca ())
```

returns a structure containing the default property values for axes objects. If `set` is called with two arguments (a graphic handle and a property name) then only the defaults for the requested property are returned.

`val = get (h)`  [Built-in Function]
`val = get (h, p)`  [Built-in Function]

Return the value of the named property *p* from the graphics handle *h*.

If *p* is omitted, return the complete property list for *h*.

If *h* is a vector, return a cell array including the property values or lists respectively.

**See also:** [set], page 359.

`set (h, property, value, ...)`  [Built-in Function]
`set (h, properties, values)`  [Built-in Function]
`set (h, pv)`  [Built-in Function]
`value_list = set (h, property)`  [Built-in Function]
`all_value_list = set (h)`  [Built-in Function]

Set named property values for the graphics handle (or vector of graphics handles) *h*.

There are three ways to give the property names and values:

- as a comma separated list of *property*, *value* pairs

  Here, each *property* is a string containing the property name, each *value* is a value of the appropriate type for the property.

- as a cell array of strings *properties* containing property names and a cell array *values* containing property values.

  In this case, the number of columns of *values* must match the number of elements in *properties*. The first column of *values* contains values for the first entry in *properties*, etc. The number of rows of *values* must be 1 or match the number of elements of $h$. In the first case, each handle in $h$ will be assigned the same values. In the latter case, the first handle in $h$ will be assigned the values from the first row of *values* and so on.

- as a structure array *pv*

  Here, the field names of *pv* represent the property names, and the field values give the property values. In contrast to the previous case, all elements of *pv* will be set in all handles in $h$ independent of the dimensions of *pv*.

`set` is also used to query the list of values a named property will take. `clist = set (h, "property")` will return the list of possible values for `"property"` in the cell list *clist*. If no output variable is used then the list is formatted and printed to the screen.

If no property is specified (`slist = set (h)`) then a structure *slist* is returned where the fieldnames are the properties of the object $h$ and the fields are the list of possible values for each property. If no output variable is used then the list is formatted and printed to the screen.

For example,

```
hf = figure ();
set (hf, "paperorientation")
⇒   paperorientation:  [ landscape | {portrait} | rotated ]
```

shows the paperorientation property can take three values with the default being `"portrait"`.

**See also:** [get], page 359.

*parent* = ancestor (*h*, *type*)  [Function File]
*parent* = ancestor (*h*, *type*, "*toplevel*")  [Function File]

Return the first ancestor of handle object $h$ whose type matches *type*, where *type* is a character string.

If *type* is a cell array of strings, return the first parent whose type matches any of the given type strings.

If the handle object $h$ itself is of type *type*, return $h$.

If `"toplevel"` is given as a third argument, return the highest parent in the object hierarchy that matches the condition, instead of the first (nearest) one.

**See also:** [findobj], page 382, [findall], page 383, [allchild], page 360.

*h* = allchild (*handles*)  [Function File]

Find all children, including hidden children, of a graphics object.

Chapter 15: Plotting 361

This function is similar to get (h, "children"), but also returns hidden objects (HandleVisibility = "off").

If *handles* is a scalar, *h* will be a vector. Otherwise, *h* will be a cell matrix of the same size as *handles* and each cell will contain a vector of handles.

**See also:** [findall], page 383, [findobj], page 382, [get], page 359, [set], page 359.

findfigs ()  [Function File]
Find all visible figures that are currently off the screen and move them onto the screen.

**See also:** [allchild], page 360, [figure], page 335, [get], page 359, [set], page 359.

Figures can be printed or saved in many graphics formats with print and saveas. Occasionally, however, it may be useful to save the original Octave handle graphic directly so that further modifications can be made such as modifying a title or legend.

This can be accomplished with the following functions by

```
fig_struct = hdl2struct (gcf);
save myplot.fig -struct fig_struct;
...
fig_struct = load ("myplot.fig");
struct2hdl (fig_struct);
```

s = hdl2struct (h)  [Function File]
Return a structure, *s*, whose fields describe the properties of the object, and its children, associated with the handle, *h*.

The fields of the structure *s* are "type", "handle", "properties", "children", and "special".

**See also:** [struct2hdl], page 361, [hgsave], page 349, [findobj], page 382.

h = struct2hdl (s)  [Function File]
h = struct2hdl (s, p)  [Function File]
h = struct2hdl (s, p, hilev)  [Function File]
Construct a graphics handle object *h* from the structure *s*.

The structure must contain the fields "handle", "type", "children", "properties", and "special".

If the handle of an existing figure or axes is specified, *p*, the new object will be created as a child of that object. If no parent handle is provided then a new figure and the necessary children will be constructed using the default values from the root figure.

A third boolean argument *hilev* can be passed to specify whether the function should preserve listeners/callbacks, e.g., for legends or hggroups. The default is false.

**See also:** [hdl2struct], page 361, [hgload], page 349, [findobj], page 382.

hnew = copyobj (horig)  [Function File]
hnew = copyobj (horig, hparent)  [Function File]
Construct a copy of the graphic object associated with handle *horig* and return a handle *hnew* to the new object.

If a parent handle *hparent* (root, figure, axes, or hggroup) is specified, the copied object will be created as a child of *hparent*.

**See also:** [struct2hdl], page 361, [hdl2struct], page 361, [findobj], page 382.

### 15.3.3 Graphics Object Properties

In this Section the graphics object properties are discussed in detail, starting with the root figure properties and continuing through the objects hierarchy. The documentation about a specific graphics object can be displayed using doc function, e.g., doc ("axes properties") will show Section 15.3.3.3 [Axes Properties], page 367.

The allowed values for radio (string) properties can be retrieved programmatically or displayed using the one or two arguments call to set function. See [set], page 359.

In the following documentation, default values are enclosed in { }.

### 15.3.3.1 Root Figure Properties

The root figure properties are:

__modified__: "off" | {"on"}
beingdeleted: {"off"} | "on"
> beingdeleted is unused.

busyaction: "cancel" | {"queue"}
> busyaction is unused.

buttondownfcn: string | function handle, def. [](0x0)
> buttondownfcn is unused.

callbackobject (read-only): graphics handle, def. [](0x0)
children (read-only): vector of graphics handles, def. [](0x1)
> Graphics handles of the root's children.

clipping: "off" | {"on"}
> clipping is unused.

commandwindowsize (read-only): def. [0 0]
createfcn: string | function handle, def. [](0x0)
> createfcn is unused.

currentfigure: graphics handle, def. [](0x0)
> Graphics handle of the current figure.

deletefcn: string | function handle, def. [](0x0)
> deletefcn is unused.

diary: {"off"} | "on"
> If diary is "on", the Octave command window session is saved to file. See [diaryfile property], page 362.

diaryfile: string, def. "diary"
> The name of the diary file. See [diary function], page 34.

echo: {"off"} | "on"
> Control whether Octave displays commands executed from scripts. See [echo function], page 34.

errormessage (read-only): string, def. ""
> The last error message raised. See [lasterr function], page 209.

Chapter 15: Plotting 363

`fixedwidthfontname`: string, def. `"Courier"`
`format`: `"+"` | `"bank"` | `"bit"` | `"hex"` | `"long"` | `"longe"` | `"longeng"` | `"longg"` | `"native-bit"` | `"native-hex"` | `"none"` | `"rat"` | `{"short"}` | `"shorte"` | `"shorteng"` | `"shortg"`

>This property is a wrapper around the `format` function. See [format function], page 232.

`formatspacing`: `"compact"` | `{"loose"}`

>This property is a wrapper around the `format` function. See [format function], page 232.

`handlevisibility`: `"callback"` | `"off"` | `{"on"}`

>`handlevisibility` is unused.

`hittest`: `"off"` | `{"on"}`

>`hittest` is unused.

`interruptible`: `"off"` | `{"on"}`

>`interruptible` is unused.

`language`: string, def. `"ascii"`
`monitorpositions`:

>`monitorpositions` is unused.

`parent`: graphics handle, def. `[](0x0)`

>Root figure has no parent graphics object. `parent` is always empty.

`pointerlocation`: two-element vector, def. `[0 0]`

>`pointerlocation` is unused.

`pointerwindow` (read-only): graphics handle, def. 0

>`pointerwindow` is unused.

`recursionlimit`: double, def. 256

>The maximum number of times a function can be called recursively. See [max_recursion_depth function], page 141.

`screendepth` (read-only): double
`screenpixelsperinch` (read-only): double
`screensize` (read-only): four-element vector
`selected`: `{"off"}` | `"on"`

>`selected` is unused.

`selectionhighlight`: `"off"` | `{"on"}`

>`selectionhighlight` is unused.

`showhiddenhandles`: `{"off"}` | `"on"`

>If `showhiddenhandles` is `"on"`, all graphics objects handles are visible in their parents' children list, regardless of the value of their `handlevisibility` property.

`tag`: string, def. `""`

>A user-defined string to label the graphics object.

type (read-only): string
: Class name of the graphics object. type is always "root"

uicontextmenu: graphics handle, def. [](0x0)
: uicontextmenu is unused.

units: "centimeters" | "inches" | "normalized" | {"pixels"} | "points"
userdata: Any Octave data, def. [](0x0)
: User-defined data to associate with the graphics object.

visible: "off" | {"on"}
: visible is unused.

### 15.3.3.2 Figure Properties

The figure properties are:

__modified__: "off" | {"on"}
alphamap: def. 64-by-1 double
: Transparency is not yet implemented for figure objects. alphamap is unused.

beingdeleted: {"off"} | "on"
busyaction: "cancel" | {"queue"}
buttondownfcn: string | function handle, def. [](0x0)
children (read-only): vector of graphics handles, def. [](0x1)
: Graphics handles of the figure's children.

clipping: "off" | {"on"}
: clipping is unused.

closerequestfcn: string | function handle, def. "closereq"
color: colorspec, def. [1 1 1]
: Color of the figure background. See Section 15.4.1 [colorspec], page 384.

colormap: N-by-3 matrix, def. 64-by-3 double
: A matrix containing the RGB color map for the current axes.

createfcn: string | function handle, def. [](0x0)
: Callback function executed immediately after figure has been created. Function is set by using default property on root object, e.g., set (0, "defaultfigurecreatefcn", 'disp ("figure created!")').

currentaxes: graphics handle, def. [](0x0)
: Handle to the graphics object of the current axes.

currentcharacter (read-only): def. ""
: currentcharacter is unused.

currentobject (read-only): graphics handle, def. [](0x0)
currentpoint (read-only): two-element vector, def. [0; 0]
: A 1-by-2 matrix which holds the coordinates of the point over which the mouse pointer was when a mouse event occurred. The X and Y coordinates are in units defined by the figure's units property and their origin is the lower left corner of the plotting area.

    Events which set currentpoint are

Chapter 15: Plotting 365

>   A mouse button was pressed
>       always
>
>   A mouse button was released
>       only if the figure's callback `windowbuttonupfcn` is defined
>
>   The pointer was moved while pressing the mouse button (drag)
>       only if the figure's callback `windowbuttonmotionfcn` is defined

`deletefcn`: string | function handle, def. [] (0x0)
>   Callback function executed immediately before figure is deleted.

`dockcontrols`: {"off"} | "on"
>   `dockcontrols` is unused.

`doublebuffer`: "off" | {"on"}
`filename`: string, def. ""
>   The filename used when saving the plot figure

`handlevisibility`: "callback" | "off" | {"on"}
>   If `handlevisibility` is "off", the figure's handle is not visible in its parent's "children" property.

`hittest`: "off" | {"on"}
`integerhandle`: "off" | {"on"}
>   Assign the next lowest unused integer as the Figure number.

`interruptible`: "off" | {"on"}
`inverthardcopy`: {"off"} | "on"
`keypressfcn`: string | function handle, def. [] (0x0)
`keyreleasefcn`: string | function handle, def. [] (0x0)
>   With `keypressfcn`, the keyboard callback functions. These callback functions are called when a key is pressed/released respectively. The functions are called with two input arguments. The first argument holds the handle of the calling figure. The second argument holds an event structure which has the following members:
>
>   Character:
>       The ASCII value of the key
>
>   Key:    Lowercase value of the key
>
>   Modifier:
>       A cell array containing strings representing the modifiers pressed with the key.

`menubar`: {"figure"} | "none"
>   Control the display of the figure menu bar in the upper left of the figure.

`mincolormap`: def. 64
`name`: string, def. ""
>   Name to be displayed in the figure title bar. The name is displayed to the right of any title determined by the `numbertitle` property.

`nextplot`: {"add"} | "new" | "replace" | "replacechildren"
`numbertitle`: "off" | {"on"}
: Display "Figure" followed by the numerical figure handle value in the figure title bar.

`outerposition`: four-element vector, def. [-1 -1 -1 -1]
`paperorientation`: "landscape" | {"portrait"} | "rotated"
`paperposition`: four-element vector, def. [0.25000 2.50000 8.00000 6.00000]
: Vector [x0 y0 width height] defining the position of the figure (in `paperunits` units) on the printed page. Setting `paperposition` also forces the `paperpositionmode` property to be set to "manual".

`paperpositionmode`: "auto" | {"manual"}
: If `paperpositionmode` is set to "auto", the `paperposition` property is automatically computed: the printed figure will have the same size as the on-screen figure and will be centered on the output page.

`papersize`: two-element vector, def. [8.5000 11.0000]
: Vector [width height] defining the size of the paper for printing. Setting this property forces the `papertype` property to be set to "<custom>".

`papertype`: "<custom>" | "a" | "a0" | "a1" | "a2" | "a3" | "a4" | "a5" | "arch-a" | "arch-b" | "arch-c" | "arch-d" | "arch-e" | "b" | "b0" | "b1" | "b2" | "b3" | "b4" | "b5" | "c" | "d" | "e" | "tabloid" | "uslegal" | {"usletter"}
: Name of the paper used for printed output. Setting `papertype` also changes `papersize` accordingly.

`paperunits`: "centimeters" | {"inches"} | "normalized" | "points"
: The unit used to compute the `paperposition` property.

`parent`: graphics handle, def. 0
: Handle of the parent graphics object.

`pointer`: {"arrow"} | "botl" | "botr" | "bottom" | "circle" | "cross" | "crosshair" | "custom" | "fleur" | "fullcrosshair" | "hand" | "ibeam" | "left" | "right" | "top" | "topl" | "topr" | "watch"
: `pointer` is unused.

`pointershapecdata`: def. 16-by-16 double
: `pointershapecdata` is unused.

`pointershapehotspot`: def. [0 0]
: `pointershapehotspot` is unused.

`position`: four-element vector, def. [300 200 560 420]
`renderer`: "none" | "opengl" | {"painters"} | "zbuffer"
`renderermode`: {"auto"} | "manual"
`resize`: "off" | {"on"}
`resizefcn`: string | function handle, def. [](0x0)
`selected`: {"off"} | "on"
`selectionhighlight`: "off" | {"on"}
`selectiontype`: "alt" | "extend" | {"normal"} | "open"
: `selectiontype` is unused.

Chapter 15: Plotting                                                                                     367

tag: string, def. ""
    A user-defined string to label the graphics object.

toolbar: {"auto"} | "figure" | "none"
    toolbar is unused.

type (read-only): string
    Class name of the graphics object. type is always "figure"

uicontextmenu: graphics handle, def. [](0x0)
    Graphics handle of the uicontextmenu object that is currently associated to this figure object.

units: "centimeters" | "characters" | "inches" | "normalized" | {"pixels"} | "points"
    The unit used to compute the position and outerposition properties.

userdata: Any Octave data, def. [](0x0)
    User-defined data to associate with the graphics object.

visible: "off" | {"on"}
    If visible is "off", the figure is not rendered on screen.

windowbuttondownfcn: string | function handle, def. [](0x0)
    See [windowbuttonupfcn property], page 367.

windowbuttonmotionfcn: string | function handle, def. [](0x0)
    See [windowbuttonupfcn property], page 367.

windowbuttonupfcn: string | function handle, def. [](0x0)
    With windowbuttondownfcn and windowbuttonmotionfcn, the mouse callback functions. These callback functions are called when a mouse button is pressed, dragged, or released respectively. When these callback functions are executed, the currentpoint property holds the current coordinates of the cursor.

windowkeypressfcn: string | function handle, def. [](0x0)
windowkeyreleasefcn: string | function handle, def. [](0x0)
windowscrollwheelfcn: string | function handle, def. [](0x0)
windowstyle: "docked" | "modal" | {"normal"}
wvisual: def. ""
wvisualmode: {"auto"} | "manual"
xdisplay: def. ""
xvisual: def. ""
xvisualmode: {"auto"} | "manual"

### 15.3.3.3 Axes Properties

The axes properties are:

__modified__: "off" | {"on"}
activepositionproperty: {"outerposition"} | "position"
alim: def. [0 1]
    Transparency is not yet implemented for axes objects. alim is unused.

`alimmode:` `{"auto"}` | `"manual"`

`ambientlightcolor:` def. `[1 1 1]`
      Light is not yet implemented for axes objects. `ambientlightcolor` is unused.

`beingdeleted:` `{"off"}` | `"on"`

`box:` `"off"` | `{"on"}`
      Control whether the axes has a surrounding box.

`busyaction:` `"cancel"` | `{"queue"}`

`buttondownfcn:` string | function handle, def. `[](0x0)`

`cameraposition:` three-element vector, def. `[0.50000 0.50000 9.16025]`

`camerapositionmode:` `{"auto"}` | `"manual"`

`cameratarget:` three-element vector, def. `[0.50000 0.50000 0.50000]`

`cameratargetmode:` `{"auto"}` | `"manual"`

`cameraupvector:` three-element vector, def. `[-0 1 0]`

`cameraupvectormode:` `{"auto"}` | `"manual"`

`cameraviewangle:` scalar, def. `6.6086`

`cameraviewanglemode:` `{"auto"}` | `"manual"`

`children` (read-only): vector of graphics handles, def. `[](0x1)`
      Graphics handles of the axes's children.

`clim:` two-element vector, def. `[0 1]`
      Define the limits for the color axis of image children. Setting `clim` also forces the `climmode` property to be set to `"manual"`. See [pcolor function], page 296.

`climmode:` `{"auto"}` | `"manual"`

`clipping:` `"off"` | `{"on"}`
      `clipping` is unused.

`color:` colorspec, def. `[1 1 1]`
      Color of the axes background. See Section 15.4.1 [colorspec], page 384.

`colororder:` N-by-3 RGB matrix, def. 7-by-3 double
      RGB values used by plot function for automatic line coloring.

`createfcn:` string | function handle, def. `[](0x0)`
      Callback function executed immediately after axes has been created. Function is set by using default property on root object, e.g., `set (0, "defaultaxescreatefcn", 'disp ("axes created!")')`.

`currentpoint:` 2-by-3 matrix, def. 2-by-3 double
      Matrix `[xf, yf, zf; xb, yb, zb]` which holds the coordinates (in axes data units) of the point over which the mouse pointer was when the mouse button was pressed. If a mouse callback function is defined, `currentpoint` holds the pointer coordinates at the time the mouse button was pressed. For 3-D plots, the first row of the returned matrix specifies the point nearest to the current camera position and the second row the furthest point. The two points forms a line which is perpendicular to the screen.

`dataaspectratio:` three-element vector, def. `[1 1 1]`
      Specify the relative height and width of the data displayed in the axes. Setting `dataaspectratio` to `[1, 2]` causes the length of one unit as displayed

Chapter 15: Plotting    369

> on the x-axis to be the same as the length of 2 units on the y-axis. Setting
> `dataaspectratio` also forces the `dataaspectratiomode` property to be set to
> "manual".

`dataaspectratiomode`: {"auto"} | "manual"
`deletefcn`: string | function handle, def. [] (0x0)
> Callback function executed immediately before axes is deleted.

`drawmode`: "fast" | {"normal"}
`fontangle`: "italic" | {"normal"} | "oblique"
`fontname`: string, def. "*"
> Name of the font used for axes annotations.

`fontsize`: scalar, def. 10
> Size of the font used for axes annotations. See [fontunits property], page 369.

`fontunits`: "centimeters" | "inches" | "normalized" | "pixels" | {"points"}
> Unit used to interpret `fontsize` property.

`fontweight`: "bold" | "demi" | "light" | {"normal"}
`gridlinestyle`: "-" | "--" | "-." | {":"} | "none"
`handlevisibility`: "callback" | "off" | {"on"}
> If `handlevisibility` is "off", the axes's handle is not visible in its parent's
> "children" property.

`hittest`: "off" | {"on"}
`interpreter`: "latex" | {"none"} | "tex"
`interruptible`: "off" | {"on"}
`layer`: {"bottom"} | "top"
> Control whether the axes is drawn below child graphics objects (ticks, labels,
> etc. covered by plotted objects) or above.

`linestyleorder`: def. "-"
`linewidth`: def. 0.50000
`minorgridlinestyle`: "-" | "--" | "-." | {":"} | "none"
`mousewheelzoom`: scalar in the range (0, 1), def. 0.50000
> Fraction of axes limits to zoom for each wheel movement.

`nextplot`: "add" | {"replace"} | "replacechildren"
`outerposition`: four-element vector, def. [0 0 1 1]
> Specify the position of the plot including titles, axes, and legend. The four
> elements of the vector are the coordinates of the lower left corner and width
> and height of the plot, in units normalized to the width and height of the plot
> window. For example, [0.2, 0.3, 0.4, 0.5] sets the lower left corner of the
> axes at $(0.2, 0.3)$ and the width and height to be 0.4 and 0.5 respectively. See
> [position property], page 370.

`parent`: graphics handle
> Handle of the parent graphics object.

`plotboxaspectratio`: def. [1 1 1]
`plotboxaspectratiomode`: {"auto"} | "manual"
`position`: four-element vector, def. [0.13000 0.11000 0.77500 0.81500]
> Specify the position of the plot excluding titles, axes, and legend. The four elements of the vector are the coordinates of the lower left corner and width and height of the plot, in units normalized to the width and height of the plot window. For example, [0.2, 0.3, 0.4, 0.5] sets the lower left corner of the axes at (0.2, 0.3) and the width and height to be 0.4 and 0.5 respectively. See [outerposition property], page 369.

`projection`: {"orthographic"} | "perspective"
`selected`: {"off"} | "on"
`selectionhighlight`: "off" | {"on"}
`tag`: string, def. ""
> A user-defined string to label the graphics object.

`tickdir`: {"in"} | "out"
> Control whether axes tick marks project "in" to the plot box or "out".

`tickdirmode`: {"auto"} | "manual"
`ticklength`: two-element vector, def. [0.010000 0.025000]
> Two-element vector [2Dlen 3Dlen] specifying the length of the tickmarks relative to the longest visible axis.

`tightinset` (read-only): def. [0.042857 0.038106 0.000000 0.023810]
`title`: graphics handle
> Graphics handle of the title text object.

`type` (read-only): string
> Class name of the graphics object. `type` is always `"axes"`

`uicontextmenu`: graphics handle, def. [](0x0)
> Graphics handle of the uicontextmenu object that is currently associated to this axes object.

`units`: "centimeters" | "characters" | "inches" | {"normalized"} | "pixels" | "points"
`userdata`: Any Octave data, def. [](0x0)
> User-defined data to associate with the graphics object.

`view`: two-element vector, def. [0 90]
> Two-element vector [`azimuth elevation`] specifying the viewpoint for three-dimensional plots

`visible`: "off" | {"on"}
> If `visible` is `"off"`, the axes is not rendered on screen.

`xaxislocation`: {"bottom"} | "top" | "zero"
`xcolor`: {colorspec} | "none", def. [0 0 0]
> Color of the x-axis. See Section 15.4.1 [colorspec], page 384.

`xdir`: {"normal"} | "reverse"
`xgrid`: {"off"} | "on"
> Control whether major x grid lines are displayed.

Chapter 15: Plotting

xlabel: graphics handle
: Graphics handle of the x label text object.

xlim: two-element vector, def. [0 1]
: Two-element vector [xmin xmax] specifying the limits for the x-axis. Setting xlim also forces the xlimmode property to be set to "manual". See [xlim function], page 301.

xlimmode: {"auto"} | "manual"
xminorgrid: {"off"} | "on"
: Control whether minor x grid lines are displayed.

xminortick: {"off"} | "on"
xscale: {"linear"} | "log"
xtick: vector, def. 1-by-6 double
: Position of x tick marks. Setting xtick also forces the xtickmode property to be set to "manual".

xticklabel: string | cell array of strings, def. 1-by-6 cell
: Labels of x tick marks. Setting xticklabel also forces the xticklabelmode property to be set to "manual".

xticklabelmode: {"auto"} | "manual"
xtickmode: {"auto"} | "manual"
yaxislocation: {"left"} | "right" | "zero"
ycolor: {colorspec} | "none", def. [0 0 0]
: Color of the y-axis. See Section 15.4.1 [colorspec], page 384.

ydir: {"normal"} | "reverse"
ygrid: {"off"} | "on"
: Control whether major y grid lines are displayed.

ylabel: graphics handle
: Graphics handle of the y label text object.

ylim: two-element vector, def. [0 1]
: Two-element vector [ymin ymax] specifying the limits for the y-axis. Setting ylim also forces the ylimmode property to be set to "manual". See [ylim function], page 301.

ylimmode: {"auto"} | "manual"
yminorgrid: {"off"} | "on"
: Control whether minor y grid lines are displayed.

yminortick: {"off"} | "on"
yscale: {"linear"} | "log"
ytick: vector, def. 1-by-6 double
: Position of y tick marks. Setting ytick also forces the ytickmode property to be set to "manual".

yticklabel: string | cell array of strings, def. 1-by-6 cell
: Labels of y tick marks. Setting yticklabel also forces the yticklabelmode property to be set to "manual".

yticklabelmode: {"auto"} | "manual"
ytickmode: {"auto"} | "manual"
zcolor: {colorspec} | "none", def. [0 0 0]
: Color of the z-axis. See Section 15.4.1 [colorspec], page 384.

zdir: {"normal"} | "reverse"
zgrid: {"off"} | "on"
: Control whether major z grid lines are displayed.

zlabel: graphics handle
: Graphics handle of the z label text object.

zlim: two-element vector, def. [0 1]
: Two-element vector [zmin zmaz] specifying the limits for the z-axis. Setting zlim also forces the zlimmode property to be set to "manual". See [zlim function], page 301.

zlimmode: {"auto"} | "manual"
zminorgrid: {"off"} | "on"
: Control whether minor z grid lines are displayed.

zminortick: {"off"} | "on"
zscale: {"linear"} | "log"
ztick: vector, def. 1-by-6 double
: Position of z tick marks. Setting ztick also forces the ztickmode property to be set to "manual".

zticklabel: string | cell array of strings, def. 1-by-6 cell
: Labels of z tick marks. Setting zticklabel also forces the zticklabelmode property to be set to "manual".

zticklabelmode: {"auto"} | "manual"
ztickmode: {"auto"} | "manual"

### 15.3.3.4 Line Properties

The line properties are:

__modified__: "off" | {"on"}
beingdeleted: {"off"} | "on"
busyaction: "cancel" | {"queue"}
buttondownfcn: string | function handle, def. [](0x0)
children (read-only): vector of graphics handles, def. [](0x1)
: children is unused.

clipping: "off" | {"on"}
: If clipping is "on", the line is clipped in its parent axes limits.

color: colorspec, def. [0 0 0]
: Color of the line object. See Section 15.4.1 [colorspec], page 384.

createfcn: string | function handle, def. [](0x0)
: Callback function executed immediately after line has been created. Function is set by using default property on root object, e.g., set (0, "defaultlinecreatefcn", 'disp ("line created!")').

`deletefcn`: string | function handle, def. `[](0x0)`
> Callback function executed immediately before line is deleted.

`displayname`: string | cell array of strings, def. `""`
> Text for the legend entry corresponding to this line.

`erasemode`: `"background"` | `"none"` | `{"normal"}` | `"xor"`
> erasemode is unused.

`handlevisibility`: `"callback"` | `"off"` | `{"on"}`
> If `handlevisibility` is `"off"`, the line's handle is not visible in its parent's `"children"` property.

`hittest`: `"off"` | `{"on"}`
`interpreter`: `"latex"` | `"none"` | `{"tex"}`
`interruptible`: `"off"` | `{"on"}`
`linestyle`: `{"-"}` | `"--"` | `"-."` | `":"` | `"none"`
> See Section 15.4.2 [Line Styles], page 385.

`linewidth`: def. `0.50000`
> Width of the line object measured in points.

`marker`: `"*"` | `"+"` | `"."` | `"<"` | `">"` | `"^"` | `"d"` | `"diamond"` | `"h"` | `"hexagram"` | `{"none"}` | `"o"` | `"p"` | `"pentagram"` | `"s"` | `"square"` | `"v"` | `"x"`
> Shape of the marker for each data point. See Section 15.4.3 [Marker Styles], page 385.

`markeredgecolor`: `{"auto"}` | `"none"`
> Color of the edge of the markers. When set to `"auto"`, the marker edges have the same color as the line. If set to `"none"`, no marker edges are displayed. This property can also be set to any color. See Section 15.4.1 [colorspec], page 384.

`markerfacecolor`: `"auto"` | `{"none"}`
> Color of the face of the markers. When set to `"auto"`, the marker faces have the same color as the line. If set to `"none"`, the marker faces are not displayed. This property can also be set to any color. See Section 15.4.1 [colorspec], page 384.

`markersize`: scalar, def. `6`
> Size of the markers measured in points.

`parent`: graphics handle
> Handle of the parent graphics object.

`selected`: `{"off"}` | `"on"`
`selectionhighlight`: `"off"` | `{"on"}`
`tag`: string, def. `""`
> A user-defined string to label the graphics object.

`type` (read-only): string
> Class name of the graphics object. `type` is always `"line"`

`uicontextmenu`: graphics handle, def. `[](0x0)`
> Graphics handle of the uicontextmenu object that is currently associated to this line object.

userdata: Any Octave data, def. [](0x0)
: User-defined data to associate with the graphics object.

visible: "off" | {"on"}
: If visible is "off", the line is not rendered on screen.

xdata: vector, def. [0 1]
: Vector of x data to be plotted.

xdatasource: string, def. ""
: Name of a vector in the current base workspace to use as x data.

ydata: vector, def. [0 1]
: Vector of y data to be plotted.

ydatasource: string, def. ""
: Name of a vector in the current base workspace to use as y data.

zdata: vector, def. [](0x0)
: Vector of z data to be plotted.

zdatasource: string, def. ""
: Name of a vector in the current base workspace to use as z data.

### 15.3.3.5 Text Properties

The text properties are:

__modified__: "off" | {"on"}

backgroundcolor: colorspec, def. "none"
: Background area is not yet implemented for text objects. backgroundcolor is unused.

beingdeleted: {"off"} | "on"

busyaction: "cancel" | {"queue"}

buttondownfcn: string | function handle, def. [](0x0)

children (read-only): vector of graphics handles, def. [](0x1)
: children is unused.

clipping: "off" | {"on"}
: If clipping is "on", the text is clipped in its parent axes limits.

color: colorspec, def. [0 0 0]
: Color of the text. See Section 15.4.1 [colorspec], page 384.

createfcn: string | function handle, def. [](0x0)
: Callback function executed immediately after text has been created. Function is set by using default property on root object, e.g., set (0, "defaulttextcreatefcn", 'disp ("text created!")').

deletefcn: string | function handle, def. [](0x0)
: Callback function executed immediately before text is deleted.

displayname: def. ""

edgecolor: colorspec, def. "none"
: Background area is not yet implemented for text objects. edgecolor is unused.

Chapter 15: Plotting 375

editing: {"off"} | "on"
erasemode: "background" | "none" | {"normal"} | "xor"
    erasemode is unused.

extent (read-only): def. [0.000000 -0.005843 0.000000 0.032136]
fontangle: "italic" | {"normal"} | "oblique"
    Flag whether the font is italic or normal. fontangle is currently unused.

fontname: string, def. "*"
    The font used for the text.

fontsize: scalar, def. 10
    The font size of the text as measured in fontunits.

fontunits: "centimeters" | "inches" | "normalized" | "pixels" | {"points"}
    The units used to interpret fontsize property.

fontweight: "bold" | "demi" | "light" | {"normal"}
    Control variant of base font used: bold, light, normal, etc.

handlevisibility: "callback" | "off" | {"on"}
    If handlevisibility is "off", the text's handle is not visible in its parent's "children" property.

hittest: "off" | {"on"}
horizontalalignment: "center" | {"left"} | "right"
interpreter: "latex" | "none" | {"tex"}
interruptible: "off" | {"on"}
linestyle: {"-"} | "--" | "-." | ":" | "none"
    Background area is not yet implemented for text objects. linestyle is unused.

linewidth: scalar, def. 0.50000
    Background area is not yet implemented for text objects. linewidth is unused.

margin: scalar, def. 2
    Background area is not yet implemented for text objects. margin is unused.

parent: graphics handle
    Handle of the parent graphics object.

position: four-element vector, def. [0 0 0]
    Vector [X0 Y0 Z0] where X0, Y0 and Z0 indicate the position of the text anchor as defined by verticalalignment and horizontalalignment.

rotation: scalar, def. 0
    The angle of rotation for the displayed text, measured in degrees.

selected: {"off"} | "on"
selectionhighlight: "off" | {"on"}
string: string, def. ""
    The text object string content.

tag: string, def. ""
    A user-defined string to label the graphics object.

type (read-only): string

> Class name of the graphics object. type is always "text"

uicontextmenu: graphics handle, def. [](0x0)

> Graphics handle of the uicontextmenu object that is currently associated to this text object.

units: "centimeters" | {"data"} | "inches" | "normalized" | "pixels" | "points"

userdata: Any Octave data, def. [](0x0)

> User-defined data to associate with the graphics object.

verticalalignment: "baseline" | "bottom" | "cap" | {"middle"} | "top"

visible: "off" | {"on"}

> If visible is "off", the text is not rendered on screen.

### 15.3.3.6 Image Properties

The image properties are:

__modified__: "off" | {"on"}

alphadata: scalar | matrix, def. 1

> Transparency is not yet implemented for image objects. alphadata is unused.

alphadatamapping: "direct" | {"none"} | "scaled"

> Transparency is not yet implemented for image objects. alphadatamapping is unused.

beingdeleted: {"off"} | "on"

busyaction: "cancel" | {"queue"}

buttondownfcn: string | function handle, def. [](0x0)

cdata: matrix, def. 64-by-64 double

cdatamapping: {"direct"} | "scaled"

children (read-only): vector of graphics handles, def. [](0x1)

> children is unused.

clipping: "off" | {"on"}

> If clipping is "on", the image is clipped in its parent axes limits.

createfcn: string | function handle, def. [](0x0)

> Callback function executed immediately after image has been created. Function is set by using default property on root object, e.g., set (0, "defaultimagecreatefcn", 'disp ("image created!")').

deletefcn: string | function handle, def. [](0x0)

> Callback function executed immediately before image is deleted.

displayname: string | cell array of strings, def. ""

> Text for the legend entry corresponding to this image.

erasemode: "background" | "none" | {"normal"} | "xor"

> erasemode is unused.

handlevisibility: "callback" | "off" | {"on"}

> If handlevisibility is "off", the image's handle is not visible in its parent's "children" property.

Chapter 15: Plotting

hittest: "off" | {"on"}
interruptible: "off" | {"on"}
parent: graphics handle
> Handle of the parent graphics object.

selected: {"off"} | "on"
selectionhighlight: "off" | {"on"}
tag: string, def. ""
> A user-defined string to label the graphics object.

type (read-only): string
> Class name of the graphics object. type is always "image"

uicontextmenu: graphics handle, def. [](0x0)
> Graphics handle of the uicontextmenu object that is currently associated to this image object.

userdata: Any Octave data, def. [](0x0)
> User-defined data to associate with the graphics object.

visible: "off" | {"on"}
> If visible is "off", the image is not rendered on screen.

xdata: two-element vector, def. [1 64]
> Two-element vector [xmin xmax] specifying the x coordinates of the first and last columns of the image.
>
> Setting xdata to the empty matrix ([]) will restore the default value of [1 columns(image)].

ydata: two-element vector, def. [1 64]
> Two-element vector [ymin ymax] specifying the y coordinates of the first and last rows of the image.
>
> Setting ydata to the empty matrix ([]) will restore the default value of [1 rows(image)].

### 15.3.3.7 Patch Properties

The patch properties are:

__modified__: "off" | {"on"}
alphadatamapping: "direct" | "none" | {"scaled"}
> Transparency is not yet implemented for patch objects. alphadatamapping is unused.

ambientstrength: scalar, def. 0.30000
> Light is not yet implemented for patch objects. ambientstrength is unused.

backfacelighting: "lit" | {"reverselit"} | "unlit"
> Light is not yet implemented for patch objects. backfacelighting is unused.

beingdeleted: {"off"} | "on"
busyaction: "cancel" | {"queue"}
buttondownfcn: string | function handle, def. [](0x0)
cdata: scalar | matrix, def. [](0x0)

    Data defining the patch object color. Patch color can be defined for faces or for vertices.

    If cdata is a scalar index into the current colormap or a RGB triplet, it defines the color of all faces.

    If cdata is an N-by-1 vector of indices or an N-by-3 (RGB) matrix, it defines the color of each one of the N faces.

    If cdata is an N-by-M or an N-by-M-by-3 (RGB) matrix, it defines the color at each vertex.

cdatamapping: "direct" | {"scaled"}
children (read-only): vector of graphics handles, def. [](0x1)

    children is unused.

clipping: "off" | {"on"}

    If clipping is "on", the patch is clipped in its parent axes limits.

createfcn: string | function handle, def. [](0x0)

    Callback function executed immediately after patch has been created. Function is set by using default property on root object, e.g., set (0, "defaultpatchcreatefcn", 'disp ("patch created!")').

deletefcn: string | function handle, def. [](0x0)

    Callback function executed immediately before patch is deleted.

diffusestrength: scalar, def. 0.60000

    Light is not yet implemented for patch objects. diffusestrength is unused.

displayname: def. ""

    Text of the legend entry corresponding to this patch.

edgealpha: scalar | matrix, def. 1

    Transparency is not yet implemented for patch objects. edgealpha is unused.

edgecolor: def. [0 0 0]
edgelighting: "flat" | "gouraud" | {"none"} | "phong"

    Light is not yet implemented for patch objects. edgelighting is unused.

erasemode: "background" | "none" | {"normal"} | "xor"

    erasemode is unused.

facealpha: scalar | matrix, def. 1

    Transparency is not yet implemented for patch objects. facealpha is unused.

facecolor: {colorspec} | "none" | "flat" | "interp", def. [0 0 0]
facelighting: "flat" | "gouraud" | {"none"} | "phong"

    Light is not yet implemented for patch objects. facelighting is unused.

Chapter 15: Plotting 379

`faces`: def. [1 2 3]

`facevertexalphadata`: scalar | matrix, def. [](0x0)
> Transparency is not yet implemented for patch objects. `facevertexalphadata` is unused.

`facevertexcdata`: def. [](0x0)

`handlevisibility`: "callback" | "off" | {"on"}
> If `handlevisibility` is "off", the patch's handle is not visible in its parent's "children" property.

`hittest`: "off" | {"on"}

`interpreter`: "latex" | "none" | {"tex"}
> `interpreter` is unused.

`interruptible`: "off" | {"on"}

`linestyle`: {"-"} | "--" | "-." | ":" | "none"

`linewidth`: def. 0.50000

`marker`: "*" | "+" | "." | "<" | ">" | "^" | "d" | "diamond" | "h" | "hexagram" | {"none"} | "o" | "p" | "pentagram" | "s" | "square" | "v" | "x"
> See [line marker property], page 373.

`markeredgecolor`: {"auto"} | "flat" | "none"
> See [line markeredgecolor property], page 373.

`markerfacecolor`: "auto" | "flat" | {"none"}
> See [line markerfacecolor property], page 373.

`markersize`: scalar, def. 6
> See [line markersize property], page 373.

`normalmode`: {"auto"} | "manual"

`parent`: graphics handle
> Handle of the parent graphics object.

`selected`: {"off"} | "on"

`selectionhighlight`: "off" | {"on"}

`specularcolorreflectance`: scalar, def. 1
> Light is not yet implemented for patch objects. `specularcolorreflectance` is unused.

`specularexponent`: scalar, def. 10
> Light is not yet implemented for patch objects. `specularexponent` is unused.

`specularstrength`: scalar, def. 0.90000
> Light is not yet implemented for patch objects. `specularstrength` is unused.

`tag`: string, def. ""
> A user-defined string to label the graphics object.

`type` (read-only): string
> Class name of the graphics object. `type` is always "patch"

`uicontextmenu`: graphics handle, def. [](0x0)
> Graphics handle of the uicontextmenu object that is currently associated to this patch object.

userdata: Any Octave data, def. [](0x0)
: User-defined data to associate with the graphics object.
vertexnormals: def. [](0x0)
vertices: vector | matrix, def. 3-by-2 double
visible: "off" | {"on"}
: If visible is "off", the patch is not rendered on screen.
xdata: vector | matrix, def. [0; 1; 0]
ydata: vector | matrix, def. [1; 1; 0]
zdata: vector | matrix, def. [](0x0)

### 15.3.3.8 Surface Properties

The surface properties are:

__modified__: "off" | {"on"}
alphadata: scalar | matrix, def. 1
: Transparency is not yet implemented for surface objects. alphadata is unused.

alphadatamapping: "direct" | "none" | {"scaled"}
: Transparency is not yet implemented for surface objects. alphadatamapping is unused.

ambientstrength: def. 0.30000
: Light is not yet implemented for surface objects. ambientstrength is unused.

backfacelighting: "lit" | {"reverselit"} | "unlit"
: Light is not yet implemented for surface objects. backfacelighting is unused.

beingdeleted: {"off"} | "on"
busyaction: "cancel" | {"queue"}
buttondownfcn: string | function handle, def. [](0x0)
cdata: matrix, def. 3-by-3 double
cdatamapping: "direct" | {"scaled"}
cdatasource: def. ""
children (read-only): vector of graphics handles, def. [](0x1)
: children is unused.

clipping: "off" | {"on"}
: If clipping is "on", the surface is clipped in its parent axes limits.

createfcn: string | function handle, def. [](0x0)
: Callback function executed immediately after surface has been created. Function is set by using default property on root object, e.g., set (0, "defaultsurfacecreatefcn", 'disp ("surface created!")').

deletefcn: string | function handle, def. [](0x0)
: Callback function executed immediately before surface is deleted.

diffusestrength: def. 0.60000
: Light is not yet implemented for surface objects. diffusestrength is unused.

displayname: def. ""
: Text for the legend entry corresponding to this surface.

# Chapter 15: Plotting

`edgealpha`: scalar, def. 1
> Transparency is not yet implemented for surface objects. `edgealpha` is unused.

`edgecolor`: def. [0 0 0]
`edgelighting`: `"flat"` | `"gouraud"` | `{"none"}` | `"phong"`
> Light is not yet implemented for surface objects. `edgelighting` is unused.

`erasemode`: `"background"` | `"none"` | `{"normal"}` | `"xor"`
> `erasemode` is unused.

`facealpha`: scalar | matrix, def. 1
> Transparency is not yet implemented for surface objects. `facealpha` is unused.

`facecolor`: `{"flat"}` | `"interp"` | `"none"` | `"texturemap"`
`facelighting`: `"flat"` | `"gouraud"` | `{"none"}` | `"phong"`
> Light is not yet implemented for surface objects. `facelighting` is unused.

`handlevisibility`: `"callback"` | `"off"` | `{"on"}`
> If `handlevisibility` is `"off"`, the surface's handle is not visible in its parent's `"children"` property.

`hittest`: `"off"` | `{"on"}`
`interpreter`: `"latex"` | `"none"` | `{"tex"}`
`interruptible`: `"off"` | `{"on"}`
`linestyle`: `{"-"}` | `"--"` | `"-."` | `":"` | `"none"`
> See Section 15.4.2 [Line Styles], page 385.

`linewidth`: def. 0.50000
> See [line linewidth property], page 373.

`marker`: `"*"` | `"+"` | `"."` | `"<"` | `">"` | `"^"` | `"d"` | `"diamond"` | `"h"` | `"hexagram"` | `{"none"}` | `"o"` | `"p"` | `"pentagram"` | `"s"` | `"square"` | `"v"` | `"x"`
> See Section 15.4.3 [Marker Styles], page 385.

`markeredgecolor`: `{"auto"}` | `"flat"` | `"none"`
> See [line markeredgecolor property], page 373.

`markerfacecolor`: `"auto"` | `"flat"` | `{"none"}`
> See [line markerfacecolor property], page 373.

`markersize`: scalar, def. 6
> See [line markersize property], page 373.

`meshstyle`: `{"both"}` | `"column"` | `"row"`
`normalmode`: `{"auto"}` | `"manual"`
`parent`: graphics handle
> Handle of the parent graphics object.

`selected`: `{"off"}` | `"on"`
`selectionhighlight`: `"off"` | `{"on"}`
`specularcolorreflectance`: def. 1
> Light is not yet implemented for surface objects. `specularcolorreflectance` is unused.

specularexponent: def. 10
: Light is not yet implemented for surface objects. `specularexponent` is unused.

specularstrength: def. 0.90000
: Light is not yet implemented for surface objects. `specularstrength` is unused.

tag: string, def. ""
: A user-defined string to label the graphics object.

type (read-only): string
: Class name of the graphics object. `type` is always `"surface"`

uicontextmenu: graphics handle, def. [](0x0)
: Graphics handle of the uicontextmenu object that is currently associated to this surface object.

userdata: Any Octave data, def. [](0x0)
: User-defined data to associate with the graphics object.

vertexnormals: def. 3-by-3-by-3 double

visible: "off" | {"on"}
: If `visible` is `"off"`, the surface is not rendered on screen.

xdata: matrix, def. [1 2 3]

xdatasource: def. ""

ydata: matrix, def. [1; 2; 3]

ydatasource: def. ""

zdata: matrix, def. 3-by-3 double

zdatasource: def. ""

### 15.3.4 Searching Properties

h = findobj ()        [Function File]
h = findobj (*prop_name*, *prop_value*, ...)        [Function File]
h = findobj (*prop_name*, *prop_value*, "*-logical_op*",        [Function File]
    *prop_name*, *prop_value*)
h = findobj ("*-property*", *prop_name*)        [Function File]
h = findobj ("*-regexp*", *prop_name*, *pattern*)        [Function File]
h = findobj (*hlist*, ...)        [Function File]
h = findobj (*hlist*, "*flat*", ...)        [Function File]
h = findobj (*hlist*, "*-depth*", *d*, ...)        [Function File]

Find graphics object with specified property values.

The simplest form is

    findobj (*prop_name*, *prop_value*)

which returns the handles of all objects which have a property named *prop_name* that has the value *prop_value*. If multiple property/value pairs are specified then only objects meeting all of the conditions are returned.

The search can be limited to a particular set of objects and their descendants, by passing a handle or set of handles *hlist* as the first argument.

The depth of the object hierarchy to search can be limited with the `"-depth"` argument. An example of searching only three generations of children is:

# Chapter 15: Plotting

    `findobj (hlist, "-depth", 3, prop_name, prop_value)`

Specifying a depth $d$ of 0, limits the search to the set of objects passed in *hlist*. A depth $d$ of 0 is equivalent to the `"flat"` argument.

A specified logical operator may be applied to the pairs of *prop_name* and *prop_value*. The supported logical operators are: `"-and"`, `"-or"`, `"-xor"`, `"-not"`.

Objects may also be matched by comparing a regular expression to the property values, where property values that match `regexp (prop_value, pattern)` are returned.

Finally, objects may be matched by property name only by using the `"-property"` option.

Implementation Note: The search only includes objects with visible handles (HandleVisibility = `"on"`). See [findall], page 383, to search for all objects including hidden ones.

**See also:** [findall], page 383, [allchild], page 360, [get], page 359, [set], page 359.

| | |
|---|---|
| `h = findall ()` | [Function File] |
| `h = findall (prop_name, prop_value, ...)` | [Function File] |
| `h = findall (prop_name, prop_value, "-logical_op",`  `prop_name, prop_value)` | [Function File] |
| `h = findall ("-property", prop_name)` | [Function File] |
| `h = findall ("-regexp", prop_name, pattern)` | [Function File] |
| `h = findall (hlist, ...)` | [Function File] |
| `h = findall (hlist, "flat", ...)` | [Function File] |
| `h = findall (hlist, "-depth", d, ...)` | [Function File] |

Find graphics object, including hidden ones, with specified property values.

The return value $h$ is a list of handles to the found graphic objects.

`findall` performs the same search as `findobj`, but it includes hidden objects (HandleVisibility = `"off"`). For full documentation, see [findobj], page 382.

**See also:** [findobj], page 382, [allchild], page 360, [get], page 359, [set], page 359.

## 15.3.5 Managing Default Properties

Object properties have two classes of default values, *factory defaults* (the initial values) and *user-defined defaults*, which may override the factory defaults.

Although default values may be set for any object, they are set in parent objects and apply to child objects, of the specified object type. For example, setting the default `color` property of `line` objects to `"green"`, for the `root` object, will result in all `line` objects inheriting the `color "green"` as the default value.

    `set (0, "defaultlinecolor", "green");`

sets the default line color for all objects. The rule for constructing the property name to set a default value is

    `default + object-type + property-name`

This rule can lead to some strange looking names, for example `defaultlinelinewidth"` specifies the default `linewidth` property for `line` objects.

The example above used the root figure object, 0, so the default property value will apply to all line objects. However, default values are hierarchical, so defaults set in a figure objects override those set in the root figure object. Likewise, defaults set in axes objects override those set in figure or root figure objects. For example,

```
subplot (2, 1, 1);
set (0, "defaultlinecolor", "red");
set (1, "defaultlinecolor", "green");
set (gca (), "defaultlinecolor", "blue");
line (1:10, rand (1, 10));
subplot (2, 1, 2);
line (1:10, rand (1, 10));
figure (2)
line (1:10, rand (1, 10));
```

produces two figures. The line in first subplot window of the first figure is blue because it inherits its color from its parent axes object. The line in the second subplot window of the first figure is green because it inherits its color from its parent figure object. The line in the second figure window is red because it inherits its color from the global root figure parent object.

To remove a user-defined default setting, set the default property to the value `"remove"`. For example,

```
set (gca (), "defaultlinecolor", "remove");
```

removes the user-defined default line color setting from the current axes object. To quickly remove all user-defined defaults use the `reset` function.

`reset (h)` [Built-in Function]
    Reset the properties of the graphic object $h$ to their default values.

    For figures, the properties `"position"`, `"units"`, `"windowstyle"`, and `"paperunits"` are not affected. For axes, the properties `"position"` and `"units"` are not affected.

    The input $h$ may also be a vector of graphic handles in which case each individual object will be reset.

    **See also:** [cla], page 339, [clf], page 339, [newplot], page 337.

Getting the `"default"` property of an object returns a list of user-defined defaults set for the object. For example,

```
get (gca (), "default");
```

returns a list of user-defined default values for the current axes object.

Factory default values are stored in the root figure object. The command

```
get (0, "factory");
```

returns a list of factory defaults.

## 15.4 Advanced Plotting

### 15.4.1 Colors

Colors may be specified as RGB triplets with values ranging from zero to one, or by name. Recognized color names include `"blue"`, `"black"`, `"cyan"`, `"green"`, `"magenta"`, `"red"`, `"white"`, and `"yellow"`.

## 15.4.2 Line Styles

Line styles are specified by the following properties:

`linestyle`
>   May be one of
>
>   | | |
>   |---|---|
>   | `"-"` | Solid line. [default] |
>   | `"--"` | Dashed line. |
>   | `":"` | Dotted line. |
>   | `"-."` | A dash-dot line. |
>   | `"none"` | No line. Points will still be marked using the current Marker Style. |

`linewidth`
>   A number specifying the width of the line. The default is 1. A value of 2 is twice as wide as the default, etc.

## 15.4.3 Marker Styles

Marker styles are specified by the following properties:

`marker`
>   A character indicating a plot marker to be place at each data point, or `"none"`, meaning no markers should be displayed.

`markeredgecolor`
>   The color of the edge around the marker, or `"auto"`, meaning that the edge color is the same as the face color. See Section 15.4.1 [Colors], page 384.

`markerfacecolor`
>   The color of the marker, or `"none"` to indicate that the marker should not be filled. See Section 15.4.1 [Colors], page 384.

`markersize`
>   A number specifying the size of the marker. The default is 1. A value of 2 is twice as large as the default, etc.

The `colstyle` function will parse a `plot`-style specification and will return the color, line, and marker values that would result.

[*style, color, marker, msg*] = colstyle (*linespec*)  [Function File]
>   Parse *linespec* and return the line style, color, and markers given.
>
>   In the case of an error, the string *msg* will return the text of the error.

## 15.4.4 Callbacks

Callback functions can be associated with graphics objects and triggered after certain events occur. The basic structure of all callback function is

>   `function mycallback (src, data)`
>   `...`
>   `endfunction`

where `src` gives a handle to the source of the callback, and `code` gives some event specific data. This can then be associated with an object either at the objects creation or later with the `set` function. For example,

```
plot (x, "DeleteFcn", @(s, e) disp ("Window Deleted"))
```
where at the moment that the plot is deleted, the message "Window Deleted" will be displayed.

Additional user arguments can be passed to callback functions, and will be passed after the 2 default arguments. For example:
```
plot (x, "DeleteFcn", {@mycallback, "1"})
...
function mycallback (src, data, a1)
  fprintf ("Closing plot %d\n", a1);
endfunction
```
The basic callback functions that are available for all graphics objects are

- CreateFcn This is the callback that is called at the moment of the objects creation. It is not called if the object is altered in any way, and so it only makes sense to define this callback in the function call that defines the object. Callbacks that are added to `CreateFcn` later with the `set` function will never be executed.
- DeleteFcn This is the callback that is called at the moment an object is deleted.
- ButtonDownFcn This is the callback that is called if a mouse button is pressed while the pointer is over this object. Note, that the gnuplot interface does not respect this callback.

The object and figure that the event occurred in that resulted in the callback being called can be found with the `gcbo` and `gcbf` functions.

*h* = gcbo ()  [Function File]
[*h*, *fig*] = gcbo ()  [Function File]

Return a handle to the object whose callback is currently executing.

If no callback is executing, this function returns the empty matrix. This handle is obtained from the root object property `"CallbackObject"`.

When called with a second output argument, return the handle of the figure containing the object whose callback is currently executing. If no callback is executing the second output is also set to the empty matrix.

**See also:** [gcbf], page 386, [gco], page 358, [gca], page 358, [gcf], page 357, [get], page 359, [set], page 359.

*fig* = gcbf ()  [Function File]

Return a handle to the figure containing the object whose callback is currently executing.

If no callback is executing, this function returns the empty matrix. The handle returned by this function is the same as the second output argument of `gcbo`.

**See also:** [gcbo], page 386, [gcf], page 357, [gco], page 358, [gca], page 358, [get], page 359, [set], page 359.

Callbacks can equally be added to properties with the `addlistener` function described below.

Chapter 15: Plotting

## 15.4.5 Application-defined Data

Octave has a provision for attaching application-defined data to a graphics handle. The data can be anything which is meaningful to the application, and will be completely ignored by Octave.

setappdata (*h*, *name*, *value*) [Function File]
setappdata (*h*, *name1*, *value1*, *name2*, *value3*, ...) [Function File]

    Set the application data *name* to *value* for the graphics object with handle *h*.

    *h* may also be a vector of graphics handles. If the application data with the specified *name* does not exist, it is created. Multiple *name*/*value* pairs can be specified at a time.

    **See also:** [getappdata], page 387, [isappdata], page 387, [rmappdata], page 387, [guidata], page 737, [get], page 359, [set], page 359, [getpref], page 739, [setpref], page 740.

*value* = getappdata (*h*, *name*) [Function File]
*appdata* = getappdata (*h*) [Function File]

    Return the *value* of the application data *name* for the graphics object with handle *h*.

    *h* may also be a vector of graphics handles. If no second argument *name* is given then getappdata returns a structure, *appdata*, whose fields correspond to the appdata properties.

    **See also:** [setappdata], page 387, [isappdata], page 387, [rmappdata], page 387, [guidata], page 737, [get], page 359, [set], page 359, [getpref], page 739, [setpref], page 740.

rmappdata (*h*, *name*) [Function File]
rmappdata (*h*, *name1*, *name2*, ...) [Function File]

    Delete the application data *name* from the graphics object with handle *h*.

    *h* may also be a vector of graphics handles. Multiple application data names may be supplied to delete several properties at once.

    **See also:** [setappdata], page 387, [getappdata], page 387, [isappdata], page 387.

*valid* = isappdata (*h*, *name*) [Function File]

    Return true if the named application data, *name*, exists for the graphics object with handle *h*.

    *h* may also be a vector of graphics handles.

    **See also:** [getappdata], page 387, [setappdata], page 387, [rmappdata], page 387, [guidata], page 737, [get], page 359, [set], page 359, [getpref], page 739, [setpref], page 740.

## 15.4.6 Object Groups

A number of Octave high level plot functions return groups of other graphics objects or they return graphics objects that have their properties linked in such a way that changes to one of the properties results in changes in the others. A graphic object that groups other objects is an hggroup

hggroup ()                                                    [Function File]
hggroup (*hax*)                                               [Function File]
hggroup (..., *property*, *value*, ...)                       [Function File]
*h* = hggroup (...)                                           [Function File]

Create handle graphics group object with axes parent *hax*.

If no parent is specified, the group is created in the current axes.

Multiple property/value pairs may be specified for the hggroup, but they must appear in pairs.

The optional return value *h* is a graphics handle to the created hggroup object.

Programming Note: An hggroup is a way to group base graphics objects such as line objects or patch objects into a single unit which can react appropriately. For example, the individual lines of a contour plot are collected into a single hggroup so that they can be made visible/invisible with a single command, `set (hg_handle, "visible", "off")`.

See also: [addproperty], page 388, [addlistener], page 389.

For example a simple use of a hggroup might be

```
x = 0:0.1:10;
hg = hggroup ();
plot (x, sin (x), "color", [1, 0, 0], "parent", hg);
hold on
plot (x, cos (x), "color", [0, 1, 0], "parent", hg);
set (hg, "visible", "off");
```

which groups the two plots into a single object and controls their visibility directly. The default properties of an hggroup are the same as the set of common properties for the other graphics objects. Additional properties can be added with the addproperty function.

addproperty (*name*, *h*, *type*)                             [Built-in Function]
addproperty (*name*, *h*, *type*, *arg*, ...)                 [Built-in Function]

Create a new property named *name* in graphics object *h*.

*type* determines the type of the property to create. *args* usually contains the default value of the property, but additional arguments might be given, depending on the type of the property.

The supported property types are:

string   A string property. *arg* contains the default string value.

any      An un-typed property. This kind of property can hold any octave value. *args* contains the default value.

radio    A string property with a limited set of accepted values. The first argument must be a string with all accepted values separated by a vertical bar ('|'). The default value can be marked by enclosing it with a '{' '}' pair. The default value may also be given as an optional second string argument.

boolean  A boolean property. This property type is equivalent to a radio property with "on|off" as accepted values. *arg* contains the default property value.

Chapter 15: Plotting 389

>     double     A scalar double property. *arg* contains the default value.

>     handle     A handle property. This kind of property holds the handle of a graphics
>                object. *arg* contains the default handle value. When no default value is
>                given, the property is initialized to the empty matrix.

>     data       A data (matrix) property. *arg* contains the default data value. When no
>                default value is given, the data is initialized to the empty matrix.

>     color      A color property. *arg* contains the default color value. When no default
>                color is given, the property is set to black. An optional second string
>                argument may be given to specify an additional set of accepted string
>                values (like a radio property).

> *type* may also be the concatenation of a core object type and a valid property name
> for that object type. The property created then has the same characteristics as the
> referenced property (type, possible values, hidden state...). This allows one to clone
> an existing property into the graphics object *h*.
>
> Examples:
> ```
>     addproperty ("my_property", gcf, "string", "a string value");
>     addproperty ("my_radio", gcf, "radio", "val_1|val_2|{val_3}");
>     addproperty ("my_style", gcf, "linelinestyle", "--");
> ```
>
> **See also:** [addlistener], page 389, [hggroup], page 387.

Once a property in added to an **hggroup**, it is not linked to any other property of either
the children of the group, or any other graphics object. Add so to control the way in which
this newly added property is used, the **addlistener** function is used to define a callback
function that is executed when the property is altered.

**addlistener (*h*, *prop*, *fcn*)** [Built-in Function]
> Register *fcn* as listener for the property *prop* of the graphics object *h*.
>
> Property listeners are executed (in order of registration) when the property is set.
> The new value is already available when the listeners are executed.
>
> *prop* must be a string naming a valid property in *h*.
>
> *fcn* can be a function handle, a string or a cell array whose first element is a function
> handle. If *fcn* is a function handle, the corresponding function should accept at least
> 2 arguments, that will be set to the object handle and the empty matrix respectively.
> If *fcn* is a string, it must be any valid octave expression. If *fcn* is a cell array, the first
> element must be a function handle with the same signature as described above. The
> next elements of the cell array are passed as additional arguments to the function.
>
> Example:
> ```
>     function my_listener (h, dummy, p1)
>       fprintf ("my_listener called with p1=%s\n", p1);
>     endfunction
>
>     addlistener (gcf, "position", {@my_listener, "my string"})
> ```
> **See also:** [addproperty], page 388, [hggroup], page 387.

`dellistener (h, prop, fcn)`                                                                                [Built-in Function]

Remove the registration of *fcn* as a listener for the property *prop* of the graphics object *h*.

The function *fcn* must be the same variable (not just the same value), as was passed to the original call to `addlistener`.

If *fcn* is not defined then all listener functions of *prop* are removed.

Example:

```
function my_listener (h, dummy, p1)
  fprintf ("my_listener called with p1=%s\n", p1);
endfunction


c = {@my_listener, "my string"};
addlistener (gcf, "position", c);
dellistener (gcf, "position", c);
```

An example of the use of these two functions might be

```
x = 0:0.1:10;
hg = hggroup ();
h = plot (x, sin (x), "color", [1, 0, 0], "parent", hg);
addproperty ("linestyle", hg, "linelinestyle", get (h, "linestyle"));
addlistener (hg, "linestyle", @update_props);
hold on
plot (x, cos (x), "color", [0, 1, 0], "parent", hg);

function update_props (h, d)
  set (get (h, "children"), "linestyle", get (h, "linestyle"));
endfunction
```

that adds a `linestyle` property to the `hggroup` and propagating any changes its value to the children of the group. The `linkprop` function can be used to simplify the above to be

```
x = 0:0.1:10;
hg = hggroup ();
h1 = plot (x, sin (x), "color", [1, 0, 0], "parent", hg);
addproperty ("linestyle", hg, "linelinestyle", get (h, "linestyle"));
hold on
h2 = plot (x, cos (x), "color", [0, 1, 0], "parent", hg);
hlink = linkprop ([hg, h1, h2], "color");
```

`hlink = linkprop (h, "prop")`                                                 [Function File]
`hlink = linkprop (h, {"prop1", "prop2", ...})`                        [Function File]

Link graphic object properties, such that a change in one is propagated to the others.

The input *h* is a vector of graphic handles to link.

*prop* may be a string when linking a single property, or a cell array of strings for multiple properties. During the linking process all properties in *prop* will initially be set to the values that exist on the first object in the list *h*.

The function returns *hlink* which is a special object describing the link. As long as the reference *hlink* exists the link between graphic objects will be active. This means

Chapter 15: Plotting 391

that *hlink* must be preserved in a workspace variable, a global variable, or otherwise stored using a function such as `setappdata`, `guidata`. To unlink properties, execute `clear` *hlink*.

An example of the use of `linkprop` is

```
x = 0:0.1:10;
subplot (1,2,1);
h1 = plot (x, sin (x));
subplot (1,2,2);
h2 = plot (x, cos (x));
hlink = linkprop ([h1, h2], {"color","linestyle"});
set (h1, "color", "green");
set (h2, "linestyle", "--");
```

**See also:** [linkaxes], page 391.

`linkaxes (`*hax*`)` [Function File]
`linkaxes (`*hax*`, `*optstr*`)` [Function File]

Link the axis limits of 2-D plots such that a change in one is propagated to the others.

The axes handles to be linked are passed as the first argument *hax*.

The optional second argument is a string which defines which axis limits will be linked. The possible values for *optstr* are:

`"x"`   Link x-axes

`"y"`   Link y-axes

`"xy"` (default)
    Link both axes

`"off"` Turn off linking

If unspecified the default is to link both X and Y axes.

When linking, the limits from the first axes in *hax* are applied to the other axes in the list. Subsequent changes to any one of the axes will be propagated to the others.

**See also:** [linkprop], page 390, [addproperty], page 388.

These capabilities are used in a number of basic graphics objects. The `hggroup` objects created by the functions of Octave contain one or more graphics object and are used to:

- group together multiple graphics objects,
- create linked properties between different graphics objects, and
- to hide the nominal user data, from the actual data of the objects.

For example the `stem` function creates a stem series where each `hggroup` of the stem series contains two line objects representing the body and head of the stem. The `ydata` property of the `hggroup` of the stem series represents the head of the stem, whereas the body of the stem is between the baseline and this value. For example

```
h = stem (1:4)
get (h, "xdata")
⇒ [ 1   2   3   4]'
get (get (h, "children")(1), "xdata")
⇒ [ 1   1 NaN   2   2 NaN   3   3 NaN   4   4 NaN]'
```

shows the difference between the `xdata` of the `hggroup` of a stem series object and the underlying line.

The basic properties of such group objects is that they consist of one or more linked `hggroup`, and that changes in certain properties of these groups are propagated to other members of the group. Whereas, certain properties of the members of the group only apply to the current member.

In addition the members of the group can also be linked to other graphics objects through callback functions. For example the baseline of the `bar` or `stem` functions is a line object, whose length and position are automatically adjusted, based on changes to the corresponding hggroup elements.

### 15.4.6.1 Data Sources in Object Groups

All of the group objects contain data source parameters. There are string parameters that contain an expression that is evaluated to update the relevant data property of the group when the `refreshdata` function is called.

**refreshdata ()** [Function File]
**refreshdata (***h***)** [Function File]
**refreshdata (***h***, *workspace*)** [Function File]

Evaluate any 'datasource' properties of the current figure and update the plot if the corresponding data has changed.

If the first argument *h* is a list of graphic handles, then operate on these objects rather than the current figure returned by `gcf`.

The optional second argument *workspace* can take the following values:

"base"    Evaluate the datasource properties in the base workspace. (default).

"caller"  Evaluate the datasource properties in the workspace of the function that called `refreshdata`.

An example of the use of `refreshdata` is:

```
x = 0:0.1:10;
y = sin (x);
plot (x, y, "ydatasource", "y");
for i = 1 : 100
  pause (0.1);
  y = sin (x + 0.1*i);
  refreshdata ();
endfor
```

### 15.4.6.2 Area Series

Area series objects are created by the `area` function. Each of the `hggroup` elements contains a single patch object. The properties of the area series are

`basevalue`
The value where the base of the area plot is drawn.

linewidth
linestyle

> The line width and style of the edge of the patch objects making up the areas. See Section 15.4.2 [Line Styles], page 385.

edgecolor
facecolor

> The line and fill color of the patch objects making up the areas. See Section 15.4.1 [Colors], page 384.

xdata
ydata

> The x and y coordinates of the original columns of the data passed to `area` prior to the cumulative summation used in the `area` function.

xdatasource
ydatasource

> Data source variables.

### 15.4.6.3 Bar Series

Bar series objects are created by the `bar` or `barh` functions. Each `hggroup` element contains a single patch object. The properties of the bar series are

showbaseline
baseline
basevalue

> The property `showbaseline` flags whether the baseline of the bar series is displayed (default is `"on"`). The handle of the graphics object representing the baseline is given by the `baseline` property and the y-value of the baseline by the `basevalue` property.
>
> Changes to any of these properties are propagated to the other members of the bar series and to the baseline itself. Equally, changes in the properties of the base line itself are propagated to the members of the corresponding bar series.

barwidth
barlayout
horizontal

> The property `barwidth` is the width of the bar corresponding to the *width* variable passed to `bar` or *barh*. Whether the bar series is `"grouped"` or `"stacked"` is determined by the `barlayout` property and whether the bars are horizontal or vertical by the `horizontal` property.
>
> Changes to any of these property are propagated to the other members of the bar series.

linewidth
linestyle

> The line width and style of the edge of the patch objects making up the bars. See Section 15.4.2 [Line Styles], page 385.

edgecolor
facecolor
: The line and fill color of the patch objects making up the bars. See Section 15.4.1 [Colors], page 384.

xdata
: The nominal x positions of the bars. Changes in this property and propagated to the other members of the bar series.

ydata
: The y value of the bars in the `hggroup`.

xdatasource
ydatasource
: Data source variables.

### 15.4.6.4 Contour Groups

Contour group objects are created by the `contour`, `contourf` and `contour3` functions. The are equally one of the handles returned by the `surfc` and `meshc` functions. The properties of the contour group are

contourmatrix
: A read only property that contains the data return by `contourc` used to create the contours of the plot.

fill
: A radio property that can have the values "on" or "off" that flags whether the contours to plot are to be filled.

zlevelmode
zlevel
: The radio property `zlevelmode` can have the values "none", "auto", or "manual". When its value is "none" there is no z component to the plotted contours. When its value is "auto" the z value of the plotted contours is at the same value as the contour itself. If the value is "manual", then the z value at which to plot the contour is determined by the `zlevel` property.

levellistmode
levellist
levelstepmode
levelstep
: If `levellistmode` is "manual", then the levels at which to plot the contours is determined by `levellist`. If `levellistmode` is set to "auto", then the distance between contours is determined by `levelstep`. If both `levellistmode` and `levelstepmode` are set to "auto", then there are assumed to be 10 equal spaced contours.

textlistmode
textlist
textstepmode
textstep
: If `textlistmode` is "manual", then the labeled contours is determined by `textlist`. If `textlistmode` is set to "auto", then the distance between labeled contours is determined by `textstep`. If both `textlistmode` and `textstepmode` are set to "auto", then there are assumed to be 10 equal spaced labeled contours.

Chapter 15: Plotting 395

`showtext`   Flag whether the contour labels are shown or not.

`labelspacing`
:   The distance between labels on a single contour in points.

`linewidth`
`linestyle`
`linecolor`
:   The properties of the contour lines. The properties `linewidth` and `linestyle` are similar to the corresponding properties for lines. The property `linecolor` is a color property (see Section 15.4.1 [Colors], page 384), that can also have the values of "none" or "auto". If `linecolor` is "none", then no contour line is drawn. If `linecolor` is "auto" then the line color is determined by the colormap.

`xdata`
`ydata`
`zdata`   The original x, y, and z data of the contour lines.

`xdatasource`
`ydatasource`
`zdatasource`
:   Data source variables.

### 15.4.6.5 Error Bar Series

Error bar series are created by the `errorbar` function. Each `hggroup` element contains two line objects representing the data and the errorbars separately. The properties of the error bar series are

`color`   The RGB color or color name of the line objects of the error bars. See Section 15.4.1 [Colors], page 384.

`linewidth`
`linestyle`
:   The line width and style of the line objects of the error bars. See Section 15.4.2 [Line Styles], page 385.

`marker`
`markeredgecolor`
`markerfacecolor`
`markersize`
:   The line and fill color of the markers on the error bars. See Section 15.4.1 [Colors], page 384.

`xdata`
`ydata`
`ldata`
`udata`
`xldata`
`xudata`   The original x, y, l, u, xl, xu data of the error bars.

```
xdatasource
ydatasource
ldatasource
udatasource
xldatasource
xudatasource
```
        Data source variables.

### 15.4.6.6 Line Series

Line series objects are created by the `plot` and `plot3` functions and are of the type `line`. The properties of the line series with the ability to add data sources.

`color`      The RGB color or color name of the line objects. See Section 15.4.1 [Colors], page 384.

```
linewidth
linestyle
```
        The line width and style of the line objects. See Section 15.4.2 [Line Styles], page 385.

```
marker
markeredgecolor
markerfacecolor
markersize
```
        The line and fill color of the markers. See Section 15.4.1 [Colors], page 384.

```
xdata
ydata
zdata
```
        The original x, y and z data.

```
xdatasource
ydatasource
zdatasource
```
        Data source variables.

### 15.4.6.7 Quiver Group

Quiver series objects are created by the `quiver` or `quiver3` functions. Each `hggroup` element of the series contains three line objects as children representing the body and head of the arrow, together with a marker as the point of origin of the arrows. The properties of the quiver series are

```
autoscale
autoscalefactor
```
        Flag whether the length of the arrows is scaled or defined directly from the u, v and w data. If the arrow length is flagged as being scaled by the `autoscale` property, then the length of the autoscaled arrow is controlled by the `autoscalefactor`.

`maxheadsize`
        This property controls the size of the head of the arrows in the quiver series. The default value is 0.2.

Chapter 15: Plotting 397

showarrowhead
: Flag whether the arrow heads are displayed in the quiver plot.

color
: The RGB color or color name of the line objects of the quiver. See Section 15.4.1 [Colors], page 384.

linewidth
linestyle
: The line width and style of the line objects of the quiver. See Section 15.4.2 [Line Styles], page 385.

marker
markerfacecolor
markersize
: The line and fill color of the marker objects at the original of the arrows. See Section 15.4.1 [Colors], page 384.

xdata
ydata
zdata
: The origins of the values of the vector field.

udata
vdata
wdata
: The values of the vector field to plot.

xdatasource
ydatasource
zdatasource
udatasource
vdatasource
wdatasource
: Data source variables.

### 15.4.6.8 Scatter Group

Scatter series objects are created by the `scatter` or `scatter3` functions. A single hggroup element contains as many children as there are points in the scatter plot, with each child representing one of the points. The properties of the stem series are

linewidth
: The line width of the line objects of the points. See Section 15.4.2 [Line Styles], page 385.

marker
markeredgecolor
markerfacecolor
: The line and fill color of the markers of the points. See Section 15.4.1 [Colors], page 384.

xdata
ydata
zdata
: The original x, y and z data of the stems.

`cdata`    The color data for the points of the plot. Each point can have a separate color, or a unique color can be specified.

`sizedata`    The size data for the points of the plot. Each point can its own size or a unique size can be specified.

`xdatasource`
`ydatasource`
`zdatasource`
`cdatasource`
`sizedatasource`
   Data source variables.

### 15.4.6.9 Stair Group

Stair series objects are created by the `stair` function. Each `hggroup` element of the series contains a single line object as a child representing the stair. The properties of the stair series are

`color`    The RGB color or color name of the line objects of the stairs. See Section 15.4.1 [Colors], page 384.

`linewidth`
`linestyle`
   The line width and style of the line objects of the stairs. See Section 15.4.2 [Line Styles], page 385.

`marker`
`markeredgecolor`
`markerfacecolor`
`markersize`
   The line and fill color of the markers on the stairs. See Section 15.4.1 [Colors], page 384.

`xdata`
`ydata`    The original x and y data of the stairs.

`xdatasource`
`ydatasource`
   Data source variables.

### 15.4.6.10 Stem Series

Stem series objects are created by the `stem` or `stem3` functions. Each `hggroup` element contains a single line object as a child representing the stems. The properties of the stem series are

`showbaseline`
`baseline`
`basevalue`
   The property `showbaseline` flags whether the baseline of the stem series is displayed (default is "on"). The handle of the graphics object representing the baseline is given by the `baseline` property and the y-value (or z-value for `stem3`) of the baseline by the `basevalue` property.

Chapter 15: Plotting 399

> Changes to any of these property are propagated to the other members of the stem series and to the baseline itself. Equally changes in the properties of the base line itself are propagated to the members of the corresponding stem series.

`color`
> The RGB color or color name of the line objects of the stems. See Section 15.4.1 [Colors], page 384.

`linewidth`
`linestyle`
> The line width and style of the line objects of the stems. See Section 15.4.2 [Line Styles], page 385.

`marker`
`markeredgecolor`
`markerfacecolor`
`markersize`
> The line and fill color of the markers on the stems. See Section 15.4.1 [Colors], page 384.

`xdata`
`ydata`
`zdata`
> The original x, y and z data of the stems.

`xdatasource`
`ydatasource`
`zdatasource`
> Data source variables.

### 15.4.6.11 Surface Group

Surface group objects are created by the `surf` or `mesh` functions, but are equally one of the handles returned by the `surfc` or `meshc` functions. The surface group is of the type `surface`.

The properties of the surface group are

`edgecolor`
`facecolor`
> The RGB color or color name of the edges or faces of the surface. See Section 15.4.1 [Colors], page 384.

`linewidth`
`linestyle`
> The line width and style of the lines on the surface. See Section 15.4.2 [Line Styles], page 385.

`marker`
`markeredgecolor`
`markerfacecolor`
`markersize`
> The line and fill color of the markers on the surface. See Section 15.4.1 [Colors], page 384.

`xdata`
`ydata`
`zdata`
`cdata`    The original x, y, z and c data.

`xdatasource`
`ydatasource`
`zdatasource`
`cdatasource`
    Data source variables.

### 15.4.7 Graphics Toolkits

`name = graphics_toolkit ()`                                       [Function File]
`name = graphics_toolkit (hlist)`                                  [Function File]
`graphics_toolkit (name)`                                          [Function File]
`graphics_toolkit (hlist, name)`                                   [Function File]
    Query or set the default graphics toolkit which is assigned to new figures.

    With no inputs, return the current default graphics toolkit. If the input is a list of figure graphic handles, hlist, then return the name of the graphics toolkit in use for each figure.

    When called with a single input name set the default graphics toolkit to name. If the toolkit is not already loaded, it is initialized by calling the function `__init_name__`. If the first input is a list of figure handles, hlist, then the graphics toolkit is set to name for these figures only.

    See also: [available_graphics_toolkits], page 400.

`available_graphics_toolkits ()`                                   [Built-in Function]
    Return a cell array of registered graphics toolkits.

    See also: [graphics_toolkit], page 400, [register_graphics_toolkit], page 400.

`loaded_graphics_toolkits ()`                                      [Built-in Function]
    Return a cell array of the currently loaded graphics toolkits.

    See also: [available_graphics_toolkits], page 400.

`register_graphics_toolkit (toolkit)`                              [Built-in Function]
    List toolkit as an available graphics toolkit.

    See also: [available_graphics_toolkits], page 400.

#### 15.4.7.1 Customizing Toolkit Behavior

The specific behavior of the backend toolkit may be modified using the following utility functions. Note: Not all functions apply to every graphics toolkit.

`[prog, args] = gnuplot_binary ()`                                 [Loadable Function]
`[old_prog, old_args] = gnuplot_binary (new_prog, arg1, ...)`      [Loadable Function]
    Query or set the name of the program invoked by the plot command when the graphics toolkit is set to "gnuplot".

Additional arguments to pass to the external plotting program may also be given. The default value is `"gnuplot"` with no additional arguments. See Appendix G [Installation], page 889.

**See also:** [graphics_toolkit], page 400.

# 16 Matrix Manipulation

There are a number of functions available for checking to see if the elements of a matrix meet some condition, and for rearranging the elements of a matrix. For example, Octave can easily tell you if all the elements of a matrix are finite, or are less than some specified value. Octave can also rotate the elements, extract the upper- or lower-triangular parts, or sort the columns of a matrix.

## 16.1 Finding Elements and Checking Conditions

The functions `any` and `all` are useful for determining whether any or all of the elements of a matrix satisfy some condition. The `find` function is also useful in determining which elements of a matrix meet a specified condition.

**any** (*x*)                                                                                               [Built-in Function]
**any** (*x*, *dim*)                                                                                    [Built-in Function]

    For a vector argument, return true (logical 1) if any element of the vector is nonzero.

    For a matrix argument, return a row vector of logical ones and zeros with each element indicating whether any of the elements of the corresponding column of the matrix are nonzero. For example:

```
any (eye (2, 4))
     ⇒ [ 1, 1, 0, 0 ]
```

    If the optional argument *dim* is supplied, work along dimension *dim*. For example:

```
any (eye (2, 4), 2)
     ⇒ [ 1; 1 ]
```

    **See also:** [all], page 403.

**all** (*x*)                                                                                                 [Built-in Function]
**all** (*x*, *dim*)                                                                                   [Built-in Function]

    For a vector argument, return true (logical 1) if all elements of the vector are nonzero.

    For a matrix argument, return a row vector of logical ones and zeros with each element indicating whether all of the elements of the corresponding column of the matrix are nonzero. For example:

```
all ([2, 3; 1, 0])
     ⇒ [ 1, 0 ]
```

    If the optional argument *dim* is supplied, work along dimension *dim*.

    **See also:** [any], page 403.

Since the comparison operators (see Section 8.4 [Comparison Ops], page 145) return matrices of ones and zeros, it is easy to test a matrix for many things, not just whether the elements are nonzero. For example,

```
all (all (rand (5) < 0.9))
     ⇒ 0
```

tests a random 5 by 5 matrix to see if all of its elements are less than 0.9.

Note that in conditional contexts (like the test clause of `if` and `while` statements) Octave treats the test as if you had typed `all (all (condition))`.

z = xor (x, y)      [Function File]
z = xor (x1, x2, ...)      [Function File]

    Return the *exclusive or* of x and y.

    For boolean expressions x and y, xor (x, y) is true if and only if one of x or y is true. Otherwise, if x and y are both true or both false, xor returns false.

    The truth table for the xor operation is

```
x  y  z
-  -  -

0  0  0
1  0  1
0  1  1
1  1  0
```

    If more than two arguments are given the xor operation is applied cumulatively from left to right:

        (...((x1 XOR x2) XOR x3) XOR ...)

    **See also:** [and], page 147, [or], page 147, [not], page 147.

diff (x)      [Built-in Function]
diff (x, k)      [Built-in Function]
diff (x, k, dim)      [Built-in Function]

    If x is a vector of length n, diff (x) is the vector of first differences $x_2 - x_1, \ldots, x_n - x_{n-1}$.

    If x is a matrix, diff (x) is the matrix of column differences along the first non-singleton dimension.

    The second argument is optional. If supplied, diff (x, k), where k is a non-negative integer, returns the k-th differences. It is possible that k is larger than the first non-singleton dimension of the matrix. In this case, diff continues to take the differences along the next non-singleton dimension.

    The dimension along which to take the difference can be explicitly stated with the optional variable *dim*. In this case the k-th order differences are calculated along this dimension. In the case where k exceeds size (x, dim) an empty matrix is returned.

    **See also:** [sort], page 412, [merge], page 149.

isinf (x)      [Mapping Function]

    Return a logical array which is true where the elements of x are infinite and false where they are not.

    For example:

        isinf ([13, Inf, NA, NaN])
           ⇒ [ 0, 1, 0, 0 ]

    **See also:** [isfinite], page 405, [isnan], page 404, [isna], page 43.

isnan (x)      [Mapping Function]

    Return a logical array which is true where the elements of x are NaN values and false where they are not.

Chapter 16: Matrix Manipulation 405

NA values are also considered NaN values. For example:

```
isnan ([13, Inf, NA, NaN])
    ⇒ [ 0, 0, 1, 1 ]
```

**See also:** [isna], page 43, [isinf], page 404, [isfinite], page 405.

**isfinite** (*x*)                                                                  [Mapping Function]

Return a logical array which is true where the elements of *x* are finite values and false where they are not.

For example:

```
isfinite ([13, Inf, NA, NaN])
    ⇒ [ 1, 0, 0, 0 ]
```

**See also:** [isinf], page 404, [isnan], page 404, [isna], page 43.

**[err, y1, ...] = common_size** (*x1, ...*)                               [Function File]

Determine if all input arguments are either scalar or of common size.

If true, *err* is zero, and *yi* is a matrix of the common size with all entries equal to *xi* if this is a scalar or *xi* otherwise. If the inputs cannot be brought to a common size, *err* is 1, and *yi* is *xi*. For example:

```
[errorcode, a, b] = common_size ([1 2; 3 4], 5)
    ⇒ errorcode = 0
    ⇒ a = [ 1, 2; 3, 4 ]
    ⇒ b = [ 5, 5; 5, 5 ]
```

This is useful for implementing functions where arguments can either be scalars or of common size.

*idx* = **find** (*x*)                                                                           [Built-in Function]
*idx* = **find** (*x, n*)                                                      [Built-in Function]
*idx* = **find** (*x, n, direction*)                          [Built-in Function]
[*i, j*] = **find** (...)                                                 [Built-in Function]
[*i, j, v*] = **find** (...)                                        [Built-in Function]

Return a vector of indices of nonzero elements of a matrix, as a row if *x* is a row vector or as a column otherwise.

To obtain a single index for each matrix element, Octave pretends that the columns of a matrix form one long vector (like Fortran arrays are stored). For example:

```
find (eye (2))
    ⇒ [ 1; 4 ]
```

If two inputs are given, *n* indicates the maximum number of elements to find from the beginning of the matrix or vector.

If three inputs are given, *direction* should be one of "**first**" or "**last**", requesting only the first or last *n* indices, respectively. However, the indices are always returned in ascending order.

If two outputs are requested, **find** returns the row and column indices of nonzero elements of a matrix. For example:

```
[i, j] = find (2 * eye (2))
    ⇒ i = [ 1; 2 ]
    ⇒ j = [ 1; 2 ]
```

If three outputs are requested, `find` also returns a vector containing the nonzero values. For example:

```
[i, j, v] = find (3 * eye (2))
    ⇒ i = [ 1; 2 ]
    ⇒ j = [ 1; 2 ]
    ⇒ v = [ 3; 3 ]
```

Note that this function is particularly useful for sparse matrices, as it extracts the nonzero elements as vectors, which can then be used to create the original matrix. For example:

```
sz = size (a);
[i, j, v] = find (a);
b = sparse (i, j, v, sz(1), sz(2));
```

**See also:** [nonzeros], page 524.

*idx* = lookup (*table*, *y*)            [Built-in Function]
*idx* = lookup (*table*, *y*, *opt*)            [Built-in Function]

Lookup values in a sorted table.

This function is usually used as a prelude to interpolation.

If table is increasing and `idx = lookup (table, y)`, then `table(idx(i)) <= y(i) < table(idx(i+1))` for all `y(i)` within the table. If `y(i) < table(1)` then `idx(i)` is 0. If `y(i) >= table(end)` or `isnan (y(i))` then `idx(i)` is n.

If the table is decreasing, then the tests are reversed. For non-strictly monotonic tables, empty intervals are always skipped. The result is undefined if *table* is not monotonic, or if *table* contains a NaN.

The complexity of the lookup is $O(M*\log(N))$ where N is the size of *table* and M is the size of *y*. In the special case when *y* is also sorted, the complexity is $O(\min(M*\log(N), M+N))$.

*table* and *y* can also be cell arrays of strings (or *y* can be a single string). In this case, string lookup is performed using lexicographical comparison.

If *opts* is specified, it must be a string with letters indicating additional options.

- m      `table(idx(i)) == val(i)` if `val(i)` occurs in table; otherwise, `idx(i)` is zero.

- b      `idx(i)` is a logical 1 or 0, indicating whether `val(i)` is contained in table or not.

- l      For numeric lookups the leftmost subinterval shall be extended to infinity (i.e., all indices at least 1)

- r      For numeric lookups the rightmost subinterval shall be extended to infinity (i.e., all indices at most n-1).

If you wish to check if a variable exists at all, instead of properties its elements may have, consult Section 7.3 [Status of Variables], page 127.

# Chapter 16: Matrix Manipulation

## 16.2 Rearranging Matrices

**fliplr** (*x*) [Function File]

Flip array left to right.

Return a copy of x with the order of the columns reversed. In other words, x is flipped left-to-right about a vertical axis. For example:

```
fliplr ([1, 2; 3, 4])
    ⇒   2   1
        4   3
```

**See also:** [flipud], page 407, [flip], page 407, [rot90], page 408, [rotdim], page 408.

**flipud** (*x*) [Function File]

Flip array upside down.

Return a copy of x with the order of the rows reversed. In other words, x is flipped upside-down about a horizontal axis. For example:

```
flipud ([1, 2; 3, 4])
    ⇒   3   4
        1   2
```

**See also:** [fliplr], page 407, [flip], page 407, [rot90], page 408, [rotdim], page 408.

**flip** (*x*) [Function File]
**flip** (*x*, *dim*) [Function File]

Flip array across dimension *dim*.

Return a copy of x flipped about the dimension *dim*. *dim* defaults to the first non-singleton dimension. For example:

```
flip ([1 2 3 4])
    ⇒   4   3   2   1

flip ([1; 2; 3; 4])
    ⇒   4
        3
        2
        1

flip ([1 2; 3 4])
    ⇒   3   4
        1   2

flip ([1 2; 3 4], 2)
    ⇒   2   1
        4   3
```

**See also:** [fliplr], page 407, [flipud], page 407, [rot90], page 408, [rotdim], page 408, [permute], page 409, [transpose], page 144.

`rot90 (A)` [Function File]
`rot90 (A, k)` [Function File]

Rotate array by 90 degree increments.

Return a copy of A with the elements rotated counterclockwise in 90-degree increments.

The second argument is optional, and specifies how many 90-degree rotations are to be applied (the default value is 1). Negative values of k rotate the matrix in a clockwise direction. For example,

```
rot90 ([1, 2; 3, 4], -1)
  ⇒  3  1
     4  2
```

rotates the given matrix clockwise by 90 degrees. The following are all equivalent statements:

```
rot90 ([1, 2; 3, 4], -1)
rot90 ([1, 2; 3, 4], 3)
rot90 ([1, 2; 3, 4], 7)
```

The rotation is always performed on the plane of the first two dimensions, i.e., rows and columns. To perform a rotation on any other plane, use `rotdim`.

See also: [rotdim], page 408, [fliplr], page 407, [flipud], page 407, [flip], page 407.

`rotdim (x)` [Function File]
`rotdim (x, n)` [Function File]
`rotdim (x, n, plane)` [Function File]

Return a copy of x with the elements rotated counterclockwise in 90-degree increments.

The second argument n is optional, and specifies how many 90-degree rotations are to be applied (the default value is 1). Negative values of n rotate the matrix in a clockwise direction.

The third argument is also optional and defines the plane of the rotation. If present, plane is a two element vector containing two different valid dimensions of the matrix. When plane is not given the first two non-singleton dimensions are used.

For example,

```
rotdim ([1, 2; 3, 4], -1, [1, 2])
  ⇒  3  1
     4  2
```

rotates the given matrix clockwise by 90 degrees. The following are all equivalent statements:

```
rotdim ([1, 2; 3, 4], -1, [1, 2])
rotdim ([1, 2; 3, 4], 3, [1, 2])
rotdim ([1, 2; 3, 4], 7, [1, 2])
```

See also: [rot90], page 408, [fliplr], page 407, [flipud], page 407, [flip], page 407.

`cat (dim, array1, array2, ..., arrayN)` [Built-in Function]

Return the concatenation of N-D array objects, array1, array2, ..., arrayN along dimension dim.

# Chapter 16: Matrix Manipulation

```
A = ones (2, 2);
B = zeros (2, 2);
cat (2, A, B)
    ⇒ 1 1 0 0
      1 1 0 0
```

Alternatively, we can concatenate A and B along the second dimension in the following way:

```
[A, B]
```

*dim* can be larger than the dimensions of the N-D array objects and the result will thus have *dim* dimensions as the following example shows:

```
cat (4, ones (2, 2), zeros (2, 2))
    ⇒ ans(:,:,1,1) =

        1 1
        1 1

      ans(:,:,1,2) =

        0 0
        0 0
```

**See also:** [horzcat], page 409, [vertcat], page 409.

**horzcat (*array1*, *array2*, ..., *arrayN*)**  [Built-in Function]

Return the horizontal concatenation of N-D array objects, *array1*, *array2*, ..., *arrayN* along dimension 2.

Arrays may also be concatenated horizontally using the syntax for creating new matrices. For example:

```
hcat = [ array1, array2, ... ]
```

**See also:** [cat], page 408, [vertcat], page 409.

**vertcat (*array1*, *array2*, ..., *arrayN*)**  [Built-in Function]

Return the vertical concatenation of N-D array objects, *array1*, *array2*, ..., *arrayN* along dimension 1.

Arrays may also be concatenated vertically using the syntax for creating new matrices. For example:

```
vcat = [ array1; array2; ... ]
```

**See also:** [cat], page 408, [horzcat], page 409.

**permute (*A*, *perm*)**  [Built-in Function]

Return the generalized transpose for an N-D array object *A*.

The permutation vector *perm* must contain the elements `1:ndims (A)` (in any order, but each element must appear only once).

The Nth dimension of *A* gets remapped to dimension *PERM(N)*. For example:

```
x = zeros ([2, 3, 5, 7]);
size (x)
    ⇒  2   3   5   7

size (permute (x, [2, 1, 3, 4]))
    ⇒  3   2   5   7

size (permute (x, [1, 3, 4, 2]))
    ⇒  2   5   7   3

## The identity permutation
size (permute (x, [1, 2, 3, 4]))
    ⇒  2   3   5   7
```

See also: [ipermute], page 410.

**ipermute** (*A*, *iperm*)                                                                                    [Built-in Function]

The inverse of the **permute** function.

The expression

```
ipermute (permute (A, perm), perm)
```

returns the original array *A*.

See also: [permute], page 409.

**reshape** (*A*, *m*, *n*, ...)                                                     [Built-in Function]
**reshape** (*A*, [*m n* ...])                                                    [Built-in Function]
**reshape** (*A*, ..., [], ...)                                              [Built-in Function]
**reshape** (*A*, *size*)                                                              [Built-in Function]

Return a matrix with the specified dimensions (*m*, *n*, ...) whose elements are taken from the matrix *A*.

The elements of the matrix are accessed in column-major order (like Fortran arrays are stored).

The following code demonstrates reshaping a 1x4 row vector into a 2x2 square matrix.

```
reshape ([1, 2, 3, 4], 2, 2)
    ⇒  1   3
       2   4
```

Note that the total number of elements in the original matrix (**prod (size (A))**) must match the total number of elements in the new matrix (**prod ([m n ...])**).

A single dimension of the return matrix may be left unspecified and Octave will determine its size automatically. An empty matrix ([]) is used to flag the unspecified dimension.

See also: [resize], page 410, [vec], page 415, [postpad], page 415, [cat], page 408, [squeeze], page 46.

**resize** (*x*, *m*)                                                                      [Built-in Function]
**resize** (*x*, *m*, *n*, ...)                                                      [Built-in Function]

Chapter 16: Matrix Manipulation 411

**resize (x, [m n ...])**                                                                         [Built-in Function]

Resize x cutting off elements as necessary.

In the result, element with certain indices is equal to the corresponding element of x if the indices are within the bounds of x; otherwise, the element is set to zero.

In other words, the statement

```
y = resize (x, dv)
```

is equivalent to the following code:

```
y = zeros (dv, class (x));
sz = min (dv, size (x));
for i = 1:length (sz)
  idx{i} = 1:sz(i);
endfor
y(idx{:}) = x(idx{:});
```

but is performed more efficiently.

If only m is supplied, and it is a scalar, the dimension of the result is m-by-m. If m, n, ... are all scalars, then the dimensions of the result are m-by-n-by-.... If given a vector as input, then the dimensions of the result are given by the elements of that vector.

An object can be resized to more dimensions than it has; in such case the missing dimensions are assumed to be 1. Resizing an object to fewer dimensions is not possible.

**See also:** [reshape], page 410, [postpad], page 415, [prepad], page 415, [cat], page 408.

**y = circshift (x, n)**                                                                                     [Function File]

Circularly shift the values of the array x.

n must be a vector of integers no longer than the number of dimensions in x. The values of n can be either positive or negative, which determines the direction in which the values or x are shifted. If an element of n is zero, then the corresponding dimension of x will not be shifted. For example:

```
x = [1, 2, 3; 4, 5, 6; 7, 8, 9];
circshift (x, 1)
⇒  7, 8, 9
   1, 2, 3
   4, 5, 6
circshift (x, -2)
⇒  7, 8, 9
   1, 2, 3
   4, 5, 6
circshift (x, [0,1])
⇒  3, 1, 2
   6, 4, 5
   9, 7, 8
```

**See also:** [permute], page 409, [ipermute], page 410, [shiftdim], page 412.

shift (*x, b*)                                                                                    [Function File]
shift (*x, b, dim*)                                                                              [Function File]
    If *x* is a vector, perform a circular shift of length *b* of the elements of *x*.

    If *x* is a matrix, do the same for each column of *x*.

    If the optional *dim* argument is given, operate along this dimension.

*y* = shiftdim (*x, n*)                                                                      [Function File]
[*y, ns*] = shiftdim (*x*)                                                                 [Function File]
    Shift the dimensions of *x* by *n*, where *n* must be an integer scalar.

    When *n* is positive, the dimensions of *x* are shifted to the left, with the leading dimensions circulated to the end. If *n* is negative, then the dimensions of *x* are shifted to the right, with *n* leading singleton dimensions added.

    Called with a single argument, `shiftdim`, removes the leading singleton dimensions, returning the number of dimensions removed in the second output argument *ns*.

    For example:

```
x = ones (1, 2, 3);
size (shiftdim (x, -1))
    ⇒ [1, 1, 2, 3]
size (shiftdim (x, 1))
    ⇒ [2, 3]
[b, ns] = shiftdim (x)
    ⇒ b = [1, 1, 1; 1, 1, 1]
    ⇒ ns = 1
```

    **See also:** [reshape], page 410, [permute], page 409, [ipermute], page 410, [circshift], page 411, [squeeze], page 46.

[*s, i*] = sort (*x*)                                                                          [Built-in Function]
[*s, i*] = sort (*x, dim*)                                                                 [Built-in Function]
[*s, i*] = sort (*x, mode*)                                                             [Built-in Function]
[*s, i*] = sort (*x, dim, mode*)                                                     [Built-in Function]
    Return a copy of *x* with the elements arranged in increasing order.

    For matrices, `sort` orders the elements within columns

    For example:

```
sort ([1, 2; 2, 3; 3, 1])
    ⇒   1   1
        2   2
        3   3
```

    If the optional argument *dim* is given, then the matrix is sorted along the dimension defined by *dim*. The optional argument `mode` defines the order in which the values will be sorted. Valid values of `mode` are `"ascend"` or `"descend"`.

    The `sort` function may also be used to produce a matrix containing the original row indices of the elements in the sorted matrix. For example:

Chapter 16: Matrix Manipulation 413

```
[s, i] = sort ([1, 2; 2, 3; 3, 1])
    ⇒ s = 1   1
          2   2
          3   3
    ⇒ i = 1   3
          2   1
          3   2
```

For equal elements, the indices are such that equal elements are listed in the order in which they appeared in the original list.

Sorting of complex entries is done first by magnitude (`abs (z)`) and for any ties by phase angle (`angle (z)`). For example:

```
sort ([1+i; 1; 1-i])
    ⇒ 1 + 0i
      1 - 1i
      1 + 1i
```

NaN values are treated as being greater than any other value and are sorted to the end of the list.

The `sort` function may also be used to sort strings and cell arrays of strings, in which case ASCII dictionary order (uppercase 'A' precedes lowercase 'a') of the strings is used.

The algorithm used in `sort` is optimized for the sorting of partially ordered lists.

**See also:** [sortrows], page 413, [issorted], page 413.

`[s, i] = sortrows (A)`  [Function File]
`[s, i] = sortrows (A, c)`  [Function File]

Sort the rows of the matrix $A$ according to the order of the columns specified in $c$.

If $c$ is omitted, a lexicographical sort is used. By default ascending order is used however if elements of $c$ are negative then the corresponding column is sorted in descending order.

**See also:** [sort], page 412.

`issorted (a)`  [Built-in Function]
`issorted (a, mode)`  [Built-in Function]
`issorted (a, "rows", mode)`  [Built-in Function]

Return true if the array is sorted according to *mode*, which may be either `"ascending"`, `"descending"`, or `"either"`.

By default, *mode* is `"ascending"`. NaNs are treated in the same manner as `sort`.

If the optional argument `"rows"` is supplied, check whether the array is sorted by rows as output by the function `sortrows` (with no options).

This function does not support sparse matrices.

**See also:** [sort], page 412, [sortrows], page 413.

`nth_element (x, n)`  [Built-in Function]
`nth_element (x, n, dim)`  [Built-in Function]

Select the n-th smallest element of a vector, using the ordering defined by `sort`.

The result is equivalent to `sort(x)(n)`.

n can also be a contiguous range, either ascending `1:u` or descending `u:-1:1`, in which case a range of elements is returned.

If x is an array, `nth_element` operates along the dimension defined by *dim*, or the first non-singleton dimension if *dim* is not given.

Programming Note: nth_element encapsulates the C++ standard library algorithms nth_element and partial_sort. On average, the complexity of the operation is O(M*log(K)), where `M = size (x, dim)` and `K = length (n)`. This function is intended for cases where the ratio K/M is small; otherwise, it may be better to use `sort`.

**See also:** [sort], page 412, [min], page 443, [max], page 442.

| | |
|---|---|
| `tril (A)` | [Function File] |
| `tril (A, k)` | [Function File] |
| `tril (A, k, pack)` | [Function File] |
| `triu (A)` | [Function File] |
| `triu (A, k)` | [Function File] |
| `triu (A, k, pack)` | [Function File] |

Return a new matrix formed by extracting the lower (`tril`) or upper (`triu`) triangular part of the matrix *A*, and setting all other elements to zero.

The second argument is optional, and specifies how many diagonals above or below the main diagonal should also be set to zero.

The default value of *k* is zero, so that `triu` and `tril` normally include the main diagonal as part of the result.

If the value of *k* is nonzero integer, the selection of elements starts at an offset of *k* diagonals above or below the main diagonal; above for positive *k* and below for negative *k*.

The absolute value of *k* must not be greater than the number of subdiagonals or superdiagonals.

For example:

```
tril (ones (3), -1)
    ⇒  0  0  0
       1  0  0
       1  1  0
```

and

```
tril (ones (3), 1)
    ⇒  1  1  0
       1  1  1
       1  1  1
```

If the option "pack" is given as third argument, the extracted elements are not inserted into a matrix, but rather stacked column-wise one above other.

**See also:** [diag], page 415.

Chapter 16: Matrix Manipulation											415

v = vec (x)                                                                          [Built-in Function]
v = vec (x, dim)                                                                     [Built-in Function]
    Return the vector obtained by stacking the columns of the matrix x one above the other.

    Without *dim* this is equivalent to x(:).

    If *dim* is supplied, the dimensions of v are set to *dim* with all elements along the last dimension. This is equivalent to shiftdim (x(:), 1-dim).

    **See also:** [vech], page 415, [resize], page 410, [cat], page 408.

vech (x)                                                                              [Function File]
    Return the vector obtained by eliminating all superdiagonal elements of the square matrix x and stacking the result one column above the other.

    This has uses in matrix calculus where the underlying matrix is symmetric and it would be pointless to keep values above the main diagonal.

    **See also:** [vec], page 415.

prepad (x, l)                                                                         [Function File]
prepad (x, l, c)                                                                      [Function File]
prepad (x, l, c, dim)                                                                 [Function File]
    Prepend the scalar value c to the vector x until it is of length *l*. If c is not given, a value of 0 is used.

    If length (x) > l, elements from the beginning of x are removed until a vector of length *l* is obtained.

    If x is a matrix, elements are prepended or removed from each row.

    If the optional argument *dim* is given, operate along this dimension.

    If *dim* is larger than the dimensions of x, the result will have *dim* dimensions.

    **See also:** [postpad], page 415, [cat], page 408, [resize], page 410.

postpad (x, l)                                                                        [Function File]
postpad (x, l, c)                                                                     [Function File]
postpad (x, l, c, dim)                                                                [Function File]
    Append the scalar value c to the vector x until it is of length *l*. If c is not given, a value of 0 is used.

    If length (x) > l, elements from the end of x are removed until a vector of length *l* is obtained.

    If x is a matrix, elements are appended or removed from each row.

    If the optional argument *dim* is given, operate along this dimension.

    If *dim* is larger than the dimensions of x, the result will have *dim* dimensions.

    **See also:** [prepad], page 415, [cat], page 408, [resize], page 410.

M = diag (v)                                                                          [Built-in Function]
M = diag (v, k)                                                                       [Built-in Function]
M = diag (v, m, n)                                                                    [Built-in Function]
v = diag (M)                                                                          [Built-in Function]

`v = diag (M, k)`   [Built-in Function]

Return a diagonal matrix with vector v on diagonal k.

The second argument is optional. If it is positive, the vector is placed on the k-th superdiagonal. If it is negative, it is placed on the -k-th subdiagonal. The default value of k is 0, and the vector is placed on the main diagonal. For example:

```
diag ([1, 2, 3], 1)
⇒  0  1  0  0
   0  0  2  0
   0  0  0  3
   0  0  0  0
```

The 3-input form returns a diagonal matrix with vector v on the main diagonal and the resulting matrix being of size m rows x n columns.

Given a matrix argument, instead of a vector, `diag` extracts the k-th diagonal of the matrix.

`blkdiag (A, B, C, ...)`   [Function File]

Build a block diagonal matrix from A, B, C, ...

All arguments must be numeric and either two-dimensional matrices or scalars. If any argument is of type sparse, the output will also be sparse.

**See also:** [diag], page 415, [horzcat], page 409, [vertcat], page 409, [sparse], page 522.

## 16.3 Special Utility Matrices

`eye (n)`   [Built-in Function]
`eye (m, n)`   [Built-in Function]
`eye ([m n])`   [Built-in Function]
`eye (..., class)`   [Built-in Function]

Return an identity matrix.

If invoked with a single scalar argument n, return a square NxN identity matrix.

If supplied two scalar arguments (m, n), `eye` takes them to be the number of rows and columns. If given a vector with two elements, `eye` uses the values of the elements as the number of rows and columns, respectively. For example:

```
eye (3)
⇒  1  0  0
   0  1  0
   0  0  1
```

The following expressions all produce the same result:

```
eye (2)
≡
eye (2, 2)
≡
eye (size ([1, 2; 3, 4]))
```

The optional argument *class*, allows `eye` to return an array of the specified type, like

Chapter 16: Matrix Manipulation

```
val = zeros (n,m, "uint8")
```

Calling **eye** with no arguments is equivalent to calling it with an argument of 1. Any negative dimensions are treated as zero. These odd definitions are for compatibility with MATLAB.

**See also:** [speye], page 520, [ones], page 417, [zeros], page 417.

ones (*n*)            [Built-in Function]
ones (*m, n*)            [Built-in Function]
ones (*m, n, k, ...*)            [Built-in Function]
ones ([*m n ...*])            [Built-in Function]
ones (..., *class*)            [Built-in Function]

Return a matrix or N-dimensional array whose elements are all 1.

If invoked with a single scalar integer argument *n*, return a square NxN matrix.

If invoked with two or more scalar integer arguments, or a vector of integer values, return an array with the given dimensions.

To create a constant matrix whose values are all the same use an expression such as

```
val_matrix = val * ones (m, n)
```

The optional argument *class* specifies the class of the return array and defaults to double. For example:

```
val = ones (m,n, "uint8")
```

**See also:** [zeros], page 417.

zeros (*n*)            [Built-in Function]
zeros (*m, n*)            [Built-in Function]
zeros (*m, n, k, ...*)            [Built-in Function]
zeros ([*m n ...*])            [Built-in Function]
zeros (..., *class*)            [Built-in Function]

Return a matrix or N-dimensional array whose elements are all 0.

If invoked with a single scalar integer argument, return a square NxN matrix.

If invoked with two or more scalar integer arguments, or a vector of integer values, return an array with the given dimensions.

The optional argument *class* specifies the class of the return array and defaults to double. For example:

```
val = zeros (m,n, "uint8")
```

**See also:** [ones], page 417.

repmat (*A, m*)            [Function File]
repmat (*A, m, n*)            [Function File]
repmat (*A, m, n, p ...*)            [Function File]
repmat (*A,* [*m n*])            [Function File]
repmat (*A,* [*m n p ...*])            [Function File]

Form a block matrix of size *m* by *n*, with a copy of matrix *A* as each element.

If *n* is not specified, form an *m* by *m* block matrix. For copying along more than two dimensions, specify the number of times to copy across each dimension *m, n, p, ...*, in a vector in the second argument.

See also: [repelems], page 418.

`repelems (x, r)`  [Built-in Function]

Construct a vector of repeated elements from x.

r is a 2xN integer matrix specifying which elements to repeat and how often to repeat each element. Entries in the first row, r(1,j), select an element to repeat. The corresponding entry in the second row, r(2,j), specifies the repeat count. If x is a matrix then the columns of x are imagined to be stacked on top of each other for purposes of the selection index. A row vector is always returned.

Conceptually the result is calculated as follows:

```
y = [];
for i = 1:columns (r)
  y = [y, x(r(1,i)*ones(1, r(2,i)))];
endfor
```

See also: [repmat], page 417, [cat], page 408.

The functions `linspace` and `logspace` make it very easy to create vectors with evenly or logarithmically spaced elements. See Section 4.2 [Ranges], page 52.

`linspace (base, limit)`  [Built-in Function]
`linspace (base, limit, n)`  [Built-in Function]

Return a row vector with n linearly spaced elements between base and limit.

If the number of elements is greater than one, then the endpoints base and limit are always included in the range. If base is greater than limit, the elements are stored in decreasing order. If the number of points is not specified, a value of 100 is used.

The `linspace` function always returns a row vector if both base and limit are scalars. If one, or both, of them are column vectors, `linspace` returns a matrix.

For compatibility with MATLAB, return the second argument (limit) if fewer than two values are requested.

See also: [logspace], page 418.

`logspace (a, b)`  [Function File]
`logspace (a, b, n)`  [Function File]
`logspace (a, pi, n)`  [Function File]

Return a row vector with n elements logarithmically spaced from $10^a$ to $10^b$.

If n is unspecified it defaults to 50.

If b is equal to $\pi$, the points are between $10^a$ and $\pi$, *not* $10^a$ and $10^\pi$, in order to be compatible with the corresponding MATLAB function.

Also for compatibility with MATLAB, return the second argument b if fewer than two values are requested.

See also: [linspace], page 418.

`rand (n)`  [Built-in Function]
`rand (m, n, ...)`  [Built-in Function]
`rand ([m n ...])`  [Built-in Function]

Chapter 16: Matrix Manipulation 419

| | |
|---|---|
| $v$ = **rand** (*"state"*) | [Built-in Function] |
| **rand** (*"state"*, *v*) | [Built-in Function] |
| **rand** (*"state"*, *"reset"*) | [Built-in Function] |
| $v$ = **rand** (*"seed"*) | [Built-in Function] |
| **rand** (*"seed"*, *v*) | [Built-in Function] |
| **rand** (*"seed"*, *"reset"*) | [Built-in Function] |
| **rand** (..., *"single"*) | [Built-in Function] |
| **rand** (..., *"double"*) | [Built-in Function] |

Return a matrix with random elements uniformly distributed on the interval (0, 1).

The arguments are handled the same as the arguments for **eye**.

You can query the state of the random number generator using the form

    v = rand ("state")

This returns a column vector $v$ of length 625. Later, you can restore the random number generator to the state $v$ using the form

    rand ("state", v)

You may also initialize the state vector from an arbitrary vector of length $\leq 625$ for $v$. This new state will be a hash based on the value of $v$, not $v$ itself.

By default, the generator is initialized from **/dev/urandom** if it is available, otherwise from CPU time, wall clock time, and the current fraction of a second. Note that this differs from MATLAB, which always initializes the state to the same state at startup. To obtain behavior comparable to MATLAB, initialize with a deterministic state vector in Octave's startup files (see Section 2.1.2 [Startup Files], page 19).

To compute the pseudo-random sequence, **rand** uses the Mersenne Twister with a period of $2^{19937} - 1$ (See M. Matsumoto and T. Nishimura, *Mersenne Twister: A 623-dimensionally equidistributed uniform pseudorandom number generator*, ACM Trans. on Modeling and Computer Simulation Vol. 8, No. 1, pp. 3–30, January 1998, http://www.math.sci.hiroshima-u.ac.jp/~m-mat/MT/emt.html). Do **not** use for cryptography without securely hashing several returned values together, otherwise the generator state can be learned after reading 624 consecutive values.

Older versions of Octave used a different random number generator. The new generator is used by default as it is significantly faster than the old generator, and produces random numbers with a significantly longer cycle time. However, in some circumstances it might be desirable to obtain the same random sequences as produced by the old generators. To do this the keyword **"seed"** is used to specify that the old generators should be used, as in

    rand ("seed", val)

which sets the seed of the generator to *val*. The seed of the generator can be queried with

    s = rand ("seed")

However, it should be noted that querying the seed will not cause **rand** to use the old generators, only setting the seed will. To cause **rand** to once again use the new generators, the keyword **"state"** should be used to reset the state of the **rand**.

The state or seed of the generator can be reset to a new random value using the **"reset"** keyword.

The class of the value returned can be controlled by a trailing "`double`" or "`single`" argument. These are the only valid classes.

**See also:** [randn], page 420, [rande], page 421, [randg], page 422, [randp], page 421.

`randi (imax)`     [Function File]
`randi (imax, n)`     [Function File]
`randi (imax, m, n, ...)`     [Function File]
`randi ([imin imax], ...)`     [Function File]
`randi (..., "class")`     [Function File]

Return random integers in the range 1:*imax*.

Additional arguments determine the shape of the return matrix. When no arguments are specified a single random integer is returned. If one argument *n* is specified then a square matrix (*n* x *n*) is returned. Two or more arguments will return a multi-dimensional matrix (*m* x *n* x ...).

The integer range may optionally be described by a two element matrix with a lower and upper bound in which case the returned integers will be on the interval [*imin*, *imax*].

The optional argument *class* will return a matrix of the requested type. The default is "`double`".

The following example returns 150 integers in the range 1–10.

```
ri = randi (10, 150, 1)
```

Implementation Note: `randi` relies internally on `rand` which uses class "`double`" to represent numbers. This limits the maximum integer (*imax*) and range (*imax* - *imin*) to the value returned by the `bitmax` function. For IEEE floating point numbers this value is $2^{53} - 1$.

**See also:** [rand], page 418.

`randn (n)`     [Built-in Function]
`randn (m, n, ...)`     [Built-in Function]
`randn ([m n ...])`     [Built-in Function]
`v = randn ("state")`     [Built-in Function]
`randn ("state", v)`     [Built-in Function]
`randn ("state", "reset")`     [Built-in Function]
`v = randn ("seed")`     [Built-in Function]
`randn ("seed", v)`     [Built-in Function]
`randn ("seed", "reset")`     [Built-in Function]
`randn (..., "single")`     [Built-in Function]
`randn (..., "double")`     [Built-in Function]

Return a matrix with normally distributed random elements having zero mean and variance one.

The arguments are handled the same as the arguments for `rand`.

By default, `randn` uses the Marsaglia and Tsang "Ziggurat technique" to transform from a uniform to a normal distribution.

The class of the value returned can be controlled by a trailing "`double`" or "`single`" argument. These are the only valid classes.

Chapter 16: Matrix Manipulation

Reference: G. Marsaglia and W.W. Tsang, *Ziggurat Method for Generating Random Variables*, J. Statistical Software, vol 5, 2000, http://www.jstatsoft.org/v05/i08/

**See also:** [rand], page 418, [rande], page 421, [randg], page 422, [randp], page 421.

| | |
|---|---|
| rande ($n$) | [Built-in Function] |
| rande ($m, n, \ldots$) | [Built-in Function] |
| rande ($[m\ n\ \ldots]$) | [Built-in Function] |
| $v$ = rande ("*state*") | [Built-in Function] |
| rande ("*state*", $v$) | [Built-in Function] |
| rande ("*state*", "*reset*") | [Built-in Function] |
| $v$ = rande ("*seed*") | [Built-in Function] |
| rande ("*seed*", $v$) | [Built-in Function] |
| rande ("*seed*", "*reset*") | [Built-in Function] |
| rande ($\ldots$, "*single*") | [Built-in Function] |
| rande ($\ldots$, "*double*") | [Built-in Function] |

Return a matrix with exponentially distributed random elements.

The arguments are handled the same as the arguments for **rand**.

By default, **randn** uses the Marsaglia and Tsang "Ziggurat technique" to transform from a uniform to a normal distribution.

The class of the value returned can be controlled by a trailing `"double"` or `"single"` argument. These are the only valid classes.

Reference: G. Marsaglia and W.W. Tsang, *Ziggurat Method for Generating Random Variables*, J. Statistical Software, vol 5, 2000, http://www.jstatsoft.org/v05/i08/

**See also:** [rand], page 418, [randn], page 420, [randg], page 422, [randp], page 421.

| | |
|---|---|
| randp ($l, n$) | [Built-in Function] |
| randp ($l, m, n, \ldots$) | [Built-in Function] |
| randp ($l, [m\ n\ \ldots]$) | [Built-in Function] |
| $v$ = randp ("*state*") | [Built-in Function] |
| randp ("*state*", $v$) | [Built-in Function] |
| randp ("*state*", "*reset*") | [Built-in Function] |
| $v$ = randp ("*seed*") | [Built-in Function] |
| randp ("*seed*", $v$) | [Built-in Function] |
| randp ("*seed*", "*reset*") | [Built-in Function] |
| randp ($\ldots$, "*single*") | [Built-in Function] |
| randp ($\ldots$, "*double*") | [Built-in Function] |

Return a matrix with Poisson distributed random elements with mean value parameter given by the first argument, $l$.

The arguments are handled the same as the arguments for **rand**, except for the argument $l$.

Five different algorithms are used depending on the range of $l$ and whether or not $l$ is a scalar or a matrix.

For scalar $l \leq 12$, use direct method.

W.H. Press, et al., *Numerical Recipes in C*, Cambridge University Press, 1992.

For scalar $l > 12$, use rejection method.[1]
> W.H. Press, et al., *Numerical Recipes in C*, Cambridge University Press, 1992.

For matrix $l \leq 10$, use inversion method.[2]
> E. Stadlober, et al., WinRand source code, available via FTP.

For matrix $l > 10$, use patchwork rejection method.
> E. Stadlober, et al., WinRand source code, available via FTP, or H. Zechner, *Efficient sampling from continuous and discrete unimodal distributions*, Doctoral Dissertation, 156pp., Technical University Graz, Austria, 1994.

For $l > 1e8$, use normal approximation.
> L. Montanet, et al., *Review of Particle Properties*, Physical Review D 50 p1284, 1994.

The class of the value returned can be controlled by a trailing `"double"` or `"single"` argument. These are the only valid classes.

**See also:** [rand], page 418, [randn], page 420, [rande], page 421, [randg], page 422.

| | |
|---|---|
| `randg (n)` | [Built-in Function] |
| `randg (m, n, ...)` | [Built-in Function] |
| `randg ([m n ...])` | [Built-in Function] |
| `v = randg ("state")` | [Built-in Function] |
| `randg ("state", v)` | [Built-in Function] |
| `randg ("state", "reset")` | [Built-in Function] |
| `v = randg ("seed")` | [Built-in Function] |
| `randg ("seed", v)` | [Built-in Function] |
| `randg ("seed", "reset")` | [Built-in Function] |
| `randg (..., "single")` | [Built-in Function] |
| `randg (..., "double")` | [Built-in Function] |

Return a matrix with `gamma (a,1)` distributed random elements.

The arguments are handled the same as the arguments for `rand`, except for the argument *a*.

This can be used to generate many distributions:

`gamma (a, b)` for a > -1, b > 0
> `r = b * randg (a)`

`beta (a, b)` for a > -1, b > -1
> `r1 = randg (a, 1)`
> `r = r1 / (r1 + randg (b, 1))`

`Erlang (a, n)`
> `r = a * randg (n)`

`chisq (df)` for df > 0
> `r = 2 * randg (df / 2)`

`t (df)` for 0 < df < inf (use randn if df is infinite)
> `r = randn () / sqrt (2 * randg (df / 2) / df)`

## Chapter 16: Matrix Manipulation

F (n1, n2) for 0 < n1, 0 < n2
```
              ## r1 equals 1 if n1 is infinite
              r1 = 2 * randg (n1 / 2) / n1
              ## r2 equals 1 if n2 is infinite
              r2 = 2 * randg (n2 / 2) / n2
              r  = r1 / r2
```

negative binomial (n, p) for n > 0, 0 < p <= 1
```
              r = randp ((1 - p) / p * randg (n))
```

non-central chisq (df, L), for df >= 0 and L > 0
           (use chisq if L = 0)
```
              r = randp (L / 2)
              r(r > 0) = 2 * randg (r(r > 0))
              r(df > 0) += 2 * randg (df(df > 0)/2)
```

Dirichlet (a1, ... ak)
```
              r = (randg (a1), ..., randg (ak))
              r = r / sum (r)
```

The class of the value returned can be controlled by a trailing "double" or "single" argument. These are the only valid classes.

**See also:** [rand], page 418, [randn], page 420, [rande], page 421, [randp], page 421.

The generators operate in the new or old style together, it is not possible to mix the two. Initializing any generator with "state" or "seed" causes the others to switch to the same style for future calls.

The state of each generator is independent and calls to different generators can be interleaved without affecting the final result. For example,

```
rand ("state", [11, 22, 33]);
randn ("state", [44, 55, 66]);
u = rand (100, 1);
n = randn (100, 1);
```

and

```
rand ("state", [11, 22, 33]);
randn ("state", [44, 55, 66]);
u = zeros (100, 1);
n = zeros (100, 1);
for i = 1:100
  u(i) = rand ();
  n(i) = randn ();
end
```

produce equivalent results. When the generators are initialized in the old style with "seed" only **rand** and **randn** are independent, because the old **rande**, **randg** and **randp** generators make calls to **rand** and **randn**.

The generators are initialized with random states at start-up, so that the sequences of random numbers are not the same each time you run Octave.[1] If you really do need to reproduce a sequence of numbers exactly, you can set the state or seed to a specific value.

If invoked without arguments, `rand` and `randn` return a single element of a random sequence.

The original `rand` and `randn` functions use Fortran code from RANLIB, a library of Fortran routines for random number generation, compiled by Barry W. Brown and James Lovato of the Department of Biomathematics at The University of Texas, M.D. Anderson Cancer Center, Houston, TX 77030.

**randperm** (*n*) [Built-in Function]
**randperm** (*n, m*) [Built-in Function]

    Return a row vector containing a random permutation of `1:n`.

    If *m* is supplied, return *m* unique entries, sampled without replacement from `1:n`.

    The complexity is $O(n)$ in memory and $O(m)$ in time, unless $m < n/5$, in which case $O(m)$ memory is used as well. The randomization is performed using rand(). All permutations are equally likely.

    **See also:** [perms], page 605.

## 16.4 Famous Matrices

The following functions return famous matrix forms.

**gallery** (*name*) [Function File]
**gallery** (*name, args*) [Function File]

    Create interesting matrices for testing.

*c* = **gallery** ("*cauchy*", *x*) [Function File]
*c* = **gallery** ("*cauchy*", *x, y*) [Function File]

    Create a Cauchy matrix.

*c* = **gallery** ("*chebspec*", *n*) [Function File]
*c* = **gallery** ("*chebspec*", *n, k*) [Function File]

    Create a Chebyshev spectral differentiation matrix.

*c* = **gallery** ("*chebvand*", *p*) [Function File]
*c* = **gallery** ("*chebvand*", *m, p*) [Function File]

    Create a Vandermonde-like matrix for the Chebyshev polynomials.

*a* = **gallery** ("*chow*", *n*) [Function File]
*a* = **gallery** ("*chow*", *n, alpha*) [Function File]
*a* = **gallery** ("*chow*", *n, alpha, delta*) [Function File]

    Create a Chow matrix – a singular Toeplitz lower Hessenberg matrix.

*c* = **gallery** ("*circul*", *v*) [Function File]

    Create a circulant matrix.

---

[1] The old versions of `rand` and `randn` obtain their initial seeds from the system clock.

# Chapter 16: Matrix Manipulation

a = **gallery** ("*clement*", n)  [Function File]
a = **gallery** ("*clement*", n, k)  [Function File]
    Create a tridiagonal matrix with zero diagonal entries.

c = **gallery** ("*compar*", a)  [Function File]
c = **gallery** ("*compar*", a, k)  [Function File]
    Create a comparison matrix.

a = **gallery** ("*condex*", n)  [Function File]
a = **gallery** ("*condex*", n, k)  [Function File]
a = **gallery** ("*condex*", n, k, theta)  [Function File]
    Create a 'counterexample' matrix to a condition estimator.

a = **gallery** ("*cycol*", [m n])  [Function File]
a = **gallery** ("*cycol*", n)  [Function File]
a = **gallery** (..., k)  [Function File]
    Create a matrix whose columns repeat cyclically.

[c, d, e] = **gallery** ("*dorr*", n)  [Function File]
[c, d, e] = **gallery** ("*dorr*", n, theta)  [Function File]
a = **gallery** ("*dorr*", ...)  [Function File]
    Create a diagonally dominant, ill-conditioned, tridiagonal matrix.

a = **gallery** ("*dramadah*", n)  [Function File]
a = **gallery** ("*dramadah*", n, k)  [Function File]
    Create a (0, 1) matrix whose inverse has large integer entries.

a = **gallery** ("*fiedler*", c)  [Function File]
    Create a symmetric Fiedler matrix.

a = **gallery** ("*forsythe*", n)  [Function File]
a = **gallery** ("*forsythe*", n, alpha)  [Function File]
a = **gallery** ("*forsythe*", n, alpha, lambda)  [Function File]
    Create a Forsythe matrix (a perturbed Jordan block).

f = **gallery** ("*frank*", n)  [Function File]
f = **gallery** ("*frank*", n, k)  [Function File]
    Create a Frank matrix (ill-conditioned eigenvalues).

c = **gallery** ("*gcdmat*", n)  [Function File]
    Create a greatest common divisor matrix.

    $c$ is an $n$-by-$n$ matrix whose values correspond to the greatest common divisor of its coordinate values, i.e., $c(i,j)$ correspond `gcd (i, j)`.

a = **gallery** ("*gearmat*", n)  [Function File]
a = **gallery** ("*gearmat*", n, i)  [Function File]
a = **gallery** ("*gearmat*", n, i, j)  [Function File]
    Create a Gear matrix.

`g = gallery ("grcar", n)` [Function File]
`g = gallery ("grcar", n, k)` [Function File]
: Create a Toeplitz matrix with sensitive eigenvalues.

`a = gallery ("hanowa", n)` [Function File]
`a = gallery ("hanowa", n, d)` [Function File]
: Create a matrix whose eigenvalues lie on a vertical line in the complex plane.

`v = gallery ("house", x)` [Function File]
`[v, beta] = gallery ("house", x)` [Function File]
: Create a householder matrix.

`a = gallery ("integerdata", imax, [M N ...], j)` [Function File]
`a = gallery ("integerdata", imax, M, N, ..., j)` [Function File]
`a = gallery ("integerdata", [imin, imax], [M N ...], j)` [Function File]
`a = gallery ("integerdata", [imin, imax], M, N, ..., j)` [Function File]
`a = gallery ("integerdata", ..., "class")` [Function File]
: Create a matrix with random integers in the range [1, imax]. If imin is given then the integers are in the range [imin, imax].

The second input is a matrix of dimensions describing the size of the output. The dimensions can also be input as comma-separated arguments.

The input j is an integer index in the range [0, 2^32-1]. The values of the output matrix are always exactly the same (reproducibility) for a given size input and j index.

The final optional argument determines the class of the resulting matrix. Possible values for class: `"uint8"`, `"uint16"`, `"uint32"`, `"int8"`, `"int16"`, int32", `"single"`, `"double"`. The default is `"double"`.

`a = gallery ("invhess", x)` [Function File]
`a = gallery ("invhess", x, y)` [Function File]
: Create the inverse of an upper Hessenberg matrix.

`a = gallery ("invol", n)` [Function File]
: Create an involutory matrix.

`a = gallery ("ipjfact", n)` [Function File]
`a = gallery ("ipjfact", n, k)` [Function File]
: Create a Hankel matrix with factorial elements.

`a = gallery ("jordbloc", n)` [Function File]
`a = gallery ("jordbloc", n, lambda)` [Function File]
: Create a Jordan block.

`u = gallery ("kahan", n)` [Function File]
`u = gallery ("kahan", n, theta)` [Function File]
`u = gallery ("kahan", n, theta, pert)` [Function File]
: Create a Kahan matrix (upper trapezoidal).

`a = gallery ("kms", n)` [Function File]
`a = gallery ("kms", n, rho)` [Function File]
: Create a Kac-Murdock-Szego Toeplitz matrix.

## Chapter 16: Matrix Manipulation

*b* = **gallery** (*"krylov"*, **a**)                                                                 [Function File]
*b* = **gallery** (*"krylov"*, **a**, **x**)                                                   [Function File]
*b* = **gallery** (*"krylov"*, **a**, **x**, **j**)                                      [Function File]
    Create a Krylov matrix.

*a* = **gallery** (*"lauchli"*, **n**)                                                 [Function File]
*a* = **gallery** (*"lauchli"*, **n**, **mu**)                                   [Function File]
    Create a Lauchli matrix (rectangular).

*a* = **gallery** (*"lehmer"*, **n**)                                               [Function File]
    Create a Lehmer matrix (symmetric positive definite).

*t* = **gallery** (*"lesp"*, **n**)                                                  [Function File]
    Create a tridiagonal matrix with real, sensitive eigenvalues.

*a* = **gallery** (*"lotkin"*, **n**)                                               [Function File]
    Create a Lotkin matrix.

*a* = **gallery** (*"minij"*, **n**)                                                 [Function File]
    Create a symmetric positive definite matrix MIN(i,j).

*a* = **gallery** (*"moler"*, **n**)                                                [Function File]
*a* = **gallery** (*"moler"*, **n**, **alpha**)                                 [Function File]
    Create a Moler matrix (symmetric positive definite).

[*a*, *t*] = **gallery** (*"neumann"*, **n**)                                [Function File]
    Create a singular matrix from the discrete Neumann problem (sparse).

*a* = **gallery** (*"normaldata"*, [*M N* ...], *j*)                 [Function File]
*a* = **gallery** (*"normaldata"*, *M, N*, ..., *j*)                [Function File]
*a* = **gallery** (*"normaldata"*, ..., *"class"*)                  [Function File]
    Create a matrix with random samples from the standard normal distribution (mean = 0, std = 1).

    The first input is a matrix of dimensions describing the size of the output. The dimensions can also be input as comma-separated arguments.

    The input *j* is an integer index in the range [0, 2^32-1]. The values of the output matrix are always exactly the same (reproducibility) for a given size input and *j* index.

    The final optional argument determines the class of the resulting matrix. Possible values for *class*: `"single"`, `"double"`. The default is `"double"`.

*q* = **gallery** (*"orthog"*, **n**)                                              [Function File]
*q* = **gallery** (*"orthog"*, **n**, **k**)                                       [Function File]
    Create orthogonal and nearly orthogonal matrices.

*a* = **gallery** (*"parter"*, **n**)                                              [Function File]
    Create a Parter matrix (a Toeplitz matrix with singular values near pi).

*p* = **gallery** (*"pei"*, **n**)                                                    [Function File]
*p* = **gallery** (*"pei"*, **n**, **alpha**)                                     [Function File]
    Create a Pei matrix.

`a = gallery ("Poisson", n)`     [Function File]
    Create a block tridiagonal matrix from Poisson's equation (sparse).

`a = gallery ("prolate", n)`     [Function File]
`a = gallery ("prolate", n, w)`     [Function File]
    Create a prolate matrix (symmetric, ill-conditioned Toeplitz matrix).

`h = gallery ("randhess", x)`     [Function File]
    Create a random, orthogonal upper Hessenberg matrix.

`a = gallery ("rando", n)`     [Function File]
`a = gallery ("rando", n, k)`     [Function File]
    Create a random matrix with elements -1, 0 or 1.

`a = gallery ("randsvd", n)`     [Function File]
`a = gallery ("randsvd", n, kappa)`     [Function File]
`a = gallery ("randsvd", n, kappa, mode)`     [Function File]
`a = gallery ("randsvd", n, kappa, mode, kl)`     [Function File]
`a = gallery ("randsvd", n, kappa, mode, kl, ku)`     [Function File]
    Create a random matrix with pre-assigned singular values.

`a = gallery ("redheff", n)`     [Function File]
    Create a zero and ones matrix of Redheffer associated with the Riemann hypothesis.

`a = gallery ("riemann", n)`     [Function File]
    Create a matrix associated with the Riemann hypothesis.

`a = gallery ("ris", n)`     [Function File]
    Create a symmetric Hankel matrix.

`a = gallery ("smoke", n)`     [Function File]
`a = gallery ("smoke", n, k)`     [Function File]
    Create a complex matrix, with a 'smoke ring' pseudospectrum.

`t = gallery ("toeppd", n)`     [Function File]
`t = gallery ("toeppd", n, m)`     [Function File]
`t = gallery ("toeppd", n, m, w)`     [Function File]
`t = gallery ("toeppd", n, m, w, theta)`     [Function File]
    Create a symmetric positive definite Toeplitz matrix.

`p = gallery ("toeppen", n)`     [Function File]
`p = gallery ("toeppen", n, a)`     [Function File]
`p = gallery ("toeppen", n, a, b)`     [Function File]
`p = gallery ("toeppen", n, a, b, c)`     [Function File]
`p = gallery ("toeppen", n, a, b, c, d)`     [Function File]
`p = gallery ("toeppen", n, a, b, c, d, e)`     [Function File]
    Create a pentadiagonal Toeplitz matrix (sparse).

`a = gallery ("tridiag", x, y, z)`     [Function File]
`a = gallery ("tridiag", n)`     [Function File]
`a = gallery ("tridiag", n, c, d, e)`     [Function File]
    Create a tridiagonal matrix (sparse).

Chapter 16: Matrix Manipulation                                                              429

*t* = gallery (*"triw"*, *n*)                                                      [Function File]
*t* = gallery (*"triw"*, *n*, *alpha*)                                             [Function File]
*t* = gallery (*"triw"*, *n*, *alpha*, *k*)                                        [Function File]
: Create an upper triangular matrix discussed by Kahan, Golub, and Wilkinson.

*a* = gallery (*"uniformdata"*, [*M N* ...], *j*)                                  [Function File]
*a* = gallery (*"uniformdata"*, *M, N*, ..., *j*)                                  [Function File]
*a* = gallery (*"uniformdata"*, ..., *"class"*)                                    [Function File]
: Create a matrix with random samples from the standard uniform distribution (range [0,1]).

The first input is a matrix of dimensions describing the size of the output. The dimensions can also be input as comma-separated arguments.

The input *j* is an integer index in the range [0, 2^32-1]. The values of the output matrix are always exactly the same (reproducibility) for a given size input and *j* index.

The final optional argument determines the class of the resulting matrix. Possible values for *class*: `"single"`, `"double"`. The default is `"double"`.

*a* = gallery (*"wathen"*, *nx*, *ny*)                                             [Function File]
*a* = gallery (*"wathen"*, *nx*, *ny*, *k*)                                        [Function File]
: Create the Wathen matrix.

[*a*, *b*] = gallery (*"wilk"*, *n*)                                               [Function File]
: Create various specific matrices devised/discussed by Wilkinson.

hadamard (*n*)                                                                     [Function File]
: Construct a Hadamard matrix (Hn) of size *n*-by-*n*.

The size $n$ must be of the form $2^k * p$ in which p is one of 1, 12, 20 or 28. The returned matrix is normalized, meaning `Hn(:,1) == 1` and `Hn(1,:) == 1`.

Some of the properties of Hadamard matrices are:

- `kron (Hm, Hn)` is a Hadamard matrix of size *m*-by-*n*.
- `Hn * Hn' = n * eye (n)`.
- The rows of Hn are orthogonal.
- `det (A) <= abs (det (Hn))` for all *A* with `abs (A(i, j)) <= 1`.
- Multiplying any row or column by -1 and the matrix will remain a Hadamard matrix.

**See also:** [compan], page 638, [hankel], page 429, [toeplitz], page 431.

hankel (*c*)                                                                       [Function File]
hankel (*c*, *r*)                                                                  [Function File]
: Return the Hankel matrix constructed from the first column *c*, and (optionally) the last row *r*.

If the last element of *c* is not the same as the first element of *r*, the last element of *c* is used. If the second argument is omitted, it is assumed to be a vector of zeros with the same size as *c*.

A Hankel matrix formed from an m-vector c, and an n-vector r, has the elements

$$H(i,j) = \begin{cases} c_{i+j-1}, & i+j-1 \leq m; \\ r_{i+j-m}, & \text{otherwise.} \end{cases}$$

**See also:** [hadamard], page 429, [toeplitz], page 431.

**hilb (n)** [Function File]

Return the Hilbert matrix of order n.

The $i,j$ element of a Hilbert matrix is defined as

$$H(i,j) = \frac{1}{(i+j-1)}$$

Hilbert matrices are close to being singular which make them difficult to invert with numerical routines. Comparing the condition number of a random matrix 5x5 matrix with that of a Hilbert matrix of order 5 reveals just how difficult the problem is.

```
cond (rand (5))
   ⇒ 14.392
cond (hilb (5))
   ⇒ 4.7661e+05
```

**See also:** [invhilb], page 430.

**invhilb (n)** [Function File]

Return the inverse of the Hilbert matrix of order n.

This can be computed exactly using

$$A_{ij} = -1^{i+j}(i+j-1)\binom{n+i-1}{n-j}\binom{n+j-1}{n-i}\binom{i+j-2}{i-2}^2$$

$$= \frac{p(i)p(j)}{(i+j-1)}$$

where

$$p(k) = -1^k \binom{k+n-1}{k-1}\binom{n}{k}$$

The validity of this formula can easily be checked by expanding the binomial coefficients in both formulas as factorials. It can be derived more directly via the theory of Cauchy matrices. See J. W. Demmel, *Applied Numerical Linear Algebra*, p. 92.

Compare this with the numerical calculation of `inverse (hilb (n))`, which suffers from the ill-conditioning of the Hilbert matrix, and the finite precision of your computer's floating point arithmetic.

**See also:** [hilb], page 430.

**magic (n)** [Function File]

Create an n-by-n magic square.

A magic square is an arrangement of the integers 1:n^2 such that the row sums, column sums, and diagonal sums are all equal to the same value.

Note: n must be greater than 2 for the magic square to exist.

# Chapter 16: Matrix Manipulation

**pascal** (*n*)   [Function File]
**pascal** (*n, t*)   [Function File]

Return the Pascal matrix of order *n* if `t` = 0.

The default value of *t* is 0.

When `t` = 1, return the pseudo-lower triangular Cholesky factor of the Pascal matrix (The sign of some columns may be negative). This matrix is its own inverse, that is `pascal (n, 1) ^ 2 == eye (n)`.

If `t` = -1, return the true Cholesky factor with strictly positive values on the diagonal.

If `t` = 2, return a transposed and permuted version of `pascal (n, 1)`, which is the cube root of the identity matrix. That is, `pascal (n, 2) ^ 3 == eye (n)`.

**See also:** [chol], page 470.

**rosser** ()   [Function File]

Return the Rosser matrix.

This is a difficult test case used to evaluate eigenvalue algorithms.

**See also:** [wilkinson], page 432, [eig], page 465.

**toeplitz** (*c*)   [Function File]
**toeplitz** (*c, r*)   [Function File]

Return the Toeplitz matrix constructed from the first column *c*, and (optionally) the first row *r*.

If the first element of *r* is not the same as the first element of *c*, the first element of *c* is used. If the second argument is omitted, the first row is taken to be the same as the first column.

A square Toeplitz matrix has the form:

$$\begin{bmatrix} c_0 & r_1 & r_2 & \cdots & r_n \\ c_1 & c_0 & r_1 & \cdots & r_{n-1} \\ c_2 & c_1 & c_0 & \cdots & r_{n-2} \\ \vdots & \vdots & \vdots & \ddots & \vdots \\ c_n & c_{n-1} & c_{n-2} & \cdots & c_0 \end{bmatrix}$$

**See also:** [hankel], page 429.

**vander** (*c*)   [Function File]
**vander** (*c, n*)   [Function File]

Return the Vandermonde matrix whose next to last column is *c*.

If *n* is specified, it determines the number of columns; otherwise, *n* is taken to be equal to the length of *c*.

A Vandermonde matrix has the form:

$$\begin{bmatrix} c_1^{n-1} & \cdots & c_1^2 & c_1 & 1 \\ c_2^{n-1} & \cdots & c_2^2 & c_2 & 1 \\ \vdots & \ddots & \vdots & \vdots & \vdots \\ c_n^{n-1} & \cdots & c_n^2 & c_n & 1 \end{bmatrix}$$

**See also:** [polyfit], page 643.

`wilkinson (n)` [Function File]
: Return the Wilkinson matrix of order $n$.

    Wilkinson matrices are symmetric and tridiagonal with pairs of nearly, but not exactly, equal eigenvalues. They are useful in testing the behavior and performance of eigenvalue solvers.

    **See also:** [rosser], page 431, [eig], page 465.

# 17 Arithmetic

Unless otherwise noted, all of the functions described in this chapter will work for real and complex scalar, vector, or matrix arguments. Functions described as *mapping functions* apply the given operation individually to each element when given a matrix argument. For example:

```
sin ([1, 2; 3, 4])
    ⇒  0.84147   0.90930
       0.14112  -0.75680
```

## 17.1 Exponents and Logarithms

**exp** (*x*)                                                                     [Mapping Function]
    Compute $e^x$ for each element of x.

    To compute the matrix exponential, see Chapter 18 [Linear Algebra], page 463.

    **See also:** [log], page 433.

**expm1** (*x*)                                                                   [Mapping Function]
    Compute $e^x - 1$ accurately in the neighborhood of zero.

    **See also:** [exp], page 433.

**log** (*x*)                                                                     [Mapping Function]
    Compute the natural logarithm, $\ln(x)$, for each element of x.

    To compute the matrix logarithm, see Chapter 18 [Linear Algebra], page 463.

    **See also:** [exp], page 433, [log1p], page 433, [log2], page 433, [log10], page 433, [logspace], page 418.

**reallog** (*x*)                                                                 [Function File]
    Return the real-valued natural logarithm of each element of x.

    If any element results in a complex return value **reallog** aborts and issues an error.

    **See also:** [log], page 433, [realpow], page 434, [realsqrt], page 434.

**log1p** (*x*)                                                                   [Mapping Function]
    Compute $\ln(1 + x)$ accurately in the neighborhood of zero.

    **See also:** [log], page 433, [exp], page 433, [expm1], page 433.

**log10** (*x*)                                                                   [Mapping Function]
    Compute the base-10 logarithm of each element of x.

    **See also:** [log], page 433, [log2], page 433, [logspace], page 418, [exp], page 433.

**log2** (*x*)                                                                    [Mapping Function]
[*f*, *e*] = **log2** (*x*)                                                       [Mapping Function]
    Compute the base-2 logarithm of each element of x.

    If called with two output arguments, split x into binary mantissa and exponent so that $\frac{1}{2} \le |f| < 1$ and e is an integer. If $x = 0$, $f = e = 0$.

    **See also:** [pow2], page 434, [log], page 433, [log10], page 433, [exp], page 433.

**pow2** (*x*)   [Function File]
**pow2** (*f*, *e*)   [Function File]

    With one input argument, compute $2^x$ for each element of *x*.

    With two input arguments, return $f \cdot 2^e$.

    **See also:** [log2], page 433, [nextpow2], page 434, [power], page 144.

**nextpow2** (*x*)   [Function File]

    Compute the exponent for the smallest power of two larger than the input.

    For each element in the input array *x*, return the first integer *n* such that $2^n \geq |x|$.

    **See also:** [pow2], page 434, [log2], page 433.

**realpow** (*x*, *y*)   [Function File]

    Compute the real-valued, element-by-element power operator.

    This is equivalent to `x .^ y`, except that `realpow` reports an error if any return value is complex.

    **See also:** [power], page 144, [reallog], page 433, [realsqrt], page 434.

**sqrt** (*x*)   [Mapping Function]

    Compute the square root of each element of *x*.

    If *x* is negative, a complex result is returned.

    To compute the matrix square root, see Chapter 18 [Linear Algebra], page 463.

    **See also:** [realsqrt], page 434, [nthroot], page 434.

**realsqrt** (*x*)   [Function File]

    Return the real-valued square root of each element of *x*.

    If any element results in a complex return value `realsqrt` aborts and issues an error.

    **See also:** [sqrt], page 434, [realpow], page 434, [reallog], page 433.

**cbrt** (*x*)   [Mapping Function]

    Compute the real cube root of each element of *x*.

    Unlike `x^(1/3)`, the result will be negative if *x* is negative.

    **See also:** [nthroot], page 434.

**nthroot** (*x*, *n*)   [Function File]

    Compute the real (non-complex) *n*-th root of *x*.

    *x* must have all real entries and *n* must be a scalar. If *n* is an even integer and *x* has negative entries then `nthroot` aborts and issues an error.

    Example:

```
nthroot (-1, 3)
⇒ -1
(-1) ^ (1 / 3)
⇒ 0.50000 - 0.86603i
```

    **See also:** [realsqrt], page 434, [sqrt], page 434, [cbrt], page 434.

## 17.2 Complex Arithmetic

In the descriptions of the following functions, $z$ is the complex number $x + iy$, where $i$ is defined as $\sqrt{-1}$.

**abs** (*z*) [Mapping Function]

    Compute the magnitude of $z$.

    The magnitude is defined as $|z| = \sqrt{x^2 + y^2}$.

    For example:

```
abs (3 + 4i)
    ⇒ 5
```

**See also:** [arg], page 435.

**arg** (*z*) [Mapping Function]
**angle** (*z*) [Mapping Function]

    Compute the argument, i.e., angle of $z$.

    This is defined as, $\theta = atan2(y, x)$, in radians.

    For example:

```
arg (3 + 4i)
    ⇒ 0.92730
```

**See also:** [abs], page 435.

**conj** (*z*) [Mapping Function]

    Return the complex conjugate of $z$.

    The complex conjugate is defined as $\bar{z} = x - iy$.

    **See also:** [real], page 436, [imag], page 435.

**cplxpair** (*z*) [Function File]
**cplxpair** (*z, tol*) [Function File]
**cplxpair** (*z, tol, dim*) [Function File]

    Sort the numbers $z$ into complex conjugate pairs ordered by increasing real part.

    The negative imaginary complex numbers are placed first within each pair. All real numbers (those with `abs (imag (z) / z) < tol`) are placed after the complex pairs.

    If *tol* is unspecified the default value is 100*`eps`.

    By default the complex pairs are sorted along the first non-singleton dimension of $z$. If *dim* is specified, then the complex pairs are sorted along this dimension.

    Signal an error if some complex numbers could not be paired. Signal an error if all complex numbers are not exact conjugates (to within *tol*). Note that there is no defined order for pairs with identical real parts but differing imaginary parts.

```
cplxpair (exp(2i*pi*[0:4]'/5)) == exp(2i*pi*[3; 2; 4; 1; 0]/5)
```

**imag** (*z*) [Mapping Function]

    Return the imaginary part of $z$ as a real number.

    **See also:** [real], page 436, [conj], page 435.

`real (z)` [Mapping Function]
>   Return the real part of z.
>
>   **See also:** [imag], page 435, [conj], page 435.

## 17.3 Trigonometry

Octave provides the following trigonometric functions where angles are specified in radians. To convert from degrees to radians multiply by $\pi/180$ (e.g., `sin (30 * pi/180)` returns the sine of 30 degrees). As an alternative, Octave provides a number of trigonometric functions which work directly on an argument specified in degrees. These functions are named after the base trigonometric function with a 'd' suffix. For example, `sin` expects an angle in radians while `sind` expects an angle in degrees.

Octave uses the C library trigonometric functions. It is expected that these functions are defined by the ISO/IEC 9899 Standard. This Standard is available at: `http://www.open-std.org/jtc1/sc22/wg14/www/docs/n1124.pdf`. Section F.9.1 deals with the trigonometric functions. The behavior of most of the functions is relatively straightforward. However, there are some exceptions to the standard behavior. Many of the exceptions involve the behavior for -0. The most complex case is atan2. Octave exactly implements the behavior given in the Standard. Including $atan2(\pm 0, -0)$ returns $\pm \pi$.

It should be noted that MATLAB uses different definitions which apparently do not distinguish -0.

`sin (x)` [Mapping Function]
>   Compute the sine for each element of x in radians.
>
>   **See also:** [asin], page 437, [sind], page 438, [sinh], page 437.

`cos (x)` [Mapping Function]
>   Compute the cosine for each element of x in radians.
>
>   **See also:** [acos], page 437, [cosd], page 438, [cosh], page 437.

`tan (z)` [Mapping Function]
>   Compute the tangent for each element of x in radians.
>
>   **See also:** [atan], page 437, [tand], page 439, [tanh], page 437.

`sec (x)` [Mapping Function]
>   Compute the secant for each element of x in radians.
>
>   **See also:** [asec], page 437, [secd], page 439, [sech], page 437.

`csc (x)` [Mapping Function]
>   Compute the cosecant for each element of x in radians.
>
>   **See also:** [acsc], page 437, [cscd], page 439, [csch], page 437.

`cot (x)` [Mapping Function]
>   Compute the cotangent for each element of x in radians.
>
>   **See also:** [acot], page 437, [cotd], page 439, [coth], page 437.

# Chapter 17: Arithmetic

**asin** (*x*)     [Mapping Function]
    Compute the inverse sine in radians for each element of *x*.
    **See also:** [sin], page 436, [asind], page 439.

**acos** (*x*)     [Mapping Function]
    Compute the inverse cosine in radians for each element of *x*.
    **See also:** [cos], page 436, [acosd], page 439.

**atan** (*x*)     [Mapping Function]
    Compute the inverse tangent in radians for each element of *x*.
    **See also:** [tan], page 436, [atand], page 439.

**asec** (*x*)     [Mapping Function]
    Compute the inverse secant in radians for each element of *x*.
    **See also:** [sec], page 436, [asecd], page 439.

**acsc** (*x*)     [Mapping Function]
    Compute the inverse cosecant in radians for each element of *x*.
    **See also:** [csc], page 436, [acscd], page 439.

**acot** (*x*)     [Mapping Function]
    Compute the inverse cotangent in radians for each element of *x*.
    **See also:** [cot], page 436, [acotd], page 439.

**sinh** (*x*)     [Mapping Function]
    Compute the hyperbolic sine for each element of *x*.
    **See also:** [asinh], page 438, [cosh], page 437, [tanh], page 437.

**cosh** (*x*)     [Mapping Function]
    Compute the hyperbolic cosine for each element of *x*.
    **See also:** [acosh], page 438, [sinh], page 437, [tanh], page 437.

**tanh** (*x*)     [Mapping Function]
    Compute hyperbolic tangent for each element of *x*.
    **See also:** [atanh], page 438, [sinh], page 437, [cosh], page 437.

**sech** (*x*)     [Mapping Function]
    Compute the hyperbolic secant of each element of *x*.
    **See also:** [asech], page 438.

**csch** (*x*)     [Mapping Function]
    Compute the hyperbolic cosecant of each element of *x*.
    **See also:** [acsch], page 438.

**coth** (*x*)     [Mapping Function]
    Compute the hyperbolic cotangent of each element of *x*.
    **See also:** [acoth], page 438.

**asinh** (*x*) [Mapping Function]
    Compute the inverse hyperbolic sine for each element of *x*.

    **See also:** [sinh], page 437.

**acosh** (*x*) [Mapping Function]
    Compute the inverse hyperbolic cosine for each element of *x*.

    **See also:** [cosh], page 437.

**atanh** (*x*) [Mapping Function]
    Compute the inverse hyperbolic tangent for each element of *x*.

    **See also:** [tanh], page 437.

**asech** (*x*) [Mapping Function]
    Compute the inverse hyperbolic secant of each element of *x*.

    **See also:** [sech], page 437.

**acsch** (*x*) [Mapping Function]
    Compute the inverse hyperbolic cosecant of each element of *x*.

    **See also:** [csch], page 437.

**acoth** (*x*) [Mapping Function]
    Compute the inverse hyperbolic cotangent of each element of *x*.

    **See also:** [coth], page 437.

**atan2** (*y*, *x*) [Mapping Function]
    Compute atan ($y$ / $x$) for corresponding elements of *y* and *x*.

    *y* and *x* must match in size and orientation.

    **See also:** [tan], page 436, [tand], page 439, [tanh], page 437, [atanh], page 438.

Octave provides the following trigonometric functions where angles are specified in degrees. These functions produce true zeros at the appropriate intervals rather than the small round-off error that occurs when using radians. For example:

```
cosd (90)
    ⇒ 0
cos (pi/2)
    ⇒ 6.1230e-17
```

**sind** (*x*) [Function File]
    Compute the sine for each element of *x* in degrees.

    Returns zero for elements where `x/180` is an integer.

    **See also:** [asind], page 439, [sin], page 436.

**cosd** (*x*) [Function File]
    Compute the cosine for each element of *x* in degrees.

    Returns zero for elements where `(x-90)/180` is an integer.

    **See also:** [acosd], page 439, [cos], page 436.

Chapter 17: Arithmetic                                                                   439

tand (x)                                                                    [Function File]
    Compute the tangent for each element of x in degrees.

    Returns zero for elements where x/180 is an integer and Inf for elements where (x-90)/180 is an integer.

    **See also:** [atand], page 439, [tan], page 436.

secd (x)                                                                    [Function File]
    Compute the secant for each element of x in degrees.

    **See also:** [asecd], page 439, [sec], page 436.

cscd (x)                                                                    [Function File]
    Compute the cosecant for each element of x in degrees.

    **See also:** [acscd], page 439, [csc], page 436.

cotd (x)                                                                    [Function File]
    Compute the cotangent for each element of x in degrees.

    **See also:** [acotd], page 439, [cot], page 436.

asind (x)                                                                   [Function File]
    Compute the inverse sine in degrees for each element of x.

    **See also:** [sind], page 438, [asin], page 437.

acosd (x)                                                                   [Function File]
    Compute the inverse cosine in degrees for each element of x.

    **See also:** [cosd], page 438, [acos], page 437.

atand (x)                                                                   [Function File]
    Compute the inverse tangent in degrees for each element of x.

    **See also:** [tand], page 439, [atan], page 437.

atan2d (y, x)                                                               [Function File]
    Compute atan2 (y / x) in degrees for corresponding elements from y and x.

    **See also:** [tand], page 439, [atan2], page 438.

asecd (x)                                                                   [Function File]
    Compute the inverse secant in degrees for each element of x.

    **See also:** [secd], page 439, [asec], page 437.

acscd (x)                                                                   [Function File]
    Compute the inverse cosecant in degrees for each element of x.

    **See also:** [cscd], page 439, [acsc], page 437.

acotd (x)                                                                   [Function File]
    Compute the inverse cotangent in degrees for each element of x.

    **See also:** [cotd], page 439, [acot], page 437.

## 17.4 Sums and Products

sum (*x*)     [Built-in Function]
sum (*x*, *dim*)     [Built-in Function]
sum (..., "*native*")     [Built-in Function]
sum (..., "*double*")     [Built-in Function]
sum (..., "*extra*")     [Built-in Function]

    Sum of elements along dimension *dim*.

    If *dim* is omitted, it defaults to the first non-singleton dimension.

    The optional "`type`" input determines the class of the variable used for calculations. If the argument "`native`" is given, then the operation is performed in the same type as the original argument, rather than the default double type.

    For example:

```
sum ([true, true])
   ⇒ 2
sum ([true, true], "native")
   ⇒ true
```

    On the contrary, if "`double`" is given, the sum is performed in double precision even for single precision inputs.

    For double precision inputs, the "`extra`" option will use a more accurate algorithm than straightforward summation. For single precision inputs, "`extra`" is the same as "`double`". Otherwise, "`extra`" has no effect.

    **See also:** [cumsum], page 440, [sumsq], page 441, [prod], page 440.

prod (*x*)     [Built-in Function]
prod (*x*, *dim*)     [Built-in Function]
prod (..., "*native*")     [Built-in Function]
prod (..., "*double*")     [Built-in Function]

    Product of elements along dimension *dim*.

    If *dim* is omitted, it defaults to the first non-singleton dimension.

    The optional "`type`" input determines the class of the variable used for calculations. If the argument "`native`" is given, then the operation is performed in the same type as the original argument, rather than the default double type.

    For example:

```
prod ([true, true])
   ⇒ 1
prod ([true, true], "native")
   ⇒ true
```

    On the contrary, if "`double`" is given, the operation is performed in double precision even for single precision inputs.

    **See also:** [cumprod], page 441, [sum], page 440.

cumsum (*x*)     [Built-in Function]
cumsum (*x*, *dim*)     [Built-in Function]

Chapter 17: Arithmetic  441

cumsum (..., "*native*")  [Built-in Function]
cumsum (..., "*double*")  [Built-in Function]
cumsum (..., "*extra*")  [Built-in Function]
> Cumulative sum of elements along dimension *dim*.
>
> If *dim* is omitted, it defaults to the first non-singleton dimension.
>
> See `sum` for an explanation of the optional parameters `"native"`, `"double"`, and `"extra"`.
>
> See also: [sum], page 440, [cumprod], page 441.

cumprod (*x*)  [Built-in Function]
cumprod (*x*, *dim*)  [Built-in Function]
> Cumulative product of elements along dimension *dim*.
>
> If *dim* is omitted, it defaults to the first non-singleton dimension.
>
> See also: [prod], page 440, [cumsum], page 440.

sumsq (*x*)  [Built-in Function]
sumsq (*x*, *dim*)  [Built-in Function]
> Sum of squares of elements along dimension *dim*.
>
> If *dim* is omitted, it defaults to the first non-singleton dimension.
>
> This function is conceptually equivalent to computing
>
>     sum (x .* conj (x), dim)
>
> but it uses less memory and avoids calling `conj` if *x* is real.
>
> See also: [sum], page 440, [prod], page 440.

## 17.5 Utility Functions

ceil (*x*)  [Mapping Function]
> Return the smallest integer not less than x.
>
> This is equivalent to rounding towards positive infinity.
>
> If *x* is complex, return `ceil (real (x)) + ceil (imag (x)) * I`.
>
>     ceil ([-2.7, 2.7])
>        ⇒ -2   3
>
> See also: [floor], page 442, [round], page 442, [fix], page 441.

fix (*x*)  [Mapping Function]
> Truncate fractional portion of *x* and return the integer portion.
>
> This is equivalent to rounding towards zero. If *x* is complex, return `fix (real (x)) + fix (imag (x)) * I`.
>
>     fix ([-2.7, 2.7])
>        ⇒ -2   2
>
> See also: [ceil], page 441, [floor], page 442, [round], page 442.

floor (x)                                                    [Mapping Function]
: Return the largest integer not greater than x.

    This is equivalent to rounding towards negative infinity. If x is complex, return floor (real (x)) + floor (imag (x)) * I.

        floor ([-2.7, 2.7])
            ⇒ -3    2

    See also: [ceil], page 441, [round], page 442, [fix], page 441.

round (x)                                                    [Mapping Function]
: Return the integer nearest to x.

    If x is complex, return round (real (x)) + round (imag (x)) * I. If there are two nearest integers, return the one further away from zero.

        round ([-2.7, 2.7])
            ⇒ -3    3

    See also: [ceil], page 441, [floor], page 442, [fix], page 441, [roundb], page 442.

roundb (x)                                                   [Mapping Function]
: Return the integer nearest to x. If there are two nearest integers, return the even one (banker's rounding).

    If x is complex, return roundb (real (x)) + roundb (imag (x)) * I.

    See also: [round], page 442.

max (x)                                                      [Built-in Function]
max (x, [], dim)                                             [Built-in Function]
[w, iw] = max (x)                                            [Built-in Function]
max (x, y)                                                   [Built-in Function]
: Find maximum values in the array x.

    For a vector argument, return the maximum value. For a matrix argument, return a row vector with the maximum value of each column. For a multi-dimensional array, max operates along the first non-singleton dimension.

    If the optional third argument *dim* is present then operate along this dimension. In this case the second argument is ignored and should be set to the empty matrix.

    For two matrices (or a matrix and a scalar), return the pairwise maximum.

    Thus,

        max (max (x))

    returns the largest element of the 2-D matrix x, and

        max (2:5, pi)
            ⇒  3.1416  3.1416  4.0000  5.0000

    compares each element of the range 2:5 with pi, and returns a row vector of the maximum values.

    For complex arguments, the magnitude of the elements are used for comparison. If the magnitudes are identical, then the results are ordered by phase angle in the range (-pi, pi]. Hence,

Chapter 17: Arithmetic                                                                 443

```
max ([-1 i 1 -i])
    ⇒ -1
```

because all entries have magnitude 1, but -1 has the largest phase angle with value pi.

If called with one input and two output arguments, `max` also returns the first index of the maximum value(s). Thus,

```
[x, ix] = max ([1, 3, 5, 2, 5])
    ⇒  x = 5
       ix = 3
```

**See also:** [min], page 443, [cummax], page 444, [cummin], page 444.

min (*x*)                                                              [Built-in Function]
min (*x*, [], *dim*)                                                   [Built-in Function]
[*w*, *iw*] = min (*x*)                                                [Built-in Function]
min (*x*, *y*)                                                         [Built-in Function]

Find minimum values in the array *x*.

For a vector argument, return the minimum value. For a matrix argument, return a row vector with the minimum value of each column. For a multi-dimensional array, `min` operates along the first non-singleton dimension.

If the optional third argument *dim* is present then operate along this dimension. In this case the second argument is ignored and should be set to the empty matrix.

For two matrices (or a matrix and a scalar), return the pairwise minimum.

Thus,

```
min (min (x))
```

returns the smallest element of the 2-D matrix *x*, and

```
min (2:5, pi)
    ⇒  2.0000  3.0000  3.1416  3.1416
```

compares each element of the range `2:5` with `pi`, and returns a row vector of the minimum values.

For complex arguments, the magnitude of the elements are used for comparison. If the magnitudes are identical, then the results are ordered by phase angle in the range (-pi, pi]. Hence,

```
min ([-1 i 1 -i])
    ⇒ -i
```

because all entries have magnitude 1, but -i has the smallest phase angle with value -pi/2.

If called with one input and two output arguments, `min` also returns the first index of the minimum value(s). Thus,

```
[x, ix] = min ([1, 3, 0, 2, 0])
    ⇒  x = 0
       ix = 3
```

**See also:** [max], page 442, [cummin], page 444, [cummax], page 444.

cummax (x) [Built-in Function]
cummax (x, dim) [Built-in Function]
[w, iw] = cummax (...) [Built-in Function]

Return the cumulative maximum values along dimension *dim*.

If *dim* is unspecified it defaults to column-wise operation. For example:

```
cummax ([1 3 2 6 4 5])
   ⇒  1  3  3  6  6  6
```

If called with two output arguments the index of the maximum value is also returned.

```
[w, iw] = cummax ([1 3 2 6 4 5])
⇒
w  = 1  3  3  6  6  6
iw = 1  2  2  4  4  4
```

See also: [cummin], page 444, [max], page 442, [min], page 443.

cummin (x) [Built-in Function]
cummin (x, dim) [Built-in Function]
[w, iw] = cummin (x) [Built-in Function]

Return the cumulative minimum values along dimension *dim*.

If *dim* is unspecified it defaults to column-wise operation. For example:

```
cummin ([5 4 6 2 3 1])
   ⇒  5  4  4  2  2  1
```

If called with two output arguments the index of the minimum value is also returned.

```
[w, iw] = cummin ([5 4 6 2 3 1])
⇒
w  = 5  4  4  2  2  1
iw = 1  2  2  4  4  6
```

See also: [cummax], page 444, [min], page 443, [max], page 442.

hypot (x, y) [Built-in Function]
hypot (x, y, z, ...) [Built-in Function]

Compute the element-by-element square root of the sum of the squares of *x* and *y*.

This is equivalent to `sqrt (x.^2 + y.^2)`, but is calculated in a manner that avoids overflows for large values of *x* or *y*.

`hypot` can also be called with more than 2 arguments; in this case, the arguments are accumulated from left to right:

```
hypot (hypot (x, y), z)
hypot (hypot (hypot (x, y), z), w), etc.
```

dx = gradient (m) [Function File]
[dx, dy, dz, ...] = gradient (m) [Function File]
[...] = gradient (m, s) [Function File]
[...] = gradient (m, x, y, z, ...) [Function File]
[...] = gradient (f, x0) [Function File]
[...] = gradient (f, x0, s) [Function File]

Chapter 17: Arithmetic                                                                 445

[...] = gradient (f, x0, x, y, ...)                                        [Function File]
    Calculate the gradient of sampled data or a function.

    If *m* is a vector, calculate the one-dimensional gradient of *m*. If *m* is a matrix the gradient is calculated for each dimension.

    [dx, dy] = gradient (m) calculates the one-dimensional gradient for *x* and *y* direction if *m* is a matrix. Additional return arguments can be use for multi-dimensional matrices.

    A constant spacing between two points can be provided by the *s* parameter. If *s* is a scalar, it is assumed to be the spacing for all dimensions. Otherwise, separate values of the spacing can be supplied by the *x*, ... arguments. Scalar values specify an equidistant spacing. Vector values for the *x*, ... arguments specify the coordinate for that dimension. The length must match their respective dimension of *m*.

    At boundary points a linear extrapolation is applied. Interior points are calculated with the first approximation of the numerical gradient

        y'(i) = 1/(x(i+1)-x(i-1)) * (y(i-1)-y(i+1)).

    If the first argument *f* is a function handle, the gradient of the function at the points in *x0* is approximated using central difference. For example, gradient (@cos, 0) approximates the gradient of the cosine function in the point $x0 = 0$. As with sampled data, the spacing values between the points from which the gradient is estimated can be set via the *s* or *dx*, *dy*, ... arguments. By default a spacing of 1 is used.

    **See also:** [diff], page 404, [del2], page 446.

dot (x, y, dim)                                                            [Built-in Function]
    Compute the dot product of two vectors.

    If *x* and *y* are matrices, calculate the dot products along the first non-singleton dimension.

    If the optional argument *dim* is given, calculate the dot products along this dimension.

    This is equivalent to sum (conj (X) .* Y, dim), but avoids forming a temporary array and is faster. When *X* and *Y* are column vectors, the result is equivalent to X' * Y.

    **See also:** [cross], page 445, [divergence], page 446.

cross (x, y)                                                               [Function File]
cross (x, y, dim)                                                          [Function File]
    Compute the vector cross product of two 3-dimensional vectors *x* and *y*.

    If *x* and *y* are matrices, the cross product is applied along the first dimension with three elements.

    The optional argument *dim* forces the cross product to be calculated along the specified dimension.

    Example Code:

        cross ([1,1,0], [0,1,1])
            ⇒ [ 1; -1; 1 ]

    **See also:** [dot], page 445, [curl], page 446, [divergence], page 446.

```
div = divergence (x, y, z, fx, fy, fz)                          [Function File]
div = divergence (fx, fy, fz)                                   [Function File]
div = divergence (x, y, fx, fy)                                 [Function File]
div = divergence (fx, fy)                                       [Function File]
```
Calculate divergence of a vector field given by the arrays fx, fy, and fz or fx, fy respectively.

$$divF(x,y,z) = \partial_x F + \partial_y F + \partial_z F$$

The coordinates of the vector field can be given by the arguments x, y, z or x, y respectively.

**See also:** [curl], page 446, [gradient], page 444, [del2], page 446, [dot], page 445.

```
[cx, cy, cz, v] = curl (x, y, z, fx, fy, fz)                    [Function File]
[cz, v] = curl (x, y, fx, fy)                                   [Function File]
[...] = curl (fx, fy, fz)                                       [Function File]
[...] = curl (fx, fy)                                           [Function File]
v = curl (...)                                                  [Function File]
```
Calculate curl of vector field given by the arrays fx, fy, and fz or fx, fy respectively.

$$curlF(x,y,z) = \left(\frac{\partial d}{\partial y}F_z - \frac{\partial d}{\partial z}F_y, \frac{\partial d}{\partial z}F_x - \frac{\partial d}{\partial x}F_z, \frac{\partial d}{\partial x}F_y - \frac{\partial d}{\partial y}F_x\right)$$

The coordinates of the vector field can be given by the arguments x, y, z or x, y respectively. v calculates the scalar component of the angular velocity vector in direction of the z-axis for two-dimensional input. For three-dimensional input the scalar rotation is calculated at each grid point in direction of the vector field at that point.

**See also:** [divergence], page 446, [gradient], page 444, [del2], page 446, [cross], page 445.

```
d = del2 (M)                                                    [Function File]
d = del2 (M, h)                                                 [Function File]
d = del2 (M, dx, dy, ...)                                       [Function File]
```
Calculate the discrete Laplace operator ($\nabla^2$).

For a 2-dimensional matrix $M$ this is defined as

$$d = \frac{1}{4}\left(\frac{d^2}{dx^2}M(x,y) + \frac{d^2}{dy^2}M(x,y)\right)$$

For N-dimensional arrays the sum in parentheses is expanded to include second derivatives over the additional higher dimensions.

The spacing between evaluation points may be defined by h, which is a scalar defining the equidistant spacing in all dimensions. Alternatively, the spacing in each dimension may be defined separately by dx, dy, etc. A scalar spacing argument defines equidistant spacing, whereas a vector argument can be used to specify variable spacing. The length of the spacing vectors must match the respective dimension of $M$. The default spacing value is 1.

Chapter 17: Arithmetic

At least 3 data points are needed for each dimension. Boundary points are calculated from the linear extrapolation of interior points.

**See also:** [gradient], page 444, [diff], page 404.

`factorial (n)` [Function File]

Return the factorial of *n* where *n* is a real non-negative integer.

If *n* is a scalar, this is equivalent to `prod (1:n)`. For vector or matrix arguments, return the factorial of each element in the array.

For non-integers see the generalized factorial function `gamma`. Note that the factorial function grows large quite quickly, and even with double precision values overflow will occur if *n* > 171. For such cases consider `gammaln`.

**See also:** [prod], page 440, [gamma], page 454, [gammaln], page 456.

`pf = factor (q)` [Function File]
`[pf, n] = factor (q)` [Function File]

Return the prime factorization of *q*.

The prime factorization is defined as `prod (pf) == q` where every element of *pf* is a prime number. If `q == 1`, return 1.

With two output arguments, return the unique prime factors *pf* and their multiplicities. That is, `prod (pf .^ n) == q`.

Implementation Note: The input *q* must be less than `bitmax` (9.0072e+15) in order to factor correctly.

**See also:** [gcd], page 447, [lcm], page 447, [isprime], page 64, [primes], page 448.

`g = gcd (a1, a2, ...)` [Built-in Function]
`[g, v1, ...] = gcd (a1, a2, ...)` [Built-in Function]

Compute the greatest common divisor of *a1, a2, ...*.

If more than one argument is given then all arguments must be the same size or scalar. In this case the greatest common divisor is calculated for each element individually. All elements must be ordinary or Gaussian (complex) integers. Note that for Gaussian integers, the gcd is only unique up to a phase factor (multiplication by 1, -1, i, or -i), so an arbitrary greatest common divisor among the four possible is returned.

Optional return arguments *v1, ...*, contain integer vectors such that,

$$g = v_1 a_1 + v_2 a_2 + \cdots$$

Example code:

```
gcd ([15, 9], [20, 18])
     ⇒  5  9
```

**See also:** [lcm], page 447, [factor], page 447, [isprime], page 64.

`lcm (x, y)` [Mapping Function]
`lcm (x, y, ...)` [Mapping Function]

Compute the least common multiple of *x* and *y*, or of the list of all arguments.

All elements must be numeric and of the same size or scalar.

**See also:** [factor], page 447, [gcd], page 447, [isprime], page 64.

`chop (x, ndigits, base)` [Function File]

    Truncate elements of *x* to a length of *ndigits* such that the resulting numbers are exactly divisible by *base*.

    If *base* is not specified it defaults to 10.

```
chop (-pi, 5, 10)
    ⇒ -3.14200000000000
chop (-pi, 5, 5)
    ⇒ -3.14150000000000
```

`rem (x, y)` [Mapping Function]

    Return the remainder of the division `x / y`.

    The remainder is computed using the expression

```
x - y .* fix (x ./ y)
```

    An error message is printed if the dimensions of the arguments do not agree, or if either of the arguments is complex.

    **See also:** [mod], page 448.

`mod (x, y)` [Mapping Function]

    Compute the modulo of *x* and *y*.

    Conceptually this is given by

```
x - y .* floor (x ./ y)
```

    and is written such that the correct modulus is returned for integer types. This function handles negative values correctly. That is, `mod (-1, 3)` is 2, not -1, as `rem (-1, 3)` returns. `mod (x, 0)` returns *x*.

    An error results if the dimensions of the arguments do not agree, or if either of the arguments is complex.

    **See also:** [rem], page 448.

`primes (n)` [Function File]

    Return all primes up to *n*.

    The output data class (double, single, uint32, etc.) is the same as the input class of *n*. The algorithm used is the Sieve of Eratosthenes.

    Notes: If you need a specific number of primes you can use the fact that the distance from one prime to the next is, on average, proportional to the logarithm of the prime. Integrating, one finds that there are about $k$ primes less than $k \log(5k)$.

    See also `list_primes` if you need a specific number *n* of primes.

    **See also:** [list_primes], page 448, [isprime], page 64.

`list_primes ()` [Function File]
`list_primes (n)` [Function File]

    List the first *n* primes.

    If *n* is unspecified, the first 25 primes are listed.

    **See also:** [primes], page 448, [isprime], page 64.

Chapter 17: Arithmetic 449

sign (x) [Mapping Function]
Compute the *signum* function.
This is defined as
$$\text{sign}(x) = \begin{cases} 1, & x > 0; \\ 0, & x = 0; \\ -1, & x < 0. \end{cases}$$

For complex arguments, `sign` returns `x ./ abs (x)`.

Note that `sign (-0.0)` is 0. Although IEEE 754 floating point allows zero to be signed, 0.0 and -0.0 compare equal. If you must test whether zero is signed, use the `signbit` function.

**See also:** [signbit], page 449.

signbit (x) [Mapping Function]
Return logical true if the value of *x* has its sign bit set and false otherwise.

This behavior is consistent with the other logical functions. See Section 4.6 [Logical Values], page 60. The behavior differs from the C language function which returns nonzero if the sign bit is set.

This is not the same as `x < 0.0`, because IEEE 754 floating point allows zero to be signed. The comparison `-0.0 < 0.0` is false, but `signbit (-0.0)` will return a nonzero value.

**See also:** [sign], page 449.

## 17.6 Special Functions

[a, *ierr*] = airy (k, z, opt) [Built-in Function]
Compute Airy functions of the first and second kind, and their derivatives.

```
          K    Function   Scale factor (if "opt" is supplied)
          ---  --------   -----------------------------------
          0    Ai (Z)     exp ((2/3) * Z * sqrt (Z))
          1    dAi(Z)/dZ  exp ((2/3) * Z * sqrt (Z))
          2    Bi (Z)     exp (-abs (real ((2/3) * Z * sqrt (Z))))
          3    dBi(Z)/dZ  exp (-abs (real ((2/3) * Z * sqrt (Z))))
```

The function call `airy (z)` is equivalent to `airy (0, z)`.

The result is the same size as *z*.

If requested, *ierr* contains the following status information and is the same size as the result.

0. Normal return.

1. Input error, return `NaN`.

2. Overflow, return `Inf`.

3. Loss of significance by argument reduction results in less than half of machine accuracy.

4. Complete loss of significance by argument reduction, return `NaN`.

5. Error—no computation, algorithm termination condition not met, return `NaN`.

| | |
|---|---|
| `[j, ierr] = besselj (alpha, x, opt)` | [Built-in Function] |
| `[y, ierr] = bessely (alpha, x, opt)` | [Built-in Function] |
| `[i, ierr] = besseli (alpha, x, opt)` | [Built-in Function] |
| `[k, ierr] = besselk (alpha, x, opt)` | [Built-in Function] |
| `[h, ierr] = besselh (alpha, k, x, opt)` | [Built-in Function] |

Compute Bessel or Hankel functions of various kinds:

- `besselj` Bessel functions of the first kind. If the argument *opt* is 1 or true, the result is multiplied by `exp (-abs (imag (x)))`.

- `bessely` Bessel functions of the second kind. If the argument *opt* is 1 or true, the result is multiplied by `exp (-abs (imag (x)))`.

- `besseli`

  Modified Bessel functions of the first kind. If the argument *opt* is 1 or true, the result is multiplied by `exp (-abs (real (x)))`.

- `besselk`

  Modified Bessel functions of the second kind. If the argument *opt* is 1 or true, the result is multiplied by `exp (x)`.

- `besselh` Compute Hankel functions of the first ($k = 1$) or second ($k = 2$) kind. If the argument *opt* is 1 or true, the result is multiplied by `exp (-I*x)` for $k = 1$ or `exp (I*x)` for $k = 2$.

If *alpha* is a scalar, the result is the same size as *x*. If *x* is a scalar, the result is the same size as *alpha*. If *alpha* is a row vector and *x* is a column vector, the result is a matrix with `length (x)` rows and `length (alpha)` columns. Otherwise, *alpha* and *x* must conform and the result will be the same size.

The value of *alpha* must be real. The value of *x* may be complex.

If requested, *ierr* contains the following status information and is the same size as the result.

0. Normal return.
1. Input error, return `NaN`.
2. Overflow, return `Inf`.
3. Loss of significance by argument reduction results in less than half of machine accuracy.
4. Complete loss of significance by argument reduction, return `NaN`.
5. Error—no computation, algorithm termination condition not met, return `NaN`.

`beta (a, b)` [Mapping Function]

Compute the Beta function for real inputs *a* and *b*.

The Beta function definition is

$$B(a, b) = \frac{\Gamma(a)\Gamma(b)}{\Gamma(a + b)}.$$

The Beta function can grow quite large and it is often more useful to work with the logarithm of the output rather than the function directly. See [betaln], page 451, for computing the logarithm of the Beta function in an efficient manner.

**See also:** [betaln], page 451, [betainc], page 451, [betaincinv], page 451.

Chapter 17: Arithmetic    451

**betainc (x, a, b)**                                                                 [Mapping Function]

Compute the regularized incomplete Beta function.

The regularized incomplete Beta function is defined by

$$I(x,a,b) = \frac{1}{B(a,b)} \int_0^x t^{(a-z)}(1-t)^{(b-1)} dt.$$

If x has more than one component, both a and b must be scalars. If x is a scalar, a and b must be of compatible dimensions.

**See also:** [betaincinv], page 451, [beta], page 450, [betaln], page 451.

**betaincinv (y, a, b)**                                                              [Mapping Function]

Compute the inverse of the incomplete Beta function.

The inverse is the value x such that

    y == betainc (x, a, b)

**See also:** [betainc], page 451, [beta], page 450, [betaln], page 451.

**betaln (a, b)**                                                                     [Mapping Function]

Compute the natural logarithm of the Beta function for real inputs a and b.

`betaln` is defined as

$$\mathrm{betaln}(a,b) = \ln(B(a,b)) \equiv \ln\left(\frac{\Gamma(a)\Gamma(b)}{\Gamma(a+b)}\right).$$

and is calculated in a way to reduce the occurrence of underflow.

The Beta function can grow quite large and it is often more useful to work with the logarithm of the output rather than the function directly.

**See also:** [beta], page 450, [betainc], page 451, [betaincinv], page 451, [gammaln], page 456.

**bincoeff (n, k)**                                                                   [Mapping Function]

Return the binomial coefficient of n and k, defined as

$$\binom{n}{k} = \frac{n(n-1)(n-2)\cdots(n-k+1)}{k!}$$

For example:

    bincoeff (5, 2)
        ⇒ 10

In most cases, the **nchoosek** function is faster for small scalar integer arguments. It also warns about loss of precision for big arguments.

**See also:** [nchoosek], page 604.

**commutation_matrix (m, n)**                                                         [Function File]

Return the commutation matrix $K_{m,n}$ which is the unique $mn \times mn$ matrix such that $K_{m,n} \cdot \mathrm{vec}(A) = \mathrm{vec}(A^T)$ for all $m \times n$ matrices A.

If only one argument m is given, $K_{m,m}$ is returned.

See Magnus and Neudecker (1988), *Matrix Differential Calculus with Applications in Statistics and Econometrics.*

`duplication_matrix (n)` *[Function File]*
> Return the duplication matrix $D_n$ which is the unique $n^2 \times n(n+1)/2$ matrix such that $D_n * \text{vech}(A) = \text{vec}(A)$ for all symmetric $n \times n$ matrices $A$.
>
> See Magnus and Neudecker (1988), *Matrix Differential Calculus with Applications in Statistics and Econometrics.*

`dawson (z)` *[Mapping Function]*
> Compute the Dawson (scaled imaginary error) function.
>
> The Dawson function is defined as
>
> $$\frac{\sqrt{\pi}}{2} e^{-z^2} \text{erfi}(z) \equiv -i \frac{\sqrt{\pi}}{2} e^{-z^2} \text{erf}(iz)$$
>
> **See also:** [erfc], page 453, [erf], page 453, [erfcx], page 453, [erfi], page 453, [erfinv], page 454, [erfcinv], page 454.

`[sn, cn, dn, err] = ellipj (u, m)` *[Built-in Function]*
`[sn, cn, dn, err] = ellipj (u, m, tol)` *[Built-in Function]*
> Compute the Jacobi elliptic functions *sn*, *cn*, and *dn* of complex argument *u* and real parameter *m*.
>
> If *m* is a scalar, the results are the same size as *u*. If *u* is a scalar, the results are the same size as *m*. If *u* is a column vector and *m* is a row vector, the results are matrices with `length (u)` rows and `length (m)` columns. Otherwise, *u* and *m* must conform in size and the results will be the same size as the inputs.
>
> The value of *u* may be complex. The value of *m* must be $0 \leq m \leq 1$.
>
> The optional input *tol* is currently ignored (MATLAB uses this to allow faster, less accurate approximation).
>
> If requested, *err* contains the following status information and is the same size as the result.
>
> 0. Normal return.
>
> 1. Error—no computation, algorithm termination condition not met, return `NaN`.
>
> Reference: Milton Abramowitz and Irene A Stegun, *Handbook of Mathematical Functions*, Chapter 16 (Sections 16.4, 16.13, and 16.15), Dover, 1965.
>
> **See also:** [ellipke], page 452.

`k = ellipke (m)` *[Function File]*
`k = ellipke (m, tol)` *[Function File]*
`[k, e] = ellipke (...)` *[Function File]*
> Compute complete elliptic integrals of the first $K(m)$ and second $E(m)$ kind.
>
> *m* must be a scalar or real array with $-\text{Inf} \leq m \leq 1$.
>
> The optional input *tol* controls the stopping tolerance of the algorithm and defaults to `eps (class (m))`. The tolerance can be increased to compute a faster, less accurate approximation.
>
> When called with one output only elliptic integrals of the first kind are returned.
>
> Mathematical Note:

Chapter 17: Arithmetic

Elliptic integrals of the first kind are defined as

$$\mathrm{K}(m) = \int_0^1 \frac{dt}{\sqrt{(1-t^2)(1-m^2t^2)}}$$

Elliptic integrals of the second kind are defined as

$$\mathrm{E}(m) = \int_0^1 \frac{\sqrt{1-m^2t^2}}{\sqrt{1-t^2}} dt$$

Reference: Milton Abramowitz and Irene A. Stegun, *Handbook of Mathematical Functions*, Chapter 17, Dover, 1965.

**See also:** [ellipj], page 452.

erf (*z*) [Mapping Function]

Compute the error function.

The error function is defined as

$$\mathrm{erf}(z) = \frac{2}{\sqrt{\pi}} \int_0^z e^{-t^2} dt$$

**See also:** [erfc], page 453, [erfcx], page 453, [erfi], page 453, [dawson], page 452, [erfinv], page 454, [erfcinv], page 454.

erfc (*z*) [Mapping Function]

Compute the complementary error function.

The complementary error function is defined as $1 - \mathrm{erf}(z)$.

**See also:** [erfcinv], page 454, [erfcx], page 453, [erfi], page 453, [dawson], page 452, [erf], page 453, [erfinv], page 454.

erfcx (*z*) [Mapping Function]

Compute the scaled complementary error function.

The scaled complementary error function is defined as

$$e^{z^2}\mathrm{erfc}(z) \equiv e^{z^2}(1 - \mathrm{erf}(z))$$

**See also:** [erfc], page 453, [erf], page 453, [erfi], page 453, [dawson], page 452, [erfinv], page 454, [erfcinv], page 454.

erfi (*z*) [Mapping Function]

Compute the imaginary error function.

The imaginary error function is defined as

$$-i\,\mathrm{erf}(iz)$$

**See also:** [erfc], page 453, [erf], page 453, [erfcx], page 453, [dawson], page 452, [erfinv], page 454, [erfcinv], page 454.

**erfinv (x)**                          [Mapping Function]

Compute the inverse error function.

The inverse error function is defined such that

    `erf (y) == x`

See also: [erf], page 453, [erfc], page 453, [erfcx], page 453, [erfi], page 453, [dawson], page 452, [erfcinv], page 454.

**erfcinv (x)**                        [Mapping Function]

Compute the inverse complementary error function.

The inverse complementary error function is defined such that

    `erfc (y) == x`

See also: [erfc], page 453, [erf], page 453, [erfcx], page 453, [erfi], page 453, [dawson], page 452, [erfinv], page 454.

**expint (x)**                            [Function File]

Compute the exponential integral:

$$E_1(x) = \int_x^\infty \frac{e^{-t}}{t} dt$$

Note: For compatibility, this functions uses the MATLAB definition of the exponential integral. Most other sources refer to this particular value as $E_1(x)$, and the exponential integral as

$$\mathrm{Ei}(x) = -\int_{-x}^\infty \frac{e^{-t}}{t} dt.$$

The two definitions are related, for positive real values of x, by $E_1(-x) = -\mathrm{Ei}(x) - i\pi$.

**gamma (z)**                           [Mapping Function]

Compute the Gamma function.

The Gamma function is defined as

$$\Gamma(z) = \int_0^\infty t^{z-1} e^{-t} dt.$$

Programming Note: The gamma function can grow quite large even for small input values. In many cases it may be preferable to use the natural logarithm of the gamma function (`gammaln`) in calculations to minimize loss of precision. The final result is then `exp (result_using_gammaln)`.

See also: [gammainc], page 454, [gammaln], page 456, [factorial], page 447.

**gammainc (x, a)**                       [Mapping Function]
**gammainc (x, a, "*lower*")**             [Mapping Function]
**gammainc (x, a, "*upper*")**             [Mapping Function]

Compute the normalized incomplete gamma function.

This is defined as

$$\gamma(x,a) = \frac{1}{\Gamma(a)} \int_0^x t^{a-1} e^{-t} dt$$

Chapter 17: Arithmetic 455

with the limiting value of 1 as x approaches infinity. The standard notation is $P(a, x)$, e.g., Abramowitz and Stegun (6.5.1).

If $a$ is scalar, then `gammainc (x, a)` is returned for each element of x and vice versa.

If neither x nor $a$ is scalar, the sizes of x and $a$ must agree, and `gammainc` is applied element-by-element.

By default the incomplete gamma function integrated from 0 to x is computed. If `"upper"` is given then the complementary function integrated from x to infinity is calculated. It should be noted that

    `gammainc (x, a)` ≡ `1 - gammainc (x, a, "upper")`

**See also:** [gamma], page 454, [gammaln], page 456.

*l* = **legendre** (*n*, *x*)                                                                                         [Function File]
*l* = **legendre** (*n*, *x*, *normalization*)                                                          [Function File]

Compute the Legendre function of degree $n$ and order $m = 0 \ldots n$.

The value $n$ must be a real non-negative integer.

x is a vector with real-valued elements in the range [-1, 1].

The optional argument *normalization* may be one of `"unnorm"`, `"sch"`, or `"norm"`. The default if no normalization is given is `"unnorm"`.

When the optional argument *normalization* is `"unnorm"`, compute the Legendre function of degree $n$ and order $m$ and return all values for $m = 0 \ldots n$. The return value has one dimension more than x.

The Legendre Function of degree $n$ and order $m$:

$$P_n^m(x) = (-1)^m (1-x^2)^{m/2} \frac{d^m}{dx^m} P_n(x)$$

with Legendre polynomial of degree $n$:

$$P(x) = \frac{1}{2^n n!} \left( \frac{d^n}{dx^n} (x^2 - 1)^n \right)$$

`legendre (3, [-1.0, -0.9, -0.8])` returns the matrix:

```
    x  |   -1.0    |   -0.9    |   -0.8
    ------------------------------------
    m=0 | -1.00000  | -0.47250  | -0.08000
    m=1 |  0.00000  | -1.99420  | -1.98000
    m=2 |  0.00000  | -2.56500  | -4.32000
    m=3 |  0.00000  | -1.24229  | -3.24000
```

When the optional argument `normalization` is `"sch"`, compute the Schmidt semi-normalized associated Legendre function. The Schmidt semi-normalized associated Legendre function is related to the unnormalized Legendre functions by the following:

For Legendre functions of degree $n$ and order 0:

$$SP_n^0(x) = P_n^0(x)$$

For Legendre functions of degree n and order m:

$$SP_n^m(x) = P_n^m(x)(-1)^m \left( \frac{2(n-m)!}{(n+m)!} \right)^{0.5}$$

When the optional argument *normalization* is `"norm"`, compute the fully normalized associated Legendre function. The fully normalized associated Legendre function is related to the unnormalized Legendre functions by the following:

For Legendre functions of degree *n* and order *m*

$$NP_n^m(x) = P_n^m(x)(-1)^m \left( \frac{(n+0.5)(n-m)!}{(n+m)!} \right)^{0.5}$$

`gammaln (x)` [Mapping Function]
`lgamma (x)` [Mapping Function]

Return the natural logarithm of the gamma function of *x*.

**See also:** [gamma], page 454, [gammainc], page 454.

## 17.7 Rational Approximations

`s = rat (x, tol)` [Function File]
`[n, d] = rat (x, tol)` [Function File]

Find a rational approximation to *x* within the tolerance defined by *tol* using a continued fraction expansion.

For example:

```
rat (pi) = 3 + 1/(7 + 1/16) = 355/113
rat (e)  = 3 + 1/(-4 + 1/(2 + 1/(5 + 1/(-2 + 1/(-7))))) 
         = 1457/536
```

When called with two output arguments return the numerator and denominator separately as two matrices.

**See also:** [rats], page 456.

`rats (x, len)` [Built-in Function]

Convert *x* into a rational approximation represented as a string.

The string can be converted back into a matrix as follows:

```
r = rats (hilb (4));
x = str2num (r)
```

The optional second argument defines the maximum length of the string representing the elements of *x*. By default *len* is 9.

If the length of the smallest possible rational approximation exceeds *len*, an asterisk (*) padded with spaces will be returned instead.

**See also:** [format], page 232, [rat], page 456.

# Chapter 17: Arithmetic

## 17.8 Coordinate Transformations

[theta, r] = cart2pol (x, y)     [Function File]
[theta, r, z] = cart2pol (x, y, z)     [Function File]
[theta, r] = cart2pol (C)     [Function File]
[theta, r, z] = cart2pol (C)     [Function File]
P = cart2pol (...)     [Function File]

> Transform Cartesian coordinates to polar or cylindrical coordinates.
>
> The inputs x, y (, and z) must be the same shape, or scalar. If called with a single matrix argument then each row of C represents the Cartesian coordinate (x, y (, z)).
>
> *theta* describes the angle relative to the positive x-axis.
>
> *r* is the distance to the z-axis (0, 0, z).
>
> If only a single return argument is requested then return a matrix P where each row represents one polar/(cylindrical) coordinate (*theta*, *phi* (, z)).
>
> **See also:** [pol2cart], page 457, [cart2sph], page 457, [sph2cart], page 458.

[x, y] = pol2cart (theta, r)     [Function File]
[x, y, z] = pol2cart (theta, r, z)     [Function File]
[x, y] = pol2cart (P)     [Function File]
[x, y, z] = pol2cart (P)     [Function File]
C = pol2cart (...)     [Function File]

> Transform polar or cylindrical coordinates to Cartesian coordinates.
>
> The inputs *theta*, *r*, (and z) must be the same shape, or scalar. If called with a single matrix argument then each row of P represents the polar/(cylindrical) coordinate (*theta*, *r* (, z)).
>
> *theta* describes the angle relative to the positive x-axis.
>
> *r* is the distance to the z-axis (0, 0, z).
>
> If only a single return argument is requested then return a matrix C where each row represents one Cartesian coordinate (x, y (, z)).
>
> **See also:** [cart2pol], page 457, [sph2cart], page 458, [cart2sph], page 457.

[theta, phi, r] = cart2sph (x, y, z)     [Function File]
[theta, phi, r] = cart2sph (C)     [Function File]
S = cart2sph (...)     [Function File]

> Transform Cartesian coordinates to spherical coordinates.
>
> The inputs x, y, and z must be the same shape, or scalar. If called with a single matrix argument then each row of C represents the Cartesian coordinate (x, y, z).
>
> *theta* describes the angle relative to the positive x-axis.
>
> *phi* is the angle relative to the xy-plane.
>
> *r* is the distance to the origin (0, 0, 0).
>
> If only a single return argument is requested then return a matrix S where each row represents one spherical coordinate (*theta*, *phi*, *r*).
>
> **See also:** [sph2cart], page 458, [cart2pol], page 457, [pol2cart], page 457.

`[x, y, z] = sph2cart (theta, phi, r)` [Function File]
`[x, y, z] = sph2cart (S)` [Function File]
`C = sph2cart (...)` [Function File]

    Transform spherical coordinates to Cartesian coordinates.

    The inputs *theta*, *phi*, and *r* must be the same shape, or scalar. If called with a single matrix argument then each row of *S* represents the spherical coordinate (*theta*, *phi*, *r*).

    *theta* describes the angle relative to the positive x-axis.

    *phi* is the angle relative to the xy-plane.

    *r* is the distance to the origin (0, 0, 0).

    If only a single return argument is requested then return a matrix *C* where each row represents one Cartesian coordinate (*x*, *y*, *z*).

    **See also:** [cart2sph], page 457, [pol2cart], page 457, [cart2pol], page 457.

## 17.9 Mathematical Constants

`e` [Built-in Function]
`e (n)` [Built-in Function]
`e (n, m)` [Built-in Function]
`e (n, m, k, ...)` [Built-in Function]
`e (..., class)` [Built-in Function]

    Return a scalar, matrix, or N-dimensional array whose elements are all equal to the base of natural logarithms.

    The constant $e$ satisfies the equation $\log(e) = 1$.

    When called with no arguments, return a scalar with the value $e$.

    When called with a single argument, return a square matrix with the dimension specified.

    When called with more than one scalar argument the first two arguments are taken as the number of rows and columns and any further arguments specify additional matrix dimensions.

    The optional argument *class* specifies the return type and may be either `"double"` or `"single"`.

    **See also:** [log], page 433, [exp], page 433, [pi], page 458, [I], page 459.

`pi` [Built-in Function]
`pi (n)` [Built-in Function]
`pi (n, m)` [Built-in Function]
`pi (n, m, k, ...)` [Built-in Function]
`pi (..., class)` [Built-in Function]

    Return a scalar, matrix, or N-dimensional array whose elements are all equal to the ratio of the circumference of a circle to its diameter($\pi$).

    Internally, `pi` is computed as '`4.0 * atan (1.0)`'.

    When called with no arguments, return a scalar with the value of $\pi$.

Chapter 17: Arithmetic 459

When called with a single argument, return a square matrix with the dimension specified.

When called with more than one scalar argument the first two arguments are taken as the number of rows and columns and any further arguments specify additional matrix dimensions.

The optional argument *class* specifies the return type and may be either `"double"` or `"single"`.

**See also:** [e], page 458, [I], page 459.

I   [Built-in Function]
I (*n*)   [Built-in Function]
I (*n*, *m*)   [Built-in Function]
I (*n*, *m*, *k*, ...)   [Built-in Function]
I (..., *class*)   [Built-in Function]

Return a scalar, matrix, or N-dimensional array whose elements are all equal to the pure imaginary unit, defined as $\sqrt{-1}$.

I, and its equivalents i, j, and J, are functions so any of the names may be reused for other purposes (such as i for a counter variable).

When called with no arguments, return a scalar with the value $i$.

When called with a single argument, return a square matrix with the dimension specified.

When called with more than one scalar argument the first two arguments are taken as the number of rows and columns and any further arguments specify additional matrix dimensions.

The optional argument *class* specifies the return type and may be either `"double"` or `"single"`.

**See also:** [e], page 458, [pi], page 458, [log], page 433, [exp], page 433.

Inf   [Built-in Function]
Inf (*n*)   [Built-in Function]
Inf (*n*, *m*)   [Built-in Function]
Inf (*n*, *m*, *k*, ...)   [Built-in Function]
Inf (..., *class*)   [Built-in Function]

Return a scalar, matrix or N-dimensional array whose elements are all equal to the IEEE representation for positive infinity.

Infinity is produced when results are too large to be represented using the IEEE floating point format for numbers. Two common examples which produce infinity are division by zero and overflow.

```
[ 1/0 e^800 ]
⇒ Inf    Inf
```

When called with no arguments, return a scalar with the value '`Inf`'.

When called with a single argument, return a square matrix with the dimension specified.

When called with more than one scalar argument the first two arguments are taken as the number of rows and columns and any further arguments specify additional matrix dimensions.

The optional argument *class* specifies the return type and may be either `"double"` or `"single"`.

**See also:** [isinf], page 404, [NaN], page 460.

NaN                              [Built-in Function]
NaN (*n*)                        [Built-in Function]
NaN (*n, m*)                     [Built-in Function]
NaN (*n, m, k, . . .*)           [Built-in Function]
NaN (*. . ., class*)             [Built-in Function]

Return a scalar, matrix, or N-dimensional array whose elements are all equal to the IEEE symbol NaN (Not a Number).

NaN is the result of operations which do not produce a well defined numerical result. Common operations which produce a NaN are arithmetic with infinity ($\infty - \infty$), zero divided by zero (0/0), and any operation involving another NaN value (5 + NaN).

Note that NaN always compares not equal to NaN (NaN != NaN). This behavior is specified by the IEEE standard for floating point arithmetic. To find NaN values, use the `isnan` function.

When called with no arguments, return a scalar with the value 'NaN'.

When called with a single argument, return a square matrix with the dimension specified.

When called with more than one scalar argument the first two arguments are taken as the number of rows and columns and any further arguments specify additional matrix dimensions.

The optional argument *class* specifies the return type and may be either `"double"` or `"single"`.

**See also:** [isnan], page 404, [Inf], page 459.

eps                              [Built-in Function]
eps (*x*)                        [Built-in Function]
eps (*n, m*)                     [Built-in Function]
eps (*n, m, k, . . .*)           [Built-in Function]
eps (*. . ., class*)             [Built-in Function]

Return a scalar, matrix or N-dimensional array whose elements are all eps, the machine precision.

More precisely, `eps` is the relative spacing between any two adjacent numbers in the machine's floating point system. This number is obviously system dependent. On machines that support IEEE floating point arithmetic, `eps` is approximately $2.2204 \times 10^{-16}$ for double precision and $1.1921 \times 10^{-7}$ for single precision.

When called with no arguments, return a scalar with the value `eps (1.0)`.

Given a single argument *x*, return the distance between *x* and the next largest value.

When called with more than one argument the first two arguments are taken as the number of rows and columns and any further arguments specify additional matrix

Chapter 17: Arithmetic 461

dimensions. The optional argument *class* specifies the return type and may be either `"double"` or `"single"`.

**See also:** [realmax], page 461, [realmin], page 461, [intmax], page 55, [bitmax], page 58.

`realmax`                                                    [Built-in Function]
`realmax (n)`                                                [Built-in Function]
`realmax (n, m)`                                             [Built-in Function]
`realmax (n, m, k, ...)`                                     [Built-in Function]
`realmax (..., class)`                                       [Built-in Function]

Return a scalar, matrix, or N-dimensional array whose elements are all equal to the largest floating point number that is representable.

The actual value is system dependent. On machines that support IEEE floating point arithmetic, `realmax` is approximately $1.7977 \times 10^{308}$ for double precision and $3.4028 \times 10^{38}$ for single precision.

When called with no arguments, return a scalar with the value `realmax ("double")`.

When called with a single argument, return a square matrix with the dimension specified.

When called with more than one scalar argument the first two arguments are taken as the number of rows and columns and any further arguments specify additional matrix dimensions.

The optional argument *class* specifies the return type and may be either `"double"` or `"single"`.

**See also:** [realmin], page 461, [intmax], page 55, [bitmax], page 58, [eps], page 460.

`realmin`                                                    [Built-in Function]
`realmin (n)`                                                [Built-in Function]
`realmin (n, m)`                                             [Built-in Function]
`realmin (n, m, k, ...)`                                     [Built-in Function]
`realmin (..., class)`                                       [Built-in Function]

Return a scalar, matrix, or N-dimensional array whose elements are all equal to the smallest normalized floating point number that is representable.

The actual value is system dependent. On machines that support IEEE floating point arithmetic, `realmin` is approximately $2.2251 \times 10^{-308}$ for double precision and $1.1755 \times 10^{-38}$ for single precision.

When called with no arguments, return a scalar with the value `realmin ("double")`.

When called with a single argument, return a square matrix with the dimension specified.

When called with more than one scalar argument the first two arguments are taken as the number of rows and columns and any further arguments specify additional matrix dimensions.

The optional argument *class* specifies the return type and may be either `"double"` or `"single"`.

**See also:** [realmax], page 461, [intmin], page 56, [eps], page 460.

Chapter 18: Linear Algebra

# 18 Linear Algebra

This chapter documents the linear algebra functions provided in Octave. Reference material for many of these functions may be found in Golub and Van Loan, *Matrix Computations*, *2nd Ed.*, Johns Hopkins, 1989, and in the LAPACK *Users' Guide*, SIAM, 1992. The LAPACK *Users' Guide* is available at: *http://www.netlib.org/lapack/lug/*

A common text for engineering courses is G. Strang, *Linear Algebra and Its Applications*, *4th Edition*. It has become a widespread reference for linear algebra. An alternative is P. Lax *Linear Algebra and Its Applications*, and also is a good choice. It claims to be suitable for high school students with substantial mathematical interests as well as first-year undergraduates.

## 18.1 Techniques Used for Linear Algebra

Octave includes a polymorphic solver that selects an appropriate matrix factorization depending on the properties of the matrix itself. Generally, the cost of determining the matrix type is small relative to the cost of factorizing the matrix itself. In any case the matrix type is cached once it is calculated so that it is not re-determined each time it is used in a linear equation.

The selection tree for how the linear equation is solved or a matrix inverse is formed is given by:

1. If the matrix is upper or lower triangular sparse use a forward or backward substitution using the LAPACK xTRTRS function, and goto 4.

2. If the matrix is square, Hermitian with a real positive diagonal, attempt Cholesky factorization using the LAPACK xPOTRF function.

3. If the Cholesky factorization failed or the matrix is not Hermitian with a real positive diagonal, and the matrix is square, factorize using the LAPACK xGETRF function.

4. If the matrix is not square, or any of the previous solvers flags a singular or near singular matrix, find a least squares solution using the LAPACK xGELSD function.

The user can force the type of the matrix with the `matrix_type` function. This overcomes the cost of discovering the type of the matrix. However, it should be noted that identifying the type of the matrix incorrectly will lead to unpredictable results, and so `matrix_type` should be used with care.

It should be noted that the test for whether a matrix is a candidate for Cholesky factorization, performed above, and by the `matrix_type` function, does not make certain that the matrix is Hermitian. However, the attempt to factorize the matrix will quickly detect a non-Hermitian matrix.

## 18.2 Basic Matrix Functions

$AA$ = balance ($A$)     [Built-in Function]
$AA$ = balance ($A$, *opt*)     [Built-in Function]
[$DD$, $AA$] = balance ($A$, *opt*)     [Built-in Function]
[$D$, $P$, $AA$] = balance ($A$, *opt*)     [Built-in Function]
[$CC$, $DD$, $AA$, $BB$] = balance ($A$, $B$, *opt*)     [Built-in Function]

    Balance the matrix $A$ to reduce numerical errors in future calculations.

Compute $AA = DD \backslash A * DD$ in which $AA$ is a matrix whose row and column norms are roughly equal in magnitude, and $DD = P * D$, in which $P$ is a permutation matrix and $D$ is a diagonal matrix of powers of two. This allows the equilibration to be computed without round-off. Results of eigenvalue calculation are typically improved by balancing first.

If two output values are requested, `balance` returns the diagonal $D$ and the permutation $P$ separately as vectors. In this case, $DD =$ `eye(n)(:,P) * diag (D)`, where $n$ is the matrix size.

If four output values are requested, compute $AA = CC*A*DD$ and $BB = CC*B*DD$, in which $AA$ and $BB$ have nonzero elements of approximately the same magnitude and $CC$ and $DD$ are permuted diagonal matrices as in $DD$ for the algebraic eigenvalue problem.

The eigenvalue balancing option *opt* may be one of:

"noperm", "S"
    Scale only; do not permute.

"noscal", "P"
    Permute only; do not scale.

Algebraic eigenvalue balancing uses standard LAPACK routines.

Generalized eigenvalue problem balancing uses Ward's algorithm (SIAM Journal on Scientific and Statistical Computing, 1981).

`bw = bandwidth (A, type)`     [Function File]
`[lower, upper] = bandwidth (A)`     [Function File]
    Compute the bandwidth of $A$.

The *type* argument is the string "lower" for the lower bandwidth and "upper" for the upper bandwidth. If no *type* is specified return both the lower and upper bandwidth of $A$.

The lower/upper bandwidth of a matrix is the number of subdiagonals/superdiagonals with nonzero entries.

**See also:** [isbanded], page 64, [isdiag], page 64, [istril], page 64, [istriu], page 64.

`cond (A)`     [Function File]
`cond (A, p)`     [Function File]
    Compute the *p*-norm condition number of a matrix.

`cond (A)` is defined as $\| A \|_p * \| A^{-1} \|_p$.

By default, $p = 2$ is used which implies a (relatively slow) singular value decomposition. Other possible selections are $p = 1$, `Inf`, `"fro"` which are generally faster. See `norm` for a full discussion of possible $p$ values.

The condition number of a matrix quantifies the sensitivity of the matrix inversion operation when small changes are made to matrix elements. Ideally the condition number will be close to 1. When the number is large this indicates small changes (such as underflow or round-off error) will produce large changes in the resulting output. In such cases the solution results from numerical computing are not likely to be accurate.

**See also:** [condest], page 540, [rcond], page 470, [norm], page 468, [svd], page 479.

Chapter 18: Linear Algebra

det (A) [Built-in Function]
[d, rcond] = det (A) [Built-in Function]

Compute the determinant of A.

Return an estimate of the reciprocal condition number if requested.

Programming Notes: Routines from LAPACK are used for full matrices and code from UMFPACK is used for sparse matrices.

The determinant should not be used to check a matrix for singularity. For that, use any of the condition number functions: cond, condest, rcond.

**See also:** [cond], page 464, [condest], page 540, [rcond], page 470.

lambda = eig (A) [Built-in Function]
lambda = eig (A, B) [Built-in Function]
[V, lambda] = eig (A) [Built-in Function]
[V, lambda] = eig (A, B) [Built-in Function]

Compute the eigenvalues (and optionally the eigenvectors) of a matrix or a pair of matrices

The algorithm used depends on whether there are one or two input matrices, if they are real or complex, and if they are symmetric (Hermitian if complex) or non-symmetric.

The eigenvalues returned by eig are not ordered.

**See also:** [eigs], page 543, [svd], page 479.

G = givens (x, y) [Built-in Function]
[c, s] = givens (x, y) [Built-in Function]

Compute the Givens rotation matrix G.

The Givens matrix is a 2 × 2 orthogonal matrix

$$G = \begin{bmatrix} c & s \\ -s' & c \end{bmatrix}$$

such that

$$G \begin{bmatrix} x \\ y \end{bmatrix} = \begin{bmatrix} * \\ 0 \end{bmatrix}$$

with $x$ and $y$ scalars.

If two output arguments are requested, return the factors c and s rather than the Givens rotation matrix.

For example:

```
givens (1, 1)
    ⇒   0.70711   0.70711
       -0.70711   0.70711
```

**See also:** [planerot], page 465.

[G, y] = planerot (x) [Function File]

Given a two-element column vector, return the 2 × 2 orthogonal matrix G such that y = g * x and y(2) = 0.

**See also:** [givens], page 465.

`x = inv (A)`    [Built-in Function]
`[x, rcond] = inv (A)`    [Built-in Function]

    Compute the inverse of the square matrix $A$.

    Return an estimate of the reciprocal condition number if requested, otherwise warn of an ill-conditioned matrix if the reciprocal condition number is small.

    In general it is best to avoid calculating the inverse of a matrix directly. For example, it is both faster and more accurate to solve systems of equations ($A*x = b$) with `y = A \ b`, rather than `y = inv (A) * b`.

    If called with a sparse matrix, then in general x will be a full matrix requiring significantly more storage. Avoid forming the inverse of a sparse matrix if possible.

    **See also:** [ldivide], page 143, [rdivide], page 144.

`x = linsolve (A, b)`    [Function File]
`x = linsolve (A, b, opts)`    [Function File]
`[x, R] = linsolve (...)`    [Function File]

    Solve the linear system `A*x = b`.

    With no options, this function is equivalent to the left division operator (`x = A \ b`) or the matrix-left-divide function (`x = mldivide (A, b)`).

    Octave ordinarily examines the properties of the matrix $A$ and chooses a solver that best matches the matrix. By passing a structure *opts* to `linsolve` you can inform Octave directly about the matrix $A$. In this case Octave will skip the matrix examination and proceed directly to solving the linear system.

    **Warning:** If the matrix $A$ does not have the properties listed in the *opts* structure then the result will not be accurate AND no warning will be given. When in doubt, let Octave examine the matrix and choose the appropriate solver as this step takes little time and the result is cached so that it is only done once per linear system.

    Possible *opts* fields (set value to true/false):

| | |
|---|---|
| LT | $A$ is lower triangular |
| UT | $A$ is upper triangular |
| UHESS | $A$ is upper Hessenberg (currently makes no difference) |
| SYM | $A$ is symmetric or complex Hermitian (currently makes no difference) |
| POSDEF | $A$ is positive definite |
| RECT | $A$ is general rectangular (currently makes no difference) |
| TRANSA | Solve `A'*x = b` by `transpose (A) \ b` |

    The optional second output $R$ is the inverse condition number of $A$ (zero if matrix is singular).

    **See also:** [mldivide], page 143, [matrix_type], page 466, [rcond], page 470.

`type = matrix_type (A)`    [Built-in Function]
`type = matrix_type (A, "nocompute")`    [Built-in Function]
`A = matrix_type (A, type)`    [Built-in Function]
`A = matrix_type (A, "upper", perm)`    [Built-in Function]

# Chapter 18: Linear Algebra

*A* = **matrix_type** (*A*, "*lower*", *perm*) [Built-in Function]
*A* = **matrix_type** (*A*, "*banded*", *nl*, *nu*) [Built-in Function]

Identify the matrix type or mark a matrix as a particular type.

This allows more rapid solutions of linear equations involving *A* to be performed.

Called with a single argument, `matrix_type` returns the type of the matrix and caches it for future use.

Called with more than one argument, `matrix_type` allows the type of the matrix to be defined.

If the option `"nocompute"` is given, the function will not attempt to guess the type if it is still unknown. This is useful for debugging purposes.

The possible matrix types depend on whether the matrix is full or sparse, and can be one of the following

"unknown"
: Remove any previously cached matrix type, and mark type as unknown.

"full"
: Mark the matrix as full.

"positive definite"
: Probable full positive definite matrix.

"diagonal"
: Diagonal matrix. (Sparse matrices only)

"permuted diagonal"
: Permuted Diagonal matrix. The permutation does not need to be specifically indicated, as the structure of the matrix explicitly gives this. (Sparse matrices only)

"upper"
: Upper triangular. If the optional third argument *perm* is given, the matrix is assumed to be a permuted upper triangular with the permutations defined by the vector *perm*.

"lower"
: Lower triangular. If the optional third argument *perm* is given, the matrix is assumed to be a permuted lower triangular with the permutations defined by the vector *perm*.

"banded"
"banded positive definite"
: Banded matrix with the band size of *nl* below the diagonal and *nu* above it. If *nl* and *nu* are 1, then the matrix is tridiagonal and treated with specialized code. In addition the matrix can be marked as probably a positive definite. (Sparse matrices only)

"singular"
: The matrix is assumed to be singular and will be treated with a minimum norm solution.

Note that the matrix type will be discovered automatically on the first attempt to solve a linear equation involving *A*. Therefore `matrix_type` is only useful to give Octave hints of the matrix type. Incorrectly defining the matrix type will result in incorrect results from solutions of linear equations; it is entirely **the responsibility of the user** to correctly identify the matrix type.

Also, the test for positive definiteness is a low-cost test for a Hermitian matrix with a real positive diagonal. This does not guarantee that the matrix is positive definite, but only that it is a probable candidate. When such a matrix is factorized, a Cholesky factorization is first attempted, and if that fails the matrix is then treated with an LU factorization. Once the matrix has been factorized, `matrix_type` will return the correct classification of the matrix.

norm (*A*) [Built-in Function]
norm (*A, p*) [Built-in Function]
norm (*A, p, opt*) [Built-in Function]

Compute the p-norm of the matrix *A*.

If the second argument is missing, p = 2 is assumed.

If *A* is a matrix (or sparse matrix):

$p = 1$      1-norm, the largest column sum of the absolute values of *A*.

$p = 2$      Largest singular value of *A*.

$p =$ Inf or "inf"
     Infinity norm, the largest row sum of the absolute values of *A*.

$p =$ "fro"
     Frobenius norm of *A*, `sqrt (sum (diag (A' * A)))`.

other *p*, *p* > 1
     maximum `norm (A*x, p)` such that `norm (x, p) == 1`

If *A* is a vector or a scalar:

$p =$ Inf or "inf"
     `max (abs (A))`.

$p =$ -Inf      `min (abs (A))`.

$p =$ "fro"
     Frobenius norm of *A*, `sqrt (sumsq (abs (A)))`.

$p = 0$      Hamming norm - the number of nonzero elements.

other *p*, *p* > 1
     p-norm of *A*, `(sum (abs (A) .^ p)) ^ (1/p)`.

other *p* *p* < 1
     the p-pseudonorm defined as above.

If *opt* is the value "rows", treat each row as a vector and compute its norm. The result is returned as a column vector. Similarly, if *opt* is "columns" or "cols" then compute the norms of each column and return a row vector.

**See also:** [cond], page 464, [svd], page 479.

Chapter 18: Linear Algebra 469

**null** (*A*)                                                         [Function File]
**null** (*A, tol*)                                    [Function File]

Return an orthonormal basis of the null space of *A*.

The dimension of the null space is taken as the number of singular values of *A* not greater than *tol*. If the argument *tol* is missing, it is computed as

```
max (size (A)) * max (svd (A)) * eps
```

**See also:** [orth], page 469.

**orth** (*A*)                                                     [Function File]
**orth** (*A, tol*)                                     [Function File]

Return an orthonormal basis of the range space of *A*.

The dimension of the range space is taken as the number of singular values of *A* greater than *tol*. If the argument *tol* is missing, it is computed as

```
max (size (A)) * max (svd (A)) * eps
```

**See also:** [null], page 469.

**[y, h] = mgorth** (*x, v*)                              [Built-in Function]

Orthogonalize a given column vector *x* with respect to a set of orthonormal vectors comprising the columns of *v* using the modified Gram-Schmidt method.

On exit, *y* is a unit vector such that:

```
norm (y) = 1
v' * y = 0
x = [v, y]*h'
```

**pinv** (*x*)                                                [Built-in Function]
**pinv** (*x, tol*)                                   [Built-in Function]

Return the pseudoinverse of *x*.

Singular values less than *tol* are ignored.

If the second argument is omitted, it is taken to be

```
tol = max (size (x)) * sigma_max (x) * eps,
```

where `sigma_max (x)` is the maximal singular value of *x*.

**rank** (*A*)                                                [Function File]
**rank** (*A, tol*)                                  [Function File]

Compute the rank of matrix *A*, using the singular value decomposition.

The rank is taken to be the number of singular values of *A* that are greater than the specified tolerance *tol*. If the second argument is omitted, it is taken to be

```
tol = max (size (A)) * sigma(1) * eps;
```

where `eps` is machine precision and `sigma(1)` is the largest singular value of *A*.

The rank of a matrix is the number of linearly independent rows or columns and determines how many particular solutions exist to a system of equations. Use `null` for finding the remaining homogenous solutions.

Example:

```
x = [1 2 3
     4 5 6
     7 8 9];
rank (x)
   ⇒ 2
```

The number of linearly independent rows is only 2 because the final row is a linear combination of -1*row1 + 2*row2.

**See also:** [null], page 469, [sprank], page 541, [svd], page 479.

`c = rcond (A)` [Built-in Function]

Compute the 1-norm estimate of the reciprocal condition number as returned by LAPACK.

If the matrix is well-conditioned then $c$ will be near 1 and if the matrix is poorly conditioned it will be close to 0.

The matrix $A$ must not be sparse. If the matrix is sparse then `condest (A)` or `rcond (full (A))` should be used instead.

**See also:** [cond], page 464, [condest], page 540.

`trace (A)` [Function File]

Compute the trace of $A$, the sum of the elements along the main diagonal.

The implementation is straightforward: `sum (diag (A))`.

**See also:** [eig], page 465.

`rref (A)` [Function File]
`rref (A, tol)` [Function File]
`[r, k] = rref (...)` [Function File]

Return the reduced row echelon form of $A$.

*tol* defaults to `eps * max (size (A)) * norm (A, inf)`.

The optional return argument $k$ contains the vector of "bound variables", which are those columns on which elimination has been performed.

## 18.3 Matrix Factorizations

`R = chol (A)` [Loadable Function]
`[R, p] = chol (A)` [Loadable Function]
`[R, p, Q] = chol (S)` [Loadable Function]
`[R, p, Q] = chol (S, "vector")` [Loadable Function]
`[L, ...] = chol (..., "lower")` [Loadable Function]
`[L, ...] = chol (..., "upper")` [Loadable Function]

Compute the Cholesky factor, $R$, of the symmetric positive definite matrix $A$.

The Cholesky factor is defined by $R^T R = A$.

Called with one output argument `chol` fails if $A$ or $S$ is not positive definite. With two or more output arguments $p$ flags whether the matrix was positive definite and `chol` does not fail. A zero value indicated that the matrix was positive definite and the $R$ gives the factorization, and $p$ will have a positive value otherwise.

Chapter 18: Linear Algebra 471

If called with 3 outputs then a sparsity preserving row/column permutation is applied to $A$ prior to the factorization. That is $R$ is the factorization of `A(Q,Q)` such that $R^T R = Q^T A Q$.

The sparsity preserving permutation is generally returned as a matrix. However, given the flag `"vector"`, $Q$ will be returned as a vector such that $R^T R = A(Q, Q)$.

Called with either a sparse or full matrix and using the `"lower"` flag, `chol` returns the lower triangular factorization such that $LL^T = A$.

For full matrices, if the `"lower"` flag is set only the lower triangular part of the matrix is used for the factorization, otherwise the upper triangular part is used.

In general the lower triangular factorization is significantly faster for sparse matrices.

**See also:** [hess], page 472, [lu], page 473, [qr], page 474, [qz], page 477, [schur], page 478, [svd], page 479, [ichol], page 550, [cholinv], page 471, [chol2inv], page 471, [cholupdate], page 471, [cholinsert], page 472, [choldelete], page 472, [cholshift], page 472.

`cholinv (A)`                                                          [Loadable Function]

Compute the inverse of the symmetric positive definite matrix $A$ using the Cholesky factorization.

**See also:** [chol], page 470, [chol2inv], page 471, [inv], page 466.

`chol2inv (U)`                                                         [Loadable Function]

Invert a symmetric, positive definite square matrix from its Cholesky decomposition, $U$.

Note that $U$ should be an upper-triangular matrix with positive diagonal elements. `chol2inv (U)` provides `inv (U'*U)` but it is much faster than using `inv`.

**See also:** [chol], page 470, [cholinv], page 471, [inv], page 466.

`[R1, info] = cholupdate (R, u, op)`                                   [Loadable Function]

Update or downdate a Cholesky factorization.

Given an upper triangular matrix $R$ and a column vector $u$, attempt to determine another upper triangular matrix $R1$ such that

- $R1'*R1 = R'*R + u*u'$ if *op* is `"+"`
- $R1'*R1 = R'*R - u*u'$ if *op* is `"-"`

If *op* is `"-"`, *info* is set to

- 0 if the downdate was successful,
- 1 if $R'*R - u*u'$ is not positive definite,
- 2 if $R$ is singular.

If *info* is not present, an error message is printed in cases 1 and 2.

**See also:** [chol], page 470, [cholinsert], page 472, [choldelete], page 472, [cholshift], page 472.

*R1* = cholinsert (*R, j, u*) [Loadable Function]
[*R1, info*] = cholinsert (*R, j, u*) [Loadable Function]

Given a Cholesky factorization of a real symmetric or complex Hermitian positive definite matrix $A = R'*R$, R upper triangular, return the Cholesky factorization of *A1*, where A1(p,p) = A, A1(:,j) = A1(j,:)' = u and p = [1:j-1,j+1:n+1]. u(j) should be positive.

On return, *info* is set to

- 0 if the insertion was successful,
- 1 if *A1* is not positive definite,
- 2 if *R* is singular.

If *info* is not present, an error message is printed in cases 1 and 2.

**See also:** [chol], page 470, [cholupdate], page 471, [choldelete], page 472, [cholshift], page 472.

*R1* = choldelete (*R, j*) [Loadable Function]

Given a Cholesky factorization of a real symmetric or complex Hermitian positive definite matrix $A = R'*R$, R upper triangular, return the Cholesky factorization of A(p,p), where p = [1:j-1,j+1:n+1].

**See also:** [chol], page 470, [cholupdate], page 471, [cholinsert], page 472, [cholshift], page 472.

*R1* = cholshift (*R, i, j*) [Loadable Function]

Given a Cholesky factorization of a real symmetric or complex Hermitian positive definite matrix $A = R'*R$, R upper triangular, return the Cholesky factorization of A(p,p), where p is the permutation
p = [1:i-1, shift(i:j, 1), j+1:n] if $i < j$
or
p = [1:j-1, shift(j:i,-1), i+1:n] if $j < i$.

**See also:** [chol], page 470, [cholupdate], page 471, [cholinsert], page 472, [choldelete], page 472.

*H* = hess (*A*) [Built-in Function]
[*P, H*] = hess (*A*) [Built-in Function]

Compute the Hessenberg decomposition of the matrix *A*.

The Hessenberg decomposition is

$$A = PHP^T$$

where $P$ is a square unitary matrix ($P^T P = I$), and $H$ is upper Hessenberg ($H_{i,j} = 0, \forall i \geq j+1$).

The Hessenberg decomposition is usually used as the first step in an eigenvalue computation, but has other applications as well (see Golub, Nash, and Van Loan, IEEE Transactions on Automatic Control, 1979).

**See also:** [eig], page 465, [chol], page 470, [lu], page 473, [qr], page 474, [qz], page 477, [schur], page 478, [svd], page 479.

| | |
|---|---|
| [L, U] = lu (*A*) | [Built-in Function] |
| [L, U, P] = lu (*A*) | [Built-in Function] |
| [L, U, P, Q] = lu (*S*) | [Built-in Function] |
| [L, U, P, Q, R] = lu (*S*) | [Built-in Function] |
| [...] = lu (*S, thres*) | [Built-in Function] |
| y = lu (...) | [Built-in Function] |
| [...] = lu (..., "*vector*") | [Built-in Function] |

Compute the LU decomposition of $A$.

If $A$ is full subroutines from LAPACK are used and if $A$ is sparse then UMFPACK is used.

The result is returned in a permuted form, according to the optional return value $P$. For example, given the matrix a = [1, 2; 3, 4],

```
[l, u, p] = lu (a)
```

returns

```
l =

   1.00000   0.00000
   0.33333   1.00000

u =

   3.00000   4.00000
   0.00000   0.66667

p =

   0   1
   1   0
```

The matrix is not required to be square.

When called with two or three output arguments and a spare input matrix, lu does not attempt to perform sparsity preserving column permutations. Called with a fourth output argument, the sparsity preserving column transformation $Q$ is returned, such that $P * A * Q = L * U$.

Called with a fifth output argument and a sparse input matrix, lu attempts to use a scaling factor $R$ on the input matrix such that $P * (R \backslash A) * Q = L * U$. This typically leads to a sparser and more stable factorization.

An additional input argument *thres*, that defines the pivoting threshold can be given. *thres* can be a scalar, in which case it defines the UMFPACK pivoting tolerance for both symmetric and unsymmetric cases. If *thres* is a 2-element vector, then the first element defines the pivoting tolerance for the unsymmetric UMFPACK pivoting strategy and the second for the symmetric strategy. By default, the values defined by **spparms** are used ([0.1, 0.001]).

Given the string argument "**vector**", lu returns the values of $P$ and $Q$ as vector values, such that for full matrix, $A\ (P,:) = L * U$, and $R(P,:) * A\ (:,Q) = L * U$.

With two output arguments, returns the permuted forms of the upper and lower triangular matrices, such that A = L * U. With one output argument y, then the matrix returned by the LAPACK routines is returned. If the input matrix is sparse then the matrix L is embedded into U to give a return value similar to the full case. For both full and sparse matrices, lu loses the permutation information.

**See also:** [luupdate], page 474, [ilu], page 552, [chol], page 470, [hess], page 472, [qr], page 474, [qz], page 477, [schur], page 478, [svd], page 479.

[L, U] = luupdate (L, U, x, y)      [Built-in Function]
[L, U, P] = luupdate (L, U, P, x, y)      [Built-in Function]

Given an LU factorization of a real or complex matrix $A = L*U$, L lower unit trapezoidal and U upper trapezoidal, return the LU factorization of $A + x*y.'$, where x and y are column vectors (rank-1 update) or matrices with equal number of columns (rank-k update).

Optionally, row-pivoted updating can be used by supplying a row permutation (pivoting) matrix P; in that case, an updated permutation matrix is returned. Note that if L, U, P is a pivoted LU factorization as obtained by lu:

    [L, U, P] = lu (A);

then a factorization of A+x*y.' can be obtained either as

    [L1, U1] = lu (L, U, P*x, y)

or

    [L1, U1, P1] = lu (L, U, P, x, y)

The first form uses the unpivoted algorithm, which is faster, but less stable. The second form uses a slower pivoted algorithm, which is more stable.

The matrix case is done as a sequence of rank-1 updates; thus, for large enough k, it will be both faster and more accurate to recompute the factorization from scratch.

**See also:** [lu], page 473, [cholupdate], page 471, [qrupdate], page 475.

[Q, R, P] = qr (A)      [Loadable Function]
[Q, R, P] = qr (A, '0')      [Loadable Function]
[C, R] = qr (A, B)      [Loadable Function]
[C, R] = qr (A, B, '0')      [Loadable Function]

Compute the QR factorization of A, using standard LAPACK subroutines.

For example, given the matrix A = [1, 2; 3, 4],

    [Q, R] = qr (A)

returns

```
Q =

  -0.31623  -0.94868
  -0.94868   0.31623

R =

  -3.16228  -4.42719
   0.00000  -0.63246
```

# Chapter 18: Linear Algebra

The qr factorization has applications in the solution of least squares problems

$$\min_x \|Ax - b\|_2$$

for overdetermined systems of equations (i.e., $A$ is a tall, thin matrix). The QR factorization is $QR = A$ where $Q$ is an orthogonal matrix and $R$ is upper triangular.

If given a second argument of '0', qr returns an economy-sized QR factorization, omitting zero rows of $R$ and the corresponding columns of $Q$.

If the matrix $A$ is full, the permuted QR factorization [Q, R, P] = qr (A) forms the QR factorization such that the diagonal entries of $R$ are decreasing in magnitude order. For example, given the matrix a = [1, 2; 3, 4],

```
[Q, R, P] = qr (A)
```

returns

```
Q =

  -0.44721  -0.89443
  -0.89443   0.44721

R =

  -4.47214  -3.13050
   0.00000   0.44721

P =

  0  1
  1  0
```

The permuted qr factorization [Q, R, P] = qr (A) factorization allows the construction of an orthogonal basis of span (A).

If the matrix $A$ is sparse, then compute the sparse QR factorization of $A$, using CSPARSE. As the matrix $Q$ is in general a full matrix, this function returns the $Q$-less factorization $R$ of $A$, such that R = chol (A' * A).

If the final argument is the scalar 0 and the number of rows is larger than the number of columns, then an economy factorization is returned. That is $R$ will have only size (A,1) rows.

If an additional matrix $B$ is supplied, then qr returns $C$, where C = Q' * B. This allows the least squares approximation of A \ B to be calculated as

```
[C, R] = qr (A, B)
x = R \ C
```

**See also:** [chol], page 470, [hess], page 472, [lu], page 473, [qz], page 477, [schur], page 478, [svd], page 479, [qrupdate], page 475, [qrinsert], page 476, [qrdelete], page 476, [qrshift], page 476.

[Q1, R1] = qrupdate (Q, R, u, v)  [Loadable Function]

Given a QR factorization of a real or complex matrix $A = Q*R$, $Q$ unitary and $R$ upper trapezoidal, return the QR factorization of $A + u*v'$, where $u$ and $v$ are column vectors (rank-1 update) or matrices with equal number of columns (rank-k update). Notice that the latter case is done as a sequence of rank-1 updates; thus, for k large enough, it will be both faster and more accurate to recompute the factorization from scratch.

The QR factorization supplied may be either full (Q is square) or economized (R is square).

**See also:** [qr], page 474, [qrinsert], page 476, [qrdelete], page 476, [qrshift], page 476.

[Q1, R1] = qrinsert (Q, R, j, x, orient)  [Loadable Function]
Given a QR factorization of a real or complex matrix $A = Q*R$, $Q$ unitary and $R$ upper trapezoidal, return the QR factorization of $[A(:,1:j-1)$ x $A(:,j:n)]$, where $u$ is a column vector to be inserted into $A$ (if *orient* is "col"), or the QR factorization of $[A(1:j-1,:);x;A(:,j:n)]$, where $x$ is a row vector to be inserted into $A$ (if *orient* is "row").

The default value of *orient* is "col". If *orient* is "col", $u$ may be a matrix and $j$ an index vector resulting in the QR factorization of a matrix $B$ such that $B(:,j)$ gives $u$ and $B(:,j) = []$ gives $A$. Notice that the latter case is done as a sequence of k insertions; thus, for k large enough, it will be both faster and more accurate to recompute the factorization from scratch.

If *orient* is "col", the QR factorization supplied may be either full (Q is square) or economized (R is square).

If *orient* is "row", full factorization is needed.

**See also:** [qr], page 474, [qrupdate], page 475, [qrdelete], page 476, [qrshift], page 476.

[Q1, R1] = qrdelete (Q, R, j, orient)  [Loadable Function]
Given a QR factorization of a real or complex matrix $A = Q*R$, $Q$ unitary and $R$ upper trapezoidal, return the QR factorization of $[A(:,1:j-1)\ A(:,j+1:n)]$, i.e., $A$ with one column deleted (if *orient* is "col"), or the QR factorization of $[A(1:j-1,:);A(j+1:n,:)]$, i.e., $A$ with one row deleted (if *orient* is "row").

The default value of *orient* is "col".

If *orient* is "col", $j$ may be an index vector resulting in the QR factorization of a matrix $B$ such that $A(:,j) = []$ gives $B$. Notice that the latter case is done as a sequence of k deletions; thus, for k large enough, it will be both faster and more accurate to recompute the factorization from scratch.

If *orient* is "col", the QR factorization supplied may be either full (Q is square) or economized (R is square).

If *orient* is "row", full factorization is needed.

**See also:** [qr], page 474, [qrupdate], page 475, [qrinsert], page 476, [qrshift], page 476.

[Q1, R1] = qrshift (Q, R, i, j)  [Loadable Function]
Given a QR factorization of a real or complex matrix $A = Q*R$, $Q$ unitary and $R$ upper trapezoidal, return the QR factorization of $A(:,p)$, where p is the permutation
p = [1:i-1, shift(i:j, 1), j+1:n] if $i < j$
or

Chapter 18: Linear Algebra 477

```
p = [1:j-1, shift(j:i,-1), i+1:n] if j < i.
```

**See also:** [qr], page 474, [qrupdate], page 475, [qrinsert], page 476, [qrdelete], page 476.

*lambda* = qz (*A, B*)      [Built-in Function]
*lambda* = qz (*A, B, opt*)      [Built-in Function]

QZ decomposition of the generalized eigenvalue problem ($Ax = sBx$).

There are three ways to call this function:

1. *lambda* = qz (*A, B*)

    Computes the generalized eigenvalues $\lambda$ of $(A - sB)$.

2. [AA, BB, Q, Z, V, W, *lambda*] = qz (*A, B*)

    Computes QZ decomposition, generalized eigenvectors, and generalized eigenvalues of $(A - sB)$

    $$AV = BV \operatorname{diag}(\lambda)$$

    $$W^T A = \operatorname{diag}(\lambda) W^T B$$

    $$AA = Q^T A Z, BB = Q^T B Z$$

    with $Q$ and $Z$ orthogonal (unitary)$= I$

3. [AA,BB,Z{, *lambda*}] = qz (*A, B, opt*)

    As in form [2], but allows ordering of generalized eigenpairs for, e.g., solution of discrete time algebraic Riccati equations. Form 3 is not available for complex matrices, and does not compute the generalized eigenvectors $V$, $W$, nor the orthogonal matrix $Q$.

    | | |
    |---|---|
    | *opt* | for ordering eigenvalues of the GEP pencil. The leading block of the revised pencil contains all eigenvalues that satisfy: |
    | "N" | = unordered (default) |
    | "S" | = small: leading block has all \|lambda\| $\leq 1$ |
    | "B" | = big: leading block has all \|lambda\| $\geq 1$ |
    | "-" | = negative real part: leading block has all eigenvalues in the open left half-plane |
    | "+" | = non-negative real part: leading block has all eigenvalues in the closed right half-plane |

Note: **qz** performs permutation balancing, but not scaling (see [XREFbalance], page 463). The order of output arguments was selected for compatibility with MATLAB.

**See also:** [eig], page 465, [balance], page 463, [lu], page 473, [chol], page 470, [hess], page 472, [qr], page 474, [qzhess], page 478, [schur], page 478, [svd], page 479.

[aa, bb, q, z] = qzhess (A, B)  [Function File]

Compute the Hessenberg-triangular decomposition of the matrix pencil (A, B), returning aa = q * A * z, bb = q * B * z, with q and z orthogonal.

For example:

```
[aa, bb, q, z] = qzhess ([1, 2; 3, 4], [5, 6; 7, 8])
     ⇒ aa = [ -3.02244, -4.41741;  0.92998,  0.69749 ]
     ⇒ bb = [ -8.60233, -9.99730;  0.00000, -0.23250 ]
     ⇒  q = [ -0.58124, -0.81373; -0.81373,  0.58124 ]
     ⇒  z = [ 1, 0; 0, 1 ]
```

The Hessenberg-triangular decomposition is the first step in Moler and Stewart's QZ decomposition algorithm.

Algorithm taken from Golub and Van Loan, *Matrix Computations*, 2nd edition.

**See also:** [lu], page 473, [chol], page 470, [hess], page 472, [qr], page 474, [qz], page 477, [schur], page 478, [svd], page 479.

S = schur (A)  [Built-in Function]
S = schur (A, "*real*")  [Built-in Function]
S = schur (A, "*complex*")  [Built-in Function]
S = schur (A, opt)  [Built-in Function]
[U, S] = schur (...)  [Built-in Function]

Compute the Schur decomposition of A.

The Schur decomposition is defined as

$$S = U^T A U$$

where $U$ is a unitary matrix ($U^T U$ is identity) and $S$ is upper triangular. The eigenvalues of $A$ (and $S$) are the diagonal elements of $S$. If the matrix $A$ is real, then the real Schur decomposition is computed, in which the matrix $U$ is orthogonal and $S$ is block upper triangular with blocks of size at most $2 \times 2$ along the diagonal. The diagonal elements of $S$ (or the eigenvalues of the $2 \times 2$ blocks, when appropriate) are the eigenvalues of $A$ and $S$.

The default for real matrices is a real Schur decomposition. A complex decomposition may be forced by passing the flag `"complex"`.

The eigenvalues are optionally ordered along the diagonal according to the value of opt. opt = "a" indicates that all eigenvalues with negative real parts should be moved to the leading block of S (used in are), opt = "d" indicates that all eigenvalues with magnitude less than one should be moved to the leading block of S (used in dare), and opt = "u", the default, indicates that no ordering of eigenvalues should occur. The leading $k$ columns of $U$ always span the $A$-invariant subspace corresponding to the $k$ leading eigenvalues of $S$.

The Schur decomposition is used to compute eigenvalues of a square matrix, and has applications in the solution of algebraic Riccati equations in control (see are and dare).

**See also:** [rsf2csf], page 479, [ordschur], page 479, [lu], page 473, [chol], page 470, [hess], page 472, [qr], page 474, [qz], page 477, [svd], page 479.

# Chapter 18: Linear Algebra

[U, T] = rsf2csf (UR, TR)  [Function File]

Convert a real, upper quasi-triangular Schur form TR to a complex, upper triangular Schur form T.

Note that the following relations hold:

$UR \cdot TR \cdot UR^T = UTU^\dagger$ and $U^\dagger U$ is the identity matrix I.

Note also that U and T are not unique.

**See also:** [schur], page 478.

[UR, SR] = ordschur (U, S, select)  [Loadable Function]

Reorders the real Schur factorization (U,S) obtained with the schur function, so that selected eigenvalues appear in the upper left diagonal blocks of the quasi triangular Schur matrix.

The logical vector *select* specifies the selected eigenvalues as they appear along S's diagonal.

For example, given the matrix A = [1, 2; 3, 4], and its Schur decomposition

```
[U, S] = schur (A)
```

which returns

```
U =

  -0.82456  -0.56577
   0.56577  -0.82456

S =

  -0.37228  -1.00000
   0.00000   5.37228
```

It is possible to reorder the decomposition so that the positive eigenvalue is in the upper left corner, by doing:

```
[U, S] = ordschur (U, S, [0,1])
```

**See also:** [schur], page 478.

angle = subspace (A, B)  [Function File]

Determine the largest principal angle between two subspaces spanned by the columns of matrices A and B.

s = svd (A)  [Built-in Function]
[U, S, V] = svd (A)  [Built-in Function]
[U, S, V] = svd (A, econ)  [Built-in Function]

Compute the singular value decomposition of A

$$A = USV^\dagger$$

The function svd normally returns only the vector of singular values. When called with three return values, it computes U, S, and V. For example,

```
        svd (hilb (3))
returns
        ans =

           1.4083189
           0.1223271
           0.0026873
and
        [u, s, v] = svd (hilb (3))
returns
        u =

          -0.82704   0.54745   0.12766
          -0.45986  -0.52829  -0.71375
          -0.32330  -0.64901   0.68867

        s =

           1.40832   0.00000   0.00000
           0.00000   0.12233   0.00000
           0.00000   0.00000   0.00269

        v =

          -0.82704   0.54745   0.12766
          -0.45986  -0.52829  -0.71375
          -0.32330  -0.64901   0.68867
```

If given a second argument, **svd** returns an economy-sized decomposition, eliminating the unnecessary rows or columns of $U$ or $V$.

**See also:** [svd_driver], page 480, [svds], page 545, [eig], page 465, [lu], page 473, [chol], page 470, [hess], page 472, [qr], page 474, [qz], page 477.

*val* = svd_driver ()                                               [Built-in Function]
*old_val* = svd_driver (*new_val*)                                  [Built-in Function]
svd_driver (*new_val*, "*local*")                                   [Built-in Function]

Query or set the underlying LAPACK driver used by **svd**.

Currently recognized values are "**gesvd**" and "**gesdd**". The default is "**gesvd**".

When called from inside a function with the "**local**" option, the variable is changed locally for the function and any subroutines it calls. The original variable value is restored when exiting the function.

**See also:** [svd], page 479.

[*housv*, *beta*, *zer*] = housh (*x*, *j*, *z*)                    [Function File]

Compute Householder reflection vector *housv* to reflect *x* to be the *j*-th column of identity, i.e.,

Chapter 18: Linear Algebra

```
        (I - beta*housv*housv')x =  norm (x)*e(j) if x(j) < 0,
        (I - beta*housv*housv')x = -norm (x)*e(j) if x(j) >= 0
```

Inputs

    *x*            vector

    *j*            index into vector

    *z*            threshold for zero (usually should be the number 0)

Outputs (see Golub and Van Loan):

    *beta*      If beta = 0, then no reflection need be applied (zer set to 0)

    *housv*     householder vector

**[u, h, nu] = krylov (A, V, k, eps1, pflg)**                           [Function File]
Construct an orthogonal basis *u* of block Krylov subspace

```
        [v a*v a^2*v ... a^(k+1)*v]
```

using Householder reflections to guard against loss of orthogonality.

If *V* is a vector, then *h* contains the Hessenberg matrix such that `a*u == u*h+rk*ek'`, in which `rk = a*u(:,k)-u*h(:,k)`, and `ek'` is the vector `[0, 0, ..., 1]` of length k. Otherwise, *h* is meaningless.

If *V* is a vector and *k* is greater than `length (A) - 1`, then *h* contains the Hessenberg matrix such that `a*u == u*h`.

The value of *nu* is the dimension of the span of the Krylov subspace (based on *eps1*).

If *b* is a vector and *k* is greater than *m-1*, then *h* contains the Hessenberg decomposition of *A*.

The optional parameter *eps1* is the threshold for zero. The default value is 1e-12.

If the optional parameter *pflg* is nonzero, row pivoting is used to improve numerical behavior. The default value is 0.

Reference: A. Hodel, P. Misra, *Partial Pivoting in the Computation of Krylov Subspaces of Large Sparse Systems*, Proceedings of the 42nd IEEE Conference on Decision and Control, December 2003.

## 18.4 Functions of a Matrix

**expm (A)**                                                                                                [Function File]
Return the exponential of a matrix.

The matrix exponential is defined as the infinite Taylor series

$$\exp(A) = I + A + \frac{A^2}{2!} + \frac{A^3}{3!} + \cdots$$

However, the Taylor series is *not* the way to compute the matrix exponential; see Moler and Van Loan, *Nineteen Dubious Ways to Compute the Exponential of a Matrix*, SIAM Review, 1978. This routine uses Ward's diagonal Padé approximation method with three step preconditioning (SIAM Journal on Numerical Analysis, 1977). Diagonal Padé approximations are rational polynomials of matrices $D_q(A)^{-1} N_q(A)$

whose Taylor series matches the first $2q + 1$ terms of the Taylor series above; direct evaluation of the Taylor series (with the same preconditioning steps) may be desirable in lieu of the Padé approximation when $D_q(A)$ is ill-conditioned.

**See also:** [logm], page 482, [sqrtm], page 482.

`s = logm (A)` [Function File]
`s = logm (A, opt_iters)` [Function File]
`[s, iters] = logm (...)` [Function File]

Compute the matrix logarithm of the square matrix $A$.

The implementation utilizes a Padé approximant and the identity

```
logm (A) = 2^k * logm (A^(1 / 2^k))
```

The optional input *opt_iters* is the maximum number of square roots to compute and defaults to 100.

The optional output *iters* is the number of square roots actually computed.

**See also:** [expm], page 481, [sqrtm], page 482.

`s = sqrtm (A)` [Built-in Function]
`[s, error_estimate] = sqrtm (A)` [Built-in Function]

Compute the matrix square root of the square matrix $A$.

Ref: N.J. Higham. *A New sqrtm for* MATLAB. Numerical Analysis Report No. 336, Manchester Centre for Computational Mathematics, Manchester, England, January 1999.

**See also:** [expm], page 481, [logm], page 482.

`kron (A, B)` [Built-in Function]
`kron (A1, A2, ...)` [Built-in Function]

Form the Kronecker product of two or more matrices.

This is defined block by block as

```
x = [ a(i,j)*b ]
```

For example:

```
kron (1:4, ones (3, 1))
    ⇒  1  2  3  4
       1  2  3  4
       1  2  3  4
```

If there are more than two input arguments $A1, A2, \ldots, An$ the Kronecker product is computed as

```
kron (kron (A1, A2), ..., An)
```

Since the Kronecker product is associative, this is well-defined.

`blkmm (A, B)` [Built-in Function]

Compute products of matrix blocks.

The blocks are given as 2-dimensional subarrays of the arrays $A$, $B$. The size of $A$ must have the form `[m,k,...]` and size of $B$ must be `[k,n,...]`. The result is then of size `[m,n,...]` and is computed as follows:

Chapter 18: Linear Algebra 483

```
for i = 1:prod (size (A)(3:end))
  C(:,:,i) = A(:,:,i) * B(:,:,i)
endfor
```

*X* = syl (*A*, *B*, *C*)  [Built-in Function]
    Solve the Sylvester equation

$$AX + XB = C$$

using standard LAPACK subroutines.

For example:

```
sylvester ([1, 2; 3, 4], [5, 6; 7, 8], [9, 10; 11, 12])
  ⇒ [ 0.50000, 0.66667; 0.66667, 0.50000 ]
```

## 18.5 Specialized Solvers

*x* = bicg (*A*, *b*, *rtol*, *maxit*, *M1*, *M2*, *x0*)  [Function File]
*x* = bicg (*A*, *b*, *rtol*, *maxit*, *P*)  [Function File]
[*x*, *flag*, *relres*, *iter*, *resvec*] = bicg (*A*, *b*, ...)  [Function File]
    Solve A x = b using the Bi-conjugate gradient iterative method.

- *rtol* is the relative tolerance, if not given or set to [] the default value 1e-6 is used.

- *maxit* the maximum number of outer iterations, if not given or set to [] the default value min (20, numel (b)) is used.

- *x0* the initial guess, if not given or set to [] the default value zeros (size (b)) is used.

*A* can be passed as a matrix or as a function handle or inline function f such that f(x, "notransp") = A*x and f(x, "transp") = A'*x.

The preconditioner *P* is given as P = M1 * M2. Both *M1* and *M2* can be passed as a matrix or as a function handle or inline function g such that g(x, "notransp") = M1 \ x or g(x, "notransp") = M2 \ x and g(x, "transp") = M1' \ x or g(x, "transp") = M2' \ x.

If called with more than one output parameter

- *flag* indicates the exit status:

    - 0: iteration converged to the within the chosen tolerance
    - 1: the maximum number of iterations was reached before convergence
    - 3: the algorithm reached stagnation

   (the value 2 is unused but skipped for compatibility).

- *relres* is the final value of the relative residual.

- *iter* is the number of iterations performed.

- *resvec* is a vector containing the relative residual at each iteration.

**See also:** [bicgstab], page 484, [cgs], page 484, [gmres], page 485, [pcg], page 546, [qmr], page 486.

`x = bicgstab (A, b, rtol, maxit, M1, M2, x0)` [Function File]
`x = bicgstab (A, b, rtol, maxit, P)` [Function File]
`[x, flag, relres, iter, resvec] = bicgstab (A, b, ...)` [Function File]

Solve `A x = b` using the stabilizied Bi-conjugate gradient iterative method.

- *rtol* is the relative tolerance, if not given or set to [] the default value 1e-6 is used.
- *maxit* the maximum number of outer iterations, if not given or set to [] the default value `min (20, numel (b))` is used.
- *x0* the initial guess, if not given or set to [] the default value `zeros (size (b))` is used.

A can be passed as a matrix or as a function handle or inline function `f` such that `f(x) = A*x`.

The preconditioner P is given as P = M1 * M2. Both *M1* and *M2* can be passed as a matrix or as a function handle or inline function `g` such that `g(x) = M1 \ x` or `g(x) = M2 \ x`.

If called with more than one output parameter

- *flag* indicates the exit status:
  - 0: iteration converged to the within the chosen tolerance
  - 1: the maximum number of iterations was reached before convergence
  - 3: the algorithm reached stagnation

  (the value 2 is unused but skipped for compatibility).
- *relres* is the final value of the relative residual.
- *iter* is the number of iterations performed.
- *resvec* is a vector containing the relative residual at each iteration.

**See also:** [bicg], page 483, [cgs], page 484, [gmres], page 485, [pcg], page 546, [qmr], page 486.

`x = cgs (A, b, rtol, maxit, M1, M2, x0)` [Function File]
`x = cgs (A, b, rtol, maxit, P)` [Function File]
`[x, flag, relres, iter, resvec] = cgs (A, b, ...)` [Function File]

Solve `A x = b`, where A is a square matrix, using the Conjugate Gradients Squared method.

- *rtol* is the relative tolerance, if not given or set to [] the default value 1e-6 is used.
- *maxit* the maximum number of outer iterations, if not given or set to [] the default value `min (20, numel (b))` is used.
- *x0* the initial guess, if not given or set to [] the default value `zeros (size (b))` is used.

A can be passed as a matrix or as a function handle or inline function `f` such that `f(x) = A*x`.

The preconditioner P is given as P = M1 * M2. Both *M1* and *M2* can be passed as a matrix or as a function handle or inline function `g` such that `g(x) = M1 \ x` or `g(x) = M2 \ x`.

If called with more than one output parameter

Chapter 18: Linear Algebra 485

- *flag* indicates the exit status:
  - 0: iteration converged to the within the chosen tolerance
  - 1: the maximum number of iterations was reached before convergence
  - 3: the algorithm reached stagnation

  (the value 2 is unused but skipped for compatibility).
- *relres* is the final value of the relative residual.
- *iter* is the number of iterations performed.
- *resvec* is a vector containing the relative residual at each iteration.

**See also:** [pcg], page 546, [bicgstab], page 484, [bicg], page 483, [gmres], page 485, [qmr], page 486.

x = gmres (A, b, m, rtol, maxit, M1, M2, x0)  [Function File]
x = gmres (A, b, m, rtol, maxit, P)  [Function File]
[x, flag, relres, iter, resvec] = gmres (...)  [Function File]

Solve A x = b using the Preconditioned GMRES iterative method with restart, a.k.a. PGMRES(m).

- *rtol* is the relative tolerance, if not given or set to [] the default value 1e-6 is used.
- *maxit* is the maximum number of outer iterations, if not given or set to [] the default value `min (10, numel (b) / restart)` is used.
- *x0* is the initial guess, if not given or set to [] the default value `zeros (size (b))` is used.
- *m* is the restart parameter, if not given or set to [] the default value `numel (b)` is used.

Argument A can be passed as a matrix, function handle, or inline function f such that `f(x) = A*x`.

The preconditioner P is given as P = M1 * M2. Both *M1* and *M2* can be passed as a matrix, function handle, or inline function g such that `g(x) = M1\x` or `g(x) = M2\x`.

Besides the vector x, additional outputs are:

- *flag* indicates the exit status:

  0 : iteration converged to within the specified tolerance
  1 : maximum number of iterations exceeded
  2 : unused, but skipped for compatibility
  3 : algorithm reached stagnation (no change between iterations)
- *relres* is the final value of the relative residual.
- *iter* is a vector containing the number of outer iterations and total iterations performed.
- *resvec* is a vector containing the relative residual at each iteration.

**See also:** [bicg], page 483, [bicgstab], page 484, [cgs], page 484, [pcg], page 546, [pcr], page 548, [qmr], page 486.

*x* = qmr (*A*, *b*, *rtol*, *maxit*, *M1*, *M2*, *x0*)  [Function File]
*x* = qmr (*A*, *b*, *rtol*, *maxit*, *P*)  [Function File]
[*x*, *flag*, *relres*, *iter*, *resvec*] = qmr (*A*, *b*, ...)  [Function File]

Solve `A x = b` using the Quasi-Minimal Residual iterative method (without look-ahead).

- *rtol* is the relative tolerance, if not given or set to [] the default value 1e-6 is used.
- *maxit* the maximum number of outer iterations, if not given or set to [] the default value `min (20, numel (b))` is used.
- *x0* the initial guess, if not given or set to [] the default value `zeros (size (b))` is used.

*A* can be passed as a matrix or as a function handle or inline function `f` such that `f(x, "notransp") = A*x` and `f(x, "transp") = A'*x`.

The preconditioner *P* is given as `P = M1 * M2`. Both *M1* and *M2* can be passed as a matrix or as a function handle or inline function g such that `g(x, "notransp") = M1 \ x` or `g(x, "notransp") = M2 \ x` and `g(x, "transp") = M1' \ x` or `g(x, "transp") = M2' \ x`.

If called with more than one output parameter

- *flag* indicates the exit status:
    - 0: iteration converged to the within the chosen tolerance
    - 1: the maximum number of iterations was reached before convergence
    - 3: the algorithm reached stagnation

    (the value 2 is unused but skipped for compatibility).
- *relres* is the final value of the relative residual.
- *iter* is the number of iterations performed.
- *resvec* is a vector containing the residual norms at each iteration.

References:

1. R. Freund and N. Nachtigal, *QMR: a quasi-minimal residual method for non-Hermitian linear systems*, Numerische Mathematik, 1991, 60, pp. 315-339.
2. R. Barrett, M. Berry, T. Chan, J. Demmel, J. Donato, J. Dongarra, V. Eijkhour, R. Pozo, C. Romine, and H. van der Vorst, *Templates for the solution of linear systems: Building blocks for iterative methods*, SIAM, 2nd ed., 1994.

**See also:** [bicg], page 483, [bicgstab], page 484, [cgs], page 484, [gmres], page 485, [pcg], page 546.

# 19 Vectorization and Faster Code Execution

Vectorization is a programming technique that uses vector operations instead of element-by-element loop-based operations. Besides frequently producing more succinct Octave code, vectorization also allows for better optimization in the subsequent implementation. The optimizations may occur either in Octave's own Fortran, C, or C++ internal implementation, or even at a lower level depending on the compiler and external numerical libraries used to build Octave. The ultimate goal is to make use of your hardware's vector instructions if possible or to perform other optimizations in software.

Vectorization is not a concept unique to Octave, but it is particularly important because Octave is a matrix-oriented language. Vectorized Octave code will see a dramatic speed up (10X–100X) in most cases.

This chapter discusses vectorization and other techniques for writing faster code.

## 19.1 Basic Vectorization

To a very good first approximation, the goal in vectorization is to write code that avoids loops and uses whole-array operations. As a trivial example, consider

```
for i = 1:n
  for j = 1:m
    c(i,j) = a(i,j) + b(i,j);
  endfor
endfor
```

compared to the much simpler

```
c = a + b;
```

This isn't merely easier to write; it is also internally much easier to optimize. Octave delegates this operation to an underlying implementation which, among other optimizations, may use special vector hardware instructions or could conceivably even perform the additions in parallel. In general, if the code is vectorized, the underlying implementation has more freedom about the assumptions it can make in order to achieve faster execution.

This is especially important for loops with "cheap" bodies. Often it suffices to vectorize just the innermost loop to get acceptable performance. A general rule of thumb is that the "order" of the vectorized body should be greater or equal to the "order" of the enclosing loop.

As a less trivial example, instead of

```
for i = 1:n-1
  a(i) = b(i+1) - b(i);
endfor
```

write

```
a = b(2:n) - b(1:n-1);
```

This shows an important general concept about using arrays for indexing instead of looping over an index variable. See Section 8.1 [Index Expressions], page 135. Also use boolean indexing generously. If a condition needs to be tested, this condition can also be written as a boolean index. For instance, instead of

```
for i = 1:n
  if (a(i) > 5)
    a(i) -= 20
  endif
endfor
```

write

```
a(a>5) -= 20;
```

which exploits the fact that `a > 5` produces a boolean index.

Use elementwise vector operators whenever possible to avoid looping (operators like `.*` and `.^`). See Section 8.3 [Arithmetic Ops], page 141. For simple inline functions, the `vectorize` function can do this automatically.

`vectorize (fun)`                                                              [Built-in Function]

Create a vectorized version of the inline function *fun* by replacing all occurrences of `*`, `/`, etc., with `.*`, `./`, etc.

This may be useful, for example, when using inline functions with numerical integration or optimization where a vector-valued function is expected.

```
fcn = vectorize (inline ("x^2 - 1"))
    ⇒ fcn = f(x) = x.^2 - 1
quadv (fcn, 0, 3)
    ⇒ 6
```

**See also:** [inline], page 202, [formula], page 202, [argnames], page 202.

Also exploit broadcasting in these elementwise operators both to avoid looping and unnecessary intermediate memory allocations. See Section 19.2 [Broadcasting], page 489.

Use built-in and library functions if possible. Built-in and compiled functions are very fast. Even with an m-file library function, chances are good that it is already optimized, or will be optimized more in a future release.

For instance, even better than

```
a = b(2:n) - b(1:n-1);
```

is

```
a = diff (b);
```

Most Octave functions are written with vector and array arguments in mind. If you find yourself writing a loop with a very simple operation, chances are that such a function already exists. The following functions occur frequently in vectorized code:

- Index manipulation
  - find
  - sub2ind
  - ind2sub
  - sort
  - unique
  - lookup
  - ifelse / merge

# Chapter 19: Vectorization and Faster Code Execution

- Repetition
  - repmat
  - repelems
- Vectorized arithmetic
  - sum
  - prod
  - cumsum
  - cumprod
  - sumsq
  - diff
  - dot
  - cummax
  - cummin
- Shape of higher dimensional arrays
  - reshape
  - resize
  - permute
  - squeeze
  - deal

## 19.2 Broadcasting

Broadcasting refers to how Octave binary operators and functions behave when their matrix or array operands or arguments differ in size. Since version 3.6.0, Octave now automatically broadcasts vectors, matrices, and arrays when using elementwise binary operators and functions. Broadly speaking, smaller arrays are "broadcast" across the larger one, until they have a compatible shape. The rule is that corresponding array dimensions must either

1. be equal, or
2. one of them must be 1.

In case all dimensions are equal, no broadcasting occurs and ordinary element-by-element arithmetic takes place. For arrays of higher dimensions, if the number of dimensions isn't the same, then missing trailing dimensions are treated as 1. When one of the dimensions is 1, the array with that singleton dimension gets copied along that dimension until it matches the dimension of the other array. For example, consider

```
x = [1 2 3;
     4 5 6;
     7 8 9];

y = [10 20 30];

x + y
```

Without broadcasting, `x + y` would be an error because the dimensions do not agree. However, with broadcasting it is as if the following operation were performed:

```
x = [1 2 3
     4 5 6
     7 8 9];

y = [10 20 30
     10 20 30
     10 20 30];

x + y
⇒   11   22   33
    14   25   36
    17   28   39
```

That is, the smaller array of size [1 3] gets copied along the singleton dimension (the number of rows) until it is [3 3]. No actual copying takes place, however. The internal implementation reuses elements along the necessary dimension in order to achieve the desired effect without copying in memory.

Both arrays can be broadcast across each other, for example, all pairwise differences of the elements of a vector with itself:

```
y - y'
⇒    0   10   20
   -10    0   10
   -20  -10    0
```

Here the vectors of size [1 3] and [3 1] both get broadcast into matrices of size [3 3] before ordinary matrix subtraction takes place.

A special case of broadcasting that may be familiar is when all dimensions of the array being broadcast are 1, i.e., the array is a scalar. Thus for example, operations like `x - 42` and `max (x, 2)` are basic examples of broadcasting.

For a higher-dimensional example, suppose `img` is an RGB image of size [m n 3] and we wish to multiply each color by a different scalar. The following code accomplishes this with broadcasting,

```
img .*= permute ([0.8, 0.9, 1.2], [1, 3, 2]);
```

Note the usage of permute to match the dimensions of the [0.8, 0.9, 1.2] vector with `img`.

For functions that are not written with broadcasting semantics, `bsxfun` can be useful for coercing them to broadcast.

**bsxfun** (*f, A, B*)                                                    [Built-in Function]

> The binary singleton expansion function performs broadcasting, that is, it applies a binary function *f* element-by-element to two array arguments *A* and *B*, and expands as necessary singleton dimensions in either input argument.
>
> *f* is a function handle, inline function, or string containing the name of the function to evaluate. The function *f* must be capable of accepting two column-vector arguments of equal length, or one column vector argument and a scalar.
>
> The dimensions of *A* and *B* must be equal or singleton. The singleton dimensions of the arrays will be expanded to the same dimensionality as the other array.

# Chapter 19: Vectorization and Faster Code Execution

**See also:** [arrayfun], page 492, [cellfun], page 494.

Broadcasting is only applied if either of the two broadcasting conditions hold. As usual, however, broadcasting does not apply when two dimensions differ and neither is 1:

```
x = [1 2 3
     4 5 6];
y = [10 20
     30 40];
x + y
```

This will produce an error about nonconformant arguments.

Besides common arithmetic operations, several functions of two arguments also broadcast. The full list of functions and operators that broadcast is

```
plus      +    .+
minus     -    .-
times          .*
rdivide        ./
ldivide        .\
power          .^   .**
lt        <
le        <=
eq        ==
gt        >
ge        >=
ne        !=   ~=
and       &
or        |
atan2
hypot
max
min
mod
rem
xor

+=  -=  .+=  .-=  .*=  ./=  .\=  .^=  .**=  &=  |=
```

Beware of resorting to broadcasting if a simpler operation will suffice. For matrices *a* and *b*, consider the following:

```
c = sum (permute (a, [1, 3, 2]) .* permute (b, [3, 2, 1]), 3);
```

This operation broadcasts the two matrices with permuted dimensions across each other during elementwise multiplication in order to obtain a larger 3-D array, and this array is then summed along the third dimension. A moment of thought will prove that this operation is simply the much faster ordinary matrix multiplication, `c = a*b;`.

A note on terminology: "broadcasting" is the term popularized by the Numpy numerical environment in the Python programming language. In other programming languages and environments, broadcasting may also be known as *binary singleton expansion* (BSX, in

MATLAB, and the origin of the name of the `bsxfun` function), *recycling* (R programming language), *single-instruction multiple data* (SIMD), or *replication*.

### 19.2.1 Broadcasting and Legacy Code

The new broadcasting semantics almost never affect code that worked in previous versions of Octave. Consequently, all code inherited from MATLAB that worked in previous versions of Octave should still work without change in Octave. The only exception is code such as

```
try
  c = a.*b;
catch
  c = a.*a;
end_try_catch
```

that may have relied on matrices of different size producing an error. Due to how broadcasting changes semantics with older versions of Octave, by default Octave warns if a broadcasting operation is performed. To disable this warning, refer to its ID (see [warning_ids], page 214):

```
warning ("off", "Octave:broadcast");
```

If you want to recover the old behavior and produce an error, turn this warning into an error:

```
warning ("error", "Octave:broadcast");
```

For broadcasting on scalars that worked in previous versions of Octave, this warning will not be emitted.

## 19.3 Function Application

As a general rule, functions should already be written with matrix arguments in mind and should consider whole matrix operations in a vectorized manner. Sometimes, writing functions in this way appears difficult or impossible for various reasons. For those situations, Octave provides facilities for applying a function to each element of an array, cell, or struct.

arrayfun (*func*, *A*)  [Function File]
x = arrayfun (*func*, *A*)  [Function File]
x = arrayfun (*func*, *A*, *b*, ...)  [Function File]
[x, y, ...] = arrayfun (*func*, *A*, ...)  [Function File]
arrayfun (..., "*UniformOutput*", *val*)  [Function File]
arrayfun (..., "*ErrorHandler*", *errfunc*)  [Function File]

Execute a function on each element of an array.

This is useful for functions that do not accept array arguments. If the function does accept array arguments it is better to call the function directly.

The first input argument *func* can be a string, a function handle, an inline function, or an anonymous function. The input argument *A* can be a logic array, a numeric array, a string array, a structure array, or a cell array. By a call of the function `arrayfun` all elements of *A* are passed on to the named function *func* individually.

The named function can also take more than two input arguments, with the input arguments given as third input argument *b*, fourth input argument *c*, ... If given

## Chapter 19: Vectorization and Faster Code Execution

more than one array input argument then all input arguments must have the same sizes, for example:

```
arrayfun (@atan2, [1, 0], [0, 1])
    ⇒ [ 1.5708   0.0000 ]
```

If the parameter *val* after a further string input argument `"UniformOutput"` is set `true` (the default), then the named function *func* must return a single element which then will be concatenated into the return value and is of type matrix. Otherwise, if that parameter is set to `false`, then the outputs are concatenated in a cell array. For example:

```
arrayfun (@(x,y) x:y, "abc", "def", "UniformOutput", false)
    ⇒
    {
      [1,1] = abcd
      [1,2] = bcde
      [1,3] = cdef
    }
```

If more than one output arguments are given then the named function must return the number of return values that also are expected, for example:

```
[A, B, C] = arrayfun (@find, [10; 0], "UniformOutput", false)
    ⇒
    A =
    {
      [1,1] =   1
      [2,1] = [](0x0)
    }
    B =
    {
      [1,1] =   1
      [2,1] = [](0x0)
    }
    C =
    {
      [1,1] =   10
      [2,1] = [](0x0)
    }
```

If the parameter *errfunc* after a further string input argument `"ErrorHandler"` is another string, a function handle, an inline function, or an anonymous function, then *errfunc* defines a function to call in the case that *func* generates an error. The definition of the function must be of the form

```
function [...] = errfunc (s, ...)
```

where there is an additional input argument to *errfunc* relative to *func*, given by *s*. This is a structure with the elements `"identifier"`, `"message"`, and `"index"` giving, respectively, the error identifier, the error message, and the index of the array elements that caused the error. The size of the output argument of *errfunc* must have the same size as the output argument of *func*, otherwise a real error is thrown. For example:

```
function y = ferr (s, x), y = "MyString"; endfunction
arrayfun (@str2num, [1234],
         "UniformOutput", false, "ErrorHandler", @ferr)
⇒
  {
    [1,1] = MyString
  }
```

**See also:** [spfun], page 494, [cellfun], page 494, [structfun], page 496.

*y* = **spfun** (*f*, *S*)　　　　　　　　　　　　　　　　　　　　　　　　[Function File]
　　Compute *f(S)* for the nonzero values of *S*.

　　This results in a sparse matrix with the same structure as *S*. The function *f* can be passed as a string, a function handle, or an inline function.

**See also:** [arrayfun], page 492, [cellfun], page 494, [structfun], page 496.

**cellfun** (*name*, *C*)　　　　　　　　　　　　　　　　　　　　　　　　[Built-in Function]
**cellfun** ("*size*", *C*, *k*)　　　　　　　　　　　　　　　　　　　　　　[Built-in Function]
**cellfun** ("*isclass*", *C*, *class*)　　　　　　　　　　　　　　　　　　　[Built-in Function]
**cellfun** (*func*, *C*)　　　　　　　　　　　　　　　　　　　　　　　　[Built-in Function]
**cellfun** (*func*, *C*, *D*)　　　　　　　　　　　　　　　　　　　　　　[Built-in Function]
[*a*, ...] = **cellfun** (...)　　　　　　　　　　　　　　　　　　　　　　[Built-in Function]
**cellfun** (..., "*ErrorHandler*", *errfunc*)　　　　　　　　　　　　　　　[Built-in Function]
**cellfun** (..., "*UniformOutput*", *val*)　　　　　　　　　　　　　　　　[Built-in Function]
　　Evaluate the function named *name* on the elements of the cell array *C*.

　　Elements in *C* are passed on to the named function individually. The function *name* can be one of the functions

- **isempty**　　Return 1 for empty elements.
- **islogical**　　Return 1 for logical elements.
- **isnumeric**　　Return 1 for numeric elements.
- **isreal**　　Return 1 for real elements.
- **length**　　Return a vector of the lengths of cell elements.
- **ndims**　　Return the number of dimensions of each element.
- **numel**
- **prodofsize**　　Return the number of elements contained within each cell element. The number is the product of the dimensions of the object at each cell element.
- **size**　　Return the size along the *k*-th dimension.
- **isclass**　　Return 1 for elements of *class*.

Additionally, `cellfun` accepts an arbitrary function *func* in the form of an inline function, function handle, or the name of a function (in a character string). The function can take one or more arguments, with the inputs arguments given by *C*, *D*, etc. Equally the function can return one or more output arguments. For example:

```
cellfun ("atan2", {1, 0}, {0, 1})
    ⇒ [ 1.57080   0.00000 ]
```

The number of output arguments of `cellfun` matches the number of output arguments of the function. The outputs of the function will be collected into the output arguments of `cellfun` like this:

```
function [a, b] = twoouts (x)
  a = x;
  b = x*x;
endfunction
[aa, bb] = cellfun (@twoouts, {1, 2, 3})
    ⇒
        aa =
          1 2 3
        bb =
          1 4 9
```

Note that per default the output argument(s) are arrays of the same size as the input arguments. Input arguments that are singleton (1x1) cells will be automatically expanded to the size of the other arguments.

If the parameter `"UniformOutput"` is set to true (the default), then the function must return scalars which will be concatenated into the return array(s). If `"UniformOutput"` is false, the outputs are concatenated into a cell array (or cell arrays). For example:

```
cellfun ("tolower", {"Foo", "Bar", "FooBar"},
        "UniformOutput", false)
    ⇒ {"foo", "bar", "foobar"}
```

Given the parameter `"ErrorHandler"`, then *errfunc* defines a function to call in case *func* generates an error. The form of the function is

```
function [...] = errfunc (s, ...)
```

where there is an additional input argument to *errfunc* relative to *func*, given by *s*. This is a structure with the elements `"identifier"`, `"message"` and `"index"`, giving respectively the error identifier, the error message, and the index into the input arguments of the element that caused the error. For example:

```
function y = foo (s, x), y = NaN; endfunction
cellfun ("factorial", {-1,2}, "ErrorHandler", @foo)
    ⇒ [NaN 2]
```

Use `cellfun` intelligently. The `cellfun` function is a useful tool for avoiding loops. It is often used with anonymous function handles; however, calling an anonymous function involves an overhead quite comparable to the overhead of an m-file function. Passing a handle to a built-in function is faster, because the interpreter is not involved in the internal loop. For example:

```
a = {...}
v = cellfun (@(x) det (x), a);  # compute determinants
v = cellfun (@det, a);  # faster
```

See also: [arrayfun], page 492, [structfun], page 496, [spfun], page 494.

structfun (*func*, *S*)      [Function File]
[*A*, ...] = structfun (...)      [Function File]
structfun (..., "*ErrorHandler*", *errfunc*)      [Function File]
structfun (..., "*UniformOutput*", *val*)      [Function File]

Evaluate the function named *name* on the fields of the structure *S*. The fields of *S* are passed to the function *func* individually.

**structfun** accepts an arbitrary function *func* in the form of an inline function, function handle, or the name of a function (in a character string). In the case of a character string argument, the function must accept a single argument named *x*, and it must return a string value. If the function returns more than one argument, they are returned as separate output variables.

If the parameter "UniformOutput" is set to true (the default), then the function must return a single element which will be concatenated into the return value. If "UniformOutput" is false, the outputs are placed into a structure with the same fieldnames as the input structure.

```
s.name1 = "John Smith";
s.name2 = "Jill Jones";
structfun (@(x) regexp (x, '(\w+)$', "matches"){1}, s,
          "UniformOutput", false)
⇒
  {
    name1 = Smith
    name2 = Jones
  }
```

Given the parameter "ErrorHandler", *errfunc* defines a function to call in case *func* generates an error. The form of the function is

```
function [...] = errfunc (se, ...)
```

where there is an additional input argument to *errfunc* relative to *func*, given by *se*. This is a structure with the elements "identifier", "message" and "index", giving respectively the error identifier, the error message, and the index into the input arguments of the element that caused the error. For an example on how to use an error handler, see [cellfun], page 494.

See also: [cellfun], page 494, [arrayfun], page 492, [spfun], page 494.

Consistent with earlier advice, seek to use Octave built-in functions whenever possible for the best performance. This advice applies especially to the four functions above. For example, when adding two arrays together element-by-element one could use a handle to the built-in addition function @plus or define an anonymous function @(x,y) x + y. But, the anonymous function is 60% slower than the first method. See Section 34.4.2 [Operator Overloading], page 727, for a list of basic functions which might be used in place of anonymous ones.

# Chapter 19: Vectorization and Faster Code Execution

## 19.4 Accumulation

Whenever it's possible to categorize according to indices the elements of an array when performing a computation, accumulation functions can be useful.

**accumarray** (*subs*, *vals*, *sz*, *func*, *fillval*, *issparse*)      [Function File]
**accumarray** (*subs*, *vals*, ...)      [Function File]

    Create an array by accumulating the elements of a vector into the positions defined by their subscripts.

    The subscripts are defined by the rows of the matrix *subs* and the values by *vals*. Each row of *subs* corresponds to one of the values in *vals*. If *vals* is a scalar, it will be used for each of the row of *subs*. If *subs* is a cell array of vectors, all vectors must be of the same length, and the subscripts in the $k$th vector must correspond to the $k$th dimension of the result.

    The size of the matrix will be determined by the subscripts themselves. However, if *sz* is defined it determines the matrix size. The length of *sz* must correspond to the number of columns in *subs*. An exception is if *subs* has only one column, in which case *sz* may be the dimensions of a vector and the subscripts of *subs* are taken as the indices into it.

    The default action of `accumarray` is to sum the elements with the same subscripts. This behavior can be modified by defining the *func* function. This should be a function or function handle that accepts a column vector and returns a scalar. The result of the function should not depend on the order of the subscripts.

    The elements of the returned array that have no subscripts associated with them are set to zero. Defining *fillval* to some other value allows these values to be defined. This behavior changes, however, for certain values of *func*. If *func* is `min` (respectively, `max`) then the result will be filled with the minimum (respectively, maximum) integer if *vals* is of integral type, logical false (respectively, logical true) if *vals* is of logical type, zero if *fillval* is zero and all values are non-positive (respectively, non-negative), and NaN otherwise.

    By default `accumarray` returns a full matrix. If *issparse* is logically true, then a sparse matrix is returned instead.

    The following `accumarray` example constructs a frequency table that in the first column counts how many occurrences each number in the second column has, taken from the vector *x*. Note the usage of `unique` for assigning to all repeated elements of *x* the same index (see [unique], page 633).

```
x = [91, 92, 90, 92, 90, 89, 91, 89, 90, 100, 100, 100];
[u, ~, j] = unique (x);
[accumarray(j', 1), u']
   ⇒  2    89
      3    90
      2    91
      2    92
      3   100
```

Another example, where the result is a multi-dimensional 3-D array and the default value (zero) appears in the output:

```
accumarray ([1, 1, 1;
             2, 1, 2;
             2, 3, 2;
             2, 1, 2;
             2, 3, 2], 101:105)
⇒ ans(:,:,1) = [101, 0, 0; 0, 0, 0]
⇒ ans(:,:,2) = [0, 0, 0; 206, 0, 208]
```

The sparse option can be used as an alternative to the **sparse** constructor (see [sparse], page 522). Thus

```
sparse (i, j, sv)
```

can be written with **accumarray** as

```
accumarray ([i, j], sv', [], [], 0, true)
```

For repeated indices, **sparse** adds the corresponding value. To take the minimum instead, use **min** as an accumulator function:

```
accumarray ([i, j], sv', [], @min, 0, true)
```

The complexity of accumarray in general for the non-sparse case is generally O(M+N), where N is the number of subscripts and M is the maximum subscript (linearized in multi-dimensional case). If *func* is one of @sum (default), @max, @min or @(x) {x}, an optimized code path is used. Note that for general reduction function the interpreter overhead can play a major part and it may be more efficient to do multiple accumarray calls and compute the results in a vectorized manner.

**See also:** [accumdim], page 498, [unique], page 633, [sparse], page 522.

**accumdim** (*subs, vals, dim, n, func, fillval*)  [Function File]

Create an array by accumulating the slices of an array into the positions defined by their subscripts along a specified dimension.

The subscripts are defined by the index vector *subs*. The dimension is specified by *dim*. If not given, it defaults to the first non-singleton dimension. The length of *subs* must be equal to **size** (*vals, dim*).

The extent of the result matrix in the working dimension will be determined by the subscripts themselves. However, if *n* is defined it determines this extent.

The default action of **accumdim** is to sum the subarrays with the same subscripts. This behavior can be modified by defining the *func* function. This should be a function or function handle that accepts an array and a dimension, and reduces the array along this dimension. As a special exception, the built-in **min** and **max** functions can be used directly, and **accumdim** accounts for the middle empty argument that is used in their calling.

The slices of the returned array that have no subscripts associated with them are set to zero. Defining *fillval* to some other value allows these values to be defined.

An example of the use of **accumdim** is:

```
accumdim ([1, 2, 1, 2, 1], [ 7, -10,   4;
                            -5, -12,   8;
                           -12,   2,   8;
                           -10,   9,  -3;
                            -5,  -3, -13])
```
⇒ [-10,-11,-1;-15,-3,5]

**See also:** [accumarray], page 497.

## 19.5 JIT Compiler

Vectorization is the preferred technique for eliminating loops and speeding up code. Nevertheless, it is not always possible to replace every loop. In such situations it may be worth trying Octave's **experimental** Just-In-Time (JIT) compiler.

A JIT compiler works by analyzing the body of a loop, translating the Octave statements into another language, compiling the new code segment into an executable, and then running the executable and collecting any results. The process is not simple and there is a significant amount of work to perform for each step. It can still make sense, however, if the number of loop iterations is large. Because Octave is an interpreted language every time through a loop Octave must parse the statements in the loop body before executing them. With a JIT compiler this is done just once when the body is translated to another language.

The JIT compiler is a very new feature in Octave and not all valid Octave statements can currently be accelerated. However, if no other technique is available it may be worth benchmarking the code with JIT enabled. The function `jit_enable` is used to turn compilation on or off. The function `jit_startcnt` sets the threshold for acceleration. Loops with iteration counts above `jit_startcnt` will be accelerated. The functions `jit_failcnt` and `debug_jit` are not likely to be of use to anyone not working directly on the implementation of the JIT compiler.

*val* = jit_enable ()  [Built-in Function]
*old_val* = jit_enable (*new_val*)  [Built-in Function]
jit_enable (*new_val*, "*local*")  [Built-in Function]
    Query or set the internal variable that enables Octave's JIT compiler.

    When called from inside a function with the "`local`" option, the variable is changed locally for the function and any subroutines it calls. The original variable value is restored when exiting the function.

    **See also:** [jit_startcnt], page 499, [debug_jit], page 500.

*val* = jit_startcnt ()  [Built-in Function]
*old_val* = jit_startcnt (*new_val*)  [Built-in Function]
jit_startcnt (*new_val*, "*local*")  [Built-in Function]
    Query or set the internal variable that determines whether JIT compilation will take place for a specific loop.

    Because compilation is a costly operation it does not make sense to employ JIT when the loop count is low. By default only loops with greater than 1000 iterations will be accelerated.

When called from inside a function with the `"local"` option, the variable is changed locally for the function and any subroutines it calls. The original variable value is restored when exiting the function.

**See also:** [jit_enable], page 499, [jit_failcnt], page 500, [debug_jit], page 500.

`val = jit_failcnt ()` [Built-in Function]
`old_val = jit_failcnt (new_val)` [Built-in Function]
`jit_failcnt (new_val, "local")` [Built-in Function]

Query or set the internal variable that counts the number of JIT fail exceptions for Octave's JIT compiler.

When called from inside a function with the `"local"` option, the variable is changed locally for the function and any subroutines it calls. The original variable value is restored when exiting the function.

**See also:** [jit_enable], page 499, [jit_startcnt], page 499, [debug_jit], page 500.

`val = debug_jit ()` [Built-in Function]
`old_val = debug_jit (new_val)` [Built-in Function]
`debug_jit (new_val, "local")` [Built-in Function]

Query or set the internal variable that determines whether debugging/tracing is enabled for Octave's JIT compiler.

When called from inside a function with the `"local"` option, the variable is changed locally for the function and any subroutines it calls. The original variable value is restored when exiting the function.

**See also:** [jit_enable], page 499, [jit_startcnt], page 499.

## 19.6 Miscellaneous Techniques

Here are some other ways of improving the execution speed of Octave programs.

- Avoid computing costly intermediate results multiple times. Octave currently does not eliminate common subexpressions. Also, certain internal computation results are cached for variables. For instance, if a matrix variable is used multiple times as an index, checking the indices (and internal conversion to integers) is only done once.
- Be aware of lazy copies (copy-on-write). When a copy of an object is created, the data is not immediately copied, but rather shared. The actual copying is postponed until the copied data needs to be modified. For example:

```
a = zeros (1000); # create a 1000x1000 matrix
b = a; # no copying done here
b(1) = 1; # copying done here
```

Lazy copying applies to whole Octave objects such as matrices, cells, struct, and also individual cell or struct elements (not array elements).

Additionally, index expressions also use lazy copying when Octave can determine that the indexed portion is contiguous in memory. For example:

```
a = zeros (1000); # create a 1000x1000 matrix
b = a(:,10:100); # no copying done here
b = a(10:100,:); # copying done here
```

# Chapter 19: Vectorization and Faster Code Execution

This applies to arrays (matrices), cell arrays, and structs indexed using '()'. Index expressions generating comma-separated lists can also benefit from shallow copying in some cases. In particular, when a is a struct array, expressions like {a.x}, {a(:,2).x} will use lazy copying, so that data can be shared between a struct array and a cell array.

Most indexing expressions do not live longer than their parent objects. In rare cases, however, a lazily copied slice outlasts its parent, in which case it becomes orphaned, still occupying unnecessarily more memory than needed. To provide a remedy working in most real cases, Octave checks for orphaned lazy slices at certain situations, when a value is stored into a "permanent" location, such as a named variable or cell or struct element, and possibly economizes them. For example:

```
a = zeros (1000); # create a 1000x1000 matrix
b = a(:,10:100);  # lazy slice
a = [];  # the original "a" array is still allocated
c{1} = b; # b is reallocated at this point
```

- Avoid deep recursion. Function calls to m-file functions carry a relatively significant overhead, so rewriting a recursion as a loop often helps. Also, note that the maximum level of recursion is limited.

- Avoid resizing matrices unnecessarily. When building a single result matrix from a series of calculations, set the size of the result matrix first, then insert values into it. Write

```
result = zeros (big_n, big_m)
for i = over:and_over
  ridx = ...
  cidx = ...
  result(ridx, cidx) = new_value ();
endfor
```

instead of

```
result = [];
for i = ever:and_ever
  result = [ result, new_value() ];
endfor
```

Sometimes the number of items can not be computed in advance, and stack-like operations are needed. When elements are being repeatedly inserted or removed from the end of an array, Octave detects it as stack usage and attempts to use a smarter memory management strategy by pre-allocating the array in bigger chunks. This strategy is also applied to cell and struct arrays.

```
a = [];
while (condition)
  ...
  a(end+1) = value; # "push" operation
  ...
  a(end) = []; # "pop" operation
  ...
endwhile
```

- Avoid calling `eval` or `feval` excessively. Parsing input or looking up the name of a function in the symbol table are relatively expensive operations.

  If you are using `eval` merely as an exception handling mechanism, and not because you need to execute some arbitrary text, use the `try` statement instead. See Section 10.9 [The try Statement], page 168.

- Use `ignore_function_time_stamp` when appropriate. If you are calling lots of functions, and none of them will need to change during your run, set the variable `ignore_function_time_stamp` to "all". This will stop Octave from checking the time stamp of a function file to see if it has been updated while the program is being run.

## 19.7 Examples

The following are examples of vectorization questions asked by actual users of Octave and their solutions.

- For a vector A, the following loop

  ```
  n = length (A);
  B = zeros (n, 2);
  for i = 1:length (A)
    ## this will be two columns, the first is the difference and
    ## the second the mean of the two elements used for the diff.
    B(i,:) = [A(i+1)-A(i), (A(i+1) + A(i))/2];
  endfor
  ```

  can be turned into the following one-liner:

  ```
  B = [diff(A)(:), 0.5*(A(1:end-1)+A(2:end))(:)]
  ```

  Note the usage of colon indexing to flatten an intermediate result into a column vector. This is a common vectorization trick.

# 20 Nonlinear Equations

## 20.1 Solvers

Octave can solve sets of nonlinear equations of the form

$$f(x) = 0$$

using the function `fsolve`, which is based on the MINPACK subroutine `hybrd`. This is an iterative technique so a starting point must be provided. This also has the consequence that convergence is not guaranteed even if a solution exists.

**fsolve** (*fcn, x0, options*)  [Function File]
**[x, fvec, info, output, fjac] = fsolve** (*fcn, ...*)  [Function File]
    Solve a system of nonlinear equations defined by the function *fcn*.

*fcn* should accept a vector (array) defining the unknown variables, and return a vector of left-hand sides of the equations. Right-hand sides are defined to be zeros. In other words, this function attempts to determine a vector x such that `fcn (x)` gives (approximately) all zeros.

*x0* determines a starting guess. The shape of *x0* is preserved in all calls to *fcn*, but otherwise it is treated as a column vector.

*options* is a structure specifying additional options. Currently, `fsolve` recognizes these options: "FunValCheck", "OutputFcn", "TolX", "TolFun", "MaxIter", "MaxFunEvals", "Jacobian", "Updating", "ComplexEqn" "TypicalX", "AutoScaling" and "FinDiffType".

If "Jacobian" is "on", it specifies that *fcn*, called with 2 output arguments also returns the Jacobian matrix of right-hand sides at the requested point. "TolX" specifies the termination tolerance in the unknown variables, while "TolFun" is a tolerance for equations. Default is `1e-7` for both "TolX" and "TolFun".

If "AutoScaling" is on, the variables will be automatically scaled according to the column norms of the (estimated) Jacobian. As a result, TolF becomes scaling-independent. By default, this option is off because it may sometimes deliver unexpected (though mathematically correct) results.

If "Updating" is "on", the function will attempt to use Broyden updates to update the Jacobian, in order to reduce the amount of Jacobian calculations. If your user function always calculates the Jacobian (regardless of number of output arguments) then this option provides no advantage and should be set to false.

"ComplexEqn" is "on", `fsolve` will attempt to solve complex equations in complex variables, assuming that the equations possess a complex derivative (i.e., are holomorphic). If this is not what you want, you should unpack the real and imaginary parts of the system to get a real system.

For description of the other options, see `optimset`.

On return, *fval* contains the value of the function *fcn* evaluated at x.

*info* may be one of the following values:

| | |
|---|---|
| 1 | Converged to a solution point. Relative residual error is less than specified by TolFun. |
| 2 | Last relative step size was less that TolX. |
| 3 | Last relative decrease in residual was less than TolF. |
| 0 | Iteration limit exceeded. |
| -3 | The trust region radius became excessively small. |

Note: If you only have a single nonlinear equation of one variable, using `fzero` is usually a much better idea.

Note about user-supplied Jacobians: As an inherent property of the algorithm, a Jacobian is always requested for a solution vector whose residual vector is already known, and it is the last accepted successful step. Often this will be one of the last two calls, but not always. If the savings by reusing intermediate results from residual calculation in Jacobian calculation are significant, the best strategy is to employ OutputFcn: After a vector is evaluated for residuals, if OutputFcn is called with that vector, then the intermediate results should be saved for future Jacobian evaluation, and should be kept until a Jacobian evaluation is requested or until OutputFcn is called with a different vector, in which case they should be dropped in favor of this most recent vector. A short example how this can be achieved follows:

```
function [fvec, fjac] = user_func (x, optimvalues, state)
persistent sav = [], sav0 = [];
if (nargin == 1)
  ## evaluation call
  if (nargout == 1)
    sav0.x = x; # mark saved vector
    ## calculate fvec, save results to sav0.
  elseif (nargout == 2)
    ## calculate fjac using sav.
  endif
else
  ## outputfcn call.
  if (all (x == sav0.x))
    sav = sav0;
  endif
  ## maybe output iteration status, etc.
endif
endfunction

## ...

fsolve (@user_func, x0, optimset ("OutputFcn", @user_func, ...))
```

See also: [fzero], page 505, [optimset], page 595.

The following is a complete example. To solve the set of equations
$$-2x^2 + 3xy + 4\sin(y) - 6 = 0$$
$$3x^2 - 2xy^2 + 3\cos(x) + 4 = 0$$

you first need to write a function to compute the value of the given function. For example:

```
function y = f (x)
  y = zeros (2, 1);
  y(1) = -2*x(1)^2 + 3*x(1)*x(2)   + 4*sin(x(2)) - 6;
  y(2) =  3*x(1)^2 - 2*x(1)*x(2)^2 + 3*cos(x(1)) + 4;
endfunction
```

Then, call `fsolve` with a specified initial condition to find the roots of the system of equations. For example, given the function `f` defined above,

```
[x, fval, info] = fsolve (@f, [1; 2])
```

results in the solution

```
x =

  0.57983
  2.54621

fval =

  -5.7184e-10
   5.5460e-10

info = 1
```

A value of `info = 1` indicates that the solution has converged.

When no Jacobian is supplied (as in the example above) it is approximated numerically. This requires more function evaluations, and hence is less efficient. In the example above we could compute the Jacobian analytically as

$$\begin{bmatrix} \frac{\partial f_1}{\partial x_1} & \frac{\partial f_1}{\partial x_2} \\ \frac{\partial f_2}{\partial x_1} & \frac{\partial f_2}{\partial x_2} \end{bmatrix} = \begin{bmatrix} 3x_2 - 4x_1 & 4\cos(x_2) + 3x_1 \\ -2x_2^2 - 3\sin(x_1) + 6x_1 & -4x_1 x_2 \end{bmatrix}$$

and compute it with the following Octave function

```
function [y, jac] = f (x)
  y = zeros (2, 1);
  y(1) = -2*x(1)^2 + 3*x(1)*x(2)   + 4*sin(x(2)) - 6;
  y(2) =  3*x(1)^2 - 2*x(1)*x(2)^2 + 3*cos(x(1)) + 4;
  if (nargout == 2)
    jac = zeros (2, 2);
    jac(1,1) =  3*x(2) - 4*x(1);
    jac(1,2) =  4*cos(x(2)) + 3*x(1);
    jac(2,1) = -2*x(2)^2 - 3*sin(x(1)) + 6*x(1);
    jac(2,2) = -4*x(1)*x(2);
  endif
endfunction
```

The Jacobian can then be used with the following call to `fsolve`:

```
[x, fval, info] = fsolve (@f, [1; 2], optimset ("jacobian", "on"));
```

which gives the same solution as before.

fzero (fun, x0)                                           [Function File]
fzero (fun, x0, options)                                  [Function File]
[x, fval, info, output] = fzero (...)                     [Function File]

> Find a zero of a univariate function.
>
> *fun* is a function handle, inline function, or string containing the name of the function to evaluate.
>
> *x0* should be a two-element vector specifying two points which bracket a zero. In other words, there must be a change in sign of the function between *x0*(1) and *x0*(2). More mathematically, the following must hold
>
> > sign (fun(x0(1))) * sign (fun(x0(2))) <= 0
>
> If *x0* is a single scalar then several nearby and distant values are probed in an attempt to obtain a valid bracketing. If this is not successful, the function fails.
>
> *options* is a structure specifying additional options. Currently, fzero recognizes these options: "FunValCheck", "OutputFcn", "TolX", "MaxIter", "MaxFunEvals". For a description of these options, see [optimset], page 595.
>
> On exit, the function returns *x*, the approximate zero point and *fval*, the function value thereof.
>
> *info* is an exit flag that can have these values:
>
> - 1 The algorithm converged to a solution.
> - 0 Maximum number of iterations or function evaluations has been reached.
> - -1 The algorithm has been terminated from user output function.
> - -5 The algorithm may have converged to a singular point.
>
> *output* is a structure containing runtime information about the fzero algorithm. Fields in the structure are:
>
> - iterations Number of iterations through loop.
> - nfev Number of function evaluations.
> - bracketx A two-element vector with the final bracketing of the zero along the x-axis.
> - brackety A two-element vector with the final bracketing of the zero along the y-axis.
>
> See also: [optimset], page 595, [fsolve], page 503.

## 20.2 Minimizers

Often it is useful to find the minimum value of a function rather than just the zeroes where it crosses the x-axis. fminbnd is designed for the simpler, but very common, case of a univariate function where the interval to search is bounded. For unbounded minimization of a function with potentially many variables use fminunc or fminsearch. The two functions use different internal algorithms and some knowledge of the objective function is required. For functions which can be differentiated, fminunc is appropriate. For functions with discontinuities, or for which a gradient search would fail, use fminsearch. See Chapter 25 [Optimization], page 583, for minimization with the presence of constraint functions. Note that searches can be made for maxima by simply inverting the objective function ($F_{max} = -F_{min}$).

## Chapter 20: Nonlinear Equations

**[x, fval, info, output] = fminbnd (*fun*, a, b, *options*)**         [Function File]
Find a minimum point of a univariate function.

*fun* should be a function handle or name. *a*, *b* specify a starting interval. *options* is a structure specifying additional options. Currently, fminbnd recognizes these options: "FunValCheck", "OutputFcn", "TolX", "MaxIter", "MaxFunEvals". For a description of these options, see [optimset], page 595.

On exit, the function returns *x*, the approximate minimum point and *fval*, the function value thereof.

*info* is an exit flag that can have these values:

- 1 The algorithm converged to a solution.
- 0 Maximum number of iterations or function evaluations has been exhausted.
- -1 The algorithm has been terminated from user output function.

Notes: The search for a minimum is restricted to be in the interval bound by *a* and *b*. If you only have an initial point to begin searching from you will need to use an unconstrained minimization algorithm such as `fminunc` or `fminsearch`. fminbnd internally uses a Golden Section search strategy.

**See also:** [fzero], page 505, [fminunc], page 507, [fminsearch], page 508, [optimset], page 595.

**fminunc (*fcn*, *x0*)**         [Function File]
**fminunc (*fcn*, *x0*, *options*)**         [Function File]
**[x, fval, info, output, grad, hess] = fminunc (*fcn*, ...)**         [Function File]
Solve an unconstrained optimization problem defined by the function *fcn*.

*fcn* should accept a vector (array) defining the unknown variables, and return the objective function value, optionally with gradient. fminunc attempts to determine a vector *x* such that `fcn (x)` is a local minimum.

*x0* determines a starting guess. The shape of *x0* is preserved in all calls to *fcn*, but otherwise is treated as a column vector.

*options* is a structure specifying additional options. Currently, fminunc recognizes these options: "FunValCheck", "OutputFcn", "TolX", "TolFun", "MaxIter", "MaxFunEvals", "GradObj", "FinDiffType", "TypicalX", "AutoScaling".

If "GradObj" is "on", it specifies that *fcn*, when called with 2 output arguments, also returns the Jacobian matrix of partial first derivatives at the requested point. TolX specifies the termination tolerance for the unknown variables *x*, while TolFun is a tolerance for the objective function value *fval*. The default is `1e-7` for both options.

For a description of the other options, see `optimset`.

On return, *x* is the location of the minimum and *fval* contains the value of the objective function at *x*.

*info* may be one of the following values:

1          Converged to a solution point. Relative gradient error is less than specified by `TolFun`.

2          Last relative step size was less than `TolX`.

| | |
|---|---|
| 3 | Last relative change in function value was less than `TolFun`. |
| 0 | Iteration limit exceeded—either maximum number of algorithm iterations `MaxIter` or maximum number of function evaluations `MaxFunEvals`. |
| -1 | Algorithm terminated by `OutputFcn`. |
| -3 | The trust region radius became excessively small. |

Optionally, `fminunc` can return a structure with convergence statistics (*output*), the output gradient (*grad*) at the solution *x*, and approximate Hessian (*hess*) at the solution *x*.

Application Notes: If have only a single nonlinear equation of one variable then using `fminbnd` is usually a better choice.

The algorithm used by `fminsearch` is a gradient search which depends on the objective function being differentiable. If the function has discontinuities it may be better to use a derivative-free algorithm such as `fminsearch`.

**See also:** [fminbnd], page 506, [fminsearch], page 508, [optimset], page 595.

x = **fminsearch** (*fun, x0*)  [Function File]
x = **fminsearch** (*fun, x0, options*)  [Function File]
[x, fval] = **fminsearch** (...)  [Function File]

Find a value of *x* which minimizes the function *fun*.

The search begins at the point *x0* and iterates using the Nelder & Mead Simplex algorithm (a derivative-free method). This algorithm is better-suited to functions which have discontinuities or for which a gradient-based search such as `fminunc` fails.

Options for the search are provided in the parameter *options* using the function `optimset`. Currently, `fminsearch` accepts the options: `"TolX"`, `"MaxFunEvals"`, `"MaxIter"`, `"Display"`. For a description of these options, see `optimset`.

On exit, the function returns *x*, the minimum point, and *fval*, the function value thereof.

Example usages:

```
fminsearch (@(x) (x(1)-5).^2+(x(2)-8).^4, [0;0])

fminsearch (inline ("(x(1)-5).^2+(x(2)-8).^4", "x"), [0;0])
```

**See also:** [fminbnd], page 506, [fminunc], page 507, [optimset], page 595.

# Appendix J  GNU GENERAL PUBLIC LICENSE

Version 3, 29 June 2007

Copyright © 2007 Free Software Foundation, Inc. http://fsf.org/

Everyone is permitted to copy and distribute verbatim copies of this license document, but changing it is not allowed.

## Preamble

The GNU General Public License is a free, copyleft license for software and other kinds of works.

The licenses for most software and other practical works are designed to take away your freedom to share and change the works. By contrast, the GNU General Public License is intended to guarantee your freedom to share and change all versions of a program—to make sure it remains free software for all its users. We, the Free Software Foundation, use the GNU General Public License for most of our software; it applies also to any other work released this way by its authors. You can apply it to your programs, too.

When we speak of free software, we are referring to freedom, not price. Our General Public Licenses are designed to make sure that you have the freedom to distribute copies of free software (and charge for them if you wish), that you receive source code or can get it if you want it, that you can change the software or use pieces of it in new free programs, and that you know you can do these things.

To protect your rights, we need to prevent others from denying you these rights or asking you to surrender the rights. Therefore, you have certain responsibilities if you distribute copies of the software, or if you modify it: responsibilities to respect the freedom of others.

For example, if you distribute copies of such a program, whether gratis or for a fee, you must pass on to the recipients the same freedoms that you received. You must make sure that they, too, receive or can get the source code. And you must show them these terms so they know their rights.

Developers that use the GNU GPL protect your rights with two steps: (1) assert copyright on the software, and (2) offer you this License giving you legal permission to copy, distribute and/or modify it.

For the developers' and authors' protection, the GPL clearly explains that there is no warranty for this free software. For both users' and authors' sake, the GPL requires that modified versions be marked as changed, so that their problems will not be attributed erroneously to authors of previous versions.

Some devices are designed to deny users access to install or run modified versions of the software inside them, although the manufacturer can do so. This is fundamentally incompatible with the aim of protecting users' freedom to change the software. The systematic pattern of such abuse occurs in the area of products for individuals to use, which is precisely where it is most unacceptable. Therefore, we have designed this version of the GPL to prohibit the practice for those products. If such problems arise substantially in other domains, we stand ready to extend this provision to those domains in future versions of the GPL, as needed to protect the freedom of users.

Finally, every program is threatened constantly by software patents. States should not allow patents to restrict development and use of software on general-purpose computers, but in those that do, we wish to avoid the special danger that patents applied to a free program could make it effectively proprietary. To prevent this, the GPL assures that patents cannot be used to render the program non-free.

The precise terms and conditions for copying, distribution and modification follow.

## TERMS AND CONDITIONS

0. Definitions.

    "This License" refers to version 3 of the GNU General Public License.

    "Copyright" also means copyright-like laws that apply to other kinds of works, such as semiconductor masks.

    "The Program" refers to any copyrightable work licensed under this License. Each licensee is addressed as "you". "Licensees" and "recipients" may be individuals or organizations.

    To "modify" a work means to copy from or adapt all or part of the work in a fashion requiring copyright permission, other than the making of an exact copy. The resulting work is called a "modified version" of the earlier work or a work "based on" the earlier work.

    A "covered work" means either the unmodified Program or a work based on the Program.

    To "propagate" a work means to do anything with it that, without permission, would make you directly or secondarily liable for infringement under applicable copyright law, except executing it on a computer or modifying a private copy. Propagation includes copying, distribution (with or without modification), making available to the public, and in some countries other activities as well.

    To "convey" a work means any kind of propagation that enables other parties to make or receive copies. Mere interaction with a user through a computer network, with no transfer of a copy, is not conveying.

    An interactive user interface displays "Appropriate Legal Notices" to the extent that it includes a convenient and prominently visible feature that (1) displays an appropriate copyright notice, and (2) tells the user that there is no warranty for the work (except to the extent that warranties are provided), that licensees may convey the work under this License, and how to view a copy of this License. If the interface presents a list of user commands or options, such as a menu, a prominent item in the list meets this criterion.

1. Source Code.

    The "source code" for a work means the preferred form of the work for making modifications to it. "Object code" means any non-source form of a work.

    A "Standard Interface" means an interface that either is an official standard defined by a recognized standards body, or, in the case of interfaces specified for a particular programming language, one that is widely used among developers working in that language.

The "System Libraries" of an executable work include anything, other than the work as a whole, that (a) is included in the normal form of packaging a Major Component, but which is not part of that Major Component, and (b) serves only to enable use of the work with that Major Component, or to implement a Standard Interface for which an implementation is available to the public in source code form. A "Major Component", in this context, means a major essential component (kernel, window system, and so on) of the specific operating system (if any) on which the executable work runs, or a compiler used to produce the work, or an object code interpreter used to run it.

The "Corresponding Source" for a work in object code form means all the source code needed to generate, install, and (for an executable work) run the object code and to modify the work, including scripts to control those activities. However, it does not include the work's System Libraries, or general-purpose tools or generally available free programs which are used unmodified in performing those activities but which are not part of the work. For example, Corresponding Source includes interface definition files associated with source files for the work, and the source code for shared libraries and dynamically linked subprograms that the work is specifically designed to require, such as by intimate data communication or control flow between those subprograms and other parts of the work.

The Corresponding Source need not include anything that users can regenerate automatically from other parts of the Corresponding Source.

The Corresponding Source for a work in source code form is that same work.

2. Basic Permissions.

   All rights granted under this License are granted for the term of copyright on the Program, and are irrevocable provided the stated conditions are met. This License explicitly affirms your unlimited permission to run the unmodified Program. The output from running a covered work is covered by this License only if the output, given its content, constitutes a covered work. This License acknowledges your rights of fair use or other equivalent, as provided by copyright law.

   You may make, run and propagate covered works that you do not convey, without conditions so long as your license otherwise remains in force. You may convey covered works to others for the sole purpose of having them make modifications exclusively for you, or provide you with facilities for running those works, provided that you comply with the terms of this License in conveying all material for which you do not control copyright. Those thus making or running the covered works for you must do so exclusively on your behalf, under your direction and control, on terms that prohibit them from making any copies of your copyrighted material outside their relationship with you.

   Conveying under any other circumstances is permitted solely under the conditions stated below. Sublicensing is not allowed; section 10 makes it unnecessary.

3. Protecting Users' Legal Rights From Anti-Circumvention Law.

   No covered work shall be deemed part of an effective technological measure under any applicable law fulfilling obligations under article 11 of the WIPO copyright treaty adopted on 20 December 1996, or similar laws prohibiting or restricting circumvention of such measures.

When you convey a covered work, you waive any legal power to forbid circumvention of technological measures to the extent such circumvention is effected by exercising rights under this License with respect to the covered work, and you disclaim any intention to limit operation or modification of the work as a means of enforcing, against the work's users, your or third parties' legal rights to forbid circumvention of technological measures.

4. Conveying Verbatim Copies.

   You may convey verbatim copies of the Program's source code as you receive it, in any medium, provided that you conspicuously and appropriately publish on each copy an appropriate copyright notice; keep intact all notices stating that this License and any non-permissive terms added in accord with section 7 apply to the code; keep intact all notices of the absence of any warranty; and give all recipients a copy of this License along with the Program.

   You may charge any price or no price for each copy that you convey, and you may offer support or warranty protection for a fee.

5. Conveying Modified Source Versions.

   You may convey a work based on the Program, or the modifications to produce it from the Program, in the form of source code under the terms of section 4, provided that you also meet all of these conditions:

   a. The work must carry prominent notices stating that you modified it, and giving a relevant date.

   b. The work must carry prominent notices stating that it is released under this License and any conditions added under section 7. This requirement modifies the requirement in section 4 to "keep intact all notices".

   c. You must license the entire work, as a whole, under this License to anyone who comes into possession of a copy. This License will therefore apply, along with any applicable section 7 additional terms, to the whole of the work, and all its parts, regardless of how they are packaged. This License gives no permission to license the work in any other way, but it does not invalidate such permission if you have separately received it.

   d. If the work has interactive user interfaces, each must display Appropriate Legal Notices; however, if the Program has interactive interfaces that do not display Appropriate Legal Notices, your work need not make them do so.

   A compilation of a covered work with other separate and independent works, which are not by their nature extensions of the covered work, and which are not combined with it such as to form a larger program, in or on a volume of a storage or distribution medium, is called an "aggregate" if the compilation and its resulting copyright are not used to limit the access or legal rights of the compilation's users beyond what the individual works permit. Inclusion of a covered work in an aggregate does not cause this License to apply to the other parts of the aggregate.

6. Conveying Non-Source Forms.

   You may convey a covered work in object code form under the terms of sections 4 and 5, provided that you also convey the machine-readable Corresponding Source under the terms of this License, in one of these ways:

a. Convey the object code in, or embodied in, a physical product (including a physical distribution medium), accompanied by the Corresponding Source fixed on a durable physical medium customarily used for software interchange.
b. Convey the object code in, or embodied in, a physical product (including a physical distribution medium), accompanied by a written offer, valid for at least three years and valid for as long as you offer spare parts or customer support for that product model, to give anyone who possesses the object code either (1) a copy of the Corresponding Source for all the software in the product that is covered by this License, on a durable physical medium customarily used for software interchange, for a price no more than your reasonable cost of physically performing this conveying of source, or (2) access to copy the Corresponding Source from a network server at no charge.
c. Convey individual copies of the object code with a copy of the written offer to provide the Corresponding Source. This alternative is allowed only occasionally and noncommercially, and only if you received the object code with such an offer, in accord with subsection 6b.
d. Convey the object code by offering access from a designated place (gratis or for a charge), and offer equivalent access to the Corresponding Source in the same way through the same place at no further charge. You need not require recipients to copy the Corresponding Source along with the object code. If the place to copy the object code is a network server, the Corresponding Source may be on a different server (operated by you or a third party) that supports equivalent copying facilities, provided you maintain clear directions next to the object code saying where to find the Corresponding Source. Regardless of what server hosts the Corresponding Source, you remain obligated to ensure that it is available for as long as needed to satisfy these requirements.
e. Convey the object code using peer-to-peer transmission, provided you inform other peers where the object code and Corresponding Source of the work are being offered to the general public at no charge under subsection 6d.

A separable portion of the object code, whose source code is excluded from the Corresponding Source as a System Library, need not be included in conveying the object code work.

A "User Product" is either (1) a "consumer product", which means any tangible personal property which is normally used for personal, family, or household purposes, or (2) anything designed or sold for incorporation into a dwelling. In determining whether a product is a consumer product, doubtful cases shall be resolved in favor of coverage. For a particular product received by a particular user, "normally used" refers to a typical or common use of that class of product, regardless of the status of the particular user or of the way in which the particular user actually uses, or expects or is expected to use, the product. A product is a consumer product regardless of whether the product has substantial commercial, industrial or non-consumer uses, unless such uses represent the only significant mode of use of the product.

"Installation Information" for a User Product means any methods, procedures, authorization keys, or other information required to install and execute modified versions of a covered work in that User Product from a modified version of its Corresponding Source.

The information must suffice to ensure that the continued functioning of the modified object code is in no case prevented or interfered with solely because modification has been made.

If you convey an object code work under this section in, or with, or specifically for use in, a User Product, and the conveying occurs as part of a transaction in which the right of possession and use of the User Product is transferred to the recipient in perpetuity or for a fixed term (regardless of how the transaction is characterized), the Corresponding Source conveyed under this section must be accompanied by the Installation Information. But this requirement does not apply if neither you nor any third party retains the ability to install modified object code on the User Product (for example, the work has been installed in ROM).

The requirement to provide Installation Information does not include a requirement to continue to provide support service, warranty, or updates for a work that has been modified or installed by the recipient, or for the User Product in which it has been modified or installed. Access to a network may be denied when the modification itself materially and adversely affects the operation of the network or violates the rules and protocols for communication across the network.

Corresponding Source conveyed, and Installation Information provided, in accord with this section must be in a format that is publicly documented (and with an implementation available to the public in source code form), and must require no special password or key for unpacking, reading or copying.

7. Additional Terms.

   "Additional permissions" are terms that supplement the terms of this License by making exceptions from one or more of its conditions. Additional permissions that are applicable to the entire Program shall be treated as though they were included in this License, to the extent that they are valid under applicable law. If additional permissions apply only to part of the Program, that part may be used separately under those permissions, but the entire Program remains governed by this License without regard to the additional permissions.

   When you convey a copy of a covered work, you may at your option remove any additional permissions from that copy, or from any part of it. (Additional permissions may be written to require their own removal in certain cases when you modify the work.) You may place additional permissions on material, added by you to a covered work, for which you have or can give appropriate copyright permission.

   Notwithstanding any other provision of this License, for material you add to a covered work, you may (if authorized by the copyright holders of that material) supplement the terms of this License with terms:

   a. Disclaiming warranty or limiting liability differently from the terms of sections 15 and 16 of this License; or

   b. Requiring preservation of specified reasonable legal notices or author attributions in that material or in the Appropriate Legal Notices displayed by works containing it; or

   c. Prohibiting misrepresentation of the origin of that material, or requiring that modified versions of such material be marked in reasonable ways as different from the original version; or

- d. Limiting the use for publicity purposes of names of licensors or authors of the material; or
- e. Declining to grant rights under trademark law for use of some trade names, trademarks, or service marks; or
- f. Requiring indemnification of licensors and authors of that material by anyone who conveys the material (or modified versions of it) with contractual assumptions of liability to the recipient, for any liability that these contractual assumptions directly impose on those licensors and authors.

All other non-permissive additional terms are considered "further restrictions" within the meaning of section 10. If the Program as you received it, or any part of it, contains a notice stating that it is governed by this License along with a term that is a further restriction, you may remove that term. If a license document contains a further restriction but permits relicensing or conveying under this License, you may add to a covered work material governed by the terms of that license document, provided that the further restriction does not survive such relicensing or conveying.

If you add terms to a covered work in accord with this section, you must place, in the relevant source files, a statement of the additional terms that apply to those files, or a notice indicating where to find the applicable terms.

Additional terms, permissive or non-permissive, may be stated in the form of a separately written license, or stated as exceptions; the above requirements apply either way.

8. Termination.

   You may not propagate or modify a covered work except as expressly provided under this License. Any attempt otherwise to propagate or modify it is void, and will automatically terminate your rights under this License (including any patent licenses granted under the third paragraph of section 11).

   However, if you cease all violation of this License, then your license from a particular copyright holder is reinstated (a) provisionally, unless and until the copyright holder explicitly and finally terminates your license, and (b) permanently, if the copyright holder fails to notify you of the violation by some reasonable means prior to 60 days after the cessation.

   Moreover, your license from a particular copyright holder is reinstated permanently if the copyright holder notifies you of the violation by some reasonable means, this is the first time you have received notice of violation of this License (for any work) from that copyright holder, and you cure the violation prior to 30 days after your receipt of the notice.

   Termination of your rights under this section does not terminate the licenses of parties who have received copies or rights from you under this License. If your rights have been terminated and not permanently reinstated, you do not qualify to receive new licenses for the same material under section 10.

9. Acceptance Not Required for Having Copies.

   You are not required to accept this License in order to receive or run a copy of the Program. Ancillary propagation of a covered work occurring solely as a consequence of using peer-to-peer transmission to receive a copy likewise does not require acceptance.

However, nothing other than this License grants you permission to propagate or modify any covered work. These actions infringe copyright if you do not accept this License. Therefore, by modifying or propagating a covered work, you indicate your acceptance of this License to do so.

10. Automatic Licensing of Downstream Recipients.

    Each time you convey a covered work, the recipient automatically receives a license from the original licensors, to run, modify and propagate that work, subject to this License. You are not responsible for enforcing compliance by third parties with this License.

    An "entity transaction" is a transaction transferring control of an organization, or substantially all assets of one, or subdividing an organization, or merging organizations. If propagation of a covered work results from an entity transaction, each party to that transaction who receives a copy of the work also receives whatever licenses to the work the party's predecessor in interest had or could give under the previous paragraph, plus a right to possession of the Corresponding Source of the work from the predecessor in interest, if the predecessor has it or can get it with reasonable efforts.

    You may not impose any further restrictions on the exercise of the rights granted or affirmed under this License. For example, you may not impose a license fee, royalty, or other charge for exercise of rights granted under this License, and you may not initiate litigation (including a cross-claim or counterclaim in a lawsuit) alleging that any patent claim is infringed by making, using, selling, offering for sale, or importing the Program or any portion of it.

11. Patents.

    A "contributor" is a copyright holder who authorizes use under this License of the Program or a work on which the Program is based. The work thus licensed is called the contributor's "contributor version".

    A contributor's "essential patent claims" are all patent claims owned or controlled by the contributor, whether already acquired or hereafter acquired, that would be infringed by some manner, permitted by this License, of making, using, or selling its contributor version, but do not include claims that would be infringed only as a consequence of further modification of the contributor version. For purposes of this definition, "control" includes the right to grant patent sublicenses in a manner consistent with the requirements of this License.

    Each contributor grants you a non-exclusive, worldwide, royalty-free patent license under the contributor's essential patent claims, to make, use, sell, offer for sale, import and otherwise run, modify and propagate the contents of its contributor version.

    In the following three paragraphs, a "patent license" is any express agreement or commitment, however denominated, not to enforce a patent (such as an express permission to practice a patent or covenant not to sue for patent infringement). To "grant" such a patent license to a party means to make such an agreement or commitment not to enforce a patent against the party.

    If you convey a covered work, knowingly relying on a patent license, and the Corresponding Source of the work is not available for anyone to copy, free of charge and under the terms of this License, through a publicly available network server or other readily accessible means, then you must either (1) cause the Corresponding Source to be so available, or (2) arrange to deprive yourself of the benefit of the patent license for this particular work, or (3) arrange, in a manner consistent with the requirements of this License, to extend the patent license to downstream recipients. "Knowingly relying" means you have actual knowledge that, but for the patent license, your conveying the covered work in a country, or your recipient's use of the covered work in a country, would infringe one or more identifiable patents in that country that you have reason to believe are valid.

If, pursuant to or in connection with a single transaction or arrangement, you convey, or propagate by procuring conveyance of, a covered work, and grant a patent license to some of the parties receiving the covered work authorizing them to use, propagate, modify or convey a specific copy of the covered work, then the patent license you grant is automatically extended to all recipients of the covered work and works based on it.

A patent license is "discriminatory" if it does not include within the scope of its coverage, prohibits the exercise of, or is conditioned on the non-exercise of one or more of the rights that are specifically granted under this License. You may not convey a covered work if you are a party to an arrangement with a third party that is in the business of distributing software, under which you make payment to the third party based on the extent of your activity of conveying the work, and under which the third party grants, to any of the parties who would receive the covered work from you, a discriminatory patent license (a) in connection with copies of the covered work conveyed by you (or copies made from those copies), or (b) primarily for and in connection with specific products or compilations that contain the covered work, unless you entered into that arrangement, or that patent license was granted, prior to 28 March 2007.

Nothing in this License shall be construed as excluding or limiting any implied license or other defenses to infringement that may otherwise be available to you under applicable patent law.

12. No Surrender of Others' Freedom.

    If conditions are imposed on you (whether by court order, agreement or otherwise) that contradict the conditions of this License, they do not excuse you from the conditions of this License. If you cannot convey a covered work so as to satisfy simultaneously your obligations under this License and any other pertinent obligations, then as a consequence you may not convey it at all. For example, if you agree to terms that obligate you to collect a royalty for further conveying from those to whom you convey the Program, the only way you could satisfy both those terms and this License would be to refrain entirely from conveying the Program.

13. Use with the GNU Affero General Public License.

    Notwithstanding any other provision of this License, you have permission to link or combine any covered work with a work licensed under version 3 of the GNU Affero General Public License into a single combined work, and to convey the resulting work. The terms of this License will continue to apply to the part which is the covered work, but the special requirements of the GNU Affero General Public License, section 13, concerning interaction through a network will apply to the combination as such.

14. Revised Versions of this License.

The Free Software Foundation may publish revised and/or new versions of the GNU General Public License from time to time. Such new versions will be similar in spirit to the present version, but may differ in detail to address new problems or concerns.

Each version is given a distinguishing version number. If the Program specifies that a certain numbered version of the GNU General Public License "or any later version" applies to it, you have the option of following the terms and conditions either of that numbered version or of any later version published by the Free Software Foundation. If the Program does not specify a version number of the GNU General Public License, you may choose any version ever published by the Free Software Foundation.

If the Program specifies that a proxy can decide which future versions of the GNU General Public License can be used, that proxy's public statement of acceptance of a version permanently authorizes you to choose that version for the Program.

Later license versions may give you additional or different permissions. However, no additional obligations are imposed on any author or copyright holder as a result of your choosing to follow a later version.

15. Disclaimer of Warranty.

    THERE IS NO WARRANTY FOR THE PROGRAM, TO THE EXTENT PERMITTED BY APPLICABLE LAW. EXCEPT WHEN OTHERWISE STATED IN WRITING THE COPYRIGHT HOLDERS AND/OR OTHER PARTIES PROVIDE THE PROGRAM "AS IS" WITHOUT WARRANTY OF ANY KIND, EITHER EXPRESSED OR IMPLIED, INCLUDING, BUT NOT LIMITED TO, THE IMPLIED WARRANTIES OF MERCHANTABILITY AND FITNESS FOR A PARTICULAR PURPOSE. THE ENTIRE RISK AS TO THE QUALITY AND PERFORMANCE OF THE PROGRAM IS WITH YOU. SHOULD THE PROGRAM PROVE DEFECTIVE, YOU ASSUME THE COST OF ALL NECESSARY SERVICING, REPAIR OR CORRECTION.

16. Limitation of Liability.

    IN NO EVENT UNLESS REQUIRED BY APPLICABLE LAW OR AGREED TO IN WRITING WILL ANY COPYRIGHT HOLDER, OR ANY OTHER PARTY WHO MODIFIES AND/OR CONVEYS THE PROGRAM AS PERMITTED ABOVE, BE LIABLE TO YOU FOR DAMAGES, INCLUDING ANY GENERAL, SPECIAL, INCIDENTAL OR CONSEQUENTIAL DAMAGES ARISING OUT OF THE USE OR INABILITY TO USE THE PROGRAM (INCLUDING BUT NOT LIMITED TO LOSS OF DATA OR DATA BEING RENDERED INACCURATE OR LOSSES SUSTAINED BY YOU OR THIRD PARTIES OR A FAILURE OF THE PROGRAM TO OPERATE WITH ANY OTHER PROGRAMS), EVEN IF SUCH HOLDER OR OTHER PARTY HAS BEEN ADVISED OF THE POSSIBILITY OF SUCH DAMAGES.

17. Interpretation of Sections 15 and 16.

    If the disclaimer of warranty and limitation of liability provided above cannot be given local legal effect according to their terms, reviewing courts shall apply local law that most closely approximates an absolute waiver of all civil liability in connection with the Program, unless a warranty or assumption of liability accompanies a copy of the Program in return for a fee.

# END OF TERMS AND CONDITIONS

## How to Apply These Terms to Your New Programs

If you develop a new program, and you want it to be of the greatest possible use to the public, the best way to achieve this is to make it free software which everyone can redistribute and change under these terms.

To do so, attach the following notices to the program. It is safest to attach them to the start of each source file to most effectively state the exclusion of warranty; and each file should have at least the "copyright" line and a pointer to where the full notice is found.

```
one line to give the program's name and a brief idea of what it does.
Copyright (C) year name of author

This program is free software: you can redistribute it and/or modify
it under the terms of the GNU General Public License as published by
the Free Software Foundation, either version 3 of the License, or (at
your option) any later version.

This program is distributed in the hope that it will be useful, but
WITHOUT ANY WARRANTY; without even the implied warranty of
MERCHANTABILITY or FITNESS FOR A PARTICULAR PURPOSE.  See the GNU
General Public License for more details.

You should have received a copy of the GNU General Public License
along with this program.  If not, see http://www.gnu.org/licenses/.
```

Also add information on how to contact you by electronic and paper mail.

If the program does terminal interaction, make it output a short notice like this when it starts in an interactive mode:

```
program Copyright (C) year name of author
This program comes with ABSOLUTELY NO WARRANTY; for details type 'show w'.
This is free software, and you are welcome to redistribute it
under certain conditions; type 'show c' for details.
```

The hypothetical commands 'show w' and 'show c' should show the appropriate parts of the General Public License. Of course, your program's commands might be different; for a GUI interface, you would use an "about box".

You should also get your employer (if you work as a programmer) or school, if any, to sign a "copyright disclaimer" for the program, if necessary. For more information on this, and how to apply and follow the GNU GPL, see http://www.gnu.org/licenses/.

The GNU General Public License does not permit incorporating your program into proprietary programs. If your program is a subroutine library, you may consider it more useful to permit linking proprietary applications with the library. If this is what you want to do, use the GNU Lesser General Public License instead of this License. But first, please read http://www.gnu.org/philosophy/why-not-lgpl.html.

Printed in Great Britain
by Amazon